New Frontiers in Meteorology

Volume I

New Frontiers in Meteorology
Volume I

Edited by **Dorothy Rambola**

R CALLISTO
EFERENCE

New York

Published by Callisto Reference,
106 Park Avenue, Suite 200,
New York, NY 10016, USA
www.callistoreference.com

New Frontiers in Meteorology: Volume I
Edited by Dorothy Rambola

International Standard Book Number: 978-1-63239-475-0 (Hardback)

This book contains information obtained from authentic and highly regarded sources. Copyright for all individual chapters remain with the respective authors as indicated. A wide variety of references are listed. Permission and sources are indicated; for detailed attributions, please refer to the permissions page. Reasonable efforts have been made to publish reliable data and information, but the authors, editors and publisher cannot assume any responsibility for the validity of all materials or the consequences of their use.

The publisher's policy is to use permanent paper from mills that operate a sustainable forestry policy. Furthermore, the publisher ensures that the text paper and cover boards used have met acceptable environmental accreditation standards.

Trademark Notice: Registered trademark of products or corporate names are used only for explanation and identification without intent to infringe.

Printed in the United States of America.

Contents

Preface

The origins of the word "meteorology" can be traced to its Greek roots. The term 'meteor' refers to something lofty or high in the sky and 'logy' means study. Thus, Meteorology is an area of research which studies about different aspects of the atmosphere. Though this field has been developing since the eighteenth century, it was only in the 19th century that progress in meteorology researches gained pace. This was due to the formation of numerous observational networks across many countries. With the advancement in computer technologies in the second half of 20th century, Meteorology achieved significant new heights.

Considering meteorological phenomena, the study of observable weather events such as: temperature, air pressure, water vapor, and the gradients and interactions of each variable, and how they vary in time is studied under the science of meteorology. The subject includes studies of local, regional, and global levels impact on weather and climatology. This subject has numerous sub-disciplines like climatology, atmospheric physics and atmospheric chemistry. Hydrology and meteorology combine to form another interdisciplinary field of meteorology, called hydrometeorology. The applications of meteorology can be found in numerous fields which include agriculture, climate studies, military and energy production among others.

Moreover, with the growing concern about global warming and climatic changes, meteorology has gained more prominence. Studying the changing seasonal patterns and rainfalls contribute to environmental studies. Thus, meteorology and environmental impact studies happen to be intertwined in that sense.

I wish to thank the contributors for their efforts and time. Without their timely submissions and patience, this book wouldn't have been possible. I also bid gratitude to the publishing team for all the efforts they provided in the publishing process of this book.

Editor

Characteristics of Air-Sea Fluxes Based on In Situ Observations from a Platform in the Bohai Gulf during Early Mid-August 2011

Bingui Wu,[1] Yiyang Xie,[1] Yi Lin,[1] Xinxin Ye,[2] Jing Chen,[1] Xiaobing Qiu,[1] and Yanan Wang[1]

[1] *Tianjin Municipal Meteorological Bureau, Tianjin 300074, China*
[2] *Laboratory for Climate and Ocean-Atmosphere Studies, Department of Atmospheric and Oceanic Sciences, School of Physics, Peking University, Beijing 100871, China*

Correspondence should be addressed to Yiyang Xie; tjqxjs@126.com

Academic Editor: Bin Liu

An eddy covariance system and other atmospheric and oceanic parameters were measured simultaneously from a fixed Platform-A in the Bohai Gulf during early mid-August 2011. One of the main goals of the comprehensive observation was to reveal the basic meteorological and hydrological characteristics of the Bohai Gulf. The results indicated that the diurnal characteristic curve for the air temperature (AT) was steeply unimodal, while the curve of the SST was a bimodal valley type and mainly influenced by tides with its valley value corresponding to the high water level during the observation period. Southeasterly winds dominated and the wind speed was generally lower than 8 m/s, and the atmospheric stability over the Bohai Gulf was generally unstable. The wave strength levels were generally below level 3, with a greater number of swell waves than wind waves. The latter were usually associated with more momentum transport, a larger difference between AT and SST, and less heat transport. During the observational period, the mean momentum, sensible, and latent heat turbulent fluxes were 0.21, 21.6, and 27.8 W/m^2, respectively. The ratio of the mean latent and sensible turbulent fluxes was about 1.3 and much lower than that in the South China Sea during the summer.

1. Introduction

The exchanges of heat, mass, and momentum between the atmosphere and the ocean are very influential on the structure of the marine boundary layer. They also influence the atmosphere both locally and globally, as well as ocean circulation. However, sensible heat and water vapor fluxes have depend on many parameters which are still considered to be unclear [1–3]. Such as whitecap element, which begin to appear on the ocean surface at wind speed as low as 3 m/s and cover a significant fraction (>1%) of the ocean surface at wind speed of about 10 m/s [4], increasing at a rate approximately proportional to the cube of the surface layer wind speed [5]. Again such as breaking waves, which alter the surface roughness characteristics of the ocean and produce sea spray droplets and affect the air-sea exchange process [6], and so on. The air-sea energy and mass exchange process becomes more complicated [3, 7]. In spite of a large amount of experimental data its uncertainty is unacceptably high for calculations of climatological heat transports. Since

a comprehensive understanding of the mechanics of air-sea interaction is fundamentally important for the study of ocean circulation models, ocean-atmosphere coupled models, and the energy and water cycle, in order to estimate the air-sea fluxes, air-sea exchange observation had been treated as the main content in Tropical Ocean and Global Atmosphere, Coupled Ocean-Atmosphere Response Experiment (TOGA COARE) [8]. It has helped to improve our understanding of the characteristics of air-sea exchange in many ocean areas including the warm pool in the western Pacific Ocean, the Kuroshio region, and the South China Sea monsoon region [9, 10]. Along Chinese coastal waters, most studies on air-sea fluxes have been focused on the South China Sea [7]. The characteristics of heat fluxes before and after the start of the South China Sea summer monsoon have been identified. The equations for bulk transport coefficients and the bulk transport improved method in the region of the South China Sea have been also uncovered [11]. These jobs have given us a better understanding of the air-sea heat fluxes during the eruption period of the South China Sea summer monsoon.

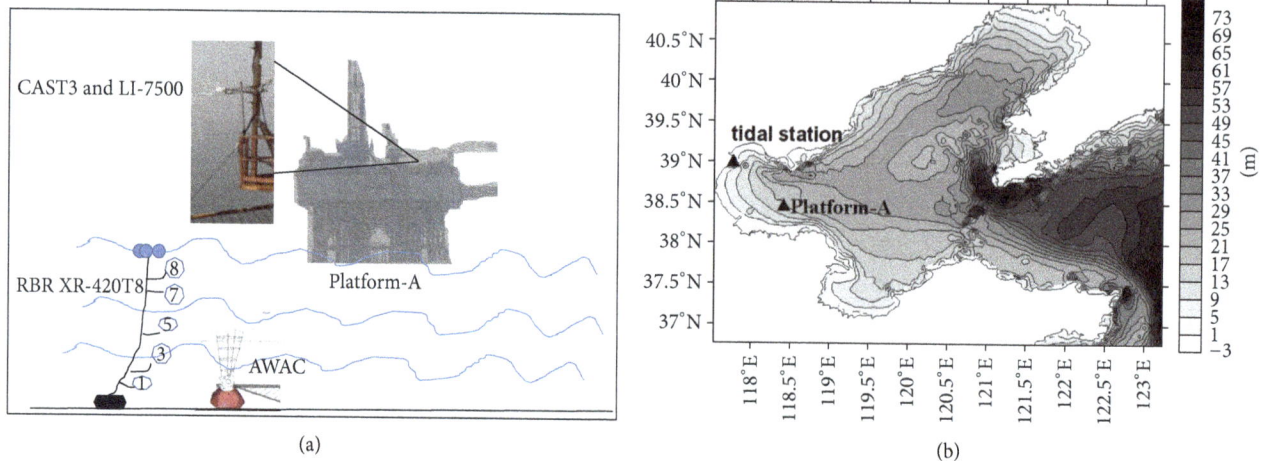

(a)

(b)

FIGURE 1: Platform-A in the Bohai Gulf and related observation instruments (a). Tide gauge station and Platform-A and water level (b).

The variations in heat fluxes at different seasons, latitudes, and coastal topographies over Chinese coastal waters have been noted by Yan et al. [12]. The heat, momentum, and water vapor fluxes under different weather conditions over the sea surface have been calculated by using the observational dataset obtained over the Xi'sha marine region in the South China Sea. Overall, research on air-sea fluxes over the South China Sea has focused on the air-sea fluxes before and after the start of the South China Sea summer monsoon. There has also been research into variations in radiation and turbulence during different weather conditions, the diurnal variation of air-sea fluxes, the bulk transport coefficients, and the influences of flux variations on the lower atmosphere and upper ocean mixing layer. According to Vickers and Mahrt [13], the momentum and sensible heat fluxes over middle-latitude coastal waters are lower than those over the open sea surfaces, and latent heat fluxes are even lower, often by 20%–50%. Compared to the South China Sea, the study of the Bohai Gulf has been fairly limited so far. Furthermore, the knowledge acquired about the South China Sea may not be valid for the Bohai Gulf. Therefore, it is necessary to do further research on air-sea interactions over the Bohai Gulf.

An observation campaign was conducted from the end of July 2011 on the Platform-A (38°27'N, 118°25'E) in the Bohai Gulf. In addition to conventional meteorological observations like air temperature (AT), wind, humidity, precipitation, cloud cover, and visibility, additional observations were included. The sea surface temperature (SST), ocean current profile, AT changes, specific humidity, and three-dimensional wind components were carried out. Marine observations are limited by many factors, such as damage done to the floating observation instrumentation by currents and waves. Since the data from early mid-August 2011 was comparably complete, data of this period were chosen for analysis for the sake of understanding of atmospheric-ocean interactions in middle-latitude continental seas. It would be beneficial to lay the foundation of further getting the quantitative description of the exchanges of heat, mass, and momentum between the atmosphere and the ocean. We present results from the comprehensive observation and show evidence of low correlation between the mean momentum flux and the wind speed. And we find that the daily variation of SST is M-type and mainly influenced by an irregular semidiurnal tide with the minimum SST corresponding to the high water level during the observation period. We also find that the sensible and latent heat fluxes have the same comparable importance in the Bohai Gulf.

2. Data and Methods

2.1. Marine Observation Platform-A and Instrumentation. The air-sea fluxes over the ocean surface are generally observed by stand or mobile platforms settled in the sea [2, 14]. The observation platform used in our experiment was the Chengbei Platform-A (Station Identity no. 54646) located in the Bohai Gulf (38°27'N, 118°25'E) where water level is about 19 m (Figure 1(b)). It is a fundamental meteorological observation station managed by the China Meteorological Administration. The platform is located at 70 km off the west coast and 40 km off the south coast of the Bohai Gulf. It is an excellent platform for conducting marine observations and 30.3 m height above mean sea level. The barometer and anemometer are located 2 m and 6.5 m above the Platform-A, respectively. In 2005, automatic observation instruments replaced the artificial observation techniques which dated back to January 1988, with observers monitoring the instrument operations. The observation data has been kept consecutively to this day.

An eddy covariance system was used to acquire supporting meteorological and boundary layer flux data. The eddy covariance system included a three-axis anemometer/thermometer (CAMPBELL CSAT3, USA) and an open-path infrared hygrometer CO_2/H_2O sensor (Licor-7500). The eddy covariance system was installed on an abandoned mast at the height of 5 m above the Platform-A. The semigirders were set up the north-south direction and the instruments faced the north. The sampling frequency was 10 Hz. The sections of wave and current speed were measured with an Acoustic Wave and Current (AWAC) from the Nortek Company based in Norway. In addition to having

TABLE 1: List of instruments used during the intensive observations on air-sea fluxes.

Instrument	Precision	Temporal resolution/frequency	Sample layer
Ultrasonic anemometer (CSAT3; Campbell, USA)	u, v, 1 mm/s; w, 0.5 mm/s, sound velocity, 1 mm/s	10 Hz	5 m above the Platform-A
CO_2/H_2O analyzer (Li7500; LiCor, USA)	0.1 μmol/mol	10 Hz	5 m above the Platform-A
Thermistor chain (XR-420 T8; RBR, Canada)	0.005°C	20 min	Undersea 1, 3, 5, 7.5, 10, 12.5, 15 m and the seabed
Acoustic wave and current (AWAC)	Current speed, 1% ± 0.5 cm/s; wave height, <1%/cm wave direction, 0.1°/2°	0.05 Hz	Surface wave height, vertical section current speed

the traditional features of an Acoustic Doppler Current Profiler (ADCP) used to measure the waves and current speeds, it can also detect wave height with Acoustic Surface Tracking (AST). The XR-420 T8 (RBR Company) was used to detect the sea water temperature section and vertical gradients. It was placed eight meters west of the observation platform and included eight temperature sensors equipped on the vertical chain (Figure 1(a)). Due to the loss of a bearing rope (and abrasion on the main one), the electric cable became overloaded, which caused the temperature section sensors to break down on August 17, 2011. Therefore, data from the August 3rd to 16th was selected for analysis. Tide gauge station used was located at the western coast of the Bohai Gulf (38°59′N, 117°46′E). It is a facility of flood control departments of Municipal government and about 91 km northwest of Platform-A (Figure 1(b)).

During the observation period, the platform was out of the monitoring of observers from 04:00 Beijing standard time (BJT, the same later) on August 7th to 09:00 BJT on August 9th because of typhoon, "Meihua," which caused missing data for August 7th and August 8th, including present weather, visibility, total amount of clouds, amount of low clouds, and cloud classifications. They were artificially and conventionally observed at 08:00, 14:00, and 20:00 BJT. In addition, due to a power failure from 01:32 to 12:21 BJT on August 8th, the data detected per minute with the automatic meteorological weather station is missing, including atmospheric pressure, air temperature, relative humidity, wind direction, and wind speed. The ultrasonic anemometer and CO_2/H_2O infrared analyzer also lost power. In terms of the hydrological data, the data of wave height and current speed from 19:45 to 22:20 BJT on August 7th is missing due to the power failure. The parameters of the instruments are listed in Table 1.

2.2. Data Processing. The TK2 Software developed by Bayreuth University in Germany was used for processing the high-frequency data to calculate heat and momentum fluxes [15, 16]. The turbulent fluxes of momentum, heat, and moisture were calculated over 30 min averages. The main formulas used are listed as follows:

friction velocity:

$$u_* = \left(\overline{u'w'}^2 + \overline{v'w'}^2 \right)^{1/4}, \tag{1}$$

Monin-Obukhov Length:

$$L = \frac{-u_*^3}{\left(k \left(g/T \right) \overline{w'T_s'} \right)}, \tag{2}$$

momentum flux:

$$\tau = -\rho u_*^2, \tag{3}$$

sensible heat flux:

$$Q_h = \rho C_{pd} \left(1 + 0.84q \right) \overline{w'\theta'}, \tag{4}$$

latent heat flux:

$$Q_e = \rho L_v \overline{w'q'}. \tag{5}$$

In the previous equations, u, v, and w are the horizontal and vertical component of turbulent velocity vector. T_s is virtual temperature and θ is the potential temperature. q is specific humidity measured by the CO_2/H_2O analyzer. The primes indicate turbulent quantities taken as deviations from the mean and the overbars represent ensemble averages. Here, ρ stands for air density, k for the Von-Karman constant (0.4 generally), g for acceleration due to gravity (9.8 m/s^2 generally), C_{pd} for the specific heat of dry air at constant pressure, L_v for latent heat for evaporation, and z for the observation height.

Considering the observation instruments (the barometer, anemometer, and the eddy covariance system) that are located 2 m to 5 m over the Platform, heat, moisture, and momentum fluxes discussed in the paper are almost still constant compared to the height 10 m. So attitude difference correction is not necessary. Strict data quality control methods were applied during the data processing with the TK2 software. Except for rejecting those data with quality labels greater than 7 marked by the TK2, some abnormal data for periods with rain, thunder, and weak winds was also rejected for the sake of data quality control.

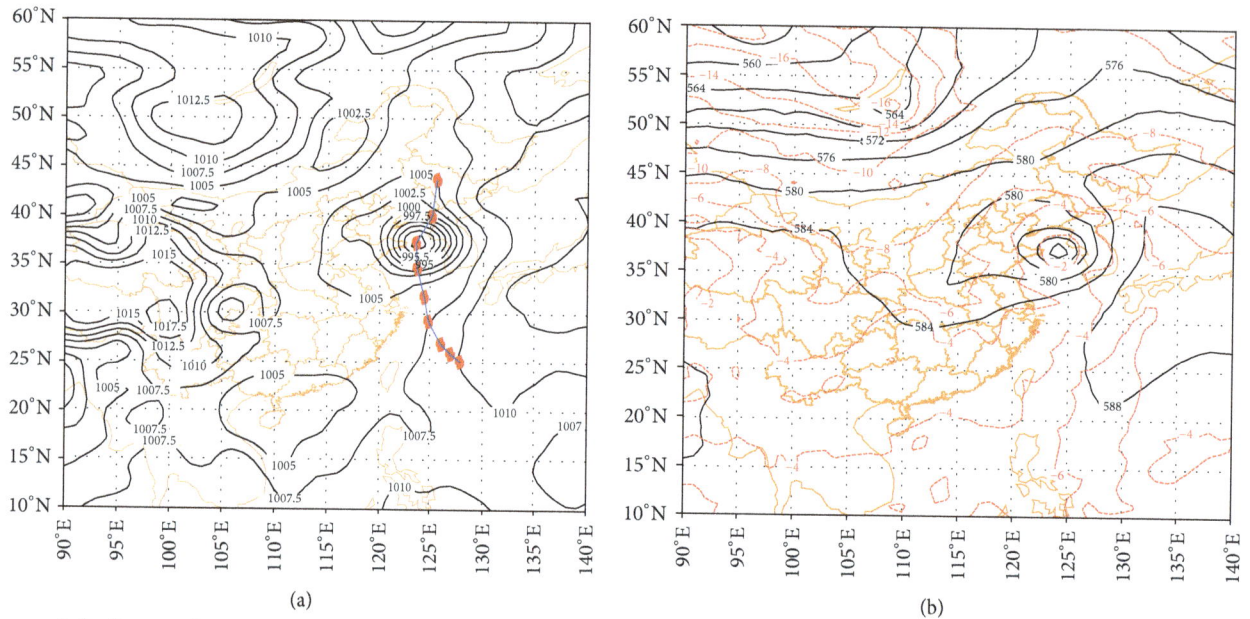

FIGURE 2: Surface weather map at 08BJT on August 8th along with typhoon mark symbol every 12 hours on the drawing from south at 08BJT on August 5th to north at 08BJT on August 9th of 2011 (a) and 500 hPa temperature (red line) and geopotential height (black line) at 08BJT on August 8th of 2011 (b).

3. Results and Discussion

3.1. Main Synoptic Processes. In early August, Northern China endures its main flood season. The region is dominantly controlled by the subtropical anticyclone system and the "westerlies." The tropical cyclones proceeding northwards bring frequent precipitation. During the observation period, the subtropical anticyclone maintained its eastern direction to the Sea of Japan. From August 3rd to 8th, the severe typhoon (No. 201109, "Meihua") passed through the East China Sea and headed into the northern Yellow Sea. According to the synoptic weather map, it was clear at the observation site caused by the continental high pressure system from August 3rd to 5th with 500 hPa isobaric surface. On August 6th, heavy rain fell over the Bohai Gulf in relation to the east-moving upper air trough. After passing the Bohai Gulf, typhoon "Meihua" made landfall on the western coast of North Korea and turned into a low pressure system (Figure 2). Although typhoon system caused surface pressure descending, it did not lead to any precipitation near the observation site. However, the weakened cyclone did make it cloudy. The weather was then clear, and it was controlled once again by the continental anticyclone form on August 9th-10th. There were showers on August 11th when a west-moving cyclone formed near the Yellow River (referred to as Yellow River cyclone) passed by. On August 12th-13th, it was cloudy weather again due to surface low pressure system influenced by Yellow River cyclone passing by. It turned clear on the 14th and conditions were once again influenced by the continental anticyclone. On the 15th, steady western flows occurred in the upper air and, because of the vortex in the lower atmosphere and the inversed trough at the surface, a short-time thunderstorm with strong winds occurred (Figure 3). In conjunction with the variations in wind direction, every precipitation

event occurred before the transition of a southeasterly wind into a northeasterly wind. Here, the northeasterly wind was always maintained for a short time. The phenomena were in accordance with the generally held rule in which precipitation always occurs when the summer monsoon system retreats and northern high-pressure systems strengthen during early mid-August.

3.2. Ocean Surface Humidity and Wind Speed. During early-mid August, the relative humidity (RH) at the ocean surface was generally above 60% without any significant interdiurnal variations (Figure 3(a)). The RH reached about 82%, when northeasterly or southeasterly winds dominated, and 76% during southwesterly winds. The atmosphere was comparably dry when the northwesterly winds prevailed, and here the RH was about 73%.

In the summer, the winds are moderate over the Bohai Gulf. They generally held between weak and medium strength. Winds with speeds below 8 m/s accounted for 97.6% of conditions. Here, winds with speeds below 3 m/s accounted for 39.3%, 3–6 m/s for 51%, and 6–8 m/s for 7.3%, and the remaining 2.4% were between 8 and 14 m/s. The latter speeds generally occurred during the thunderstorm periods (Figure 3(b)). During the three periods of precipitation, the RH reached nearly 85%, and southeasterly winds with speeds of 3 m/s prevailed. During the strong convection period on August 15th, the wind speed was greater than 6 m/s several times (Figure 3(b)).

3.3. Air Temperature, Ocean Surface Temperature, and Their Discrepancies. The air temperature (AT) was primarily affected by the amount of clouds, precipitation, and the invasion of warm or cold airflows (Figure 4).

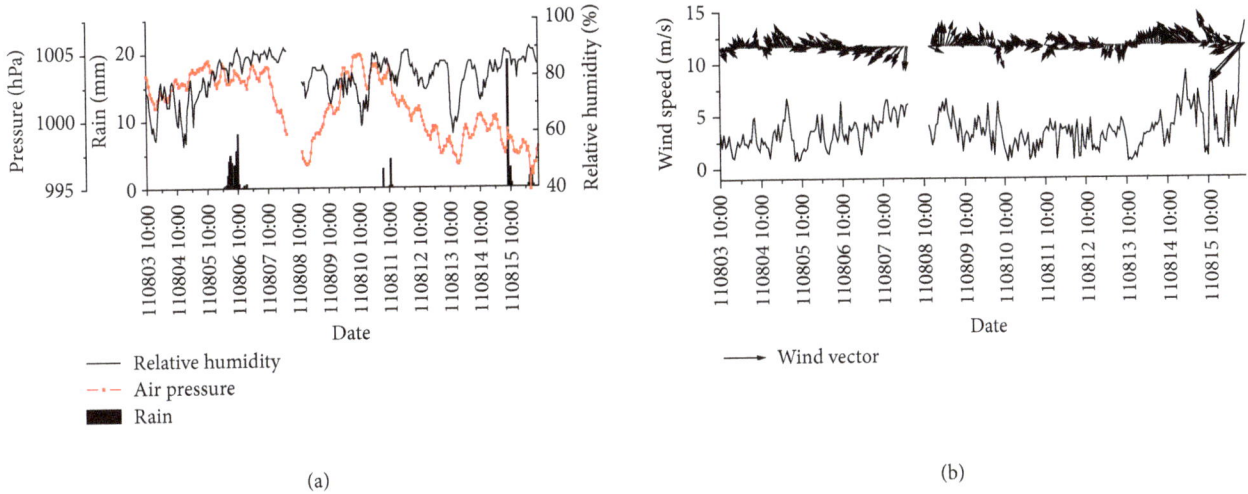

(a)

(b)

FIGURE 3: Inter-diurnal variation of meteorological elements ((a) relative humidity, precipitation, surface pressure; (b) wind speed and wind direction).

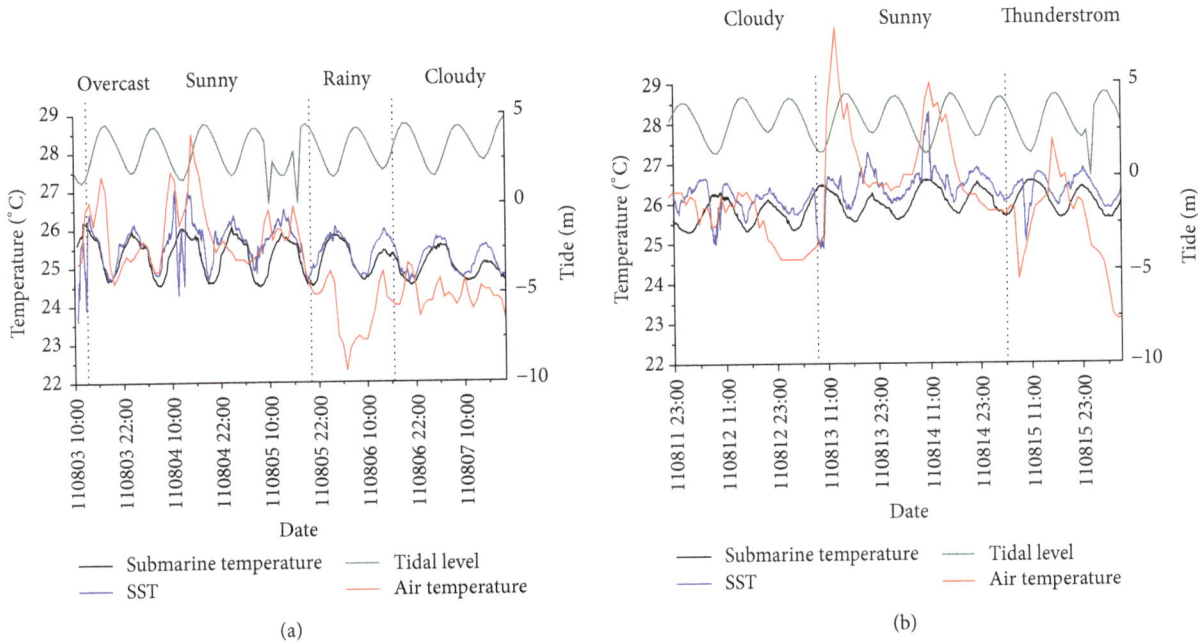

(a)

(b)

FIGURE 4: Time series for air temperature, SST, and tide level.

During clear weather, the diurnal variation of AT was large, reaching the maximum average of 26.2°C. It generally occurred around 15:00 BJT. During cloudy weather or during precipitation, the incoming solar radiation was decreased. Here, the AT decreased remarkably, especially during precipitation. The mean AT was about 24.3°C. It was 1.9°C lower than that during the clear conditions. Synoptic weather backgrounds can be representative with wind directions. Here, among the northern winds presented cold airflows from the systems to the north, with the average temperature of 25.5°C. Meanwhile, southern winds were created by the warm advection in relation to the southern systems, and the average temperature here was 26.7°C.

During the period of observation, the mean sea surface temperature (SST) was 25.9°C, and the mean bottom water temperature was 25.6°C. The variation tendencies in temperature at several depths were generally similar. The difference between the surface and bottom water temperatures averaged only 0.3°C, which indicated very strong vertical mixing of sea water and the removal of the thermocline. Although the variation tendency of SST was almost the same as the bottom water temperature, the former varied with more fluctuations and was similar to the variations in the AT. Sometimes, the SST varied in accordance with the AT, leading to large difference between the SST and bottom water temperature. Here, there was the maximum of 3.3°C and the minimum of −3.9°C. This indicated the fact that surface water is less resistant to changes in meteorological conditions such as radiation and winds.

The sea water temperature showed notable diurnal variation as well. In contrast to the unimodal-type curve of AT, the SST varied as an M-type curve. Here, two SST peaks appeared in the afternoon and near the midnight. The peaks were higher during the daytime than at nighttime, representing the heating effect of insolation. In comparison with the hourly recorded tide level from the Hydrometric Station, it was shown that an inverse correlation existed between the tide level and the sea water temperature (Figure 4). Since strong tidal currents occur in the Bohai Gulf (mainly an irregular semidiurnal tide), sea water temperature was closely related to tidal current activity. It is consistent with the observational results shown by Tang et al. [17].

When the observation experiment was conducted during early-mid August, the difference between SST and AT was obvious. On clear afternoons, the maximum AT was greater than SST by 1.5°C. During other periods, the AT was generally less than SST by about 0.5°C. Overall, AT over the Bohai Gulf exceeded SST during 62.9% of the total period. Furthermore, 63.3% of the data showed the air-sea temperature difference within 1°C. The results were different those that obtained in the South China Sea, where the daily SST average was almost always greater than the AT [12]. The one of possible reasons about the difference is because the ocean current temperature decreases with the latitude zone from the south to the north, while air temperature at both latitude zones is comparative in early-mid August. However, regardless of the weather conditions, the AT varied more evidently than the SST. It has the similar characteristic to the others as a semienclosed shallow sea, whose SST and AT are determined by the physical properties of both the sea water and the air.

3.4. Marine Boundary Layer Stratification. The stability of the lower atmosphere is an important factor for determining turbulence strength and diffusion conditions. It plays a primary role in the study of turbulence structure and for the estimation of surface fluxes.

Several common methods used to determine atmospheric stability have been previously pointed out. For instance, the temperature difference method [18] and the Monin-Obukhov (M-O) length method [19] are commonly used during observation experiments. Different criteria can draw different conclusions. When analyzing the Bohai Gulf, if we assumed positive air-sea temperature difference as an indicator of unstable marine boundary layer (MBL), then 62.9% unstable conditions would be yielded during the experiment period. Meanwhile, with M-O length method as a criterion, unstable conditions accounted for 70.2%. The judgments acquired by using two kinds of methods were often inconsistent. So the way to choose a reasonable indication for atmospheric stability is imperative.

In order to determine which indicator was better, two days of August 4th (clear) and August 12th (cloudy) with typical weather cases were selected. From 08:00 BJT to 08:00 BJT (August 4th-5th), it was clear with some dense cirrus clouds. From 08:00 BJT to 08:00 BJT (August 12th-13th), the sky contained 80% low-middle clouds. Figure 5 demonstrates

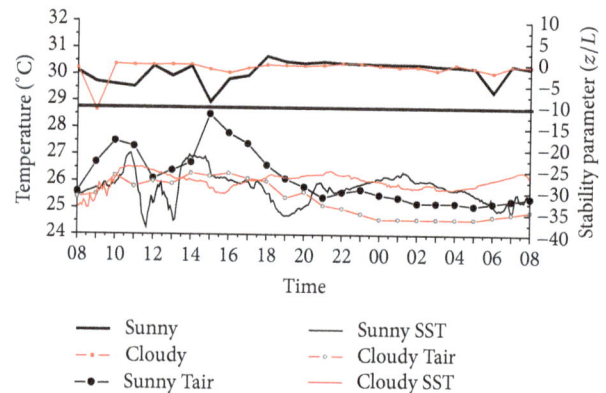

FIGURE 5: Air temperature and sea surface temperature on clear and cloudy days. The thin lines denote z/L on clear day (in black) and cloudy day (in red).

the temporal variation of the AT, SST, and stability parameter z/L. It can be seen that, on the clear day, the value of z/L was negative after sunrise, denoting unstable stratification. Meanwhile, after sunset, the MBL was neutral with z/L near zero. On the cloudy day, the values of z/L were approaching zero almost the entire day, which indicated the stratification to be neutral. Nevertheless, considering SST minus AT, when the AT increased during the day and suppressed the SST, the MBL was supposed to be stable. The opposite was considered true when the AT decreased at night. The differences between the two methods are shown clearly by the two selected cases.

Atmospheric stability is related to discrepancies in density, temperature, and the velocity of a certain air parcel within the surrounding air (especially the vertical gradient). The vertical motion of an air parcel is caused thermodynamically through buoyancy or mechanical force by wind-stress. The criterion of SST minus AT considers only the thermodynamic factor, leaving out the influences of ocean surface conditions and wind stress. Therefore, it is considered to be comparably less reliable than the M-O length criterion. The M-O length takes both thermal and mechanical factors into account at the same time. In this way, the M-O length is a better indicator of atmospheric stability.

From this point on, the dimensionless form of M-O length (i.e., z/L) was used as the criterion for the stability of the atmosphere over the Bohai Gulf. The negative z/L values denoted unstable stratifications, which accounted for 70.2% of the overall conditions. The positive values for stable conditions accounted for 29.8%. Here, if we take z/L within −0.1 to 0.1 as the neutral condition, 25.6% of all data was satisfied. The results showed that the Bohai Gulf presented unstable stratification in the lower atmosphere the majority of the time during early-mid August. It was different from the conditions over the South China Sea, where there was lower atmospheric instability almost all of the time [20].

3.5. Wave Shapes and Wave-Coherent Dynamic and Thermodynamic Features. Over the last decade it has become apparent that surface wave process can play an important role in the kinematics and dynamics of the boundary layers

TABLE 2: Wave-coherent features of wind, air temperature, and SST.

Category	Grouped by wave shape		Grouped by wave height		
	Wind wave	Swell wave	Level-2 (0.1–0.4 m)	Level-3 (0.5–1.2 m)	Level-4 (1.3–2.4 m)
Average wind speed (m/s)	3.6	1.5	1.8	3.0	9.6
Maximum wind speed (m/s)	13.8	4.1	7.8	9.2	13.8
Average SST (m/s)	26.1	25.9	25.9	26.1	26.0
Average AT (°C)	25.7	25.8	26.0	25.4	23.8
Average air-sea temperature difference (°C)	0.4	0.1	−0.1	0.7	2.2
Percentage among total samples (%)	43	57	73.3	23.6	2.1

[2, 21]. The shape and height of waves directly determine the roughness of the ocean surface. In turn, the wave also influences air-sea fluxes like heat, momentum, and mass fluxes. In order to reveal temperature difference of air-sea and stratification of the MBL, the wave shapes first need to be categorized. By referring to Wang and Chang [20], the wave shapes can be categorized in accordance with the following three conditions:

$$\frac{U^2}{H} \leq 62.0,$$

$$\delta \leq 0.0177, \tag{6}$$

$$\beta \geq 1.07.$$

For formulas (6), U is the wind speed, H is the wave height, δ is the wave steepness (height-length ratio), and β is the wave age. Wave is considered to be a swell wave if it satisfies at least two of the formulas; otherwise it is considered a wind wave. The waves in the Bohai Gulf during the experiment period were categorized according to the method. The statistical results for wind speed, AT, SST, and atmospheric stability (with different wave types) are listed in Table 2. Here, we can see that the average wind speed was 3.6 m/s under the condition of wind waves, with the maximum of 13.8 m/s. Meanwhile, for swell waves, the mean wind speed was 1.5 m/s with the maximum of only 4.1 m/s. By considering the distinct determining factors for wind and swell waves, the latter were formed outside of windy areas, generally with light and gentle winds. The differences between the average and minimum wind speeds for wind and swell waves indicated that the wave types were reasonably categorized according to the method presented earlier.

Data for the wave features was recorded every 20 s. After calculating over 60 min average and data quality control, there were 195 sets of records. In accordance with the categorization method mentioned earlier, there were 82 sets of wind waves and their ratio was 43%, and swell waves accounted for 57%. For the wind waves, the SST was higher than the AT by the average of 0.4°C. Here, unstable stratification appeared more (52%). For the swell waves, the SST was higher than the AT by just 0.1°C, and unstable stratification accounted for 78%.

Furthermore, the features of SST, AT, and atmospheric stratification were analyzed according to wave height. On the basis of the standard of wave height categorization provided by the State Oceanic Administration, the Level-2 wave height is between 0.1 and 0.4 m, the Level-3 between 0.5 and 1.2 m, and Level-4 between 1.3 and 2.4 m. During the experiment period, the wave heights in the Bohai Gulf were generally less than 2.2 m. Statistics indicated that there were 143 samples of Level-2 waves, accounting for 73.3%, 46 Level-3 waves for 23.6%, and 6 Level-4 waves for only 2.1%. Since the low number of Level-4 wave samples limited their statistical reliability, they were omitted in the following analysis. The average AT during Level-2 waves was 26.0°C, and unstable conditions amounted to 72%. For Level-3 waves, the average AT was 25.4°C (less than during the Level-2 waves). However, the mean value of SST minus AT was larger, and it reached 0.7°C. Here, the probability of unstable stratifications decreased to 56%. Therefore, we can conclude that wave height is positively correlated with wind speed as to the observation period, and that strong winds lead to AT reducing while not the same as SST. Therefore, the value of SST minus AT increased with wind speed. The variation of AT and SST with wind speed effectively demonstrates the air-sea interaction.

3.6. Wave-Coherent Momentum Flux. The air-sea fluxes exchange is one of the primary relationships in air-sea interaction. The momentum flux is closely related to current and wave features. It is also related to wind strength. For instance, during the strong convection process on August 15th, the momentum flux increased significantly in relation to the temporarily strengthened surface winds. However, the correlation coefficient of the wind speed and momentum flux was 0.37 and far less than that of the South China Sea [12].

Based on the definition of momentum flux as (3), the momentum flux is proportional to the square of friction velocity, but not with wind speed. The wind speed can determine the intensity of waves. Therefore, the wave height is directly related to wind speed and results in the variations of sea surface roughness as well as friction velocity. The air-sea fluxes as momentum and heat fluxes (grouped by wave type and wave level) are listed in Table 3. The Level-2, Level-3, and Level-4 waves correspond to average wind speeds of 1.8, 3.0, and 9.6 m/s, maximum wind speeds of 7.8, 9.2, and 13.8 m/s, and friction velocities of 0.29, 0.4, and 0.59 m/s, respectively. The friction velocity above the sea surface and momentum

TABLE 3: Wave-coherent momentum flux and heat fluxes.

Category	Grouped by wave shape		Grouped by wave level	
	Wind wave	Swell wave	Level-2 (0.1–0.4 m)	Level-3 (0.5–1.2 m)
Average momentum flux (W/m^2)	0.2	0.1	0.1	0.2
Friction velocity (m/s)	0.4	0.3	0.3	0.4
Average sensible heat flux (W/m^2)	19.7	29.0	28.2	13.9
Average latent heat flux (W/m^2)	26.3	37.1	30.6	42.2

fluxes increased with wave height, with values of 0.14, 0.23, and 0.45 W/m^2, respectively. However, they were not linearly related to wind speed. Therefore, it is not appropriate to the linear dependence of momentum flux on wind speed in the Bohai Gulf. The wind speed may roughly represent the strength of momentum exchange; however, stratification, wave features (height, steepness, and age), breaking, and current speed can also lead to variations in the friction velocity at the ocean surface [22]. So, we can conclude that the air-sea momentum flux is synthetically affected by many factors.

In addition, Table 3 shows that wind speed effectively influenced both wave shape and wave height. With the wind increasing, wind waves developed. The stronger the wind becomes, the higher the waves, and the greater the friction velocity in the range of observation data. The mean momentum flux for wind waves was 0.21 W/m^2 and 0.14 W/m^2 for swell waves. The mean friction velocity for wind waves was 0.38 m/s, and 0.29 m/s for swell waves. It indicated a stronger momentum exchange under the former conditions.

3.7. Wave-Coherent Heat Fluxes. Heat fluxes over ocean surface are in related to difference of air-sea temperature, low-atmosphere stability, humidity, wind speed, and whitecap. The diurnal variation magnitudes of AT and humidity over the sea are lower than over the land. Therefore, the heat fluxes are also comparatively lower. During the experiment period, the mean sensible heat flux was 21.6 W/m^2, and the median value was 5.9 W/m^2. It peaked in the afternoon, and decreased rapidly to nearly zero afterwards. The mean latent heat flux was 27.8 W/m^2, with a median of 21.8 W/m^2. In the daytime, the latent heat flux increased, while at night it decreased and approached zero without a notable peak. From August 6th to August 10th (with the influence of cold inflow from the northern area) the difference of air-sea temperature increased, and so did the sensible heat flux. The variations of latent heat flux and relative humidity showed an apparent anti-phase. The latent heat flux was lower during high humidity. The opposite was true during low humidity. For example, from August 10th to August 13th, northern winds prevailed in relation to the invasion of cold airflow. Here, the AT decreased with the lower humidity from 90% to 50%, and the latent heat flux was greater during this period than on other days. Overall, the ratio of sensible and latent heat flux was 1.3 during the experiment period. It indicated their comparable importance for the air-sea exchange of heat. And there was distinct from the dominant position of latent

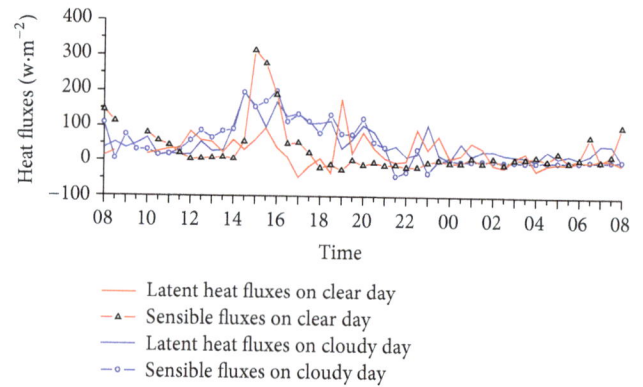

FIGURE 6: Diurnal variations in sensible and latent heat fluxes on clear and cloudy days.

heat flux found in the southern tropical sea area (the mean ratio of sensible and latent heat flux was 79 in July) [23]. In terms of diurnal variation, the magnitude of sensible heat flux was larger than that of the latent heat flux. However, the average daily latent heat flux was generally greater here than that of sensible heat flux in the location.

On clear and cloudy days (Figure 6), the heat fluxes were higher during the daytime afternoon and lower during the nighttime and early morning. The maximum sensible flux was greater on clear days than on cloudy days, with a sub-peak around 0600-0800 BJT. The mean sensible heat fluxes on clear and cloudy days were 44.8 W/m^2 and 24.7 W/m^2, respectively. The differentiation was due to Air-sea temperature difference under the condition of sunny and cloudy However, the average daily latent heat fluxes were similar under two kinds of weather conditions, 36.5 W/m^2 and 35.6 W/m^2, respectively. It was because latent heat fluxes are generally related to variations in air humidity.

4. Discussion and Conclusions

The paper used observational data collected during the composite observation experiment from early-mid August, 2011. Combining with conventional observations, meteorological and hydrological characteristics and their related air-sea fluxes were analyzed. Meanwhile, the limitation of the air-sea temperature difference for determining atmospheric stability was presented. The main conclusions are as follows.

(1) During early-mid August over the Bohai Gulf, southeaster predominates, followed by the northeaster. The winds

are light to gentle breeze. The relative humidity was maintained above 60% without any significant diurnal variations. The AT was 26.2°C on average and had unimodal diurnal curve. It was determined by the amount of cloud cover, precipitation, cold airflows, and so forth. The SST was 25.9°C on average, and its daily variation was M-type. The variation of SST was mainly influenced by the tides with the minimum SST corresponding to the high water level during the observation period.

(2) The wave height was positively related to the wind speed. During the experiment, the proportion of Level-2 waves was 73.3% and Level-3 waves 23.6%. Meanwhile, Level-4 waves account only for 2.1%. The mean wind speed during wind waves was 3.6 m/s and 1.5 m/s for swell waves. The swell waves appeared more frequently than the wind waves.

(3) The stability of the MBL over the Bohai Gulf was generally unstable, accounting for 70.2% overall in the background of the summer monsoon during early-mid August. Meanwhile, stability was related to wave shape and height. In the presence of wind waves, unstable stratification accounted for 52%, and for swell waves the proportion was 78%.

(4) Air-sea exchange was strengthened with the wind speed increasing. The mean surface momentum flux was $0.21\,\text{W/m}^2$ with the friction velocity of 0.38 m/s for wind waves and greater than that for swell waves.

(5) When compared to the momentum flux, the diurnal variations of heat fluxes were weaker. The sensible heat was transported from the sea surface into the air with the average of $21.6\,\text{W/m}^2$ during the daytime and decreased to nearly zero during the night and affected greatly by insolation and cold advection. In addition, the latent heat flux was $27.8\,\text{W/m}^2$ on average and affected greatly by relative humidity. The ratio of latent heat to sensible heat flux was average 1.3, far lower than that found in the South China Sea.

Finally, it should be noted that the results presented were based on wind speeds mostly less than 8 m/s (accounted for 97.6%), so air-sea fluxes in the paper may be only limited to the condition of light to gentle breeze. And being a semienclosed inland sea, the Bohai Sea has complex terrain, and our results about meteorological and hydrological characteristics only represented an in situ sea area in the Bohai Gulf. Whether or not the results can be extended to the whole Bohai Sea area, more investigations were needed.

Acknowledgments

This work was supported by the financial support of the Meteorological Research Project of China (GYHY201006034; GYHY201106006), the National Natural Science Foundation of China (41075004), Natural Science Foundation of Tianjin (13JCYBJC20000), Science and Technology Xinghai Project of Tianjin (KJXH2012-25).

References

[1] I. Rivin and E. Tziperman, "Sensitivity of air-sea fluxes to SST perturbations," *Journal of Climate*, vol. 10, no. 10, pp. 2431–2446, 1997.

[2] J. B. Edson, A. A. Hinton, K. E. Prada, J. E. Hare, and C. W. Fairall, "Direct covariance flux estimates from mobile platforms at sea," *Journal of Atmospheric and Oceanic Technology*, vol. 15, no. 2, pp. 547–562, 1998.

[3] B. Liu, C. L. Guan, and L. A. Xie, "The wave state and sea spray related parameterization of wind stress applicable from low to extreme winds," *Journal of Geophysical Research*, vol. 117, pp. 1–10, 2012.

[4] E. C. Monahan and I. O. Muircheartaigh, "Optimal power-law description of oceanic whitecap coverage dependence on wind speed," *Journal of Physical Oceanography*, vol. 10, no. 12, pp. 2094–2099, 1980.

[5] E. C. Monahan, C. W. Fairall, K. L. Davidson, and P. J. Boyle, "Observed inter-relations between 10m winds, ocean whitecaps and marine aerosols," *Quarterly Journal of the Royal Meteorological Society*, vol. 109, no. 460, pp. 379–392, 1983.

[6] M. A. Donelan, F. W. Dobson, S. D. Smith, and R. J. Anderson, "On the dependence of sea surface roughness on wave development," *Journal of Physical Oceanography*, vol. 23, no. 9, pp. 2143–2149, 1993.

[7] J. Liu, T. Xiao, and L. Chen, "Intercomparisons of air-sea heat fluxes over the Southern Ocean," *Journal of Climate*, vol. 24, no. 4, pp. 1198–1211, 2011.

[8] C. W. Fairall, E. F. Bradley, J. E. Hare, A. A. Grachev, and J. B. Edson, "Bulk parameterization of air-sea fluxes: updates and verification for the COARE algorithm," *Journal of Climate*, vol. 16, no. 4, pp. 571–591, 2003.

[9] B. Sun, L. Yu, and R. A. Weller, "Comparisons of surface meteorology and turbulent heat fluxes over the Atlantic: NWP model analyses versus moored buoy observations," *Journal of Climate*, vol. 16, no. 4, pp. 679–695, 2003.

[10] K. Sopkin, C. Mizak, S. Gilbert, V. Subramanian, M. Luther, and N. Poor, "Modeling air/sea flux parameters in a coastal area: a comparative study of results from the TOGA COARE model and the NOAA Buoy model," *Atmospheric Environment*, vol. 41, no. 20, pp. 4291–4303, 2007.

[11] J. T. Chuj, J. N. Chen, and L. Y. Xu, "Improved calculation of turbulent heat fluxes at air-sea interface in maritime China," *Oceanologia et Limnologia Sinica*, vol. 37, no. 6, pp. 481–487, 2006.

[12] J. Yan, H. Yao, J. Li et al., "Air-sea heat flux exchange over the South China Sea under different weather conditions before and after southwest monsoon onset in 2000," *Acta Oceanologica Sinica*, vol. 22, no. 3, pp. 369–383, 2003.

[13] D. Vickers and L. Mahrt, "Sea-surface roughness lengths in the midlatitude coastal zone," *Quarterly Journal of the Royal Meteorological Society*, vol. 136, no. 649, pp. 1089–1093, 2010.

[14] F. Anctil, M. A. Donelan, W. M. Drennan, and H. C. Graber, "Eddy-correlation measurements of air-sea fluxes from a discus buoy," *Journal of Atmospheric & Oceanic Technology*, vol. 11, no. 4, pp. 1144–1150, 1994.

[15] H. Liu, G. Peters, and T. Foken, "New equations for sonic temperature variance and buoyancy heat flux with an omnidirectional sonic anemometer," *Boundary-Layer Meteorology*, vol. 100, no. 3, pp. 459–468, 2001.

[16] M. Matthias and F. Thomas, *Documentation and Instruction Manual of the Eddy Covariance Software Package Tk2*, Universität Bayreuth, Abteilung Mikrometeorologie, 2004.

[17] M. Y. Tang, Y. Z. Liu, H. H. Li et al., "Seasonal variation characteristics of average surface sea temperature and its causes in the Bohai gulf, yellow sea and the north of the East China Sea," *Acta Oceanologica Sinica*, vol. 11, no. 5, pp. 544–553, 1989.

[18] M. Årthun and C. Schrum, "Ocean surface heat flux variability in the Barents Sea," *Journal of Marine Systems*, vol. 83, no. 1-2, pp. 88–98, 2010.

[19] E. L. Andreas, "A relationship between the aerodynamic and physical roughness of winter sea ice," *Quarterly Journal of the Royal Meteorological Society*, vol. 137, no. 659, pp. 1581–1588, 2011.

[20] B. X. Wang and R. F. Chang, "Criteria of differentiating swell from wind waves," *Journal of Oceanography of Huang Hai & Bohai Gulfs*, vol. 8, no. 1, pp. 16–24, 1990.

[21] A. Semedo, Ø. Saetra, A. Rutgersson, K. K. Kahma, and H. Pettersson, "Wave-induced wind in the marine boundary layer," *Journal of the Atmospheric Sciences*, vol. 66, no. 8, pp. 2256–2271, 2009.

[22] M. A. Donelan, B. K. Haus, N. Reul et al., "On the limiting aerodynamic roughness of the ocean in very strong winds," *Geophysical Research Letters*, vol. 31, no. 18, pp. 1–5, 2004.

[23] Q. Z. Sun, J. N. Cheng, J. Y. Yan et al., "The variation characteristics of air-sea fluxes over the Xi'sha area before and after the onset of the South China Sea monsoon in 2008," *Acta Oceanologica Sinica*, vol. 32, no. 4, pp. 12–23, 2010.

Relationship between Monthly Rainfall in NW Peru and Tropical Sea Surface Temperature

Juan Bazo,[1,2] **María de las Nieves Lorenzo,**[1] **and Rosmeri Porfirio da Rocha**[3]

[1] *Faculty of Sciences, Campus de Ourense, University of Vigo, 32004 Ourense, Spain*
[2] *Peruvian National Meteorological and Hydrological Service (SENAMHI), Casilla 11 1308, Lima 11, Peru*
[3] *Department of Atmospheric Sciences, Institute of Astronomy, Geophysics and Atmospheric Sciences, University of São Paulo, São Paulo, SP, Brazil*

Correspondence should be addressed to Juan Bazo; jbazo@senamhi.gob.pe

Academic Editor: Klaus Dethloff

This study assesses the relationship between global sea surface temperature (SST) and a regional index of rainfall (NWPR) in Piura-Tumbes, a coastal region in northwestern Peru, over the period 1965–2008 by means of the Pearson product-moment correlation. The results show that this area is strongly influenced by three indices: El Niño-Southern Oscillation (ENSO) Niño3.4 region, the Indian Ocean Dipole (IOD), and the equatorial Atlantic Oscillation (ATL3). In particular, a positive correlation has been found with the two first indices (Niño3.4 and IOD) and a negative one with ATL3 with several months of delay. This allows developing a forecast regression model for monthly rainfall in NW Peru with months in advance. The results show that linear regression model is not enough to provide satisfactory results; however, a nonlinear regression model improves considerably the prediction of rainfall anomalies in NW Peru.

1. Introduction

Sea surface temperature (SST) is a reliable variable to be used as a forecast tool. The high inertia of the sea makes the analysis of SST anomalies useful in monthly and seasonal rainfall prediction in some areas of the globe [1].

Previous works have shown the influence of anomalies in sea level pressure and precipitation variability on changes in the SST [2], in the Pacific area. Other works have analyzed the influence of global scale SST patterns on temperature and rainfall [3–6]. In particular, North Atlantic SST is related to precipitation anomaly in different European areas: Italy, Iceland, England, Iberian Peninsula, and Africa [7–10]. Also, prior researches have indicated some relationship between different areas of SST and precipitation in Peru and South America: Lagos et al. [11] employed lag-correlation analysis for statistical precipitation forecast from SST anomalies in the commonly used El Niño regions and precipitation in Peru; Woodman [12] found high correlation between rainfall in Piura and SSTs in various sectors of the eastern Pacific, allowing the development of a nonlinear regression model

to forecast (seasonal or monthly) precipitation in that region; Vuille et al. [13] observed that SST anomalies in the tropical Atlantic are closely related to the rainfall in the eastern Andes; González and Vera [14] found significant lagged correlations between SST anomalies over the tropical Indian Ocean and rainfall variability in Southern Andes.

Some researches explained that rainfall variability in South America is dominated by patterns such as El Niño-Southern Oscillation (ENSO), Intertropical Convergence Zone (ITCZ), Atlantic SST [15, 16], Indian Ocean Dipole (IOD) [17–19], and Madden-Julian Oscilation [20], with distinctive time scales, that is, intraseasonal, interannual, decadal, and long-term variations. Each pattern is closely related to SST variations of the corresponding time scales, implying the significance of the atmospheric ocean coupling to the variability of the hydrological cycle.

The climate on the northern coast of Peru is very special for two reasons: its high rainfall variability in time and its high correlation with the ENSO phenomenon. On the other hand, Tumbes and Piura regions are of great importance for the Peruvian economy due to the large fishing industry and

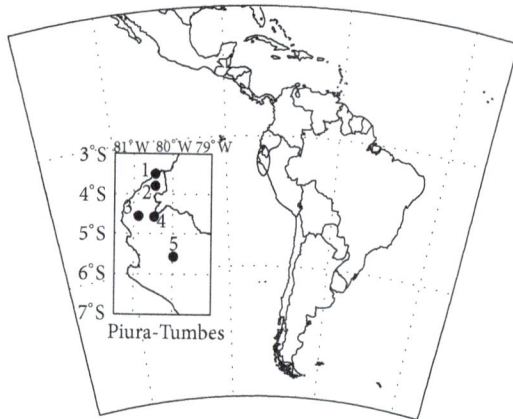

FIGURE 1: Location of Tumbes-Piura region and of the meteorological stations (numbers from 1 to 5).

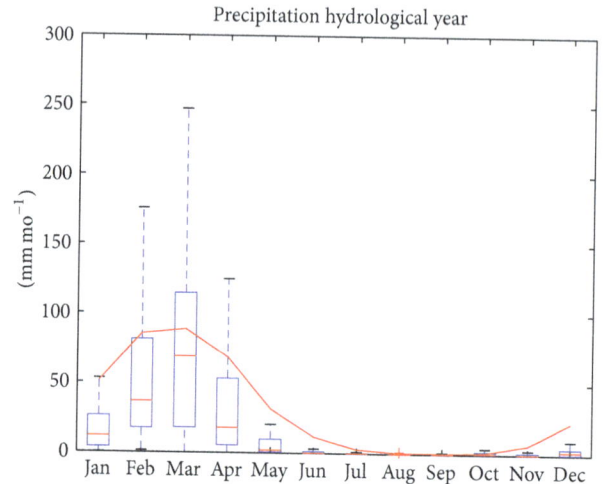

FIGURE 2: Annual hydrologic cycle corresponding to the monthly mean precipitation (red solid line) in the period 1965–2008. See text for further details.

agriculture. In addition, Tumbes and Piura are important tourist centers, generating important profits to the country. However, this region is set out to significant climatic variability, mainly due to the ENSO phenomenon with recurrent droughts and direct impacts on the biotic and physical coastal environments, both marine and terrestrial [12, 21–24]. This makes the region an area particularly sensitive to climate change impacts.

In this study, we investigate the association between global SST anomalies for 1965–2008 and monthly rainfall over Tumbes and Piura regions, in the northwestern of Peru. The main objective of this work is to investigate the potential for statistical precipitation forecasts based on global SST anomalies. Additionally, forecast regression models are proposed.

The paper is organized as follows. Details regarding the study area are provided in Section 2. Data and methodology used are described in Section 3. Results are presented and discussed in Section 4. Finally, conclusions are drawn in Section 5.

2. Study Area

Tumbes-Piura is a coastal region in northwestern Peru (from 3.50°S to 6.37°S and from 79.21°W to 81.33°W) (Figure 1). This region is located within a sharp climatic boundary between a warm and humid tropical area to the north (western Ecuador), and a cool desert land which borders the South Pacific Ocean southwards along some 3,000 km (5°–32°S). The north Peruvian coast lies also a major oceanographic limit which shows a strong thermal gradient between warm equatorial and the cold southwestern seawaters of South Pacific coast. This water is cooled down by cold currents and active upwelling processes: it is the area where the Peruvian (Humboldt) current and the Equatorial counter current are both deflected westwards, towards the central Pacific Ocean [25, 26].

A semitropical and tropical savanna in the center and the north coast and semiarid in the southern coast climates characterize the Tumbes-Piura region. Piura has tropical-dry climate monsoon weather; the average temperature is 27°C

throughout the whole year. Rainfall is scarce from May to November, and rainy season occurs from December to April at discontinuous rates. Rainfall varies in normal years from 100 to 300 mm, and it has a much defined dry season due to normal influence of ENSO. During the El Niño phenomenon, the rainfall is copious and makes the dry ravine becomes alive; the amount of precipitation in these areas exceeds more than 500% its normal value. In Figure 2, we have shown the mean annual hydrologic cycle and its variability in Tumbes-Piura region from 1965 to 2008. In this figure, the solid line represents the monthly mean values, solid line inside each box indicates the monthly median, the lower/upper whisker represents the minimum/maximum rainfall, and the lower/upper boxes represent the first/third quartiles, respectively. The period considered includes three strong El Niño events (1971–72, 1982–1983, and 1997–1998). Two clearly defined seasons (wet and dry) can be distinguished in Figure 2, with the wet season occurring from December to May and the dry one from June to November. In Figure 2, it is evident that the wet season is characterized by large rainfall interannual variability. Three synoptic situations can temporarily produce rainfall in the form of the light rain showers or isolated thundershowers. All the three are related to an El Niño event: large-scale changes in the atmospheric circulation and Pacific Ocean currents favor the development near the equatorial trough, cyclonic activity, and/or diurnal heating [27].

2.1. Teleconnection Influences on Rainfall in the Study Area. The impacts of ENSO over Peru are well documented [11–16], Waylen and Poveda [28], 2002, Douglas et al. [29], Tapley and Waylen [30], Goldberg et al. [31], and Horel and Cornejo-Garrido [32], but the effects of the Indian Ocean and Atlantic Ocean variability onto Peru rainfall are less understood. Previous studies show that Indian Ocean can remotely affect the atmospheric circulation over South America [33] leading to a significant impact on precipitation [17]. Drumond and

TABLE 1: Names and locations of the meteorological stations with rainfall data from 1965 to 2008.

Station	Altitude (m)	Latitude (S)	Longitude (W)
(1) Puerto Pizarro	1	$-3°30'$	$-80°27'$
(2) Rica Playa	98	$-3°48'$	$-80°27'$
(3) Pananga	440	$-4°33'$	$-80°53'$
(4) Lancones	135	$-4°34'$	$-80°29'$
(5) Virrey	275	$-5°34'$	$-79°58'$

TABLE 2: Monthly mean (mm), standard deviation (mm), and the coefficient of variation (CV) of the mean rainfall from the five stations (see Figure 1) in NW Peru.

Month	Mean	Standard deviation	CV (%)
January	52	36	70
February	88	41	47
March	117	42	36
April	60	35	59
May	27	57	209
June	7	14	190
July	1	10	828
August	0	1	325
October	1	1	63
November	2	10	440
December	16	30	187

Ambri [18] reported that warm SST anomalies in the Indian Ocean reach South America via teleconnection pattern, which potentially induces changes onto the circulation and rainfall over the continent during austral summer. They have showed composite of DJF precipitation anomalies for the warm IOD extreme events impacting southern Brazil and northwestern Peru. Likewise, Taschetto and Ambrizzi [19] explained a remote teleconnection pattern between the warming of tropical Indian Ocean and the atmospheric circulation/precipitation over South America. In this study, they presented results from observations, reanalysis, and simulations providing evidences of the modulation of South America summer rainfall by Indian Ocean SST variability via modifications of the Walker circulation pattern and wave-train teleconnection. In these experiments, they have calculated a partial correlation excluding the influence of El Niño index, obtaining a positive signal in seasonal precipitation over northwestern Peru.

In the same way, Vuille et al. [34] and Yoon and Zeng [35] explained a different remote mechanism associated with Atlantic Ocean affecting rainfall over South America. Besides, teleconnections of the Atlantic, Indian, and Pacific Oceans were identified by Wang et al. [36] to affect the South America. Associated with the interbasins, SST variability showed two zonal anomalous Walker circulation cells, one over tropical Pacific Ocean and another one over tropical Atlantic Ocean [36].

3. Data and Methods

Monthly precipitation data from 1965 to 2008 were obtained from the database of Peruvian National Meteorological and Hydrological Service (SENAMHI). These data underwent a quality control procedure with substitutions made for poor quality and some missing data, similar to the one used in the NCDC (National Climate Data Center, NOAA) for GHCN (Global Historical Climate Network) database [37]. Quality control for these series gave a result of only 0.01% of missing data and 90% of correlation with neighbor stations.

Table 1 summarizes the geographic characteristics of the meteorological station used in this study, and their locations are given in Figure 1. Averaging rainfalls corresponding to the five stations can constitute a valid procedure in our study because they are highly correlated. The correlations for monthly rainfall among them vary from 0.84 to 0.93 with a statistical significance of 90%.

The rainfall regime shows important interannual variations (Table 2), which can be characterized in terms of the mean, standard deviation, and the coefficient of variation (CV):

$$CV = 100 * \frac{\sigma Ri}{Ri}, \tag{1}$$

where R is the rainfall, the subscript (i) refers to the month, and (Ri) and (σRi) are the mean and the standard deviation, respectively. The CV is a dimensionless number, which ranges from 36% in March to 828% in July, which is higher during the dry season.

Monthly rainfall was expressed as anomaly relative to the mean rainfall of the period 1965–2008. The *nondimensional* rainfall anomaly index NWPR (North Western Peru Rainfall index; Phillips and McGregor [8–10, 38]), adopted in the present study is defined as:

$$NWPR = 100 \sum_{1}^{N} \left(\frac{X}{\overline{\overline{X}}} \right), \tag{2}$$

where X is the monthly rainfall anomaly at one station in mm, $\overline{X}$ is the station's mean annual rainfall in mm, and N is the number of stations. This index provides an adequate first approximation of monthly variations of the rainfall in the studied area.

SST was provided by NOAA/OAR/ESRL PSD (Smith and Reynolds [39]), and it is an extended reconstructed SST (ERSST; NOAA, ERSST-V3) obtained using the most recently available International Comprehensive Ocean-Atmosphere Data Set (ICOADS) SST and improved statistical methods that allow stable reconstruction using sparse data. This monthly analysis begins in January 1854, but due to the sparsity of data, the analyzed signal is heavily damped before 1880. Afterwards, the strength of the signal is more consistent over time. SST monthly averaged data are in a $2° \times 2°$ horizontal grid. In the present study, data were taken from January 1, 1965, through December 2008.

The values of the indices used, IOD (Indian Ocean Dipole), ATL3 (equatorial Atlantic Oscillation), and Niño3.4 (ENSO), were obtained from different sources.

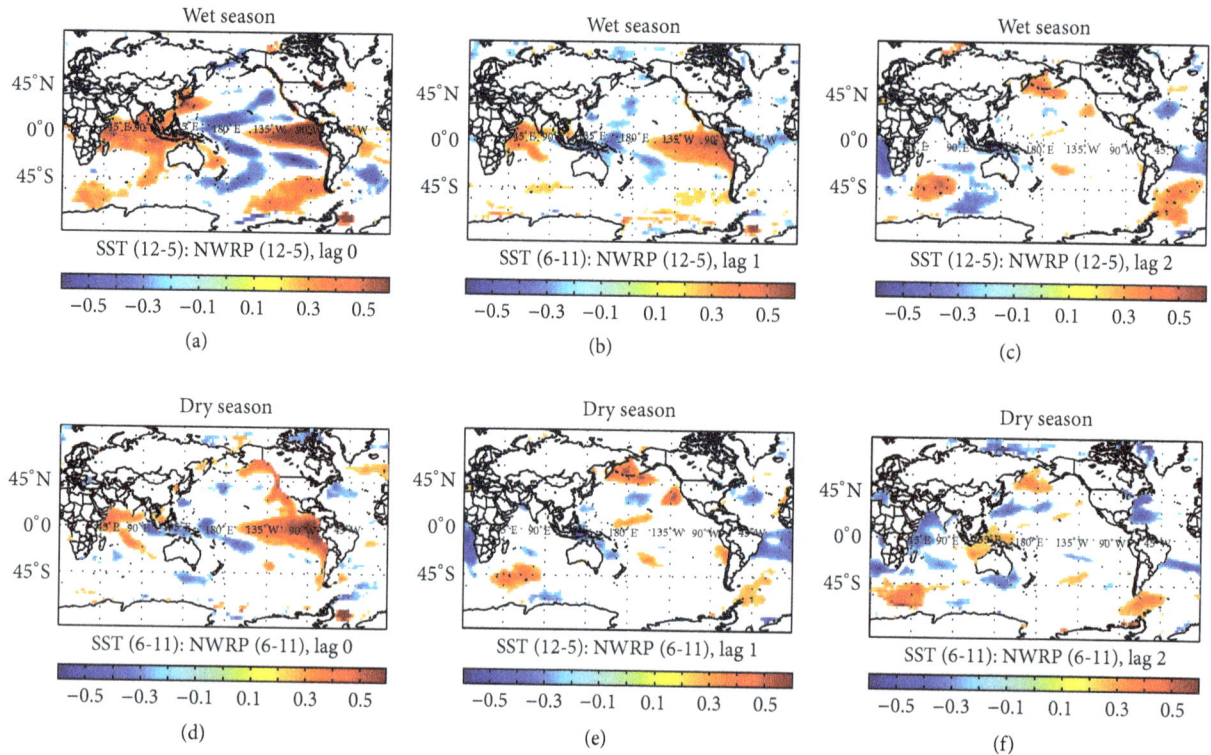

FIGURE 3: Correlation coefficient maps between SST and NWRP for wet (a, b, and c) and dry (d, e, and f) and different season lags.

The intensity of the IOD is represented by anomalous SST gradient between western equatorial Indian Ocean (50°E–70°E and 10°S–10°N) and southeastern equatorial Indian Ocean (90°E–110°E and 10°S–0°N). This gradient is named as Dipole Mode Index (DMI). When the DMI is positive (negative), then the phenomenon is refereed as positive (negative) IOD. The monthly values of this index were obtained from SST DMI dataset from 1958 to 2010 [33].

Following Zebiak [40], the ATL3 index was calculated as the SST anomaly in a region located in the central-eastern tropical Atlantic (3°S–3°N; 20°W–0°E).

The Niño3.4 SST anomaly index is an indicator of central tropical Pacific El Niño conditions (Trenberth [41]). It is an SST anomaly in the box 170°W–120°W, 5°S–5°N which was directly obtained from National Centers for Environmental Prediction (NCEP).

The Pearson product-moment correlation coefficient (r) was considered in a first step to quantify the linear association between the SST of each $2° \times 2°$ grid square and NWPR index. The coefficient significance was assessed to be greater than 95% by means of Student's t-test. As it is possible to obtain a statistically significant correlation by simply correlating two random series, we applied a test for field significance considering the properties of finiteness and interdependence of the spatial grid. More details of the applied test are shown in [8–10].

In a second step, the monthly correlations between the indices (Niño3.4, IOD, and ATL3) and NWPR were calculated considering 0–12 months of lags.

4. Results

4.1. Relationship between SST and Rainfall.
The concurrent and lagged correlation between SST and NWPR were calculated considering a seasonal division and 0–2 season lags. Two seasons were considered: one dry season from June–November and other wet season from December–May as shown in Figure 2.

Figure 3 shows the correlation maps between SST and NWPR for wet and dry seasons with their respective season lags. Although the correlations have a significance of only 90% in the applied field tests, it is possible to observe three major areas of influence of SST on northwestern Peru's rainfall. These areas are the ENSO region in the equatorial Pacific for lag 0–1 in the wet season and for lag 0 in the dry season; the Indian Ocean for lag 0–1 in the wet season and for lag 2 during dry season; the equatorial Atlantic Ocean for lag 1–2 in the wet season and for lag 1 in the dry season.

If we analyze the areas of Figure 3, we find that all these are related with climate indices. In particular, we find that the area observed in the tropical Pacific Ocean could correspond with that used to calculate the Niño3.4. The area found in the Indian Ocean could be related with the region considered to calculate the IOD, while that area detected in the Atlantic Ocean is similar to the area used to define the ATL3 index (see Figure 4).

ENSO also affects other ocean basins through perturbations in the Walker circulation which induce changes in cloud cover, evaporation, surface winds, and hence the net heat flux entering these remote oceans. This "atmospheric bridge"

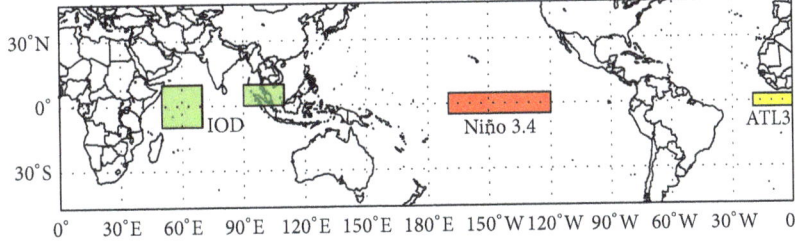

FIGURE 4: Location of SST areas used to calculate the indices IOD, ATL3, and Niño3.4.

leads to increased heat flux and positive SSTA in remote ocean basins such as the south China Sea, the Indian Ocean, and the tropical Atlantic, approximately ~3–6 months after SSTA peak in the tropical Pacific, Klein et al. [42]. The remote warming of tropical oceans leads to a coherent zonal mean warming throughout the troposphere from 30°N to 30°S during El Niño events. The peak warming however lags behind tropical Pacific SST by approximately 4 months due to the lagged response of the remote oceans [43].

The present results led us to analyze the relationship between these three indices and precipitation in the NW Peru at monthly time scale and time lags from 0 to 12 months.

Figure 5 shows the correlations between different monthly indices and monthly NWPR index (x-axis) and several lags (y-axis). The correlation coefficients showed in Figure 5 can help to elaborate a prediction model at monthly time scale.

4.2. Regression Forecast Models. In view of the findings in Figure 5, a linear regression model to predict the anomalies of rainfall was constructed. To develop the model, we must analyze Figure 5 and select the indices and months suitable to predict rainfall in each month. For example, Figure 5 indicates that the indices ATL3 of April/May, IOD of October/November, and Niño3.4 of November show strong correlations with the rainfall of December. A possible linear regression model to predict the December rainfall would be:

$$Y^{m^y} = \frac{a * SST^{m^1}_{[-3,3][3,20\,W]}}{ATL3}$$

$$+ \frac{b * SST^{m^2}_{[-10,10][50\,E,70\,E]} + c * SST^{m^2}_{[0,10][90\,E,110\,E]}}{IOD} \quad (3)$$

$$+ \frac{d * SST^{m^3}_{[-5,5][120\,W,170\,W]} + e}{Ni\tilde{n}o3.4},$$

where m^y = December, m^1 = April, m^2 = October, and m^3 = November.

During the period 1965–2008, this linear regression model to December rainfall provides a correlation of ~0.45 between the observed and predicted rainfall with a statistical significance of 95%. The same is obtained to January rainfall with a similar model. The linear model can predict the extremes episodes, but it is not a suitable predictor in the

normal years. The low values of correlation suggest that a linear regression model may not be the best solution to explain the relationship between SST indices and rainfall anomalies in NW of Peru.

In a previous work [12] assumed an exponential relationship between Piura precipitation and coastal SST. He proposed that if it were possible to predict the SST at the coast of NW Peru then would be possible to develop a rainfall prediction model. Following [12], in this present study, we assumed an exponential relationship between the NWPR rainfall and the SST of the ATL3, IOD, and Niño3.4 indices. The advantage of our study is that the influence of the indices presents several months of delay, and, in this way, knowing the temperature of these regions in particular previous months, we can elaborate a prediction of the rainfall in the next months. Therefore, considering the trend from April to November, we could make a prediction for December rainfall in NW Peru. We propose the following nonlinear model:

$$Y^{m^y} = a * \exp\left(\frac{-SST^{m^1}_{[-3,3][3,20\,W]}}{ATL3} \right.$$

$$+ \frac{SST^{m^2}_{[-10,10][50\,E,70\,E]} - SST^{m^2}_{[0,10][90\,E,110\,E]}}{IOD}$$

$$+ \left. \frac{SST^{m^3}_{[-5,5][120\,W,170\,W]}}{Ni\tilde{n}o3.4} \right) + b.$$

$$(4)$$

In (4), a minus sign is added to those areas of SST which have shown a negative correlation with the NWPR precipitation index. The superscripts indicate SST predictor month defined according Figure 5. For this model, we have achieved a correlation of ~0.85 between observed and predicted rainfall anomaly of December.

Table 3 presents the input variables (SST used to calculate ATL3, IOD, and Niño3.4 indices) and lags for linear and exponential models. These variables were selected considering the statistically significant correlations between SST and NWPR indices shown in Figure 5. Besides, Table 3 also includes the correlation coefficients between observed and modeled monthly rainfall. In order to validate our proposed models, besides the correlation coefficient (Table 3), we also calculated the mean absolute error (MAE) and the root mean

(a)

(b)

(c)

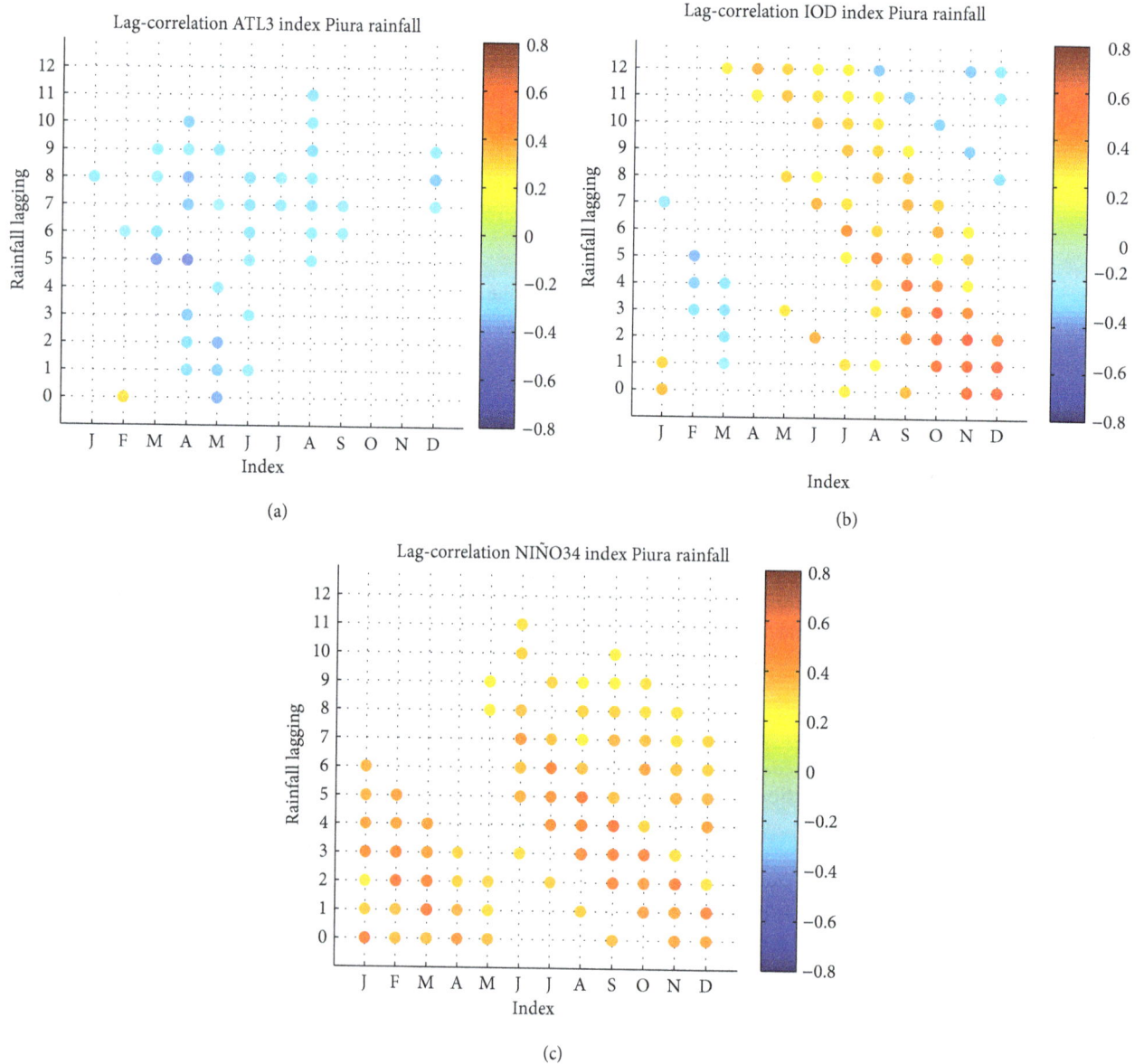

FIGURE 5: Monthly correlations between the (a) ATL3, (b) IOD, and (c) Niño3.4 indices and the NWPR index from 0 to 12 months lagged.

square error (RMSE) presented in Table 4. All statistical indices considered indicates that the nonlinear model reduces the errors in comparison with the linear model, that is, exponential model provides higher correlation and smaller MAE and RMSE. This might be explained by the nonlinear interactions between the Indian and Pacific Oceans as discussed in the recent study of Taschetto and Ambrizzi [19]. With statistical model proposed for prediction of rainfall, we can implement a system of early warning of floods in the NW Peru. The use of SST anomalies of remote areas helped us to make a forecast with several months in advance.

5. Conclusions

Many researches focus in the ENSO impact over Peru, but few studies have reported teleconnections pattern, of Indian

and Atlantic Oceans, with potential effects on rainfall in northwestern Peru.

This work has investigated links between SST variations in global oceans and rainfall index in NW Peru (NWPR). We have taken into account seasonal and monthly correlations between SST and NWPR from zero to two seasons of lag and from zero to twelve months of lag. We have applied finiteness and interdependence of the spatial grid criteria to avoid spurious correlations. We found significant lagged seasonal correlations with different tropical areas. These areas are related with the ENSO phenomenon, the Indian Ocean Dipole and the Equatorial Atlantic Oscillation. It is known that in these three areas patterns of SST variability are related with atmospheric circulation patterns.

An analysis of the relationships between SST and rainfall in Tumbes-Piura allowed us to elaborate a linear and an

TABLE 3: Variables and lags used in the linear and exponential models and correlations between modeled and observed rainfall in NW Peru.

Month	Parameters models	Correlation		Constant values	
		Exponential model	Linear model	Exponential model (a-b)	Linear model (a–e)
December	$m^1 = $ April, $m^2 = $ October, and $m^3 = $ November	0.85	0.45	(0.8527) (7.2447e − 17)	(−31.2257) (56.6475) (−18.5818) (17.2338) (−627.7589)
January	$m^1 = $ April, $m^2 = $ November, and $m^3 = $ December	0.85	0.6	(0.8517) (5.7781e − 17)	(−23.9076) (222.6269) (−114.2665) (15.4083) (−3.0392e + 03)
February	$m^1 = $ April, $m^2 = $ October, and $m^3 = $ August	0.78	0.42	(0.7845) (1.0589e − 16)	(−45.1209) (138.8208) (8.2054) (31.6697) (−3.5051e + 03)
March	$m^1 = $ July, $m^2 = $ November, and $m^3 = $ February	0.75	0.5	(0.7516) (−4.5024e − 17)	(−48.0442) (209.8999) (−59.8236) (16.5632) (−3.4925e + 03)
April	$m^1 = $ August, $m^2 = $ October, and $m^3 = $ February	0.82	0.48	(0.8219) (−3.0232e − 16)	(−35.4956) (74.0641) (39.4803) (66.5445) (−3.8771e + 03)
May	$m^1 = $ August, $m^2 = $ February, and $m^3 = $ March	0.85	0.43	(0.8493) (1.3529e − 16)	(−45.0423) (−56.2833) (78.6403) (46.9579) (−586.7617)

TABLE 4: Validation for the linear and exponential models for prediction of monthly rainfall from December to May.

Month	Linear		Exponential	
	MAE	RMSE	MAE	RMSE
December	0.76	1.03	0.28	0.52
January	0.68	0.88	0.23	0.52
February	0.88	1.07	0.45	0.61
March	0.81	0.98	0.45	0.65
April	0.80	1.01	0.36	0.56
May	0.78	1.05	0.36	0.52

exponential regression models to make monthly predictions of rainfall. We found robust and highly significant lagged correlations between SST indices and rainfall index. The linear regression model shows better skill in the prediction of January rainfall anomaly, when the correlation coefficient is the highest (0.6). However, considering all the period tested, the exponential relationship between SST and rainfall [12] provides better results. This exponential model improves the correlation in all analyzed months, especially in December,

February, and May. Considering the performance of the linear and exponential models, smaller absolute error and root mean square error are obtained with the exponential model.

In the present study, we find that despite being a remote response to ENSO events, the Indian Ocean has the potential to feedback onto the atmosphere and induce tropical and extratropical teleconnections over South America. This is in agreement with Taschetto and Ambrizzi [19] results that present evidences that South America rainfall (including Northwestern Peru) can be modulated by Indian Ocean SST variability via remote mechanisms. Furthermore, Wang et al. [36] explained teleconnetions patterns between the Indian, Atlantic, and Pacific Oceans and rainfall variability over South America.

The most important result is that the rainfall in NW of Peru is influenced by other areas besides ENSO areas, with several months in advance, which are related with known indices like the Indian Dipole Ocean or the equatorial Atlantic Oscillation. These results open new possibilities to understand and to monitor the rainfall in NW of Peru.

We consider future research studies on El Niño Modoki impacts on Peruvian rainfall variability. Tedeschi et al. [44]

presents different oceanic and atmospheric patterns when compared to Canonical ENSO. The impacts on South America precipitation indicate the importance of studying the two types separately.

Finally, we propose that further research should be conducted to study the mechanism involved in the correlations obtained in this study using atmospheric general circulation model (AGCM) and regional climate model (RCM). The first one would be used to assess the influence of SST over the circulation affecting the area under study and to provide boundary conditions to the RCM.

The RCM itself could provide the physical link between SST and rainfall, since AGCMs have difficulties with the distribution of precipitation and a dynamical downscaling may be needed [45].

Acknowledgments

This work was supported by "Xunta de Galicia" under Project 10PXIB383169PR and cofinanced by European Regional Development Fund (FEDER). This work was partially supported by Xunta de Galicia under the Programa de Consolidación e Estruturación de Unidades de Investigación (Grupos de Referencia Competitiva) funded by European Regional Development Fund (FEDER). CNPq (Grant 307202/2011-9) has also supported partially this work. The authors wish to thank Rita Ynoue for the English proofing and wish to thank the anonymous reviewers.

References

[1] A. G. Barnston, "Linear statistical short-term climate predictive skill in the Northern Hemisphere," *Journal of Climate*, vol. 7, pp. 1513–1564, 1994.

[2] D. P. Rowell, "Assessing potential seasonal predictability with an ensemble of multidecadal GCM simulations," *Journal of Climate*, vol. 11, no. 2, pp. 109–120, 1998.

[3] A. G. Barnston and T. M. Smith, "Specification and prediction of global surface temperature and precipitation from global SST using CCA," *Journal of Climate*, vol. 9, no. 11, pp. 2660–2697, 1996.

[4] M. J. Rodwell, D. P. Rowell, and C. K. Folland, "Oceanic forcing of the wintertime North Atlantic Oscillation and European climate," *Nature*, vol. 398, no. 6725, pp. 320–323, 1999.

[5] M. Drévillon, L. Terray, P. Rogel, and C. Cassou, "Mid latitude Atlantic SST influence on European winter climate variability in the NCEP reanalysis," *Climate Dynamics*, vol. 18, no. 3-4, pp. 331–344, 2001.

[6] M. N. Lorenzo, J. J. Taboada, I. Iglesias, and M. Gómez-Gesteira, "Predictability of the spring rainfall in North-west of Iberian from sea surfaces temperatures of ENSO areas," *Climatic Change*, vol. 107, no. 3-4, pp. 329–341, 2011.

[7] A. M. S. Delitala, D. Cesari, P. A. Chessa, and M. N. Ward, "Precipitation over Sardinia (Italy) during the 1946–1993 rainy seasons and associated large scale climate variations," *International Journal of Climatology*, vol. 20, pp. 519–541, 2000.

[8] I. D. Phillips and G. R. Mcgregor, "The relationship between monthly and seasonal South-West England rainfall anomalies and concurrent North Atlantic sea surface temperatures," *International Journal of Climatology*, vol. 22, no. 2, pp. 197–217, 2002.

[9] I. D. Phillips and J. Thorpe, "Icelandic precipitation—North Atlantic sea-surface temperature associations," *International Journal of Climatology*, vol. 26, no. 9, pp. 1201–1221, 2006.

[10] M. N. Lorenzo, I. Iglesias, J. J. Taboada, and M. Gómez-Gesteira, "Relationship between monthly rainfall in Northwest Iberian Peninsula and North Atlantic Sea surface temperature," *International Journal of Climatology*, vol. 30, no. 7, pp. 980–990, 2010.

[11] P. Lagos, Y. Silva, E. Nickl, and K. Mosquera, "El Niño, Climate Variability and Precipitation Extremes in Perú," *Advanced Geosciences*, vol. 14, pp. 231–237, 2008.

[12] R. Woodman, "Modelo estadístico de pronóstico de las precipitaciones en la costa norte del Peru," El Fenómeno El Niño. Investigación para una prognosis, 1er encuentro de Universidades del Pacifico Sur: Memoria 93-108, Piura-Peru, 1999.

[13] M. Vuille, R. S. Bradley, and F. Keimig, "Climate variability in the Andes of Ecuador and its relation to tropical Pacific and Atlantic Sea Surface temperature anomalies," *Journal of Climate*, vol. 13, no. 14, pp. 2520–2535, 2000.

[14] M. H. González and C. S. Vera, "On the interannual wintertime rainfall variability in the Southern Andes," *International Journal of Climatology*, vol. 30, no. 5, pp. 643–657, 2010.

[15] J. Zhou and K.-M. Lau, "Principal modes of interannual and decadal variability of summer rainfall over South America," *International Journal of Climatology*, vol. 21, no. 13, pp. 1623–1644, 2001.

[16] K. Takahashi, "The atmospheric circulation associated with extreme rainfall events in Piura, Peru, during the 1997-1998 and 2002 El Niño events," *Annales Geophysicae*, vol. 22, no. 11, pp. 3917–3926, 2004.

[17] S. C. Chan, S. K. Behera, and T. Yamagata, "Indian Ocean Dipole influence on South American rainfall," *Geophysical Research Letters*, vol. 35, no. 14, pp. 10–14, 2008.

[18] A. R. Drumond de M and T. Ambrizzi, "The role of the South Indian and Pacific oceans in South American monsoon variability. Find out how to access preview-only content," *Theoretical and Applied Climatology*, vol. 94, no. 3-4, pp. 125–137, 2008.

[19] A. S. Taschetto and T. Ambrizzi, "Can Indian Ocean SST anomalies influence South American rainfall?" *Climate Dynamics*, vol. 38, no. 7-8, pp. 1615–1628, 2012.

[20] L. M. V. Carvalho, C. Jones, A. E. Silva, B. Liebmann, and P. L. Silva Dias, "The South American Monsoon System and the 1970s climate transition," *International Journal of Climatology*, vol. 31, no. 8, pp. 1248–1256, 2011.

[21] R. C. Murphy, "Oceanic and climatic phenomena along the west coast of South America during 1925," *Geographical Review*, vol. 16, pp. 26–54, 1926.

[22] C. N. Caviedes, "El Niño 1972: its climatic, ecological, human and economic implications," *Geographical Review*, vol. 65, no. 4, pp. 494–509, 1975.

[23] C. N. Caviedes, "El Niño 1982-83," *Geographical Review*, vol. 74, no. 3, pp. 267–290, 1984.

[24] D. B. Enfield, "Progress in understanding El Niño," *Endeavour*, vol. 11, no. 4, pp. 197–204, 1987.

[25] S. Zuta and O. Guillén, "Oceanografía de las aguas costeras del Perú," *El Boletín del Instituto del Mar del Perú*, vol. 2, no. 5, pp. 157–324, 1970.

[26] R. Mugica, "Oceanografía del mar peruano," in *Historia Marítima Del Perú*, vol. I, no. I, pp. 216–474, 1972.

[27] M. T. Gilfort, M. J. Vojtesak, G. Myles, R. C Bonan, and D. L. Martens, *South America: South of the Amazon River A Climatology Study*, Environmental Technical Application Center, 1992.

[28] P. Waylen and G. Poveda, "El Nino-Southern Oscillation and aspects of western South American hydro-climatology," *Hydrological Processes*, vol. 16, no. 6, pp. 1247–1260, 2002.

[29] M. W. Douglas, M. Pena, N. Ordinola et al., "Synoptic and spatial variability of the rainfall along the northern Peruvian coast during the 1997-8 El Nino event," in *Proceedings of the Sixth International Conference on Southern Hemisphere Meteorology*, pp. 104–105, Santiago, Chile, 2000.

[30] T. D. Tapley Jr. and P. R. Waylen, "Spatial variability of annual precipitation and ENSO events in western Peru," *Hydrological Sciences Journal/Journal des Sciences Hydrologiques*, vol. 35, no. 4, pp. 429–446, 1990.

[31] R. A. Goldberg, G. Tisnado, and R. A. Scofield, "Characteristics of extreme rainfall events in north-western Peru during the 1982-1983 El Nino period," *Journal of Geophysical Research*, vol. 92, no. C13, pp. 14225–14241, 1987.

[32] J. D. Horel and A. G. Cornejo-Garrido, "Convection along the coast of northern Peru during 1983: spatial and temporal variation of clouds and rainfall," *Monthly Weather Review*, vol. 114, no. 11, pp. 2091–2105, 1986.

[33] N. H. Saji, B. N. Goswami, P. N. Vinayachandran, and T. Yamagata, "A dipole mode in the tropical Indian ocean," *Nature*, vol. 401, no. 6751, pp. 360–363, 1999.

[34] M. Vuille, R. S. Bradley, and F. Keimig, "Interannual climate variability in the Central Andes and its relation to tropical Pacific and Atlantic forcing," *Journal of Geophysical Research D*, vol. 105, no. 10, pp. 12447–12460, 2000.

[35] J. H. Yoon and N. Zeng, "An Atlantic Influence on Amazon rainfall," *Climate Dynamics*, vol. 34, no. 2-3, pp. 249–264, 2009.

[36] C. Wang, F. Kucharski, R. Barimalala, and A. Bracco, "Teleconnections of the tropical Atlantic to the tropical Indian and Pacific Oceans: a review of recent findings," *Meteorologische Zeitschrift*, vol. 18, no. 4, pp. 445–454, 2009.

[37] T. C. Peterson, R. Vose, R. Schmoyer, and V. Razuvaev, "Global historical climatology network (GHCN) quality control of monthly temperature data," *International Journal of Climatology*, vol. 18, no. 11, pp. 1169–1179, 1998.

[38] I. D. Philips and G. R. McGregor, "Western European water vapor flux-southwest England rainfall associations," *Journal of Hydrometeorology*, vol. 5, pp. 505–523, 2001.

[39] T. M. . Smith and R. W. Reynolds, "Improved Extended Reconstruction of SST, (1854-1997)," *Journal of Climate*, vol. 17, pp. 2466–2477, 2004.

[40] S. E. Zebiak, "Air-sea interaction in the equatorial Atlantic region," *Journal of Climate*, vol. 6, no. 8, pp. 1567–1586, 1993.

[41] K. E. Trenberth, "The definition of El Niño," *Bulletin of the American Meteorological Society*, vol. 78, no. 12, pp. 2771–2777, 1997.

[42] S. A. Klein, B. J. Soden, and N. C. Lau, "Remote sea surface temperature variations during ENSO: evidence for a tropical atmospheric bridge," *Journal of Climate*, vol. 12, no. 4, pp. 917–932, 1999.

[43] J. C. H. Chiang and A. H. Sobel, "Tropical tropospheric temperature variations caused by ENSO and their influence on the remote tropical climate," *Journal of Climate*, vol. 15, no. 18, pp. 2616–2631, 2002.

[44] R. G. Tedeschi, I. F. A. Cavalcanti, and A. M. Grimm, "Influences of two types of ENSO on South American precipitation," *International Journal of Climatology*, 2012.

[45] F. Giorgi and L. O. Mearns, "Introduction to special section: regional climate modeling revisited," *Journal of Geophysical Research*, vol. 104, no. D6, pp. 6335–6352, 1999.

3

A Methodical Approach to Design and Valuation of Weather Derivatives in Agriculture

Jindrich Spicka and Jiri Hnilica

Department of Business Economics, University of Economics, Prague, W. Churchill Square 4, 130 67 Prague 3, Czech Republic

Correspondence should be addressed to Jindrich Spicka; jindrich.spicka@vse.cz

Academic Editor: Ismail Gultepe

The paper deals with weather derivatives as the potentially effective risk management tool for agricultural enterprises seeking to mitigate their income exposure to variations in weather conditions. Design and valuation of the weather derivatives is an interdisciplinary approach covering agrometeorology, statistics, mathematical modeling, and financial and risk management. This paper first offers an overview of data sources and then methods of design and valuation of weather derivatives at the regional level. The accompanied case study focuses on cultivation of cereals (wheat and barley) in the Czech Republic. However, its generalizability is straightforward. The analysis of key growing phases of cereals is based on regression analysis using weather indices as the independent variables and crop yields as dependent variables. With the bootstrap tool, the burn analysis is considered as useful tool for estimating uncertainty about the payoff, option price, and statistics of probability distribution of revenues. The results show that the spatial and production basis risks reduce the efficiency of the weather derivatives. Finally, the potential for expansion of weather derivatives remains in the low income countries of Africa and Asia with systemic weather risk.

1. Introduction

Weather determines decision making over the world [1, 2]. Lazo et al. [3] quantify 3.4% interannual aggregate dollar variation in US economic activity that is attributable to the weather variability. Nevertheless, the degree of the weather sensitivity in the high income countries is relatively small because of minor share of agriculture in GDP. On the contrary, the underdeveloped world and many emerging countries with the substantial share of agriculture in GDP and employment have to face the weather risks without sufficient financial resources and infrastructure to manage them.

A financial weather contract (weather derivative, weather index insurance) is a weather contingent contract whose payoff is determined by future weather events. The contract links payments to a weather index that is the collection of weather variables measured at a stated location during an explicit period [4]. Underlying "assets" of weather derivative are most often air temperature, rainfall, wind speed, and so forth. Many relevant studies test the potential use of the weather derivatives in agriculture [5–11].

The main advantage of the index-based financial tools is their power to reduce the information asymmetry [12–16]. The derivative payoff is estimated by objective, measurable, and transparent weather variable which cannot be intentionally modified by farmers or any other subject. Alternatively, the most important disadvantage of the weather derivatives and the index insurance they most frequently advert to is the basis risk. The cause of the production basis risk is that individual yield fluctuations in general are not perfectly correlated with the relevant weather variable. The spatial basis risk arises from the difference in weather patterns at the reference point of the derivative and the location of agricultural production [17].

There is also one important prerequisite for the efficient use of the weather derivatives in agriculture. The weather derivatives are more effective in hedging the revenue risk for products with a greater likelihood for low correlations between yield and price [18, 19]. These arguments also raise the need to evaluate the price-yield concurrence.

This paper aims to assess the potential of the weather derivatives to reduce revenue risk in agriculture taking into consideration the growing conditions in the Czech Republic. The problem of risk management scheme in the Czech agriculture is that the systemic risks are not covered by insurance (drought, heat waves, and persistent rain at harvest).

The paper is organized as follows. The first part focuses on description of methodical approach for design and valuation of weather derivatives at the regional level. The second part presents results and discussion of the main findings, chances, and limitations of agricultural weather derivatives.

2. Material and Methods

Design and valuation of the weather derivatives is an interdisciplinary approach covering agrometeorology, statistics, mathematical modeling, and financial and risk management. The methodology is a modification of the procedures proposed by Jewson et al. [20], Mußhoff et al. [7], Hnilica [21], and Manfredo and Richards [22].

Argometeorological Issues

    (1) Selection of regions suitable for construction of the weather derivatives.

    (2) Selection of agricultural crops suitable for construction of the weather derivatives.

    (3) Selection of relevant available data sources.

Statistical Issues

    (4) Adjustment of data time series.

    (5) Identification of underlying index of the weather derivatives.

Mathematical Modeling

    (6) Identification of a suitable type of probability distribution of the index contract (weather). Risk layering.

    (7) Weather derivative valuation.

Risk Management Issues

    (8) Assessment of efficiency of the weather derivative contract to reduce revenue risk.

2.1. Agrometeorological Issues. Selection of regions for design of the weather derivatives is based on whether the agricultural production in the region is sufficiently important to ensure liquidity of the contract. Structural and economic data show that four Czech regions (Středočeský, Jihočeský, Vysočina, and Jihomoravský) produce more than 50% of national agricultural output. In addition, the Olomoucký and Královéhradecký regions have relatively higher risk of drought because of large lowlands. These regions are also characterized by cereal production on fertile soils. Table 1 shows basic structural and economic characteristics of selected regions.

Barnett and Mahul [5] suggest at least 20 annual data on crop yields to set a contract price for the relatively frequent adverse weather events. In this paper, the 40-year time series (1970–2009) of crop yields are obtained from the Czech

Statistical Office (CSO). The CSO is a central governmental authority in processing and publishing official statistical information. The risk analysis requires the use of data on the sown areas instead of the harvest areas [19]. Because of lack of reliable data on the sown areas, the crop yield is calculated as the share of harvested production (t) to the harvest area (ha).

Weather data are purchased from the Czech Hydrometeorological Institute (CHMI). The analysis uses the daily/monthly average air temperatures and daily/monthly cumulative rainfall. Monthly weather data are the spatial averages of data provided by the professional meteorological stations located in selected six regions. We adopt the following weather indices—air temperature (°C), rainfall (mm), and drought index S_i (combination of air temperature and rainfall, [23]) and CDD/HDD (i.e., the number of degrees that a day's average temperature is above/below a certain level). Basic air temperature of CDD is set using linear optimization to achieve the highest Pearson correlation coefficient between basic air temperature and crop yield.

2.2. Statistical Issues. The obvious technological progress in cultivation of the field crops requires detrending of the yield time series [9–11]. Linear, exponential, logarithmic, power, and polynomial (quadratic and cubic) trends are tested. The power trend is the most suitable for wheat and barley (according to the R^2). Deviations from trend are related to the 5-year average yield (2005–2009) when excluding the highest and lowest values in order to express the current average level of yields (Y_{avg}^{tr}). The detrended data are then calculated as

$$Y_t^{detr} = Y_t \frac{Y_{avg}}{Y_t^{tr}}, \qquad (1)$$

where Y_t denotes actual data in year t and Y_t^{tr} means the trend.

The regression analysis estimates the relationships between the weather index and the crop yields in critical period of vegetation (SPSS Statistics 18.0). Following types of regression functions are tested: linear, logarithmic, inverse, quadratic, cubic, power, S curve, exponential, and logistic. In order to achieve the highest possible correlation between yield and weather variables, we set weights to the critical month of vegetation. The weights are set through the highest value of correlation coefficient between yield and weather variable during the critical period of vegetation (MS Excel Solver).

The simple linear regression between the crop yield and the weather index enables to clearly find the strike level (the level of index when the contract triggers). Then we set the tick (the payoff per one index point above/below the strike level) as an expected postharvest price and the regression coefficient β. The regression function is tested for autocorrelation (Breusch-Godfrey test) and heteroskedasticity (White test) at $\alpha = 0.05$.

2.3. Mathematical Modeling. The analysis compares the effectiveness of the contracts for the two strike levels. The first strike covers risks with low frequency and high severity (strike 1). It assumes a certain degree of farmers' own contribution. At the strike 1 the probability of payoff ranges

TABLE 1: Basic characteristic of selected regions in the Czech Republic (2010).

	Středočeský	Jihomoravský	Vysočina	Jihočeský	Královéhradecký	Olomoucký
Agricultural holdings	**5,082**	**9,967**	**4,176**	**4,483**	**2,793**	**2,217**
with agricultural land up to 10 ha	2,748	8,615	2,505	2,554	1,727	1,313
Agricultural output (CZK million, current prices)	**18,939**	**12,341**	**11,408**	**11,216**	**7,637**	**7,099**
Crop output	12,162	7,670	5,337	5,028	4,056	4,008
Animal output	6,180	3,997	5,653	5,419	3,304	2,389
Agricultural services output	446	348	290	419	205	167
Utilised agricultural land (ha)	**663,524**	**362,381**	**364,600**	**422,246**	**234,041**	**244,172**
Arable land	551,096	321,537	283,052	258,465	168,653	177,223
Permanent grassland	70,978	21,153	79,990	158,948	62,278	64,358
Crop production—harvest (tonnes)						
Cereals, total	1,388,775	1,080,301	610,140	624,183	426,944	520,048
Potatoes, total	143,438	32,008	254,517	91,689	26,192	15,039
Rape	210,270	95,015	112,472	120,983	62,776	64,283
Livestock production						
Livestock intensity						
Cattle (heads/100 ha agr. land)	25.8	16.1	57.9	49.7	42.7	36.7
Pigs (heads/100 ha arable land)	60.9	80.8	99.2	67.0	73.7	62.9
Meat production (tonnes)	49,973	43,900	53,538	41,319	23,267	25,762
Milk production (thous. l)	324,039	146,827	447,592	319,395	205,659	178,612

Source: Czech Statistical Office—Regional statistical yearbooks (http://www.czso.cz).

TABLE 2: The results of the test of various weather indices.

	$T2$	$T3$	$S_i 2$	$S_i 3$	CDD 1	CDD 2	CDD 3
Wheat							
Critical month (weight)		4 (0.074)		4 (0.152)			
	5 (0.257)	5 (0.213)	5 (0.555)	5 (0.440)			
	6 (0.743)	6 (0.713)	6 (0.445)	6 (0.408)			
Pearson correlation	−0.69	−0.693	−0.617	−0.632	−0.503	−0.522	−0.569
	($P < 0.0001$)	($P < 0.0001$)	($P < 0.0001$)	($P < 0.0001$)	($P = 0.0009$)	($P = 0.0005$)	($P = 0.0001$)
R^2	0.477	0.481	0.381	0.399	0.253	0.273	0.323
Barley							
Critical month (weight)		4 (0.194)		4 (0.271)			
	5 (0.264)	5 (0.149)	5 (0.500)	5 (0.307)			
	6 (0.736)	6 (0.657)	6 (0.500)	6 (0.422)			
Pearson correlation	−0.666	−0.688	−0.592	−0.645	−0.459	−0.45	−0.497
	($P < 0.0001$)	($P < 0.0001$)	($P < 0.0001$)	($P < 0.0001$)	($P = 0.0029$)	($P = 0.0036$)	($P = 0.0011$)
R^2	0.443	0.473	0.35	0.416	0.211	0.2027	0.2472

Source: authors.

from 10% to 15%. The strike 2 represents relatively more frequent and less severe weather risks and requires no farmers' contribution.

Weather derivative pricing in the paper is based on the burn analysis. The burn analysis reflects how a contract has performed in previous years. Burn analysis in this paper is enhanced by distribution fitting and Monte Carlo simulation. The probability distribution of the weather index is estimated from the real data though the Maximum Likelihood Estimation (MLE) method [24]. The probability distribution is tested simultaneously through the Anderson-Darling test goodness-of-fit test (A-D) at $\alpha = 0.05$.

Bootstrapping [25] allows for easier estimation of uncertainty surrounding the estimate of mean and standard deviation of payoff. The parametric bootstrap [24] requires the extra information about the probability distribution.

The contract price (in this case of an option) is the average expected contract payoff. Nevertheless, the seller of the option would probably expect a reward for taking on the risk of having to pay out, and, hence, the premium would probably be slightly higher than the expected payoff by a risk loading. In the paper, the risk loading as 20% of the standard deviation of the payoff of the contract is set [20].

2.4. Risk Management Issues. Efficiency of weather derivative to reduce risk is quantified by comparing the distribution of revenues from crop sales including hedging and without hedging (option 1, option 2). If the farmer does not buy a weather derivative contract, he would realize the revenues R_0:

$$R_0 = \frac{Q \cdot P}{(1 + r_f)^n}, \quad Q = \int_{\min}^{\max} f(x)\,dx. \tag{2}$$

Q denotes crop yield (t/ha) being a function of stochastic weather variable x. P is expected postharvest crop price (CZK/t). Since the expected payoff $(Q \cdot P)$ is related to the beginning of the contract period, it is discounted by the risk-free rate r_f at the beginning of the contract period (1.90% p. a.).

If the farmer buys a weather derivative contract per 1 ha of crop, he has to pay the premium to the seller (F_0). A farmer may collect a payoff from the contract (F_T) if a weather variable exceeds the strike. The payment is a function of underlying index x. Consider

$$R_1 = R_0 + \frac{F_T}{(1 + r_f)^n} - F_0, \quad F_T = \int_{\min}^{\max} f(x)\,dx. \tag{3}$$

The simulation is processed using the Monte Carlo method with 10 000 iterations at $\alpha = 0.05$. The comparison of the simulation results without standard error (without basis risk) and including standard error in regression estimate (including basis risk) answers the question about the degree of the basis risk.

3. Results and Discussion

Tables 6 and 7 present the results of correlation analysis between yields and weather variables. A statistically significant moderate relationship occurs between the wheat/barley yields and the air temperature (and drought index S_i) in May, June, and July. Both cereals have similar sensitivity to air temperature. Precipitations are local, so the risk of lack or, conversely, excessive rainfall has not a systematic character. The inverse relationship is shown between yield and rainfall during the presowing soil preparation. However, the correlations between wheat/barley yield and precipitation at the regional level are rather weak.

The results confirm the sensitivity of wheat/barley to the lack of precipitation and higher temperatures in the spring. On the contrary, rather drier periods in some regions are beneficial during pre-sowing soil preparation. Tables 6 and 7 show Jihomoravský region as the area with highest risk of drought, where the correlation coefficient between yields and drought index S_i distinctly exceeded 0.5 and are statistically significant at $\alpha = 0.01$. The higher risk means the higher yields—the Jihomoravský region is really a growing-friendly area for high quality wheat and barley.

Sensitivity of yields to weather variables in April is also worthy of attention, but not statistically significant. The following weather indices are included in the underlying index and tested for correlation:

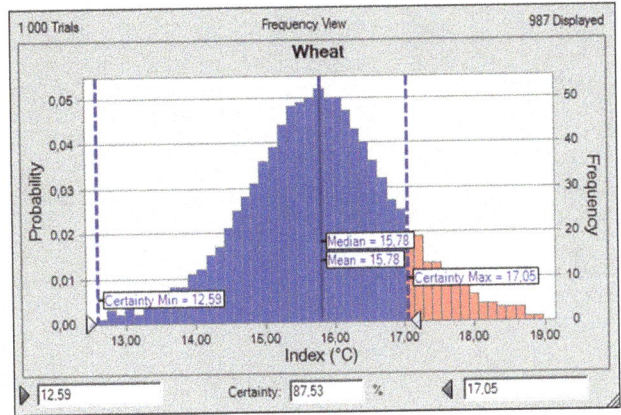

FIGURE 1: Risk layering with farmers' participation of 10% (strike 1) for wheat.

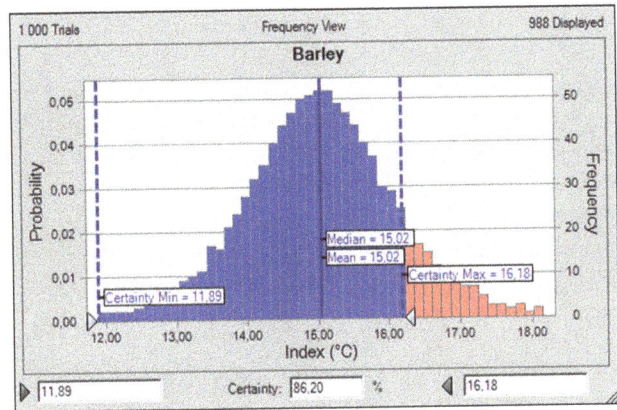

FIGURE 2: Risk layering with farmers' participation of 10% (strike 1) for barley.

(i) weighted average air temperature in the period May–June ($T2$);

(ii) weighted average air temperature in the period from April to June ($T3$);

(iii) weighted drought index S_i in the period from May to June ($S_i 2$);

(iv) weighted drought index S_i in the period April–June ($S_i 3$);

(v) number of CDDs in June assuming basic air temperature 16.5°C (wheat)/18.09°C (barley) (CDD 1);

(vi) number of CDDs in the period May–June assuming basic air temperature 13.9°C (wheat)/15.0°C (barley) (CDD 2);

(vii) number of CDDs in the period April–June assuming basic air temperature 14.0°C for both wheat and barley (CDD 3).

The highest Pearson correlation coefficient is identified between wheat/barley yields and weighted average air temperature $T3$ (Table 2). Weather index therefore consists of an average air temperature in April (weight of 7.4% for wheat/19.4% for barley), May (weight of 21.3% for

TABLE 3: Results of the regression analysis.

Crop/region	Regression	R^2	Adj. R^2	Breusch-Godfrey test (P value)	White test LM (P value)	A-D test for residuals (P value)
Wheat/Jihomoravský	$y = -0.3987x + 11.2960$	0.481	0.467	0.792216 (0.379182)	0.110964 (0.946029)	0.894
Barley/Jihomoravský	$y = -0.3623x + 9.5613$	0.473	0.460	0.778546 (0.386845)	0.593445 (0.743250)	0.957

Notes: Breusch-Godfrey test does not show the existence of autocorrelation in time series. White test did not indicate any presence of heteroskedasticity. Source: authors.

TABLE 4: Result of Anderson-Darling test of index $T3$ for wheat and barley.

Distribution	Wheat A-D	Wheat P value	Distribution	Barley A-D	Barley P value
Logistic	0.3899	0.317	Logistic	0.2230	0.776
Student's t	0.4792	×	Lognormal	0.2731	0.538
Beta	0.5144	×	Beta	0.2821	×
Normal	0.5146	0.193	Gamma	0.2867	0.55
Lognormal	0.5274	0.089	Normal	0.2889	0.623
Gamma	0.5383	0.099	Weibull	0.3096	0.524
Weibull	0.5941	0.223	Max Extreme	0.5899	0.128
Max Extreme	0.9252	0.018	Student's t	1.0695	×
Min Extreme	1.6905	0	Triangular	1.2479	×
Triangular	2.4206	×	Min Extreme	1.3435	0
Pareto	5.4825	×	Uniform	4.1707	0
Uniform	5.5950	0	Pareto	4.7336	×
BetaPERT	12.5932	×	BetaPERT	6.8939	×
Exponential	19.5335	0	Exponential	19.4727	0

Source: authors.

TABLE 5: The structure of specific-event contracts and the efficiency assessment.

	Wheat	Barley
Region	Jihomoravský	Jihomoravský
Index	Air temperature	Air temperature
Critical month for crop yield formation (weights)	April (0.074) May (0.213) June (0.713)	April (0.194) May (0.149) June (0.657)
Strike 1	17.05°C	16.18°C
Strike 2	15.80°C	15.07°C
Type of regression function	Linear	Linear
R^2	0.481	0.473
Type of option	Call	Call
Fixed price (CZK/t)	4,600	3,700
Tick	1,834 CZK/1°C	1,341 CZK/1°C
Strike 1		
Without basis risk		
Reduction of sales variability	−11.7%	−13.8%
Reduction of average sales	−0.45%	−0.47%
With basis risk		
Reduction of sales variability	−5.5%	−5.7%
Reduction of average sales	−0.41%	−0.49%
Contract price (CZK per 1 ha)	240	190
Strike 2		
Without basis risk		
Reduction of sales variability	−39.6%	−40.0%
Reduction of average sales	−1.10%	−0.88%
With basis risk		
Reduction of sales variability	−16.0%	−14.3%
Reduction of average sales	−1.00%	−1.07%
Contract price (CZK per 1 ha)	1,000	690

Source: authors.

wheat/14.9% for barley), and June (weight of 71.3% for wheat/65.7% for barley).

Table 3 describes results of regression analysis between wheat/barley yields in Jihomoravský region and weather index $T3$.

The weather index explains the variability of regional average wheat/barley yields of up to 48.1%/47.3%. The remaining part of yield variability is caused by weather conditions during other months (e.g., weather during the harvest season, hail during the vegetation) and by other influences we can term as basis risk. According to the results of Anderson-Darling goodness-of-fit test, residuals are of the normal (Gaussian) distribution value of A-D test = 0.190 (wheat)/0.152 (barley) and P value = 0.894 (wheat)/0.957 (barley). Standard error is 0.500 (wheat)/0.442 (barley). These values are the essential inputs to quantify the effectiveness of the weather derivative with the basis risk assumption.

The type and parameters of the probability distribution are estimated using the Anderson-Darling goodness-of-fit test. A histogram of weighted average air temperature $T3$ is based on historical weather data in the period of 1961–2009. Results of A-D test for both wheat and barley (Table 4) show that the most suitable approximation of real data probability distribution is logistic distribution with mean

15.783 (wheat)/15.018 (barley) and scale 0.633 (wheat)/0.616 (barley). The logistic distribution of underlying index $T3$ is therefore put in parametric bootstrapping.

Weather derivative valuation requires estimation of strike, that is, the level of index when the contract starts to pay. Figures 1 and 2 show the risk layering for wheat and barley with farmers' participation of 10%.

TABLE 6: The most significant correlation coefficients between wheat yield per hectare and the average characteristics of weather in regions of the CR (1970–2009).

Region	Air temperature	Rainfall	Drought index S_i
Středočeský	-0.43 (6, $P = 0.0059$) -0.42 (5-6, $P = 0.0068$) -0.35 (5-7, $P = 0.0290$) -0.32 (5-7, $P = 0.0422$)	0.35 (4, $P = 0.0270$) 0.33 (4-5, $P = 0.0373$) -0.38 (10-12, $t - 1$, $P = 0.0159$)	-0.33 (4-5, $P = 0.0363$) 0.43 (10-12, $t - 1$, $P = 0.0063$) 0.36 (10, $t - 1$, $P = 0.0263$) 0.36 (12, $t - 1$, $P = 0.0242$) 0.35 (10-11, $t - 1$, $P = 0.0283$)
Jihočeský	-0.48 (6, $P = 0.0016$) -0.43 (5-6, $P = 0.0060$) -0.34 (5-7, $P = 0.0339$)	0.40 (2, $P = 0.0105$) -0.34 (7, $P = 0.0301$) -0.32 (10-12, $t - 1$, $P = 0.0463$)	0.32 (1, $P = 0.0473$) 0.32 (10-12, $t - 1$, $P = 0.0449$)
Vysočina	-0.59 (6, $P < 0.0001$) -0.53 (5-6, $P = 0.0004$) -0.47 (5-7, $P = 0.0024$) -0.45 (4-6, $P = 0.0037$) -0.43 (6-7, $P = 0.0057$)	×	-0.42 (6, $P = 0.0064$) -0.38 (4-6, $P = 0.0153$) -0.36 (5-6, $P = 0.0245$)
Královéhradecký	-0.41 (6, $P = 0.0091$) -0.39 (5-6, $P = 0.0122$)	-0.43 (6-8, $P = 0.0056$) -0.41 (7-8, $P = 0.0090$) -0.35 (10-12, $t - 1$, $P = 0.0295$)	×
Jihomoravský	-0.66 (6, $P < 0.0001$) -0.66 (5-6, $P < 0.0001$) -0.60 (4-6, $P < 0.0001$) -0.52 (3-6, $P = 0.0005$) -0.50 (5-7, $P = 0.0010$)	0.34 (5, $P = 0.0304$) 0.33 (4-6, $P = 0.0398$) 0.32 (4-5, $P = 0.0418$)	-0.58 (5-6, $P < 0.0001$) -0.56 (4-6, $P = 0.0002$) -0.52 (5, $P = 0.0005$) -0.48 (3-6, $P = 0.0019$) -0.46 (6, $P = 0.0027$)
Olomoucký	-0.45 (6, $P = 0.0034$) -0.42 (5-6, $P = 0.0068$) -0.32 (5-7, $P = 0.0418$)	×	×

The data in front of round brackets are correlation coefficients. The figures in brackets denote critical months for yield formation. The P values test the two-tailed statistical significance of the correlation coefficient. The term "×" indicates no statistically significant correlation (Pearson, Spearman) at significance level 0.05. We put a maximum of 5 most statistically significant correlation coefficients.
Source: authors.

Under the assumption of the regression functions, expected crop yield, and farmers' participation (0%/10%), various levels of contract strike can be set as follows. For example, strike is 17.05°C (wheat)/16.18°C (barley) if the farmers' participation is 10% and expected yield is 5.00 t/ha (wheat)/4.12 t/ha (barley). The probability of higher underlying index $T3$ is 12.47% (wheat)/13.80% (barley). In case of no farmers' participation and the same expected yields, the strike is lower—15.80°C (wheat)/15.07°C (barley). The probability of risk occurring is then about 50%.

Simulation of average payoff and its standard deviation corresponds to the principle of actuarial pricing method as well as to the concept of fair price. Table 5 contains the assessment of the efficiency of the weather derivative contracts for wheat and barley in the Jihomoravský region.

The efficiency of weather derivative contract is relatively low—farmers can reduce the variability of revenues only by 5.5% (wheat)/5.7% (barley) at the strike 1 and under assumption of the basis risk. The analysis reveals a very high basis risk which may result in both excessive and poor payoff. If we consider no basis risk and strike 1, the contract can help reduce the variability of revenues by 11.7% (wheat)/13.8% (barley). However, the basis risk really exists. Strike 2 increases the likelihood of payoff, so the contract price and the reduction of average sales are obviously higher. Such reduction in sales could be for many farmers still acceptable if they want to manage systemic weather risk that is not insurable.

Limitation of powerful use of the weather derivatives in the Czech agriculture is both a geographical basis risk and production basis risk. Geographical basis risk can be reduced by locating the contract nearest to the reference weather station. Alternatively, the nature of such weather index contracts would increase the transaction costs (as the contract would not be standardized) and would eliminate the advantage of their transferability to the capital markets. Production basis risk cannot be influenced by the design of the weather derivative contract.

In the European conditions, the assessment of weather derivatives efficiency in agriculture was conducted in Germany in Brandenburg [7]. Brandenburg is relatively homogeneous region because it is located in large North German Lowland (Nordwestdeutsches Tiefland) with predominantly sandy soils and low water retention capacity. Weather derivative for wheat linked to total precipitation in April–June (measured at weather station Berlin—Tempelhof) refers to the regression function with $R^2 = 0.48$. Despite the relatively homogeneous conditions in the region, designed weather derivative was able to reduce the variability of wheat sales only by 11%. The extent of basis risk and efficiency of weather derivative are close to results shown in the paper.

TABLE 7: The most significant correlation coefficients between barley yield per hectare and the average characteristics of weather in regions of the CR (1970–2009).

Region	Air temperature	Rainfall	Drought index S_i
Středočeský	−0.44 (6, $P = 0.0047$) −0.43 (5-6, $P = 0.0056$) −0.41 (5–7, $P = 0.0085$) −0.37 (4-6, $P = 0.0174$) −0.37 (6-7, $P = 0.0185$)	0.31 (4, $P = 0.0479$) −0.40 (3, $P = 0.0115$) −0.33 (12, $t - 1$, $P = 0.0415$)	0.41 (3, $P = 0.0085$)
Jihočeský	−0.50 (6, $P = 0.0010$) −0.39 (6-7, $P = 0.0128$) −0.39 (5-6, $P = 0.0131$) −0.36 (5–7, $P = 0.0236$)	×	−0.31 (6, $P = 0.0479$)
Vysočina	−0.52 (6, $P = 0.0005$) −0.51 (6-7, $P = 0.0007$) −0.48 (5–7, $P = 0.0015$) −0.46 (4–7, $P = 0.0025$) −0.44 (5-6, $P = 0.0043$)	−0.35 (3, $P = 0.0268$) −0.34 (1–3, $P = 0.0337$)	0.41 (3, $P = 0.0086$) −0.40 (6, $P = 0.0010$) −0.38 (4-6, $P = 0.0162$) −0.33 (5-6, $P = 0.0382$)
Královéhradecký	−0.46 (6, $P = 0.0028$) −0.43 (5-6, $P = 0.0060$) −0.41 (5–7, $P = 0.0089$) −0.38 (6-7, $P = 0.0157$) −0.35 (4–7, $P = 0.0250$)	−0.41 (1–8, $P = 0.0081$) −0.41 (1–7, $P = 0.0082$) −0.34 (1–3, $P = 0.0302$) −0.34 (3–8, $P = 0.0317$) −0.33 (7-8, $P = 0.0391$)	0.42 (3, $P = 0.0072$) 0.38 (1–3, $P = 0.0157$) 0.36 (2-3, $P = 0.0243$) 0.33 (1–4, $P = 0.0379$) −0.34 (6, $P = 0.0345$)
Jihomoravský	−0.64 (6, $P < 0.0001$) −0.64 (5-6, $P < 0.0001$) −0.63 (4-6, $P < 0.0001$) −0.55 (4–7, $P = 0.0002$) −0.54 (5–7, $P = 0.0003$)	0.36 (4-6, $P = 0.0218$) 0.33 (4-5, $P = 0.0387$) −0.32 (3, $P = 0.0439$)	−0.60 (4-6, $P < 0.0001$) −0.56 (5-6, $P = 0.0002$) −0.49 (4-5, $P = 0.0012$) −0.47 (6, $P = 0.0020$) −0.47 (5, $P = 0.0021$)
Olomoucký	−0.46 (6, $P = 0.0027$) −0.46 (5-6, $P = 0.0029$) −0.40 (5–7, $P = 0.0103$) −0.38 (4-6, $P = 0.0171$) −0.35 (6-7, $P = 0.0279$)	−0.41 (3, $P = 0.0080$) −0.36 (1–3, $P = 0.0207$) −0.32 (1–4, $P = 0.0443$)	−0.36 (6, $P = 0.0238$) −0.34 (5-6, $P = 0.0293$) 0.33 (3, $P = 0.0376$)

The data in front of round brackets are correlation coefficients. The figures in brackets denote critical months for yield formation. The P values test the two-tailed statistical significance of the correlation coefficient. The term "×" indicates no statistically significant correlation (Pearson, Spearman) at significance level 0.05. We put a maximum of 5 most statistically significant correlation coefficients.
Source: authors.

These results confirm the findings by Manfredo and Richards [22] and Vedenov and Barnett [10] emphasizing in particular the disadvantages of weather derivatives as primary crop insurance instruments. Nevertheless, the aggregation effect suggests that the potential for weather derivatives in agriculture can be greater than previously thought, particularly for aggregators of risk, such as reinsurers [11].

4. Conclusion

The paper aims to design the weather derivative under the specific conditions of agriculture. The analysis reveals some important findings.

(i) In the Czech agriculture, the weather index explains up to 48% of the variability of the average cereal yields. More than 50% of the systemic yield risk cannot be covered by weather derivatives or weather insurance.

(ii) The main limitations on the use of the weather derivatives in the Czech Republic are heterogeneous production conditions that reduce the correlation between weather and crop yields at regional level. The Pearson correlation coefficients do not exceed ±0.7 and show weak or moderate correlation between regional yield and weather.

(iii) The analysis indicates high basis risk that can significantly distort the contract payoff. In the Czech Republic, the weather index contracts can reduce variability of the cereal revenues only by 5%-6%. If the basis risk does not exist, the contract is able to reduce variability of the cereal revenues by more than 10%.

(iv) The efficiency of weather contracts increases with higher probability of damage. At the higher strike level, the designed contract reduces the sales variability by 14%-16% (basis risk) or 40% (no basis risk). However, the contract price is higher.

(v) The higher contract price reduces revenues from −0.4% to −1.0% (basis risk). Nevertheless, this is lower than the 3% normative insurance premium rate against natural disasters. From the farmers' point of view it could be budget-wise to use weather derivatives with little efficiency but at low cost.

(vi) Assuming the potential of weather derivatives as the reinsurance instrument, it is important to clarify the legal and institutional aspects of the income risk management in agriculture using weather derivatives, especially regulation and possible areas of cooperation between the public and private sectors.

Acknowledgment

The research was supported by the long-term institutional support to the conceptual development of research organization (project of the University of Economics, Faculty of Business Administration, IGA 2, VŠE IP300040).

References

[1] J. A. Dutton, "Opportunities and priorities in a new era for weather and climate services," *Bulletin of the American Meteorological Society*, vol. 83, no. 9, pp. 1303–1311, 2002.

[2] M. Roth, Ch. Ulardic, and J. Trueb, "Critical success factors for weather risk transfer solutions in the agricultural sector: a reinsurer's view," *Agricultural Finance Review*, vol. 68, no. 1, pp. 1–7, 2008.

[3] J. K. Lazo, M. Lawson, P. H. Larsen, and D. M. Waldman, "U.S. economic sensitivity to weather variability," *Bulletin of the American Meteorological Society*, vol. 92, no. 6, pp. 709–720, 2011.

[4] R. S. Dishel, "Introduction to the weather market: dawn to mid-morning climate risk and weather market," in *Financial Risk Management With Weather Hedges*, R. S. Dishel, Ed., pp. 3–24, Risk Books, London, UK, 2002.

[5] B. J. Barnett and O. Mahul, "Weather index insurance for agriculture and rural areas in lower-income countries," *American Journal of Agricultural Economics*, vol. 89, no. 5, pp. 1241–1247, 2007.

[6] M. Miranda and D. V. Vedenov, "Innovations in agricultural and natural disaster insurance," *American Journal of Agricultural Economics*, vol. 83, no. 3, pp. 650–655, 2001.

[7] O. Mußhoff, M. Odening, and W. Xu, "Zur Quantifizierung des Basisrisikos von Wetterderivaten," Tech. Rep. 14947, German Association of Agricultural Economists, Giessen, Germany, 2006.

[8] J. R. Skees, "Innovations in index insurance for the poor in lower income countries," *Agricultural and Resource Economics Review*, vol. 37, no. 1, pp. 1–15, 2008.

[9] C. G. Turvey, "Weather derivatives for specific events risk in agriculture," *Review of Agricultural Economics*, vol. 23, no. 2, pp. 333–351, 2001.

[10] D. V. Vedenov and B. J. Barnett, "Efficiency of weather derivatives as primary crop insurance instruments," *Journal of Agricultural and Resource Economics*, vol. 29, no. 3, pp. 387–403, 2004.

[11] J. D. Woodard and P. Garcia, "Weather derivatives, spatial aggregation, and systemic risk: implications for reinsurance hedging," *Journal of Agricultural and Resource Economics*, vol. 33, no. 1, pp. 34–51, 2008.

[12] J. B. Hardaker, R. B. M. Huirne, J. R. Anderson, and G. Lien, *Coping with Risk in Agriculture*, CABI, Wallingford, UK, 2nd edition, 2004.

[13] J. K. Horowitz and E. Lichtenberg, "Insurance, moral hazard, and chemical use in agriculture," *American Journal of Agricultural Economics*, vol. 75, no. 4, pp. 926–935, 1993.

[14] S. S. Makki and A. Somwaru, "Evidence of adverse selection in crop insurance markets," *The Journal of Risk and Insurance*, vol. 68, no. 4, pp. 685–708, 2001.

[15] M. Rothschild and J. E. Stiglitz, "Equilibrium in competitive insurance markets: an essay on the economics of imperfect information," *Quarterly Journal of Economics*, vol. 90, no. 4, pp. 629–649, 1976.

[16] A. Rubinstein and M. E. Yaari, "Repeated insurance contracts and moral hazard," *Journal of Economic Theory*, vol. 30, no. 1, pp. 74–97, 1983.

[17] U. Hess, "Weather index insurance for coping with risks in agricultural production," in *Managing Weather and Climate Risks in Agriculture*, M. V. K. Sivakumar and R. P. Motha, Eds., Springer, Berlin, Germany, 2007.

[18] T. A. Fleege, T. J. Richards, M. R. Manfredo, and D. R. Sanders, "The performance of weather derivatives in managing risks of specialty crops," in *Proceedings of the NCR-134 Conference on Applied Commodity Price Analysis, Forecasting, and Market Risk Management*, St. Louis, Mo, USA, April 2004.

[19] M. P. M. Meuwissen, M. A. P. M. van Asseldonk, and R. B. M. Huirne, "The feasibility of a derivative for the potato processing industry in Netherlands," in *Proceedings of the Meeting of the Southern Association of Economics and Risk Management in Agriculture*, Gulf Shores, Ala, USA, March 2000.

[20] S. Jewson, A. Brix, and C. Ziehmann, *Weather Derivative Valuation: the Meteorological, Statistical, Financial and Mathematical Foundations*, Cambridge University Press, Cambridge, UK, 2005.

[21] J. Hnilica, "Crystal ball in weather-linked derivatives valuation," in *Proceedings of the Crystal Ball User Conference*, Denver, Colo, USA, May 2007.

[22] M. R. Manfredo and T. J. Richards, "Hedging with weather derivatives: a role for options in reducing basis risk," *Applied Financial Economics*, vol. 19, no. 2, pp. 87–97, 2009.

[23] V. Potop, L. Türkott, and V. Kožnarová, "Spatiotemporal characteristics of drought episodes in czechia," *Scientia Agriculturae Bohemica*, vol. 39, no. 3, pp. 258–268, 2008.

[24] D. Vose, *Risk Analysis: A Quantitative Guide*, John Wiley & Sons, Chichester, UK, 3rd edition, 2008.

[25] B. Efron, "Bootstrap methods: another look at the jackknife," *Annals of Statistics*, vol. 7, no. 1, pp. 1–26, 1979.

4

Trends and Variability of North Pacific Polar Lows

Fei Chen and Hans von Storch

Institute of Coastal Research, Helmholtz-Zentrum Geesthacht, Germany

Correspondence should be addressed to Fei Chen; fchen@scsio.ac.cn

Academic Editor: Lian Xie

The 6-hourly 1948–2010 NCEP 1 reanalyses have been dynamically downscaled for the region of the North Pacific. With a detecting-and-tracking algorithm, the climatology of North Pacific Polar Lows has been constructed. This derived climatology is consistent with the limited observational evidence in terms of frequency and spatial distribution. The climatology exhibits strong year-to-year variability but weak decadal variability and a small positive trend. A canonical correlation analysis describes the conditioning of the formation of Polar Lows by characteristic seasonal mean flow regimes, which favor, or limit, cold air outbreaks and upper air troughs.

1. Introduction

"Polar Low" has been defined as the generic term for all mesoscale cyclonic vortices poleward of main polar front. It should be used for intense maritime mesocyclones with scales less than 1000 km with strong wind speeds [1].

Since the availability of comprehensive observations was supported by satellite imagery, a couple of authors have dealt with space-time statistics of Polar Lows in the North Pacific [2–4]. However, these studies using satellite data cover only a few years of Polar Low occurrences. Also, a combination of subjective detections methods and inhomogeneities in data leads to the possibility that derived trends and variability may not be robust.

A number of authors have applied global reanalysis data for investigating conditions, which are favorable or unfavorable for the formation of Polar Lows and mesocyclones statistics in both Atlantic and Pacific [5, 6]. Kolstad identified low static stability and reverse-wind shear conditions in the 40-year period of ERA-40 reanalyses as favorable conditions. By computing the space-time statistics (i.e., climatology) of favorable conditions for Polar Lows over the North Atlantic, the North-West Pacific and over Southern Ocean, statistics of Polar Low occurrences could be estimated in this indirect manner.

A suitable method for constructing climatology is using long-term dynamical downscaling with regional climate models [7]. Various authors have demonstrated that Polar Lows and other mesoscale windstorms can be described by high-resolution numerical models, such as Fu et al. and Yanase et el. in Japan Sea [8–10]; Bresch et al. in Bering Sea [11]; Businger and Blier in Gulf of Alaska [2, 12]; Chen et al. in North Pacific [13]; Cavicchia and von Storch for "medicanes" in the Mediterranean Sea [14]; and Zahn and von Storch in North Atlantic [7]. Dynamical mechanisms and synoptic conditions of Polar Low formation as well as life cycles have been discussed in various case studies. Reed supported the baroclinic theory in 1979, but he did not reject the possibility of other instabilities for the Polar Low mechanism [15]. The barotropic shear instability has also been put forward to explain the genesis of meso-β-scale Polar Lows over the polar-air mass convergence zone of Japan Sea [16]. High-resolution simulations have also been applied for some idealized experiments to explain mechanisms of Polar Low development such as baroclinicity [10]. A theoretical balanced axisymmetric model has been applied to investigate the wind-induced surface heat exchange intensification mechanism in Polar Low development by Gray and Craig [17]. Various case studies indicate that usually several mechanisms together lead to the formation of Polar Lows [18–20]. A recent study also pointed out that Polar Lows influence the large-scale ocean circulation and deep water transport over the Nordic sea [21].

In this paper, we employ the method developed by Zahn and von Storch [7, 22] to dynamically downscale the gridded large-scale synoptic fields, as provided by large-scale component of reanalyses—here: NCEP 1 [23]. In a first step, we have shown in a series of cases that downscaling generates Polar Lows with sufficient accuracy [13]. Now, in this study, a regional climate model (RCM) has been run continuously for 63 years during which the NCEP 1 reanalyses are available. During the integration, the model is constrained in its large-scale components to be similar to the driving NCEP reanalysis, using the method of spectral nudging [24]. The output of this multidecadal simulation is used for investigating trends and variability of Polar Low occurrences in the North Pacific and their linkage to the seasonal mean large-scale circulation situations over the last 63 years.

We are sometimes confronted with the request that such an analysis should result in new insights into the dynamics of Polar Low formation and life cycles. However, this is not the intention of the present study. Such case studies have been done in many cases (as stated previously), and we do not intend to extend this large number of dynamical analyses. Instead, we are interested in space-time statistics, including the conditioning by large-scale dynamical configurations, of Polar Lows, in particular, on the differences between regions and years and decades, and in systematic changes. Thus, the present analysis does not contribute to dynamical meteorology but to climatology.

In Section 2, we describe briefly the model and the detection-and tracking methodology used in this study. This methodology differs a bit from the previous study of Chen et al. [13] as well as the North Atlantic study by Zahn and von Storch [22]. The derived statistics of the formation of Polar Lows in space and time is the subject of Section 3. In Section 4, we determine linkages of seasonal mean large-scale flow in the North Pacific and the number of storms formed in different subregions of the North Pacific. The paper is concluded with a discussion and the recapitulation of major results finally in Section 5.

2. Data and Methodology

The RCM we applied in this study is the COSMO-CLM 4.8 (COSMO model in CLimate Mode) [25, 26]. This model is the climate version of the operational weather prediction model of the Deutscher Wetterdienst (German National Weather Service) and the COnsortium for Small-scale MOdeling (COSMO), adapted to climate simulation purposes by the CLM-Community (http://www.clm-community.eu/). The model domain covers the whole North Pacific (Figure 3(b)). We used NCEP 1 reanalysis data as initial and lateral boundary conditions. In particular, the sea surface temperature (SST) and sea ice extent were prescribed as lower boundary according to the NCEP 1 reanalyses. Spectral nudging [24] of tropospheric wind components was applied for enforcing the NCEP large-scale situation in the model region in order to prevent COSMO-CLM from significantly deviating from the analyzed large-scale state. The nudging is applied at 850 hPa and above, with the nudging becoming stronger with

height. Only spatial scales larger than 800 km are constrained; smaller features are unconstrained. A rotated grid with 0.4° grid resolution and 220 and 80 points is employed for the longitudinal and latitudinal grid map. The 8-grid sponge zone is introduced to avoid reflection of waves at the boundaries. Boundary data is prescribed by this zone with decreasing influence for the inner grid points. The simulation results from the sponge zone are not usable for further analysis. The number of vertical levels is 40. The simulation period began on January 1, 1948 and ended on December 31, 2010. For further details about the model setup, refer to Chen et al. [13].

An automatic detection-and-tracking procedure was applied to determine the presence and tracks of Polar Lows in the North Pacific. The detection procedure searches for the minima in the band-pass filtered mean sea level pressure (MSLP) fields and concatenates the minima in consecutive time steps to tracks. Along these tracks, the fulfillment of further criteria is requested for categorizing an event as Polar Low:

(1) strength of the minimum band-pass filtered MSLP: ≤ -2 hPa once along the track;

(2) wind speed: ≥ 13.9 m/s once along the track;

(3) sea surface versus mid troposphere temperature difference: $SST-T_{500\,hPa} \geq 39$ K;

(4) average direction of the track: a north-to-south component;

(5) limits to allowable adjacent grid boxes. No land: those tracks are excluded when over 50% of the positions near the coastal grid boxes along the track.

When the minimum of the band-pass filtered MSLP along these tracks decreases below -6 hPa once and there are no coastal grid boxes close to that location, a Polar Low is recorded irrespective of the other criteria.

This is mostly identical to the North Atlantic setting of Zahn and von Storch [22], but some parameters (like the air-sea temperature difference) have been modified in order to meet the conditions in the North Pacific better [13]. However, we used now a cosine (discrete cosine transforms: DCT) band-pass filter of Denis et al. [27] instead of the digital filter used by Zahn and von Storch [22], after Xia et al. [28] pointed out the superiority of the first one. DCT is more precise in scale separation than the original digital filter.

Some Polar Lows in the North Pacific move on average in zonal direction, like the case of 22nd of March 1975 described by Chen et al. [13]. Therefore, the request for a north-to-south movement may be too strict. However, without this criterion, also smaller baroclinic storms may be categorized as Polar Lows. In order to examine the significance of this zonal movement criterion for variability and long-term trend, we did the analysis with both criteria, with and without N-S direction criterion (see Figure 1).

3. Results: Space Time Statistics of Tracks

Polar Lows are a phenomenon forming in the "cold season;" therefore, the "Polar Low season" (PLS) in the North Pacific is

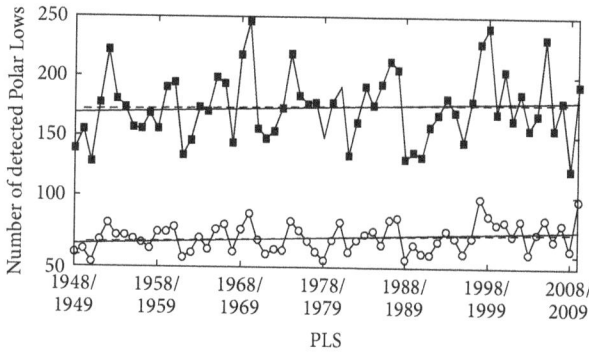

FIGURE 1: Number of detected Polar Lows in the North Pacific per Polar Low season (October to April). Top curves: the detected number of Polar Lows without N-S criterion (marked with ■). The solid line represents the trend from 62 PLSs, from 1948/1949 to 2009/2010, which is 0.17 cases/year; the dashed line represents the trend from only 60 years, from 1949/1950 to 2008/2009, which is 0.09 cases/year. Bottom curves: the detected number of Polar Lows with N-S criterion (marked with ○). The solid line represents the trend from 1948/1949 to 2009/2010, which is 0.2 cases/year; the dashed line represents the trend from 1949/1950 to 2008/2009, which is 0.14 cases/year.

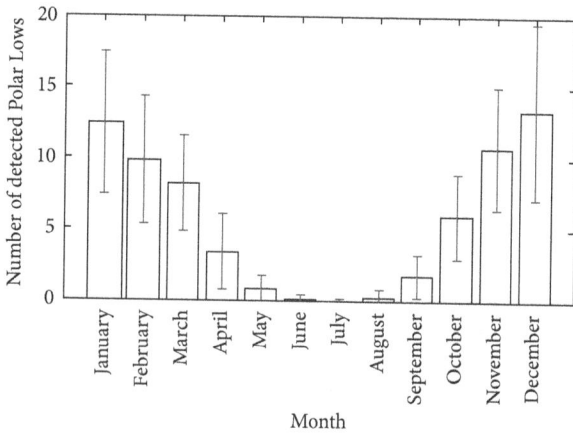

FIGURE 2: Number of detected Polar Lows in an average calendar month for the whole 62 PLSs. Tracking with N-S criterion. The error bar is the standard deviation of Polar Low variance in each month during the 62 PLSs.

defined as the time from October through April the following year. This is a bit different from Zahn and von Storch [7] in the case of the North Atlantic: they count the PLS from July to next June. As we know, there are few Polar Lows in summer (see Figure 1); the Polar Low season is addressed by the first and second year; for example, the PLS 1950/1951 begins in October 1950 and ends in April 1951. The statistical analysis excluded the first half year of 1948 and last half year of 2010, as they represent only part of a Polar Low season. So we have 62 PLSs, with the first from October 1948 to April 1949 and the last from October 2009 to April 2010.

Figure 1 shows the time series of the number of detected Polar Lows per PLS both without and with N-S criterion. In all the following analyses, the criterion of an average movement with a North-to-South component is applied.

When looking only for cases *without a directional constraint of the movement*, a total of 10812 Polar Lows were detected by the tracking algorithm during the 62 Polar Low seasons. On average, 174 Polar Lows were found per PLS, with a strong year-to-year variability indicated by a standard deviation of 29 (±17% of the long-term mean). The decadal variability is weak. The overall trend, from the first PLS in 1948/1949 to the last PLS in 2009/2010, in the frequency of Polar Low is positive with 0.17 cases/PLS, which yields about 11 Polar Lows more in the end than in the beginning of the series (11 cases correspond to 6% of the long-term mean). We have to point out that the slope of the trend depends on the number of cases of the first and last PLSs. When disregarding the last and the first PLSs, the trend of Polar Lows from PLS 1949/1950 to 2008/2009 is smaller, with only 0.09 cases/PLS.

For the result *with the N-S criterion*, fewer Polar Lows are detected, namely, only 4052 Polar Lows during the 62 Polar Low seasons. On average, 65 Polar Lows were found per PLS, with a strong year-to-year variability indicated by a standard deviation of 12 (±18% of the long-term mean). Maximum number of detected cases is found in PLS 1997/1998 with 98 cases; the minimum number is in PLS 1950/1951 with 44 cases. The overall trend, from the first PLS in 1948/1949 to the last PLS in 2009/2010, in the frequency of Polar Low is positive with, on average, additional 0.2 cases per PLS, which yields about 12 Polar Lows more in the end than in the beginning of the series (corresponding to 21% of the mean total). The trend of Polar Lows from PLS 1949/1950 to 2008/2009 is positive with 0.18 cases/PLS. When calculating 10 trends from 1948/1949–2009/2010, 1949/1950–2008/2009 to 1957/1958–2000/2001, the mean trend is 0.16 cases/PLS; the standard deviation of these 10 trends is 0.03/PLS, so that the estimate of the trend appears relatively insensitive to the early and late values.

There is no acceleration of a trend towards the end of the time series. Furthermore, the trend seems mostly uniform and rather small. According to the prewhitened Mann-Kendall trend test [29], the 62 PLSs trend of the number of detected Polar Low is significant (5% risk; result not shown) for the configurations with directional constraining, but insignificant when examining the curve derived without directional constraint.

The annual cycle of monthly numbers of detected Polar Lows with N-S criterion exhibits the highest frequency in winter with maxima in December and January and almost no Polar Low activity in summer (Figure 2), consistent with the observation that Polar Lows form in the cold season. Furthermore, we determined the annual cycle of the days with Polar Lows during the same time period as Businger [2], namely, 1975–1983, and found our results consistent with Businger's results. The results differ with respect to a secondary peak in January in our 1975–1983 climatology but in February in Businger's climatology (Figure not shown). In view of the very different methods applied by us and by Businger, namely, satellite observations versus downscaled reanalyses, the differences may be considered acceptable. When examining the set of Polar Lows derived without the directional constraint, a very similar annual cycle, apart of a uniform bias in the magnitude, is found (not shown).

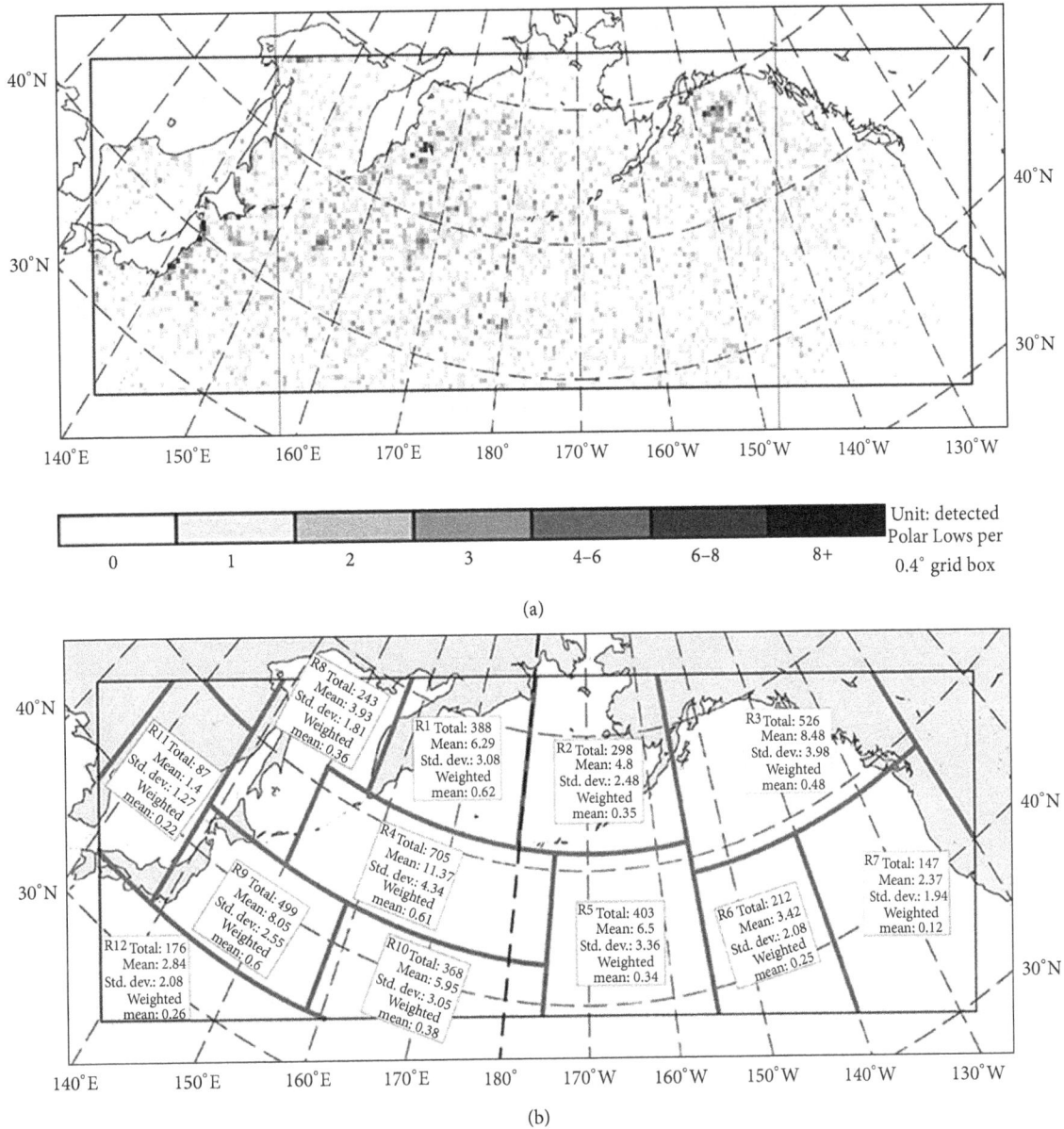

FIGURE 3: (a) Density distribution of first appearance of Polar Lows. Unit: detected Polar Lows per 0.4° grid box. (b) Subregions R1–12, for which statistics of Polar Lows were aggregated: the total number, the mean number per Polar Low season (mean), year-to-year standard deviation (std. dev.), and number of Polar Lows per number of no land grid points in the subregion (weighted mean).

There is no suggestion in the literature about the general number and trend of Polar Lows in the North Pacific that we should expect. Therefore, we find it difficult to decide if the directional constraint is really helpful or not. Clearly, the total number depends strongly on this criterion, but the overall trends are similar in both cases. This indicates that the N-S criterion does not much influence the general characteristic of long-term trend. The correlation coefficient between the two curves is 0.82, in both cases, with and without N-S criterion. Under the consideration that the N-S criterion is more reasonable from the dynamical view of Polar Low generation, we have chosen the result with N-S criterion for further discussions.

The spatial distribution of Polar Low density is shown in Figure 3(a). The Polar Low density counts the frequency per grid box of a first detection of a Polar Low. Highest densities are found in the region east of Japan in accordance with the results of Yarnal and Henderson [4]. By analysing defense meteorological satellite program (DMSP) infrared imagery from 7 winter seasons, Yarnal and Henderson concluded that the most active Polar Low cyclogenesis takes place in the western extratropical North Pacific. Our peak area of Polar Low density is just off the east coast of Japan Island, while Yarnal and Henderson's result has its peak a little north, near the island of Hokkaido. In both analyses, there is much less Polar Low activity in the eastern North Pacific than in the western part.

Yarnal and Henderson [4] pointed out that there are two bands that extend from northern Japan through the Kamchatka Peninsula into the western Bering Sea; another

(a)

(b)

CCA1

(c)

CCA2

(d)

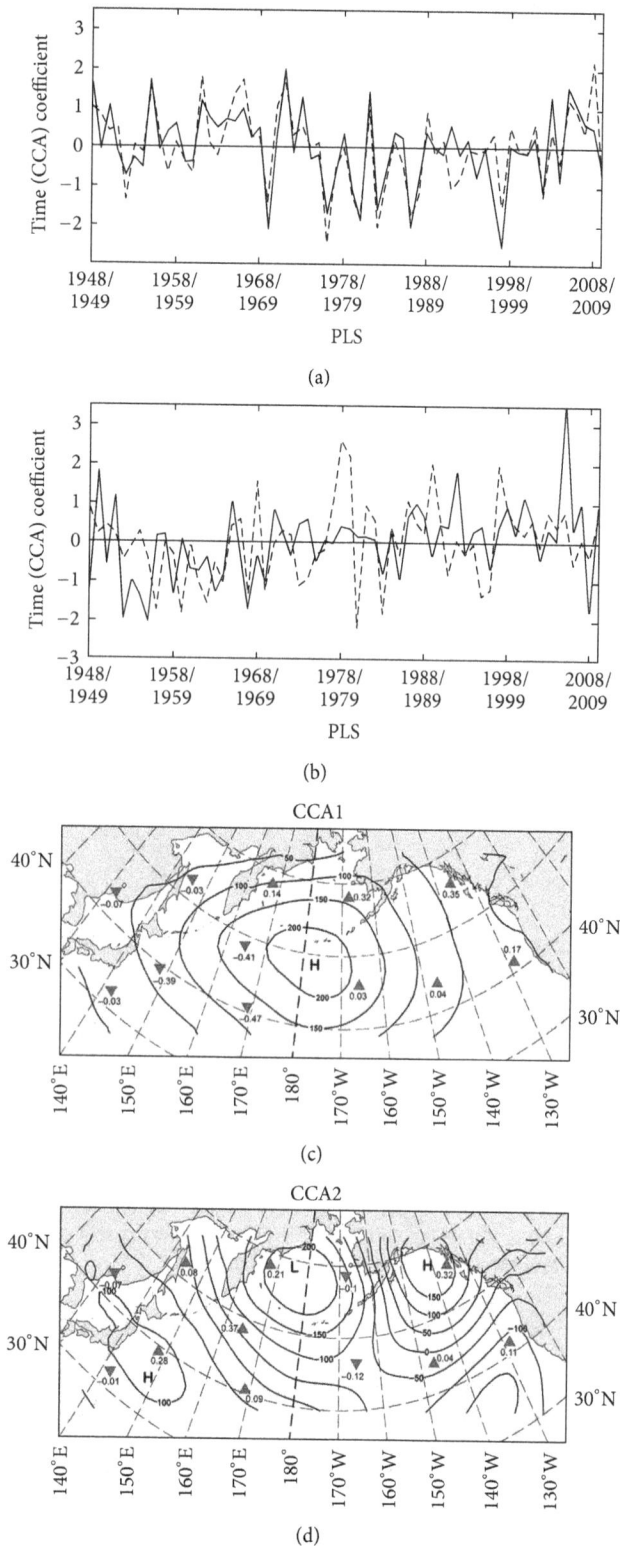

FIGURE 4: ((a), (b)) First two canonical correlation coefficient time series (dashed lines represent the Polar Low occurrence and solid lines represent the MSLP pattern) and ((c), (d)) corresponding canonical correlation patterns between regional time series of Polar Low occurrences per PLS in subregions R1–R12 ($\triangle$ for positive values, $\triangledown$ for negative values) and mean sea level pressure fields in Pa. The first CCA pair ((a) and (c)) shares a correlation coefficient of 0.89. The second CCA pair ((b) and (d)) shares a correlation of 0.72.

one extends eastward into the open waters of the North Pacific to just east of the international date line. Such two bands are also present in our climatology of the first detected position of the Polar Lows. Additionally, we find also high values for the Gulf of Alaska, where Yarnal and Henderson have detected only "a couple of small, weak pockets of formation."

In order to compare the information of Polar Lows over different regions of North Pacific, we divided the domain into 12 subregions (Figure 3(b)). Some of the subregions are based on the oceanographic feature, such as the Bering Sea (R1 and R2; they are divided by the international date line), the Gulf of Alaska (R3), the Okhotsk Sea (R8), and the Japan Sea (R11). The others are dependent on the climatology cyclogenesis by Yarnal and Henderson [4]. The two bands with the highest frequency of Polar Low occurrence are referred to as R4 and R9. The eastern North Pacific with much less activity is represented by R5, R6, and R7. R10 and R12 are the subregions in the south with seldom distribution. For the different regions in Figure 3(b), small trends and high year-to-year variability for Polar Low frequency were found. Some regions (especially for R1, R3, and R4 with high density) have a higher year-to-year standard deviation than others (R6, R7, R8, R11, and R12) (not shown).

4. Linkage to Large-Scale Pressure Patterns

We begin with investigating how *mean sea level pressure (MSLP)* fields, averaged across a Polar Low season, are related to the distribution of the numbers of Polar Low occurrences in that season. Here, we emphasize that the analysis is not about short-term synoptic situation or the instantaneous air pressure field directly related with the probability of a Polar Low to form. Instead, we compare two statistics during the same Polar Low season, namely, the geographical distribution of Polar Lows, aggregated to the twelve subregions shown in (Figure 3(b)) and the gridded time-mean MSLP field.

The link between MSLP and the number of Polar Lows is done through a canonical correlation analysis (CCA [30]). CCA is a method for calculating correlation structures between two fields of variables. To reduce noise in each set, we projected the full fields on the first 5 empirical orthogonal functions (EOFs) of the Polar Low time series (representing 77% of the variance) and of the PLS-mean MSLP (87% of the variance). Prior to the CCA, the multiyear mean field is subtracted; that is, the analysis is done with anomalies.

Figure 4 shows the resulting time series and spatial patterns of the two most important linkages between the regional distribution of Polar Lows and the time-mean MSLP field.

The first canonical pattern (CCA1, Figure 4(c)) describes a unipolar pressure distribution. When the CCA coefficient is positive, then, on average, there will be a west-eastward cold air flow across the Bering Sea and south-eastward across the Gulf of Alaska and, consistently, more Polar Lows in that region. These time-mean flows are characteristic for more or less, short-term marine cold air outbreaks in these regions. It is indicated that there is a close relationship between Polar Low formation and the presence of a trough in winter over

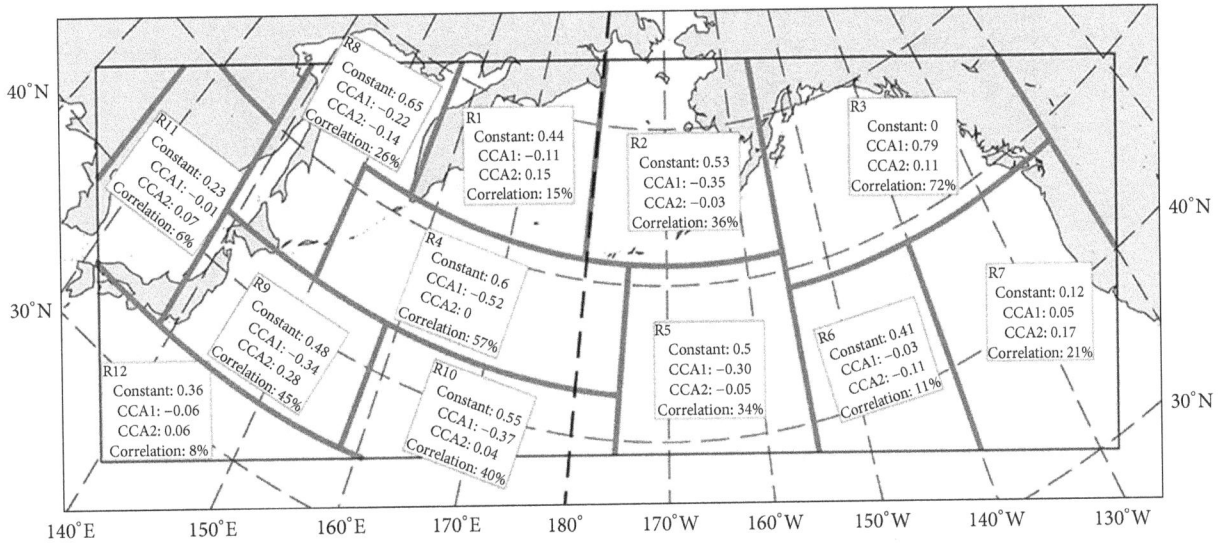

FIGURE 5: Result of multiple regression analyses of the first two CCA coefficients on the number of Polar Lows in subregions R1–12: coefficient a_0 (constant), a_1 (connected to CCA1), a_2 (connected to CCA2), and the correlation of number of Polar Lows and trough regressions estimated number of Polar Lows.

East Asia and the nearby North Pacific [2]. The strong land-sea thermal contrast along the marginal ice zone pulls cold continental polar or Arctic air over the Bering Sea and the Japan Sea. The relatively warm waters in the open ocean lead to the formation of Polar Lows through convective instability and baroclinicity.

The time series of canonical correlation coefficient (Figures 4(a) and 4(b)) explains the significance of the two fields of variables over the 62 PLSs. Dashed lines represent the pattern of Polar Low occurrence over the 12 subregions; meanwhile, the solid lines represent the MSLP pattern. When examining the time series in Figure 4(a), it is evident that CCA1 dominates the Polar Low seasons in 1954/1955, 1971/1972, and 1981/1982.

The second canonical pattern (CCA2, Figure 4(d)) shows a bipolar pressure distribution—there is a negative anomaly (below −2 hPa) on the Bering Sea at the same time as two positive anomalies on the Gulf of Alaska and Japan Island (over 1 hPa). Consequently, there will be a south-eastward time-mean flow starting from Siberia, across the Sea of Okhotsk, and then the southwest of Bering Sea where, in subregion R4, consistently more Polar Lows are detected on average. A time-mean pressure contrast of about 3 hPa between the Bering Sea and the west and east part of North Pacific (the region around Japan Island and the west coast of North American continent) is associated with 0.2 to 0.3 more Polar Lows per PLS in the corresponding regions (R1, R3, R4, and R9).

By examining the time series in Figure 4(b), we found that CCA2 dominates the Polar Low seasons of 1951/1952, 1975/1976, and 1998/1999, which means there were more Polar Lows in regions R1, R3, R4, and R9; the negative signs in 1949/1950 and 1978/1979 point to remarkably less Polar Lows in these regions.

To determine the relative importance of the two CCA patterns, we have built for each subregion R1–R12 a regression

model, of the form $\text{PL}_i(t) = a_{0,i} + a_{1,i} \times \text{CCA}_1(t) + a_{2,i} \times \text{CCA}_2(t) + \varepsilon_i$. Here, $\text{PL}_i(t)$ is the number of Polar Lows in season t and in subregion i, $a_{0,i}$, $a_{1,i}$, and $a_{2,i}$ are the regression coefficients, which are determined by a least square fit, $\text{CCA}_1(t)$ and $\text{CCA}_2(t)$ are the CCA coefficients of patterns 1 and 2 in season t, and ε_i is the residual, the unexplained part. In Figure 5, we list the regression coefficients and the correlation between the number of Polar Lows and through these regressions estimated number of Polar Lows for every subregion.

The two series, $\text{CCA}_1(t)$ and $\text{CCA}_2(t)$, are about equally important, when counting how often $|a_1(t)| > |a_2(t)|$. They are particularly successful to describe the variability in all but one of the highest frequency occurrence area which we showed in Figure 3—for subregions 3, 4, 9, and 10, the correlation is over 40%, while in the far eastern (subregion 6 and 7) and western (11 and 12), the correlations are less than 25%. When comparing with Figure 3(b), we see that the frequency of occurrence of Polar Lows in the subregions 6, 7, 11, and 12 is relatively low: the mean number of Polar Lows in 7, 11, and 12 together is 6.61 Polar Lows per PLS, while the occurrences in each of subregions 3, 4, and 9 separately are stronger than this intensity, namely, 8.48, 11.37, and 8.05 Polar Lows per PLS. In subregion 1, frequency of occurrence is relatively high (6.29 Polar Lows per PLS) but the correlation is low (15%).

Next, we examine the link to *time-mean distribution throughout the troposphere*. Therefore, we derive associated correlation patterns [30] to describe the linkage between Polar Low occurrence and geopotential height in different pressure levels (Figure 6).

Associated correlation patterns are designed to describe the relationship between time series of an "index" variable and a physical field. By calculating the correlation coefficient between the time series of CCA coefficient for Polar Low occurrence pattern (dashed lines in Figures 4(a) and 4(b))

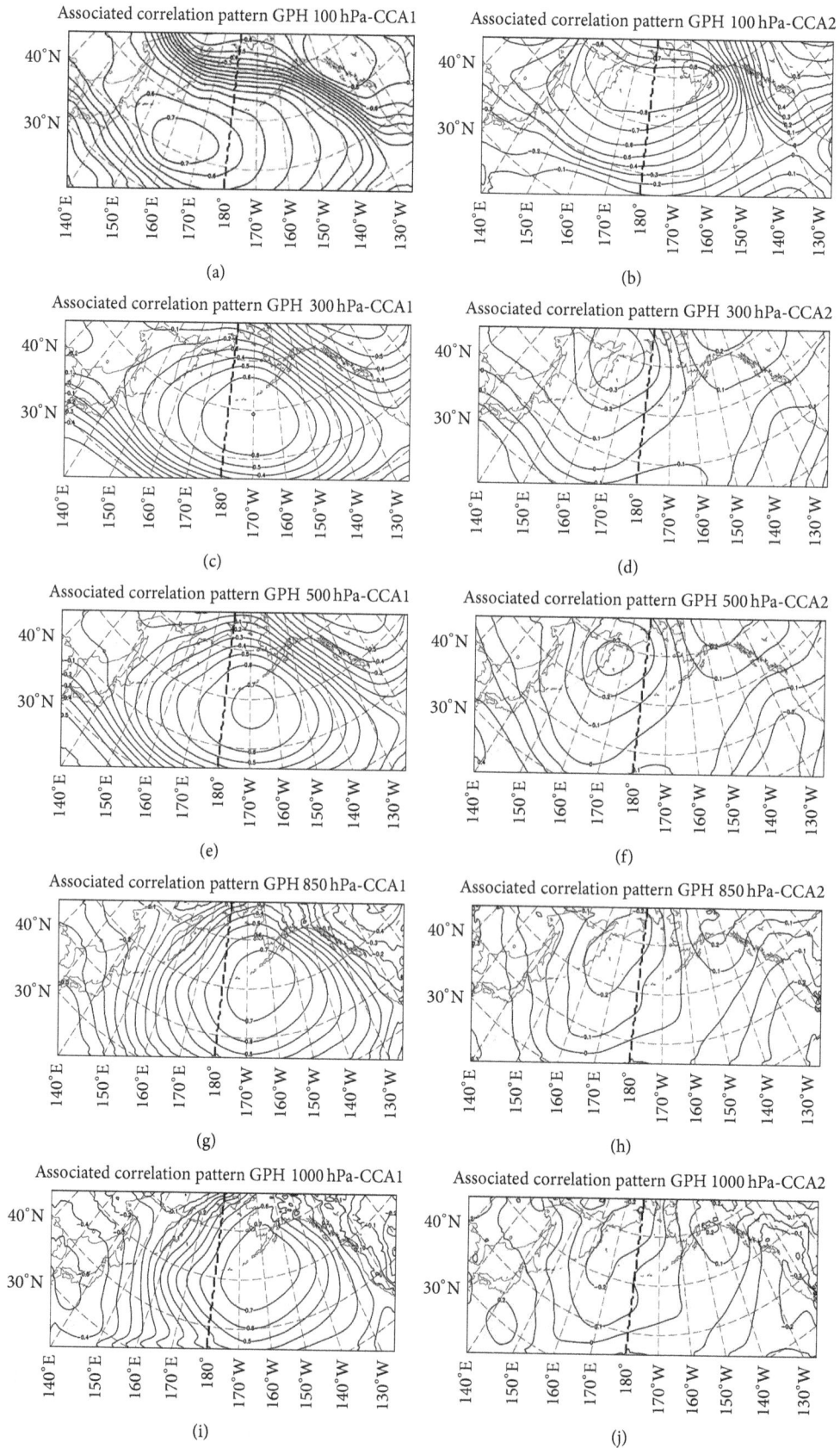

FIGURE 6: Associated correlation patterns between time series of geopotential height over 62 PLSs and time coefficient of the Polar Low occurrence from the first two CCA patterns (1st CCA pair, left column, 2nd right column). From top to bottom, 100, 300, 500, 700, 850, and 1000 hPa. All variables are averaged across a Polar Low season, that is, from October to April.

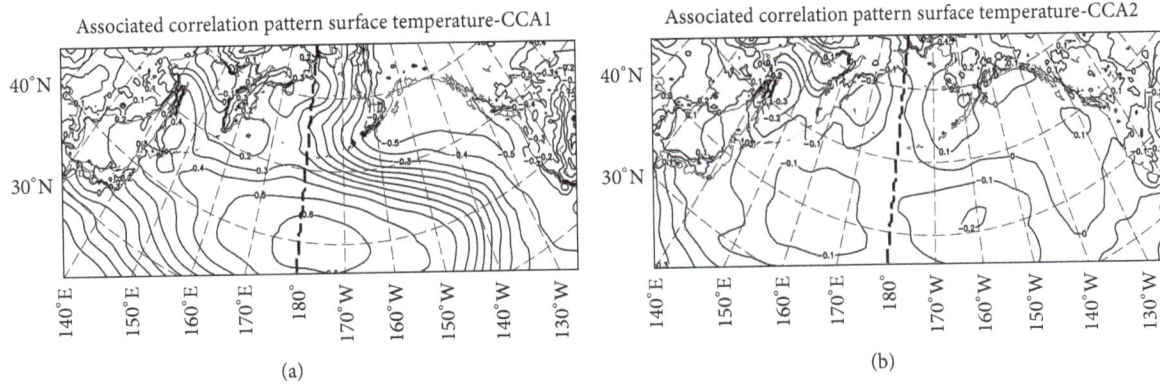

FIGURE 7: Associated correlation patterns between time series of ocean surface temperature anomalies over 62 PLSs and time coefficient of the Polar Low occurrence from the first two CCA patterns (1st CCA pair, left column, 2nd right column). All variables are averaged across a Polar Low season, that is, from October to April. The surface temperature was taken from the NCEP1 reanalysis data, over the ocean. In case of no ice, it indicates the sea surface temperature (SST); else, it indicates the ice surface temperature.

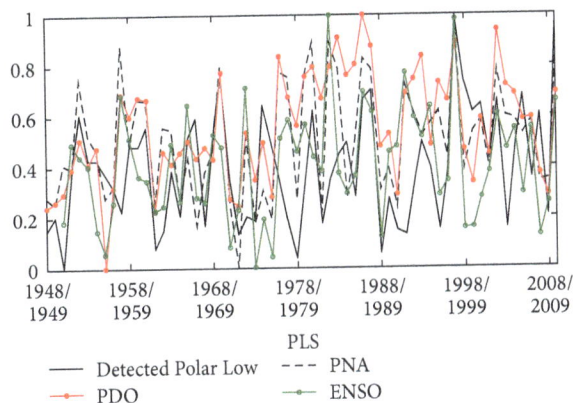

FIGURE 8: Time series of normalized detected Polar Low number (black line), PDO index (red, marked with o), PNA index (black, dashed), and ENSO index (green, marked with o). The indices are determined as average across the Polar Low seasons.

and the time series of geopotential height anomalies for each grid point on each level, we derive typical configurations on every pressure level associated with the two Polar Low patterns shown in Figures 4(c) and 4(d) (the triangles). The anomalies were formed by subtracting the time-mean fields for the 62 considered Polar Low seasons.

The maps of correlation coefficients present similar patterns to the corresponding MSLP field of the CCA results (Figures 4(c) and 4(d); isolines). It indicates that the relationship between the atmosphere circulation and Polar Low occurrence is mostly barotropic, even if in the lower stratosphere of 100 hPa the pattern is somewhat shifted. The cold flow which is inducing the Polar Low occurrence is uniform throughout the troposphere, from sea level to the upper troposphere and the lower stratosphere.

On the upper levels at 100, 300, 500, and 700 hPa, the isolines are smoother. On the lower level at 850 and 1000 hPa, the solid lines are more wiggly: the orography and sea surface temperature attain a stronger influence on the generation of more or less Polar Lows.

In order to investigate the link with *sea surface temperature (SST)* and respective sea ice temperature in case of ice, a

pair of associated correlation patterns is presented (Figure 7) to describe the linkage between the CCA-time series of Polar Low occurrence (dashed lines in Figures 4(a) and 4(b)) and the surface temperature. These temperatures were taken from the same NCEP1 reanalysis, which was used to force the regional model simulation.

Both associated correlation patterns of SST are consistent with the flow patterns of the CCA result. Temperatures tend to be lower, where more cold air is advected, and higher than on average, when the flow advects less cold or more warm air. Previous modeling studies have shown that the mechanism behind these patterns is that of an oceanic response to anomalous atmospheric flow [31, 32], first suggested by Bjerknes [33] for the Atlantic. We conclude that anomalous mean flow is responsible for both, the formation of anomalous SST as well as the formation of more, or less Polar Lows, in the North Pacific.

In order to investigate the *large-scale dynamical environment* of changing Polar Low occurrences, we analyze the correlation between the time series of detected Polar Low number in the different subregions and several climate variability indexes, namely, the Pacific decadal oscillation (PDO [34–36]), Pacific-North American teleconnection pattern (PNA [37]), and El Niño/La Niña-Southern oscillation (ENSO multivariate ENSO index; see http://www.esrl.noaa.gov/psd//people/klaus.wolter/MEI/table.html). The time series for the normalized Polar Low numbers PDO, PNA, and ENSO index are shown in Figure 8.

The correlation coefficient of detected Polar Low number in simulation area with PDO index is 0.39, with PNA index is 0.45, and with ENSO index is 0.22. It indicates that the Polar Low formation over the North Pacific has the highest relationship with PNA and PDO; there is limited influence by ENSO.

Multiple regression analyses of the number of Polar Lows in the 12 subregions (Figure 3(b)) with the time-mean circulation indices for PDO, PNA, and ENSO reveal which regions are mostly affected by the state of the three circulation systems. The regressions results are listed in Figure 9 for every subregion as well as the correlations between the number of Polar Lows and the through regression estimated number of

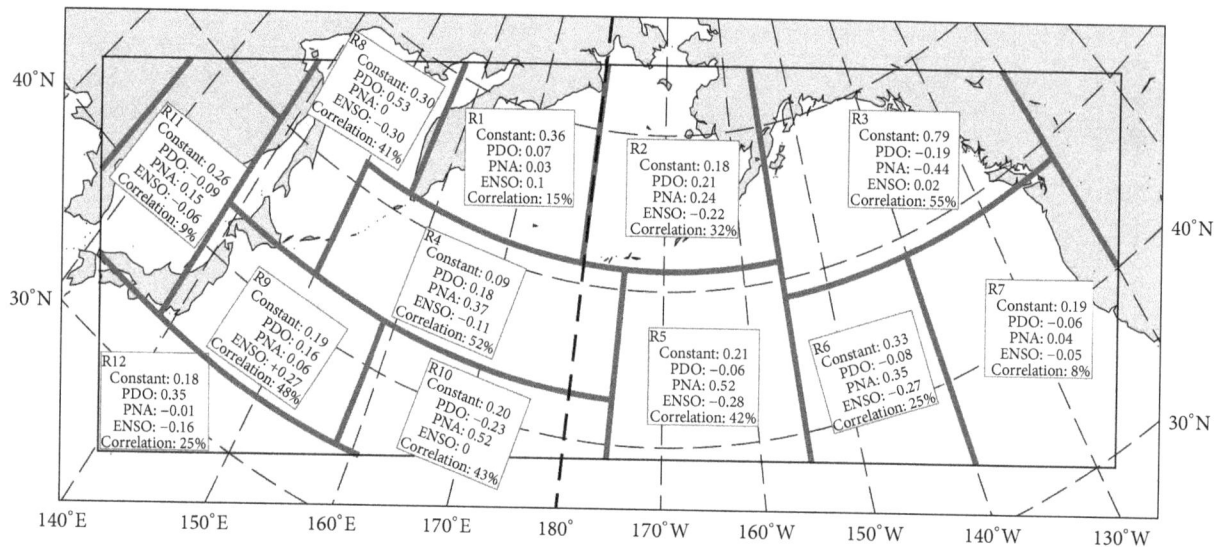

FIGURE 9: Results of multiple regression analyses of number of Polar Low subregions R1–12: the constant parts and the coefficients associated with the PLS time-mean circulation indices PDO, PNA, and ENSO.

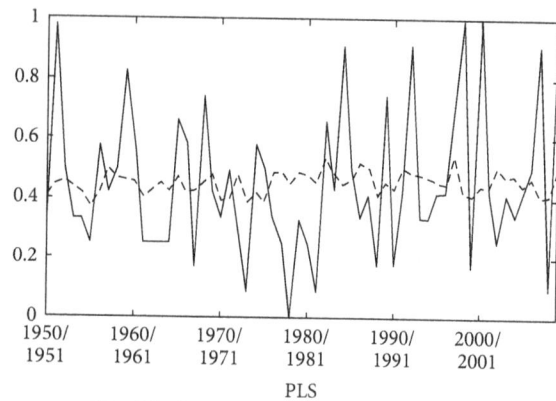

(a)

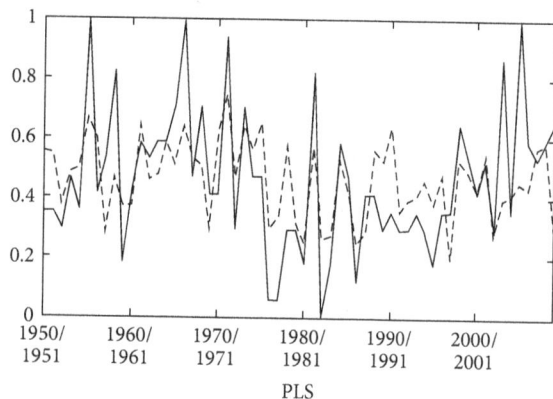

(b)

FIGURE 10: Multiple regression of PLS-time mean circulation indices PDO, PNA, and ENSO on the number of Polar Lows in two of the 12 subregions shown in Figure 3—top: R1, bottom: R4.

TABLE 1: Result of multiple regression analysis of the first two time series of MSLP patterns of CCA between MSLP and Polar Low occurrence and of PLS time mean circulation indices PDO, PNA, and ENSO.

	Constant	PDO	PNA	ENSO	Correlation
CCA1	0.07	0.19	0.54	−0.01	71%
CCA2	0.48	−0.03	0.36	−0.37	26%

Polar Lows. Obviously, the indices are not independent, so that there is no strict separation of the effect of the circulation systems. As an example, the result of the regression is shown for two subregions in Figure 10—one is R1 in the North close to the date line, with a very low correlation of only 15%, and the other is R4 south of R1 in the central part of the North Pacific, with the largest correlation, namely, 55%. Obviously, the circulation indices cannot be associated with year-to-year variability in R1, while the skill of the indices in describing this variability is remarkable in R4.

We find only two regions, R3 and R4, where the total influence of the indices amounts to a correlation of more than 50%. R3 is in the Northeast, south of Alaska, while R4 is at middle latitudes west of the date line; in both cases, PNA is associated with the largest coefficient, which is consistent with the pattern of PNA (not shown). A weak link, as expressed by a correlation of 30% or less, is found for R1, R6, R7, R11, and R12, all subregions at the boundaries of the model domain.

To characterize the variability of the CCA patterns and the large-scale state indices of PDO, PNA, and ENSO, we have also established regression models. They relate the time series of the MSLP pattern with the three indices. The success and relative importance of the three indices are summarized in Table 1. Additionally, the correlations between the time series of MSLP pattern and the one estimated through the regression are listed.

The first CCA pattern is strongly linked to, primarily, PNA and also PDO (which are of course not independent) but hardly to ENSO; this is different for the second pattern, which is described as being negatively linked to ENSO, equally strongly linked to PNA but hardly to PDO. The overall link is much stronger for the first pattern (as indicated by a correlation of 71%), while the link for the 2nd pattern is weak (correlation of 26%).

5. Summary and Outlook

For the first time, a multidecadal climatology, including trends, of Polar Low formation in the North Pacific has been constructed. Because of an insufficient database of local observations and homogeneous high-resolution analyses, a dynamical downscaling strategy has been employed, following the concept developed by Zahn and von Storch [22].

The main result is the presence of large interannual variability in 1948–2010, but little decadal variability of Polar Low occurrences and positive long-term changes in the North Pacific region. No obvious change was detected in recent decades, which indicates that the large natural variability in the region is still dominating over possible effects of global warming.

The tendency of forming more, or less, Polar lows has been related to the time-mean circulation in the North Pacific, in terms of mean sea level pressure and geopotential height throughout the troposphere. Anomalous flows from cold surfaces were found to support the large-scale synoptic environment of Polar Low formation. Two major patterns are detected, which are correlated with the variations described by the indices of PDO, PNA, and, to a lesser extent, ENSO.

Acknowledgments

The work was done with the support of the Chinese Scholarship Council CSC, and it is a contribution to the Helmholtz Climate Initiative REKLIM (Regional Climate Change), a joint research project of the Helmholtz Association of German Research Centers. The authors thank Beate Gardeike for preparing most of the diagrams. The technical and scientific support, the various comments and suggestions by Dr. Beate Geyer and Dr. Matthias Zahn have greatly improved this paper.

References

[1] G. Heinemann and C. Claud, "Meeting summary: report of a workshop on "theoretical and observational studies of polar lows" of the European Geophysical Society Polar Lows Working Group," *Bulletin of the American Meteorological Society*, vol. 78, no. 11, pp. 2643–2658, 1997.

[2] S. Businger, "The synoptic climatology of polar-low outbreaks over the Gulf of Alaska and the Bering Sea," *Tellus A*, vol. 39, no. 4, pp. 307–325, 1987.

[3] K. Ninomiya, "Polar/comma-cloud lows over the Japan Sea and the Northwestern Pacific in Winter," *Journal of the Meteorological Society of Japan*, vol. 67, pp. 83–97, 1989.

[4] B. Yarnal and K. G. Henderson, "A climatology of polar low cyclogenetic regions over the North Pacific-ocean," *Journal of Climate*, vol. 2, no. 12, pp. 1476–1491, 1989.

[5] A. Condron, G. R. Bigg, and I. A. Renfrew, "Polar mesoscale cyclones in the northeast Atlantic: comparing climatologies from ERA-40 and satellite imagery," *Monthly Weather Review*, vol. 134, no. 5, pp. 1518–1533, 2006.

[6] E. W. Kolstad, "A new climatology of favourable conditions for reverse-shear polar lows," *Tellus A*, vol. 58, no. 3, pp. 344–354, 2006.

[7] M. Zahn and H. von Storch, "A long-term climatology of North Atlantic polar lows," *Geophysical Research Letters*, vol. 35, no. 22, Article ID L22702, 2008.

[8] G. Fu, H. Niino, R. Kimura, and T. Kato, "A polar low over the Japan Sea on 21 January 1997. Part I: observational analysis," *Monthly Weather Review*, vol. 132, no. 7, pp. 1537–1551, 2004.

[9] W. Yanase, G. Fu, H. Niino, and T. Kato, "A polar low over the Japan Sea on 21 January 1997. Part II: a numerical study," *Monthly Weather Review*, vol. 132, no. 7, pp. 1552–1574, 2004.

[10] W. Yanase and H. Niino, "Dependence of polar low development on baroclinicity and physical processes: an idealized high-resolution numerical experiment," *Journal of the Atmospheric Sciences*, vol. 64, no. 9, pp. 3044–3067, 2007.

[11] J. F. Bresch, R. J. Reed, and M. D. Albright, "A polar-low development over the bering sea: analysis, numerical simulation, and sensitivity experiments," *Monthly Weather Review*, vol. 125, no. 12, pp. 3109–3130, 1997.

[12] W. Blier, "A numerical modeling investigation of a case of polar airstream cyclogenesis over the Gulf of Alaska," *Monthly Weather Review*, vol. 124, no. 12, pp. 2703–2725, 1996.

[13] F. Chen, B. Geyer, M. Zahn, and H. von Storch, "Toward a multidecadal climatology of North Pacific polar lows employing dynamical downscaling," *Terrestrial, Atmospheric and Oceanic Sciences*, vol. 23, pp. 291–301, 2012.

[14] L. Cavicchia and H. von Storch, "The simulation of medicanes in a high-resolution regional climate model," *Climate Dynamics*, vol. 39, no. 9-10, pp. 2273–2290, 2012.

[15] R. J. Reed, "Cyclogenesis in polar airstreams," *Monthly Weather Review*, vol. 107, no. 1, pp. 38–52, 1979.

[16] M. Nagata, "Meso-β-scale vortices developing along the Japan-Sea polar- airmass convergence zone (JPCZ) cloud band: numerical simulation," *Journal Meteorological Society of Japan*, vol. 71, no. 1, pp. 43–57, 1993.

[17] S. L. Gray and G. C. Craig, "A simple theoretical model for the intensification of tropical cyclones and polar lows," *Quarterly Journal of the Royal Meteorological Society*, vol. 124, no. 547, pp. 919–947, 1998.

[18] S. Businger, "The synoptic climatology of polar low outbreaks," *Tellus A*, vol. 37, no. 5, pp. 419–432, 1985.

[19] T. E. Nordeng, "A model-based diagnostic study of the development and maintenance mechanism of two polar lows," *Tellus A*, vol. 42, no. 1, pp. 92–108, 1990.

[20] J. Mailhot, D. Hanley, B. Bilodeau, and O. Hertzman, "A numerical case study of a polar low in the Labrador Sea," *Tellus A*, vol. 48, no. 3, pp. 383–402, 1996.

[21] A. Condron and I. A. Renfrew, "The impact of polar mesoscale storms on northeast Atlantic Ocean circulation," *Nature Geoscience*, vol. 6, pp. 34–37, 2013.

[22] M. Zahn and H. von Storch, "Tracking polar lows in CLM," *Meteorologische Zeitschrift*, vol. 17, no. 4, pp. 445–453, 2008.

[23] E. Kalnay, M. Kanamitsu, R. Kistler et al., "The NCEP/NCAR 40-year reanalysis project," *Bulletin of the American Meteorological Society*, vol. 77, no. 3, pp. 437–471, 1996.

[24] H. von Storch, H. Langenberg, and F. Feser, "A spectral nudging technique for dynamical downscaling purposes," *Monthly Weather Review*, vol. 128, no. 10, pp. 3664–3673, 2000.

[25] B. Rockel, A. Will, and A. Hense, "The regional climate model COSMO-CLM (CCLM)," *Meteorologische Zeitschrift*, vol. 17, no. 4, pp. 347–348, 2008.

[26] J. Steppeler, G. Doms, U. Schättler et al., "Meso-gamma scale forecasts using the nonhydrostatic model LM," *Meteorology and Atmospheric Physics*, vol. 82, no. 1–4, pp. 75–96, 2003.

[27] B. Denis, J. Côté, and R. Laprise, "Spectral decomposition of two-dimensional atmospheric fields on limited-area domains using the discrete cosine transform (DCT)," *Monthly Weather Review*, vol. 130, no. 7, pp. 1812–1829, 2002.

[28] X. Xia, M. Zahn, K. Hodges, F. Feser, and H. von Storch, "A comparison of two identification and tracking methods for polar lows," *Tellus A*, vol. 64, Article ID 17196, 2012.

[29] A. Kulkarni and H. von Storch, "Monte Carlo experiments on the effect of serial correlation on the Mann-Kendall test of trend," *Meteorologische Zeitschrift*, vol. 4, no. 2, pp. 82–85, 1995.

[30] H. von Storch and F. W. Zwiers, *Statistical Analysis in Climate Research*, Cambridge University Press, Cambridge, UK, 1999.

[31] U. Luksch, H. von Storch, and E. Maier-Reimer, "Modeling North Pacific SST anomalies as a response to anomalous atmospheric forcing," *Journal of Marine Systems*, vol. 1, no. 1-2, pp. 155–168, 1990.

[32] U. Luksch and H. von Storch, "Modeling the low-frequency sea surface temperature variability in the North Pacific," *Journal of Climate*, vol. 5, no. 9, pp. 893–906, 1992.

[33] J. Bjerknes, "Atlantic air-Sea interaction," *Advances in Geophysics C*, vol. 10, pp. 1–82, 1964.

[34] S. R. Hare and R. C. Francis, "Climate change and salmon production in the Northeast Pacific Ocean," *Canadian Special Publication of Fisheries and Aquatic Sciences*, vol. 121, pp. 357–372, 1995.

[35] Y. Zhang, J. M. Wallace, and D. S. Battisti, "ENSO-like interdecadal variability: 1900–93," *Journal of Climate*, vol. 10, no. 5, pp. 1004–1020, 1997.

[36] N. Mantua, "Comparison of typical warm PDO/El Nino SST, SLP, and wind anomalies," 2000, http://jisao.washington.edu/pdo/graphics.html.

[37] J. M. Wallace and D. S. Gutzler, "Teleconnections in the geopotential height field during the Northern Hemisphere winter," *Monthly Weather Review*, vol. 109, no. 4, pp. 784–812, 1981.

5

Intercontinental Transport and Climatic Impact of Saharan and Sahelian Dust

N'Datchoh Evelyne Touré, Abdourahamane Konaré, and Siélé Silué

Laboratoire de Physique de l'Atmosphère, Université de Cocody, 22 BP 582 Abidjan 22, Cote d'Ivoire

Correspondence should be addressed to N'Datchoh Evelyne Touré, ndatchoheve@yahoo.fr

Academic Editor: Dimitris G. Kaskaoutis

The Sahara and Sahel regions of Africa are important sources of dust particles into the atmosphere. Dust particles from these regions are transported over the Atlantic Ocean to the Eastern American Coasts. This transportation shows temporal and spatial variability and often reaches its peak during the boreal summer (June-July-August). The regional climate model (RegCM 4.0), containing a module of dust emission, transport, and deposition processes, is used in this study. Saharan and Sahelian dusts emissions, transports, and climatic impact on precipitations during the spring (March-April-May) and summer (June-July-August) were studied using this model. The results showed that the simulation were coherent with observations made by the MISR satellite and the AERONET ground stations, within the domain of Africa (Banizoumba, Cinzana, and M'Bour) and Ragged-point (Barbados Islands). The transport of dust particles was predominantly from North-East to South-West over the studied period (2005–2010). The seasonality of dust plumes' trajectories was influenced by the altitudes reached by dusts in the troposphere. The impact of dusts on climate consisted of a cooling effect both during the boreal summer and spring over West Africa (except Southern-Guinea and Northern-Liberia), Central Africa, South-America, and Caribbean where increased precipitations were observed.

1. Introduction

Numerous studies have been focused on Sahelian climate variability contrary to the tropical humid African areas for which reliable data do not exist. Servat et al. [1] showed that the tropical humid belt has been similarly affected by series of climatic episodes in comparison with those observed in the Sahelian zone. In addition, this region has been subjected to significant environmental changes due to the increase of populations and strong exploitation of natural resources such as deforestation [2]. Furthermore, this region is particularly rich in aerosols from various origins leading to coexistence in the region of maritime, desert, urban, and bushfire aerosols. Besides, the plumes of desert and Sahelian dusts ejected into the atmosphere during emission episodes are an integral part of the West African climate system [3–5]. The greater part of mineral aerosols is emitted from arid and semiarid zones on the Earth where, these surfaces are less protected from erosion because of very limited or inexistent vegetative covers, and of very low soil humidity limiting the cohesion of constituent elements [5]. Thus, the Saharan-Sahelian region has been identified as the world's first source of emission of this aerosol type [6–9]. Recent studies have made it possible to identify the most active zones in this part of the world. They revealed four major emission sources, that is, the Bodélé region, the Nubian Desert, the Libyan Desert, and the zone covering Mauritania, Mali, and Southern Algeria [10–12]. The particles, once in the troposphere, can be transported over long distances beyond the African continent. The meteorological conditions in the Sahara and surrounding region will determine the dust plume transportation characteristics (i.e., direction, speed, altitudes, trajectories, distance travel, and duration of transport). Three main trajectories for Saharan and Sahelian dust transportation have been identified, and these include the transatlantic transport toward the Gulf of Guinea, United States of America, Caribbean Island, and South America [8, 13, 14], the transport towards the Mediterranean and towards Europe [15, 16], and finally the transport towards Near East and Middle East [17]. A transcontinental transport of aerosol plumes from North Africa to Japan via the Middle East and southern Asia has been suggested by Tanaka et al. [18]. For D'Almeida [13], 60% of Saharan and Sahelian dusts are transported towards the Gulf of Guinea, 28% towards the Atlantic Ocean and 12% towards Europe. Satellite observations and direct measurements have shown that the dust plume transportation over the Arabic Peninsula, Near-East, and Middle East, is essentially responsible for the dust plume coming from Eastern Sahara (Libya, Egypt, and Sudan) [19]. This transportation is significant during three periods in the year [17], namely, the spring (March to May), the summer (July to August), and autumn (September to November).

The dust particles may also influence the local atmosphere by absorbing and reflecting solar radiations that modify the radiation toll of the atmosphere and the Earth surface [20, 21]. Their impact on the global radiation budget still remains the big uncertainty that affects the ability of models to predict climatic change [22]. However, some modeling based-studies showed that Saharan and Sahelian dusts can affect the West African monsoon [23–26]. Although all these studies agree on the fact that dusts have impact on West Africa, they did not agree on the associated atmospheric answers such as the increase and decrease of precipitations over the Sahelian region. Indeed, while Konaré et al. [24] and Solmon et al. [25] have shown that they induce a decrease in precipitations over the Sahelian band, Lau et al. [23] and Kim et al. [27] agreed on the fact that an increase of heat in the higher layers of the troposphere induces an adiabatic warming which, causes an increase of precipitations. Such a controversy perfectly illustrates the complexity of "aerosol-climate" interactions, as well as the necessity of taking them into account in climate prediction models, and the West African climate in particular. However, most of these studies were focused on the monsoon period and very few were concerned about the transitional period (March-April-May), even though intensive events of dust emissions often occur during this period (e.g., during the spring 2010).

This paper presents (1) an interseasonal analysis of dust transport from the African continent to the American coasts using regional climatic modeling and (2) assesses the impact of their radiation and associated effects during the transport with specific focus on events of the spring 2010.

2. Methodology and Data Collection

The regional climate model, RegCM 4.0, designed by the International Centre of Theoretical Physics (ICTP) in Italy, was used to do the climate simulations based on the method used by Giorgi et al. [28] and Pal et al. [29]. A dust particle module was incorporated in the version 3.1 of this model by Zakey et al. [30]. Since then, several studies on dusts have been made using the model [24, 25, 31]. The weekly data of the Optimum Intertropical Sea Surface Temperature (OISST) of the National Center for Environmental Prediction (NCEP) for the values of the Sea Surface Temperature

(SST), as well as the reanalysis data of the NCEP, were used as initial meteorological conditions while forcing the model during the simulations. Soil humidity was initialized under standard conditions of the model (RegCM) as defined by Giorgi and Bates [32]. The surface conditions and soil types used in the model were from Dickinson et al. [33] and Zobler [34], respectively. RegCM4.0 uses the radiation scheme of the NCAR CCM3, which is described in Kiehl et al. [35]. Briefly, the solar component, which accounts for the effect of O_3, H_2O, CO_2, and O_2, follows the σ-Eddington approximation of Kiehl et al. [35]. It included 18 spectral intervals from 0.2 to 5 μm. The simulations were done between November 2004–December 2010, with 48×168 grid points, at the spatial resolution of 60 km and 18 pressure levels. To validate the simulations, available observation data from Aerosol Robotic NETwork (AERONET) on the Barbados station and data from three African stations (Banizoumbou in Niger, Cinzana in Mali, and M'Bour in Sénégal) were used. In addition, monthly observation data from Terra Multi-angle Imaging Spectro-Radiometer (MISR) at the spatial resolution of 0.5×0.5 degrees were used to validate the monthly mean Aerosol Optical Depth (AOD) values over the whole domain of the simulations. The domain was chosen in manner that the particle sources, transports, and depositional processes from Africa to America could be captured in the simulations. In order to study the climatic impacts, two simulation types were done. While the first type considered dust particles, the second did not.

3. Results and Discussion

3.1. Comparison between Model and Observations. The simulations made with the climate model, RegCM 4.0, between November 2004–December 2010, showed that the model perfectly simulated the different dust emission sources in the North Africa. The four major sources (the depression of Bodélé, the Nubian Desert, the Libyan Desert, and the zone covering Mauritania, Mali, and Southern Algeria) [11, 12] were well captured (Figures not shown). Figure 1 shows the comparison between the monthly mean emissions of March, April, and May 2010 from the model and those of the available observations from Terra MISR at the resolution of 0.5 $\times$ 0.5 degrees. It appeared that the model properly captured the dust emissions of April and May (Figures 1(c), 1(d), 1(e), and 1(f)). However, values from the MISR observations were lower than the simulated results. The lower values of the observation could be due to the fact that closer to the sources, substantial amounts of dust present in the lowest atmospheric layers (below 1 km) could not be detected by the satellite [24]. In contrast, great differences were found between the observations and the model simulations for the emissions in March, 2010 (Figures 1(a) and 1(b)). These differences may be attributed to emissions from biomass fires that were not taken into account in these simulations. This month (March) coincided with the end of the fire season in this part of the Northern Hemisphere. The model captured well the AOD values in the absence of burning practices in West Africa during the months of April and May.

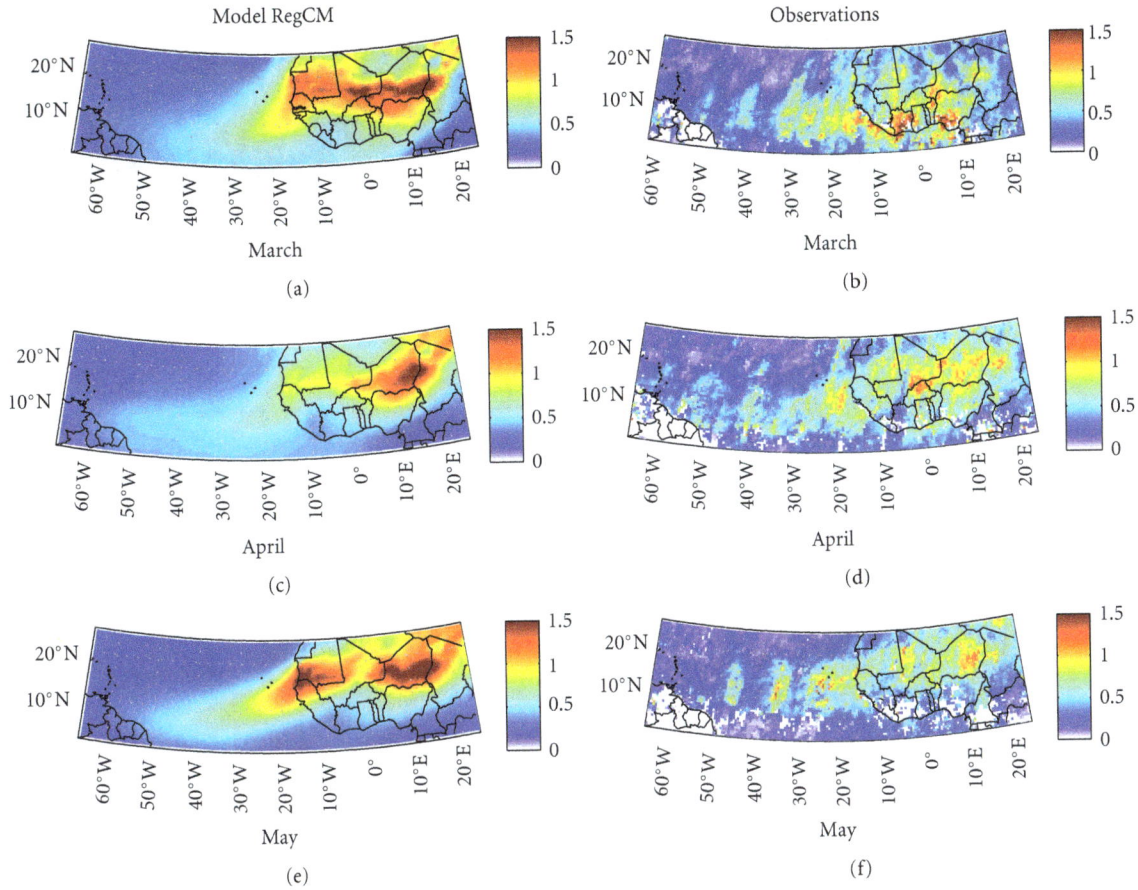

FIGURE 1: Comparison of monthly mean values of AOD during the year 2010: (a) RegCM for March, (b) MISR observations for March, (c) RegCM for April, (d) MISR observations for April, (e) RegCM for May, and (f) MISR observation for May.

However, dust sources were more extended in the model than in the observations, and these were coherent with the results from Konaré et al. [24] obtained with the version 3 of the same model. Therefore, compared to the MISR observations, the model tends to overestimate the AOD in the areas of maximum dust sources. On the other hand, close to these sources substantial amounts of dusts were present in the lower atmospheric layers (below 1 km), which present difficulties for satellite retrieval and hence not detected in the MISR data. The Bodélé Sahel was the most active region during the spring of 2010.

3.2. Dust Episodes during March 2010. This study was focused on March 2010 as a continuation of observations made at the four stations (Banizoumbou, Cinzana, and M'Bour). Modeling data obtained revealed intensive dust emission episodes for March 2010. Two important emission events were simulated by the model over Niger that has been previously identified as sources by N'Tchayi et al. [36]. The first event occurred during the first 10 days of the month (precisely during March 5–8th), whereas the second took place between the 17th and 21st of March (Figure 2).

The comparison between AOD values reproduced during these two events with the model and those found

by AERONET at the three African stations (Banizoumba, Cinzana, and M'Bour) indicated significant dust emissions (Figure 3). However, at the Banizoumba station, the model captured the first peak of March very late. In addition, observations from the stations confirmed that the AOD values were well detected in the model simulations both in terms of magnitude and variability. The second event (17th to 21st March) was well captured by the model than the event of 5th to 8th March. However, the AOD values detected with the model were slightly underestimated compared to the observed values. This underestimation might be explained firstly, by the overestimation of the duration of events (i.e., dust emissions), which was illustrated by the widening of peaks (Figure 3) and, secondly, by the interpolation method to estimate the mean of AOD values from the model.

3.3. Temporal Evolution of Dust Plumes. Although the model takes into account the emission and transport of dusts at a time, in this section we focused on dust transport in March 2010. The transportation of dust was observed from the African continent towards the Atlantic Ocean (from North-East towards South-West) (Figures 4, 5, and 6). A maximum dust plumes seen on March 6th over Mauritania and Senegal was found on March 7th over Senegal and Gambia. Then,

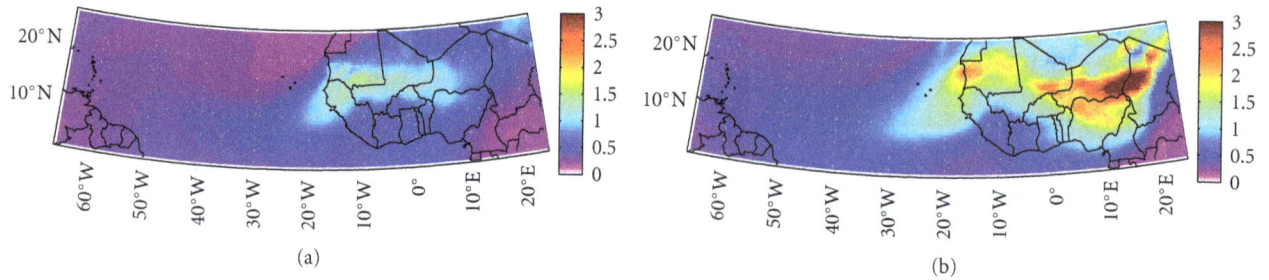

FIGURE 2: Mean AOD values during dust episodes simulated with the RegCM 4.0: (a) March 5–8 and (b) March 17–21.

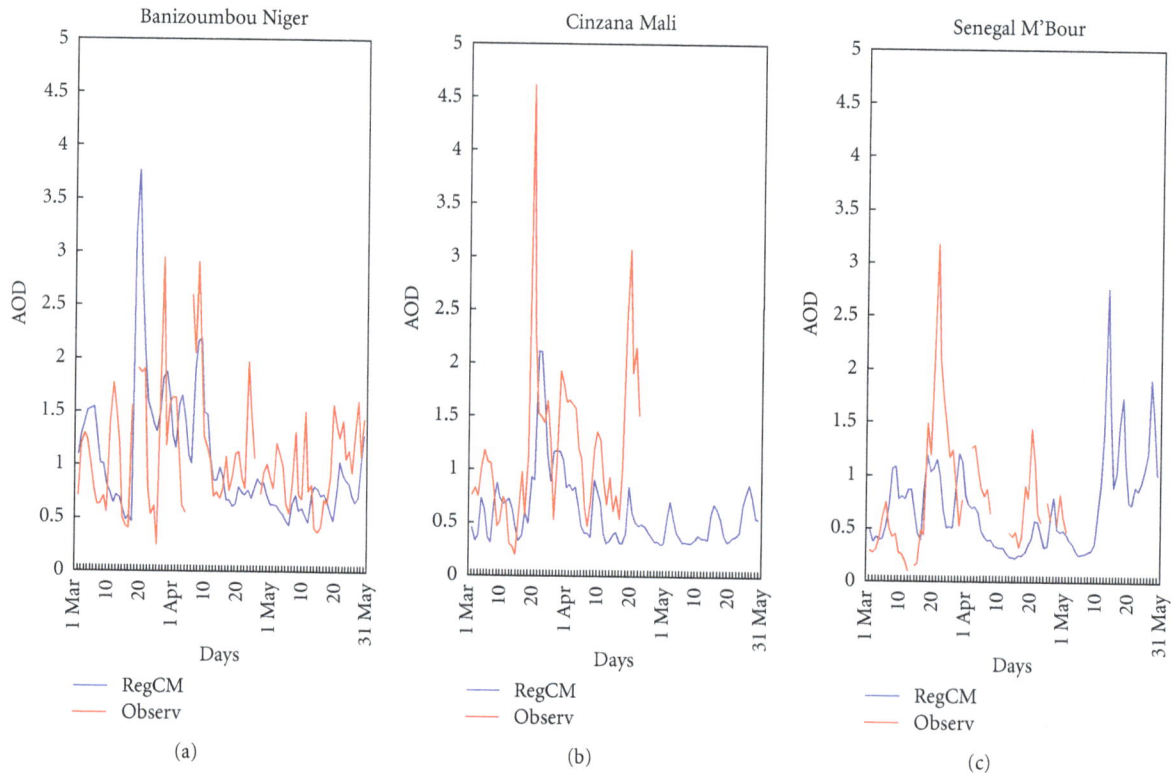

FIGURE 3: Comparison of AOD values between the RegCM 4.0 simulations and the observations made with the AERONET at 500 nm (from March to May 2010).

this continued to the Ocean beyond the Guinean Coasts on March 8th (Figure 4), before being entirely present over the Atlantic Ocean on March 9th. Dust plumes became less concentrated during the transport as a result of deposition. Thus, the particles emitted on March 6th spent 3–5 days to pass through the continent to the Atlantic Ocean (Figure 4). In contrast with these two periods, dust events on the 17th to 21st of March seemed stronger (Figure 5). In fact, the Chad and Niger sources were active simultaneously during the first three days (19th, 20th, and 21st of March) in the way that the spatial evolution was more accentuate than the temporal evolution. This may be attributed to the fact that the model captured well the events of 17th to 21st of March than that of the 5th to 8th of March. Similar paths were found for dust emissions occurring in April and May (i.e., from North-East towards South-West). This direction,

as revealed in Figures 4, 5, and 6 for the period March-May, was shown by previous studies, which suggested that dust transport from the African continent was seasonal following three trajectories: (i) the transatlantic trajectory towards the USA, Caribbean Islands, and South America [8, 15, 37], (ii) the transport towards the Mediterranean and Europe [16], and (iii) the transport to the Near East and Middle East [17]. Thus, dusts emitted from Sahara and Sahel can reach the West Indies during the summer [17], whereas during the spring, they are transported through South America to the Amazonian basin and eastern Mediterranean [8, 38].

The transatlantic trajectory of dust particles is known to be influenced by the Inter-Tropical Convergence Zone (ITCZ) [39, 40]. This trajectory is influenced by the different mechanisms of dust plume rising in the atmosphere linked to monsoon flow. The convergence of Harmattan and monsoon

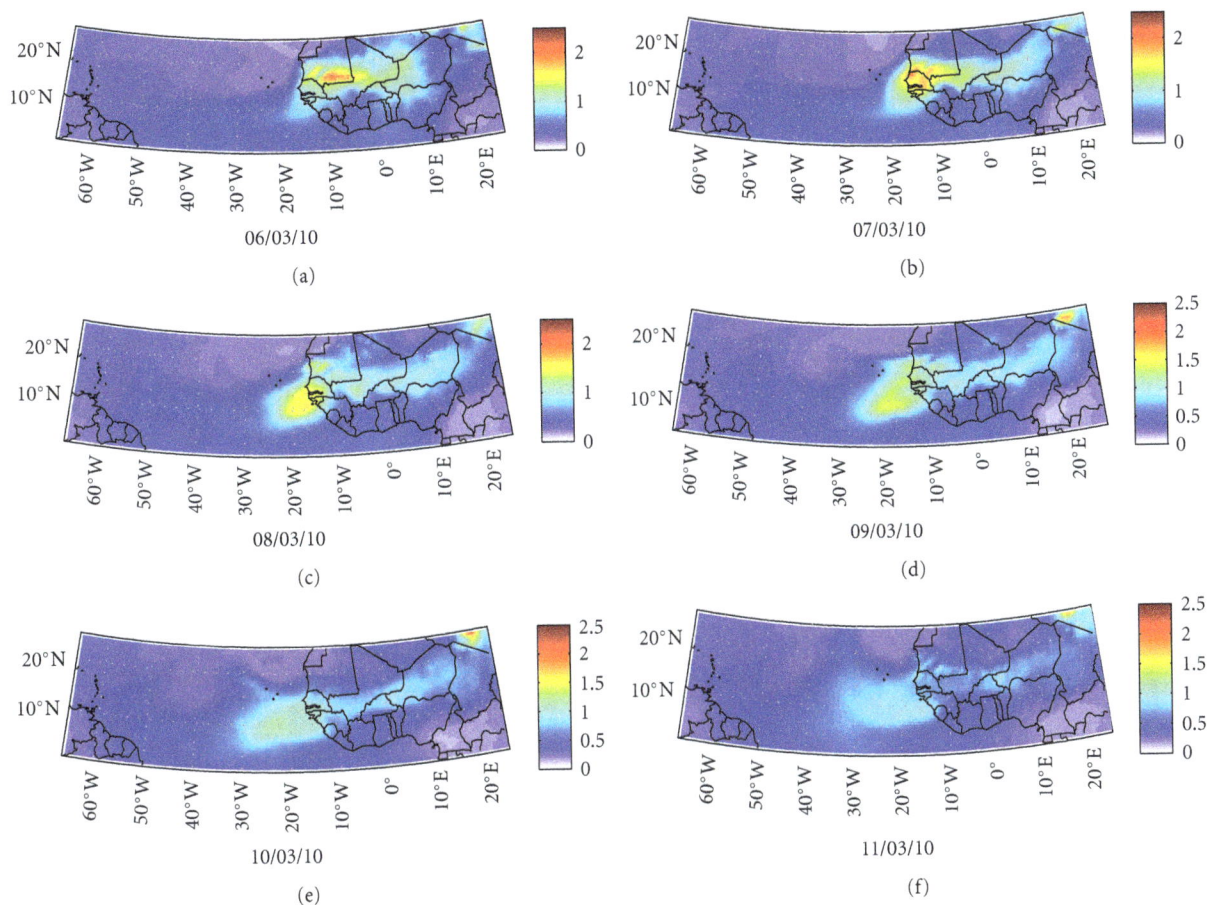

FIGURE 4: Evolution of dust particles (episode of March 5–8th simulated with the RegCM 4.0) from the African continent towards the Atlantic Ocean between March 6–11th 2010.

wind in the low pressure zone in the ITD (InterTropical Discontinuty) constitutes a favorable condition for the establishment of strong wind on either side of the ITD favoring the rising of the dust plumes [39, 40]. The directions of the dust plumes vary with the seasons linked to the ITD and the dust rising mechanisms that is governed by the West Africa Monsoon system. It is also modulated by the activity of the East African Waves (EAW), which spread from East to West [41–44]. These transport features are satisfactorily captured by the Model. Jones et al. [43] showed that 20% of dust particles ejected into the atmosphere over the major part of North Africa were associated with EAW. They suggested that EAW must have regulated the suspension of Saharan and Sahelian dust in the atmosphere. Also, 10–20% of dust seasonality concentrations over North Atlantic Ocean were linked to EAW, an indication of the modulation of dust transport by the EAW.

3.4. Dust Detection in Barbados' Islands. Studies have shown that dust particles may be detected by the AERONET in the Caribbean Island and on the American coasts [8]. We, therefore, compared the simulated AOD to those of the AERONET Ragged-point station located in the Barbados Island, with the aim of bringing out the particularity of

emissions of spring 2010, in spite of the missing observational data. The AOD values were comparable to those of the Barbados Islands, although spring was not the season of maximal detections of Saharan and Sahelian dusts in the Island. Thus, monthly AOD values found at the Ragged-point station during the months of March-April-May from 2008 to 2010 showed that the most elevated values (more than double of values observed in 2008 and 2009) were registered during March and April 2010 (Table 1). The AERONET observational AOD values showed that the mean value of March and April of 2010 was more than doubled that of March and April of 2008 and 2009 at the same station. In addition, the values registered during spring 2010 (at 500 nm) were in the same order as the monthly mean values at 870 nm, observed during the boreal summer where Saharan and Sahelian dust plumes got to the Barbados Islands in their maximal quantities [45].

The model satisfactorily reproduced the variability in most of the peaks observed by AERONET measurements (Figure 7). However, dust emissions occurring during the first 10 days of March 2010 were not detected by the model.

3.5. Zonal and Meridian Transport of Dust. Dust particles are present in the West African environment and interact

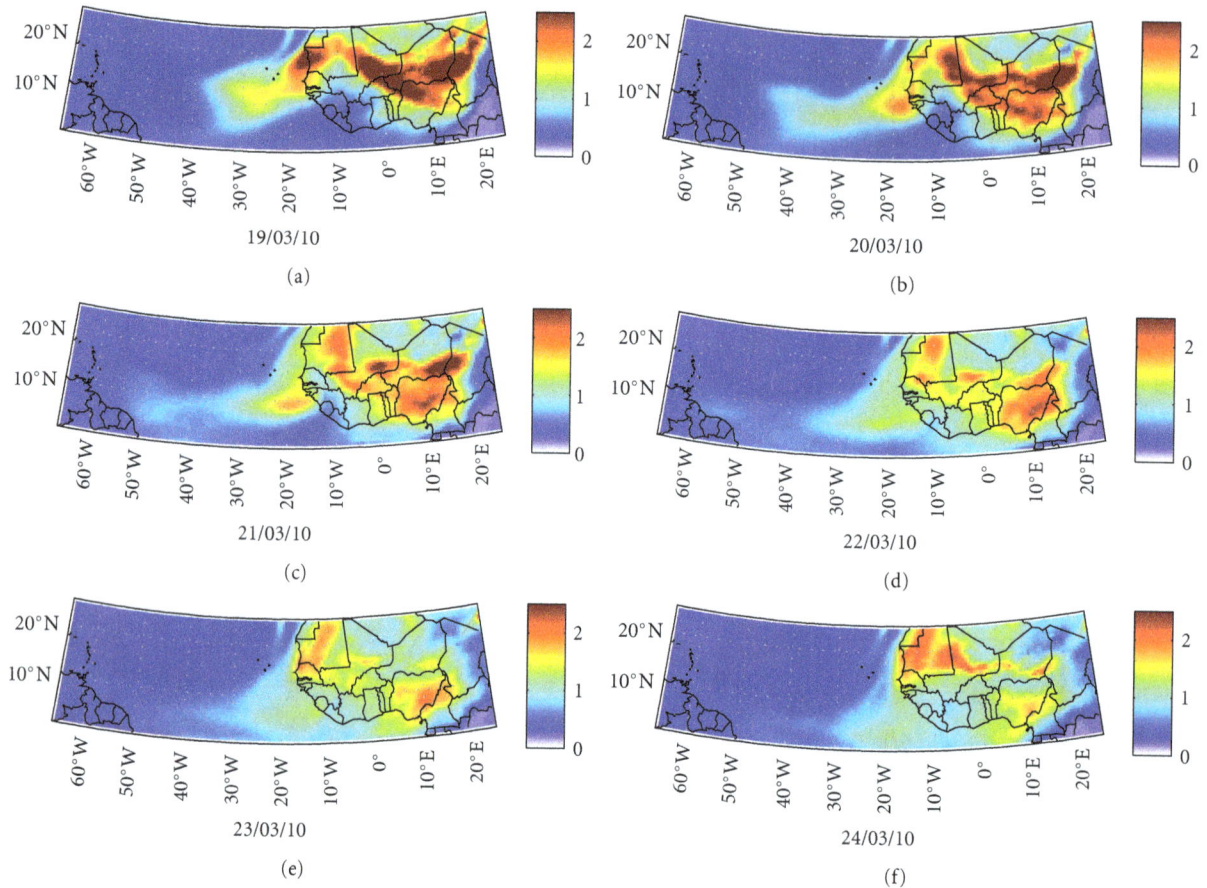

FIGURE 5: Evolution of dust particles (episode of March 17–21st, simulated with the RegCM 4.0) from the African continent towards the Atlantic Ocean between March 19–24th 2010.

TABLE 1: Comparison between monthly AOD values (at 500 nm) simulated with the RegCM 4.0 and monthly values observed at the Ragged-point station (Barbados).

	2008	2009	2010
March			
RegCM	0.11	0.14	0.21
Observation	0.09	0.08	0.19
April			
RegCM	0.15	0.17	0.21
Observation	0.18	0.09	0.31
May			
RegCM	0.19	0.22	0.18
Observation	0.16	0.23	0.24

with climate processes. Hence, they constitute an integral component of climate systems over this region. They were present in all the study areas and their concentrations progressively decreased as they moved away from their sources due to the deposition processes, which are various and cover large part of the globe with some preferential sites such as the tropical forest [8], monsoon zone [46], and the Oceans [47].

Meridional profile section (Figure 8) revealed that dust sources are located between 18°N and 22°N. The dust source around 18°N (Bodélé) possessed a quasicontinuous activity that is, during the three periods December-January-February (DJF), March-April-May (MAM), and June-July-August (JJA) with the most intensive activity occurring at DJF and MAM. During the dry season (DJF), the dusts were mostly observed in the lower troposphere where they essentially spread from the continent towards the West and from the North towards the South. During the period, MAM, they reached higher altitudes than that of DJF with elevated concentrations towards the South. During the boreal summer, there were not only elevated concentrations but also vertical rising of dust particles. This result was consistent with Chiapello et al. [48], who showed that during summer, dust particles are transported into higher altitudes than in winter. Between 1°W and 7°E, a vertical rising of dusts towards the highest atmospheric strata appeared during the period MAM (Figure 9). This may be due to the existence of a thermal anomaly, which could favor a vertical rising of dust particles. Thus, the period MAM could be a transition between the periods DJF, where the altitudes reached were minimal and JJA showing maximal altitudes. The altitudes reached during MAM were also transitional between those of DJF (low altitudes) and JJA (high altitudes).

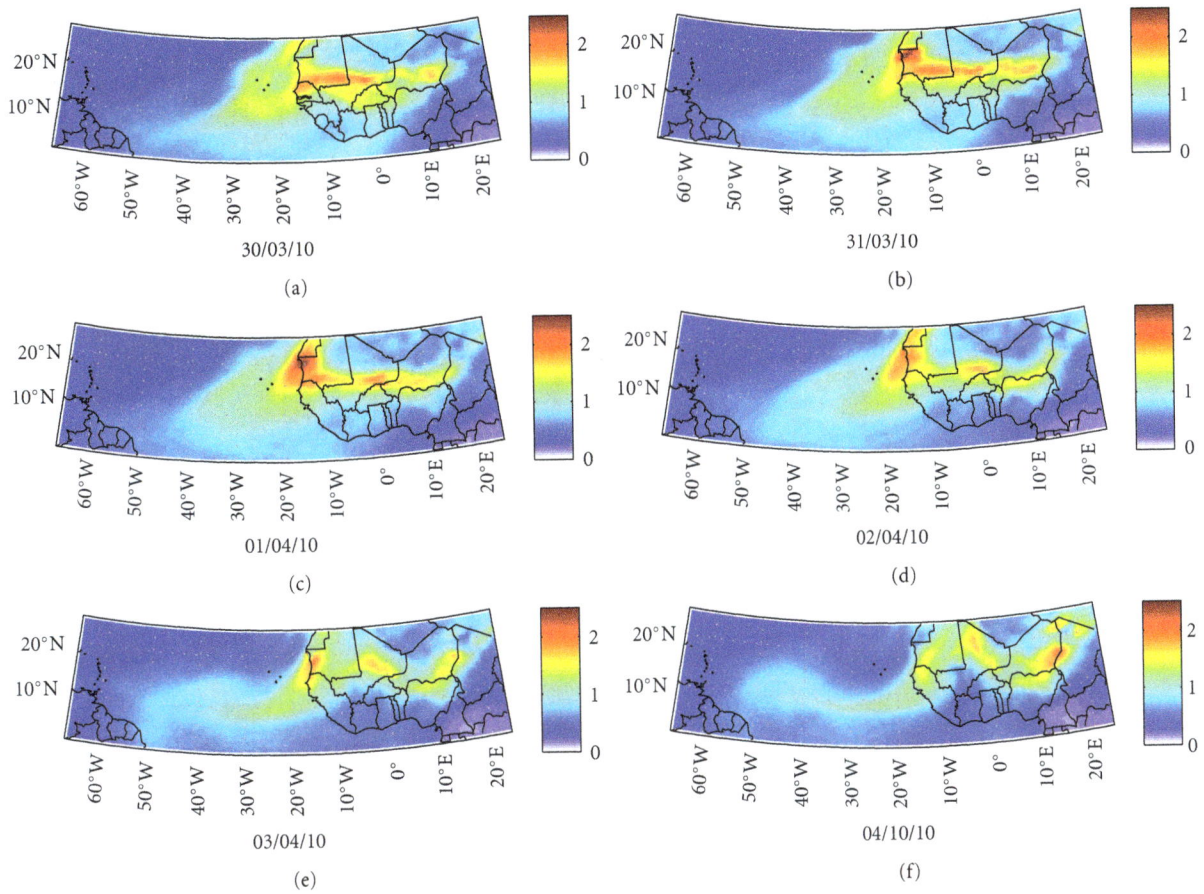

FIGURE 6: Evolution of dust particles from the African continent to the Atlantic Ocean, simulated with the RegCM 4.0 between March 30th and April 4th 2010.

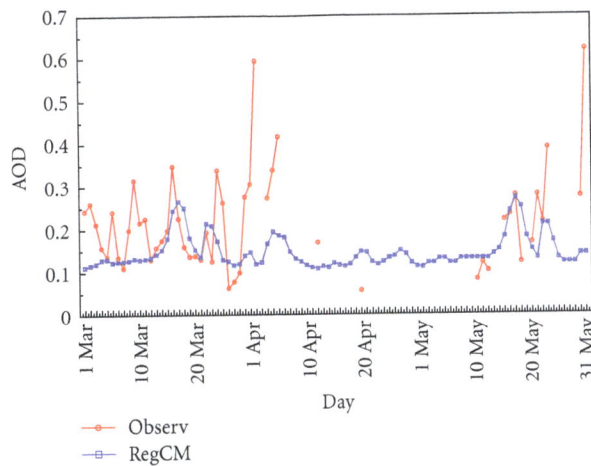

FIGURE 7: Daily simulated AOD values at the Ragged-point station, during March-April-May 2010 with the RegCM 4.0 and observations from AERONET.

The altitudes reached by the dust particles strongly impacted the trajectories of dust plumes and therefore have to be taken into account when studying the transport of dust plumes over the Atlantic Ocean. In effect, these altitudes can vary according to the season and the dust concentrations decrease progressively when they are transported from their sources towards the coasts due to the deposition processes. Dust transport is obviously strongly influenced by the regional atmospheric dynamic. Otherwise, the importance of dust concentrations in the low atmospheric strata and their propagation at such low altitudes may be explained by the occurrence of the Harmattan winds in the region during the

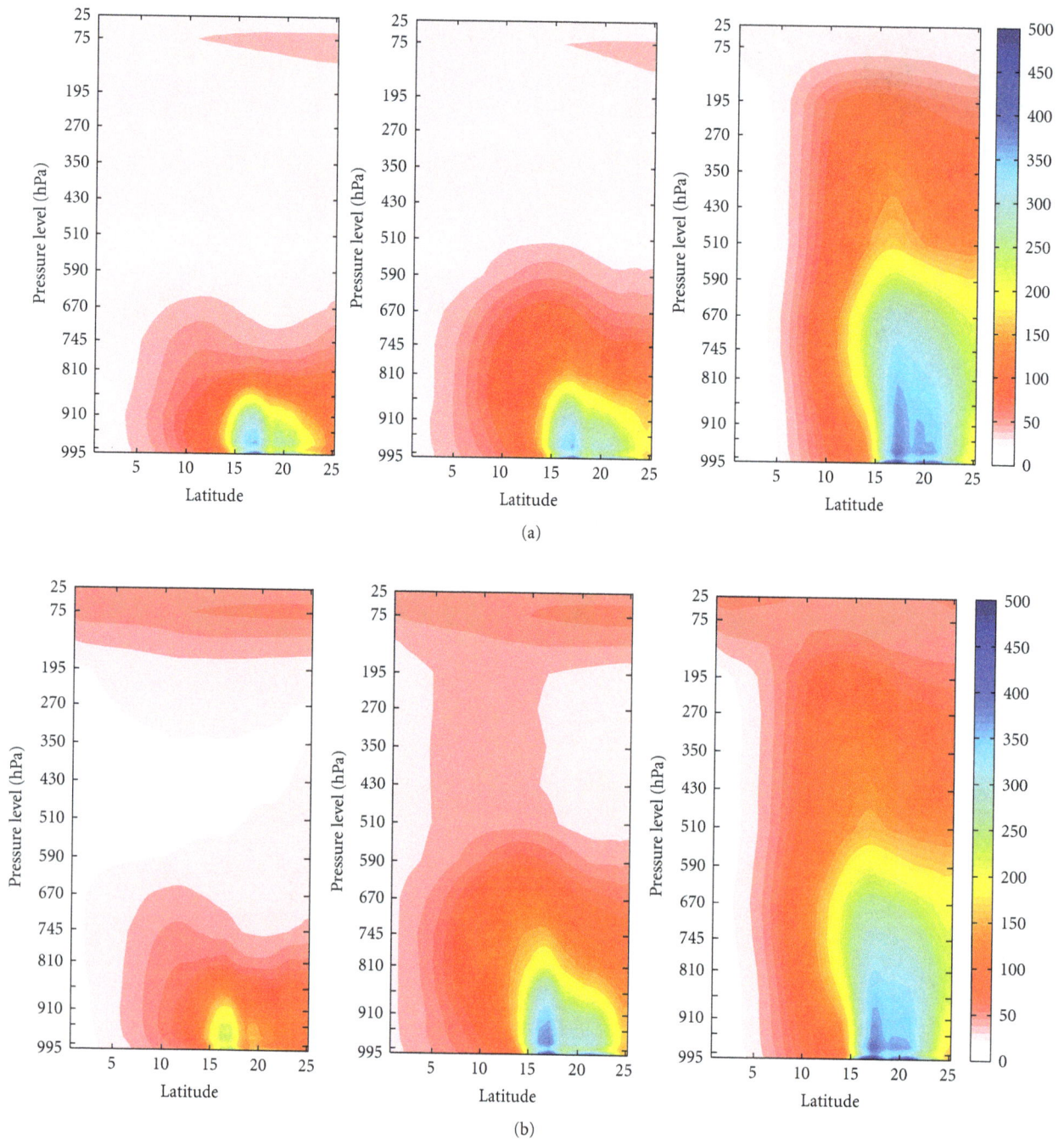

Figure 8: Meridian profile of mean desert dust concentrations in μg/kg average between 66°W and 23°E for the periods "December-January-February (DJF)," "March-April-May (MAM)," and "June-July-August (JJA)": (a) mean values for the years 2005–2010; (b) mean values for 2010. Higher concentrations are indicated with cool colors and lower concentrations with hot colors.

period, DJF. The Harmattan is a low-altitude wind with its maximal intensity during DJF; it could be assumed to aid the transportation of dust plumes. The dust transport seemed to have been enhanced by the North-East direction of the winds, before reaching the coast towards the Gulf of Guinea. As for the period MAM, the transitional period between the dry season and the monsoon of the boreal summer, a thermal effect of low strata (between 1° West and 7° East) could favor

the vertical rising of dust particles, explaining the relative importance of altitudes reached. Moreover, during the boreal summer, the regional atmospheric dynamic dominated by the West African monsoon may explain the altitude and trajectory followed by dust plumes. The migration of the Inter-Tropical Convergence Zone (ITCZ) around 10° N and the resulting dry and wet convections could favor the vertical rising of dust plumes in the troposphere. Thus, the dusts

(a)

(b)

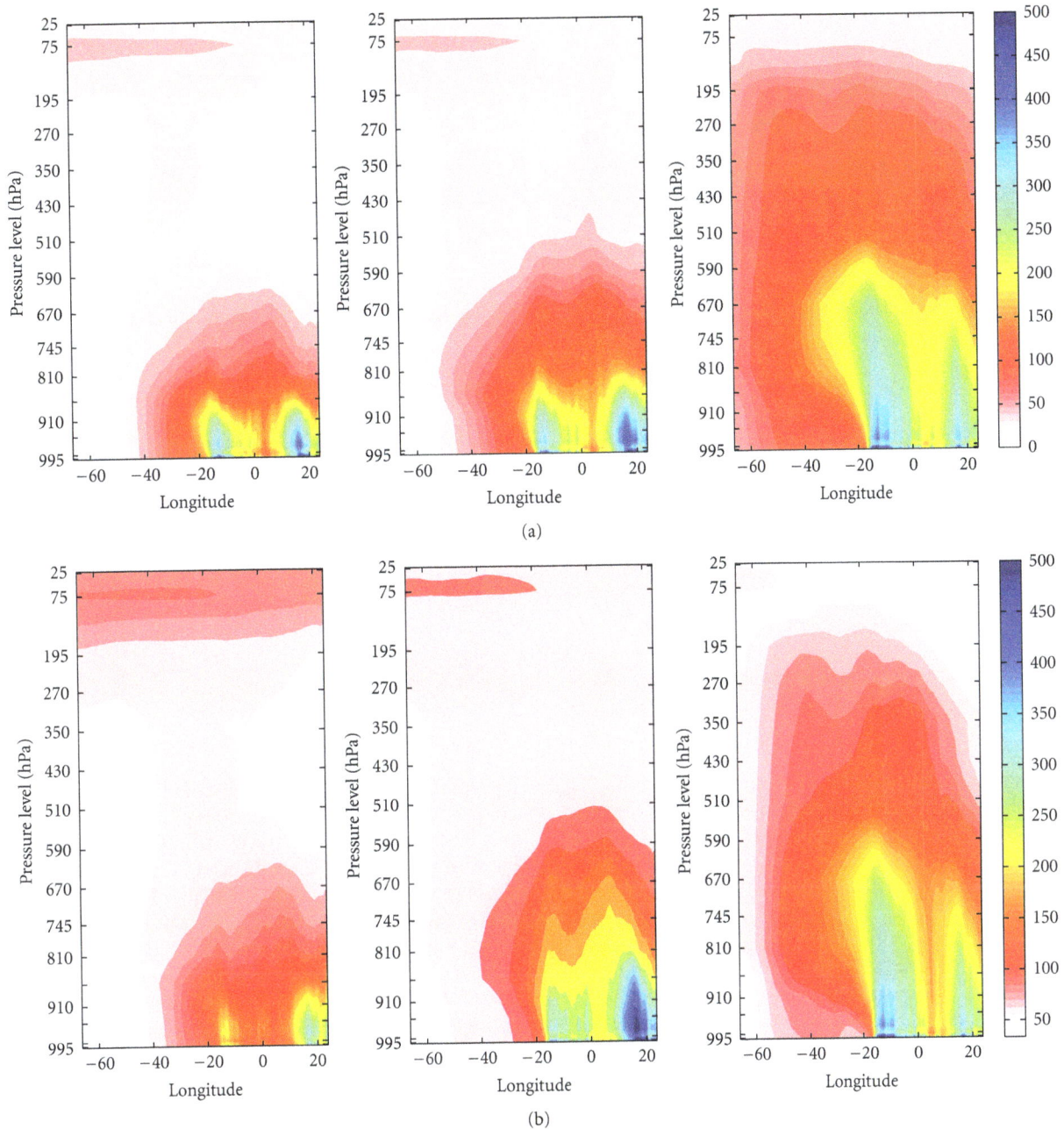

FIGURE 9: Zonal profile of mean desert dust concentrations in μg/kg in the troposphere average between 0°N and 25°N for the periods "December-January-February (DJF)," "March-April-May (MAM)," and "June-July-August (JJA)": (a) mean values for the years 2005–2010; (b) mean values for 2010. Higher concentrations are indicated with cool colors and lower concentrations with hot colors.

emitted at the surface during the summer are ejected more higher into the atmosphere where the transport towards the Atlantic Ocean is done through the Saharan Air Layer (SAL).

The model properly represented the seasonal variability of the altitudes of dust plumes, which was in agreement with the study by Chiapello et al. [48]. In addition, the influence of the ITCZ's dynamics and dust emission mechanisms during the monsoon [39, 40] were also properly represented using the model. The seasonality of sources and the importance of their activity obtained using the model simulations were coherent with the results of Laurent et al. [9] who had

showed that the dust sources located in the West have their maximum activities during June and July, whereas those located in the East have their during March and April. In addition, observations from MODIS, [49] showed that the Bodélé source is quasiactive during the year with maximum activities during January and March. In addition, the maximal dust emissions during the boreal summer coincided with the migration of the Inter-Tropical Front (IFT) towards the North (around 10°N). This migration favored dry and wet convections that modulated dust emissions [50], which were well detected with RegCM previously shown

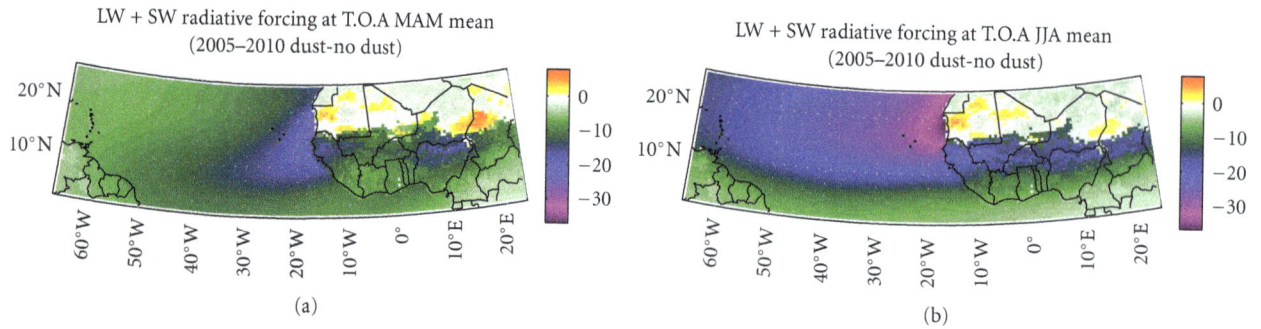

FIGURE 10: Solar radiation forcing (Longwave and shortwave) in W/m² due to the presence of dusts at the top of the atmosphere.

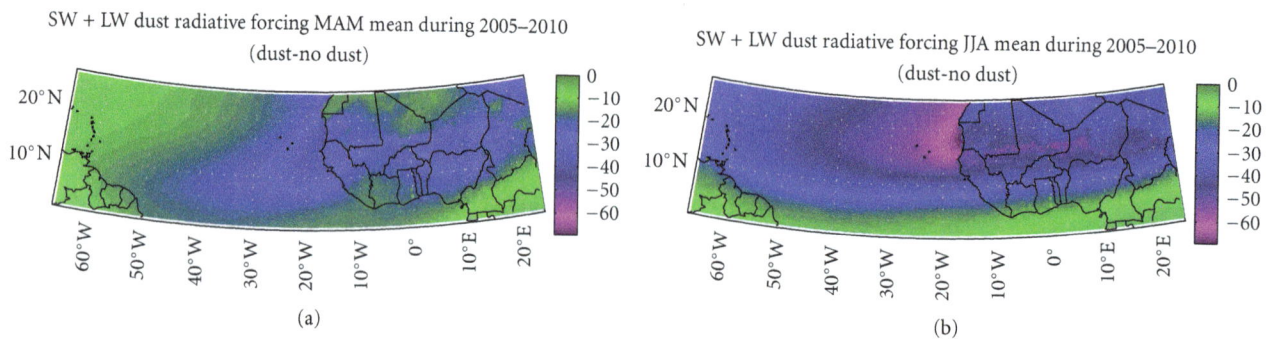

FIGURE 11: Solar radiation forcing (Longwave and shortwave) in W/m² due to the presence of dusts at the surface.

with elevated concentrations on Figures 8(b) and 9(b). The zonal and meridian vertical profile revealed very elevated dust concentrations in the troposphere during the periods DJF and MAM in 2010, when compared with the mean concentration found between 2005 and 2010 for the same periods.

Although the activity of the Bodélé source was less important than the mean activity during DJF between 2005 and 2010, dust concentrations in the troposphere and the stratosphere in particular were more important. During the period MAM 2010, a vertical rising observed in the vertical profiles suggested that the altitude at which dusts were transported represented an important factor for the study of the trajectories followed by dust plumes while crossing the Atlantic Ocean, as well as the distance they travel all over.

3.6. Climatic Impact of Dust. Dusts emitted during the monsoon period can significantly affect both its development and the precipitations, as they interfere with short and long wavelength radiations and modify the physical and radiative properties of clouds [23–27, 51]. This section was devoted to the impact of dust radiative effects on surface temperatures and precipitations both during the periods (MAM) and (JJA). It is important to underline that the results presented in this section are preliminary and required further analysis that will be done in our next work.

3.6.1. Dust Radiative Forcing. The radiative forcing caused by dusts on the top of the atmosphere is presented in Figure 10.

It measured dust-induced cooling and warming events on the Earth-Atmosphere system before the climatic processes take place, that is, before any climate process interacts with the solar radiation. The results showed that over the desert, the forcing is either positive or nil, both during MAM and JJA. This may be explained by the important elevated albedo values in the region that reduced the incident solar radiation in the short wavelengths. Change in forcing sign around 15°N was due to the change in albedo values between the desert in the North and the Sudanian savannas in the South, and also the decrease in dust quantity from sources in the South. The results showed that the radiation forcing induced by dusts at the surface was negative both during MAM and JJA, an indication that dust can exert essential cooling effect at the surface (Figure 11). The cooling effect may suppress the vertical movement of air masses due to the lack of convective processes. If the low strata cooling and the reduction in vertical movements are strong and covering large area, it may result in a reduction in ascendant air over the extended surfaces. If this type of phenomenon occurs during the monsoon over West Africa, the reduction of the convective movement on the large scale would be sufficient to reduce significantly the moisture inflow from the Gulf of Guinea in the lower and the middle strata. Such reduction in moisture inflow towards the north may induce a delay in the onset of the monsoon and also a reduction of its intensity, leading to less precipitation.

3.6.2. Effect of Dust on Temperatures. The effect of dusts included cooling the surface during the periods JJA and

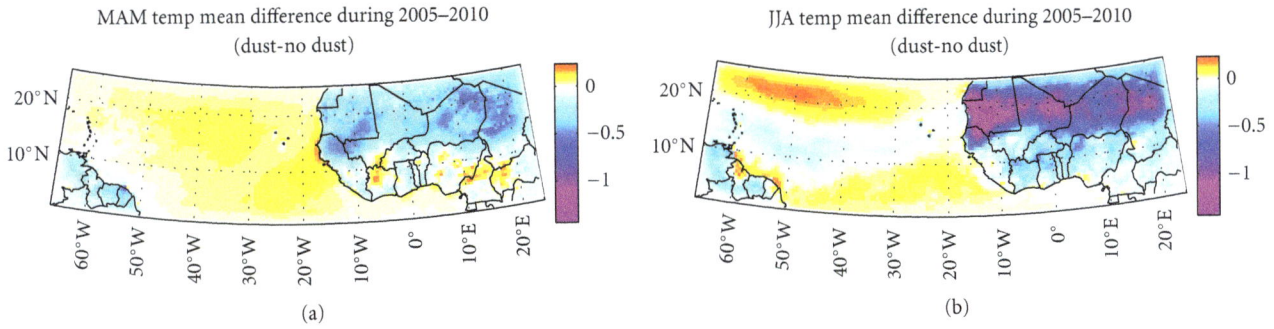

FIGURE 12: Effect of dusts on temperatures in °C at the surface during the months of (a) MAM and (b) JJA. Cooling is indicated with cool colors while warming with hot colors.

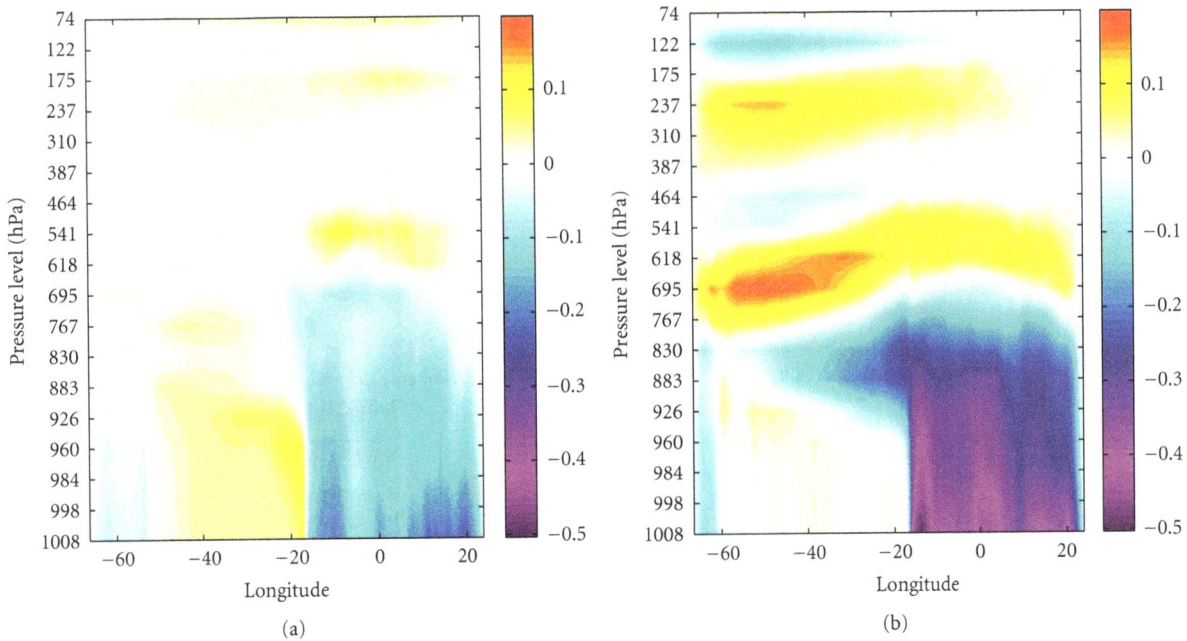

FIGURE 13: Zonal vertical profile of dust effect on temperatures in °C, averaged between 0°N and 25°N during the periods (a) MAM and (b) JJA. Cooling is indicated with cool colors while warming with hot colors.

MAM (Figure 12). This cooling seemed more important during the monsoon period than the MAM period. Results reveal also warming zones, notably along the coasts, South-Western Burkina Faso, South-Eastern Mali, Northern Côte d'Ivoire, Western Nigeria, through to Northern-Cameroon until West of Central African Republic. Over the Ocean, there was a weak dust-induced warming. This may be attributed to dust depositional phenomena. During the monsoon period, dust-induced cooling is more important. A global cooling was observed over the Southern part of the American continent during the period MAM while warming occurred over the coasts from Guyana to Brazil (Figure 12). The zonal profile section of temperatures indicated a warming at the sea surface between longitudes 18°W and 50°W, and a cooling under the domain. The cooling was observed from the surface to the middle of the troposphere around 700 hPa (Figure 13). The meridional profile section of temperatures was in conformity with the cooling observed at the surface

due to Saharan and Sahelian dust (Figure 14). The cooling at the surface was greatly accentuated and extended from the continent until the coasts during the monsoon period. However, at around 12°N, a warming seemed to occur.

3.6.3. Effect of Dust on Precipitations. Using regional climatic modeling, some authors such as Konaré et al. [24] and Solmon et al. [25] found a reduction in precipitations, but other authors [23, 27] found an increase in precipitations using a global climatic model. Most of the studies conducted were focused on the monsoon period but this study considered the intermediate period MAM by analyzing the impact of dusts at a local scale outside the monsoon system. Results revealed that the Saharan and Sahelian dusts caused drying effect in the region during the two periods: MAM and JJA (Figure 15). However, there were some zones of increased precipitations even though dust quantities were low. These included Northern-Liberia, Southern-Guinea,

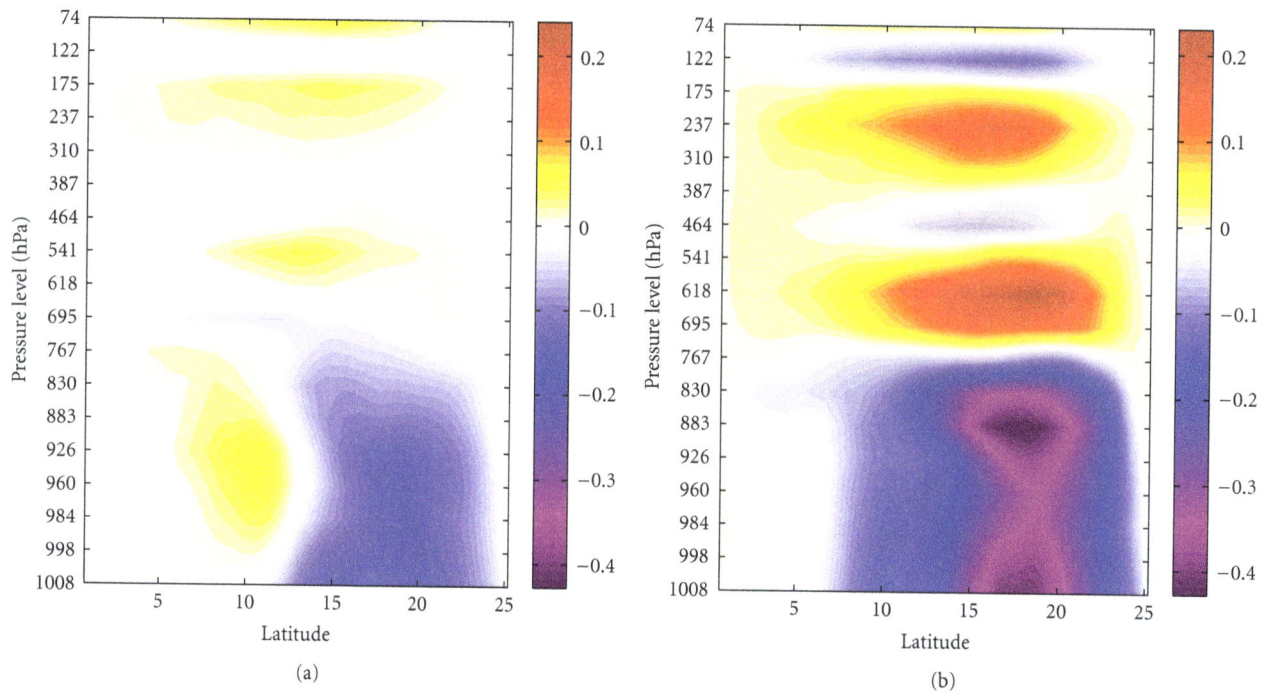

FIGURE 14: Meridian vertical profile of dust effect on temperatures in °C, averaged between 66°W and 23°E during the periods (a) MAM and (b) JJA. Cooling is indicated with cool colors while warming with hot colors.

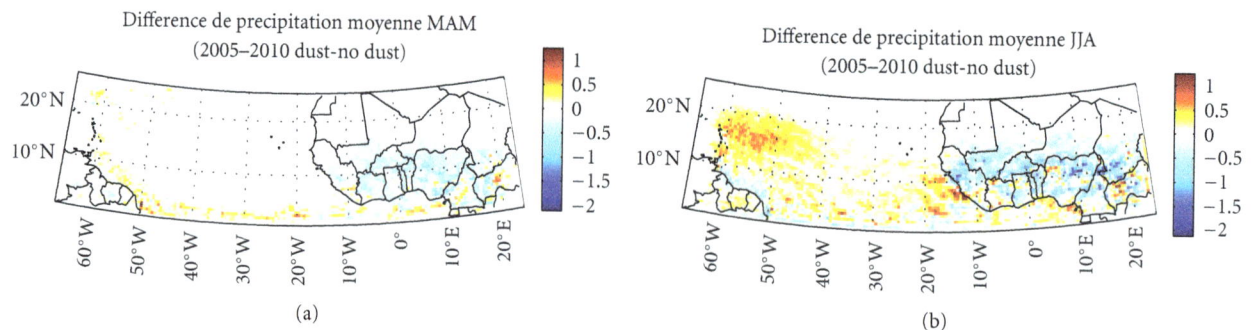

FIGURE 15: Effect of dusts on precipitations in mm/day during the period of MAM and JJA. Precipitation decrease is indicated with cool colors while increase in precipitation with hot colors.

middle of Central African Republic, Caribbean Island, and the American coasts.

During the monsoon, the dusts can induce circulation in opposite direction to the monsoon flow, which favors its weakening having the consequence of reducing precipitation. At the local level (i.e., MAM), the cooling caused can inhibit the convection and that explains the observed reduction of precipitations. These results agreed with Solmon et al. [25] who showed that dusts significantly impact on the monsoon development in West Africa by interacting with incident and reflected radiation.

4. Conclusion

Our study allowed analysis of simulated dust emissions from the Sahara and Sahel, using the regional climate model (RegCM). The trajectory of dust plumes were also analyzed

for the months March, April, and May 2010 using the same model and other observations. The analysis revealed a seasonality of trajectories associated with the seasonality of altitudes reached by dusts plumes. The vertical profile of dust concentrations revealed a progressive increase in altitude during the period, "December-January-February (DJF)" to the period "June-July-August (JJA)," with a transition altitude during the period "March-April-May (MAM)." Indeed, during the period DJF, dust plumes flow in the low altitudes while high altitudes are reached during the period JJA. The model well captured the variability of various episodes at the African stations (Banizoumba, Cinzana, and M'Bour). However, it underestimated the peaks of AOD values compared to the observations. In the Barbados Islands, it satisfactorily reproduced the variability of AOD values even though it did not capture all the peaks shown by the AERONET observations. Part of the differences between the AOD values and those of the observations from AERONET, during the month of March, is probably due to the exclusion of biomass fires from this study although they were present in the West African landscape.

The study of the impact on precipitations and local temperatures of Saharan and Sahelian dusts (outside the monsoon period) indicated that dusts could exert a cooling effect in the whole region both during the period MAM and the monsoon period (JJA). This was in agreement with the studies of Konaré et al. [24] and Solmon et al. [25] for the period JJA. However, the extension over the Atlantic Ocean did not modify the results found by these previous studies. The cooling at the surface was accompanied by a decrease in local precipitations during the period (MAM) in the studied domain with the exception of some zones in Central Africa, West Africa (Southern-Guinea and Northern-Liberia), Caribbean, and South America, where increased in precipitations were observed.

Another study is currently being undertaken to take into account the emissions coming from bush fires in order to know their contributions to observed values and analyze their roles in climatic processes in West Africa and over the Atlantic Ocean.

Acknowledgments

The work was funded by START and RIPIECSA grant. All authors acknowledge the ICTP (International Centre for the Theoretical Physics) for their RegCM model, which was used for this paper.

References

[1] E. Servat, J. E. Paturel, H. Lubès, B. Kouamé, M. Ouedraogo, and J. M. Masson, "Climatic variability in humid Africa along the Gulf of Guinea. Part I: detailed analysis of the phenomenon in Cote d'Ivoire," *Journal of Hydrology*, vol. 191, no. 1–4, pp. 1–15, 1997.

[2] X. Zheng and E. A. B. Eltahir, "The response to deforestation and desertification in a model of West African monsoons," *Geophysical Research Letters*, vol. 24, no. 2, pp. 155–158, 1997.

[3] N. Brooks and M. Legrand, "Dust variability over northern Africa and rainfall in the Sahel," in *Linking Climate Change to Landsurface Change*, S. J. McLaren and D. Kniveton, Eds., chapter 1, pp. 1–25, Springer, New York, NY, USA, 2000.

[4] M. O. Andreae, "Climatic effects of changing atmospheric aerosol levels," in *World Survey of Climatology*, A. Henderson-Seller, Ed., vol. 16 of *Future Climates of the World*, pp. 341–392, Elsevier, New York, NY, USA, 1995.

[5] R. A. Duce, "Sources, distributions, and fluxes of mineral aerosols and their relationship to climate," in *Aerosol Forcing of Climate*, R. J. Charson and J. Heintzenberg, Eds., pp. 43–72, John Wiley & Sons, New York, NY, USA, 1995.

[6] L. Schultz, Jaenicke, and H. Pieter, "Sahara dust transport over the north Atlantic ocean," in *Desert Dust: Origin, Characteristics, and Effect on Man*, T. L. Péwé, Ed., pp. 87–100, The Geological Society of America, Boulder, Colo, USA, 1981.

[7] G. A. D'almeida, "On the variability of desert aerosol radiative characteristics," *Journal of Geophysical Research*, vol. 92, no. 3, pp. 3017–3026, 1987.

[8] R. Swap, M. Garstang, S. Greco, R. Talbot, and P. Kallberg, "Saharan dust in the Amazon Basin," *Tellus B*, vol. 44, no. 2, pp. 133–149, 1992.

[9] B. Laurent, B. Marticorena, G. Bergametti, J. F. Léon, and N. M. Mahowald, "Modeling mineral dust emissions from the Sahara desert using new surface properties and soil database," *Journal of Geophysical Research D*, vol. 113, no. 14, Article ID D14218, 2008.

[10] M. Legrand, A. Plana-Fattori, and C. N'Doumé, "Satellite detection of dust using the IR imagery of Meteosat 1. Infrared difference dust index," *Journal of Geophysical Research D*, vol. 106, no. 16, pp. 18251–18274, 2001.

[11] J. M. Prospero, P. Ginoux, O. Torres, S. E. Nicholson, and T. E. Gill, "Environmental characterization of global sources of atmospheric soil dust identified with the Nimbus 7 Total Ozone Mapping Spectrometer (TOMS) absorbing aerosol product," *Reviews of Geophysics*, vol. 40, no. 1, pp. 1–31, 2002.

[12] S. Engelstaedter, I. Tegen, and R. Washington, "North African dust emissions and transport," *Earth-Science Reviews*, vol. 79, no. 1-2, pp. 73–100, 2006.

[13] G. A. D'Almeida, "A model for Saharan dust transport," *Journal of Climate & Applied Meteorology*, vol. 25, no. 7, pp. 903–916, 1986.

[14] C. Moulin, C. E. Lambert, U. Dayan et al., "Satellite climatology of African dust transport in the Mediterranean atmosphere," *Journal of Geophysical Research D*, vol. 103, no. 11, pp. 13137–13144, 1998.

[15] J. M. Prospero and P. J. Lamb, "African droughts and dust transport to the Caribbean: climate change implications," *Science*, vol. 302, no. 5647, pp. 1024–1027, 2003.

[16] I. Borbely-Kiss, A. Z. Kiss, E. Koltay, G. Szabo, and L. Bozó, "Saharan dust episodes in Hungarian aerosol: elemental signatures and transport trajectories," *Journal of Aerosol Science*, vol. 35, no. 10, pp. 1205–1224, 2004.

[17] P. L. Israelevich, E. Ganor, Z. Levin, and J. H. Joseph, "Annual variations of physical properties of desert dust over Israel," *Journal of Geophysical Research D*, vol. 108, no. 13, article 4381, pp. 1–9, 2003.

[18] T. Y. Tanaka, Y. Kurosaki, M. Chiba et al., "Possible transcontinental dust transport from North Africa and the Middle East to East Asia," *Atmospheric Environment*, vol. 39, no. 21, pp. 3901–3909, 2005.

[19] P. Alpert and E. Ganor, "Sahara mineral dust measurements from TOMS: comparison to surface observations over the Middle East for the extreme dust storm, March 14–17, 1998," *Journal of Geophysical Research D*, vol. 106, no. 16, pp. 18275–18286, 2001.

[20] H. Liao and J. H. Seinfeld, "Radiative forcing by mineral dust aerosol: sensitivity to key variables," *Journal of Geophysical Research D*, vol. 103, no. 24, pp. 31,637–31,645, 1998.

[21] D. Tanré, J. Haywood, J. Pelon et al., "Measurement and modeling of the Saharan dust radiative impact: overview of the Saharan Dust Experiment (SHADE)," *Journal of Geophysical Research D*, vol. 108, no. 18, pp. 1–12, 2003.

[22] IPCC (International Panel on Climate Change), "Fourth assessment report: climate change 2007 (AR4)," in *Contribution of Working Group I: The Physical Science Basis*, Cambridge University Press, Cambridge, UK, 2007.

[23] K. M. Lau, K. M. Kim, Y. C. Sud, and G. K. Walker, "A GCM study of the response of the atmospheric water cycle of West Africa and the Atlantic to Saharan dust radiative forcing," *Annales Geophysicae*, vol. 27, no. 10, pp. 4023–4037, 2009.

[24] A. Konaré, A. S. Zakey, F. Solmon et al., "A regional climate modeling study of the effect of desert dust on the West African

monsoon," *Journal of Geophysical Research D*, vol. 113, no. 12, Article ID D12206, 2008.

[25] F. Solmon, M. Mallet, N. Elguindi, F. Giorgi, A. Zakey, and A. Konaré, "Dust aerosol impact on regional precipitation over western Africa, mechanisms and sensitivity to absorption properties," *Geophysical Research Letters*, vol. 35, no. 24, Article ID L24705, 2008.

[26] M. Yoshioka, N. M. Mahowald, A. J. Conley et al., "Impact of desert dust radiative forcing on sahel precipitation: relative importance of dust compared to sea surface temperature variations, vegetation changes, and greenhouse gas warming," *Journal of Climate*, vol. 20, no. 8, pp. 1445–1467, 2007.

[27] K. M. Kim, W. K. M. Lau, Y. C. Sud, and G. K. Walker, "Influence of aerosol-radiative forcings on the diurnal and seasonal cycles of rainfall over West Africa and Eastern Atlantic Ocean using GCM simulations," *Climate Dynamics*, vol. 35, no. 1, pp. 115–126, 2010.

[28] F. Giorgi, M. R. Marinucci, G. T. Bates, and G. De Canio, "Development of a second-generation regional climate model (RegCM2). Part II: convective processes and assimilation of lateral boundary conditions," *Monthly Weather Review*, vol. 121, no. 10, pp. 2814–2832, 1993.

[29] J. S. Pal, F. Giorgi, X. Bi et al., "Regional climate modeling for the developing world: the ICTP RegCM3 and RegCNET," *Bulletin of the American Meteorological Society*, vol. 88, no. 9, pp. 1395–1409, 2007.

[30] A. S. Zakey, F. Solmon, and F. Giorgi, "Development and testing of a desert dust module in a regional climate model," *Atmospheric Chemistry and Physics Discussions*, vol. 6, no. 2, pp. 1749–1792, 2006.

[31] D. F. Zhang, A. S. Zakey, X. J. Gao, F. Giorgi, and F. Solmon, "Simulation of dust aerosol and its regional feedbacks over East Asia using a regional climate model," *Atmospheric Chemistry and Physics*, vol. 9, no. 4, pp. 1095–1110, 2009.

[32] F. Giorgi and G. T. Bates, "The climatological skill of a regional model over complex terrain," *Monthly Weather Review*, vol. 117, no. 11, pp. 2325–2347, 1989.

[33] R. Dickinson, A. Henderson-Sellers, and P. Kennedy, "Biosphere-atmosphere transfer scheme (bats) version 1e as coupled to the NCAR community climate model," Technical Report National Center for Atmospheric Research, 1993.

[34] L. Zobler, "A world soil file for global climate modeling," NASA—Technical Memorandum 87 802, 1986.

[35] J. T. Kiehl, J. J. Hack, G. B. Bonan et al., "Description of the ncar community climate model (ccm3)," Technical Report CAR/TN-420+STR National Center for Atmospheric Research, 1996.

[36] M. G. N'Tchayi, J. Bertrand, M. Legrand, and J. Baudet, "Temporal and spatial variations of the atmospheric dust loading throughout West Africa over the last thirty years," *Annales Geophysicae*, vol. 12, no. 2-3, pp. 265–273, 1994.

[37] K. D. Perry, T. A. Cahill, R. A. Eldred, D. D. Dutcher, and T. E. Gill, "Long-range transport of North African dust to the eastern United States," *Journal of Geophysical Research D*, vol. 102, no. 10, pp. 11225–11238, 1997.

[38] Y. J. Kaufman, I. Koren, L. A. Remer, D. Tanré, P. Ginoux, and S. Fan, "Dust transport and deposition observed from the Terra-Moderate Resolution Imaging Spectroradiometer (MODIS) spacecraft over the Atlantic Ocean," *Journal of Geophysical Research D*, vol. 110, no. 10, article D10S12, pp. 1–16, 2005.

[39] P. Tulet, M. Mallet, V. Pont, J. Pelon, and A. Boone, "The 7–13 March 2006 dust storm over West Africa: generation,

transport, and vertical stratification," *Journal of Geophysical Research D*, vol. 113, no. 23, Article ID D00C08, 2008.

[40] D. B. Karam, C. Flamant, P. Knippertz et al., "Dust emissions over the Sahel associated with the West African monsoon intertropical discontinuity region: a representative case-study," *Quarterly Journal of the Royal Meteorological Society*, vol. 134, no. 632, pp. 621–634, 2008.

[41] J. M. Prospero and T. N. Carlson, "Saharan air outbreaks over the tropical North Atlantic," *Pure and Applied Geophysics*, vol. 119, no. 3, pp. 677–691, 1981.

[42] J. M. Prospero and R. T. Nees, "Impact of the North African drought and El Nino on mineral dust in the Barbados trade winds," *Nature*, vol. 320, no. 6064, pp. 735–738, 1986.

[43] C. Jones, N. Mahowald, and C. Luo, "The role of easterly waves on African desert dust transport," *Journal of Climate*, vol. 16, pp. 3617–3628, 2003.

[44] C. Jones, N. Mahowald, and C. Luo, "Observational evidence of African desert dust intensification of easterly waves," *Geophysical Research Letters*, vol. 31, no. 17, article L17208, 4 pages, 2004.

[45] A. Smirnov, B. N. Holben, D. Savoie et al., "Relation-ship between column aerosol optical thickness and in situ ground based dust concentrations over Barbados," *Geophysical Research Letters*, vol. 27, no. 11, pp. 1643–1646, 2000.

[46] C. Flamant, C. Lavaysse, M. C. Todd, J. P. Chaboureau, and J. Pelon, "Multi-platform observations of a springtime case of Bodélé and Sudan dust emission, transport and scavenging over West Africa," *Quarterly Journal of the Royal Meteorological Society*, vol. 135, no. 639, pp. 413–430, 2009.

[47] R. A. Duce, "The impact of atmospheric nitrogen, phosphorus and iron species on marine biological productivity," in *The Role of Air-Sea Exchange in Geochemical Cycling*, P. Buat-Menard, Ed., pp. 497–529, 1986.

[48] I. Chiapello, G. Bermagetti, L. Gomes, B. Chatenet, F. Dulac, and E. S. Suares, "An additional low layer transport of Sahelian and Saharan dust over the north-eastern tropical Atlantic," *Geophysical Research Letters*, vol. 22, no. 23, pp. 3191–3194, 1995.

[49] R. Washington and M. C. Todd, "Atmospheric controls on mineral dust emission from the Bodélé depression, chad: the role of the low level jet," *Geophysical Research Letters*, vol. 32, no. 17, Article ID L17701, pp. 1–5, 2005.

[50] S. Engelstaedter and R. Whasington, "Atmospheric controls on the annual cycle of North African dust," *Journal of Geophysical Research*, vol. 112, Article ID D03103, 14 pages, 2007.

[51] R. L. Miller, I. Tegen, and J. Perlwitz, "Surface radiative forcing by soil dust aerosols and the hydrologic cycle," *Journal of Geophysical Research D*, vol. 109, no. 4, article D04203, 24 pages, 2004.

6

Temporal Forecasting with a Bayesian Spatial Predictor: Application to Ozone

Yiping Dou,[1] Nhu D. Le,[2] and James V. Zidek[3]

[1] *Finance, eBay Inc., San Jose, CA 95125, USA*
[2] *BC Cancer Agency Research Center, Vancouver, BC, Canada V5Z 4E6*
[3] *Department of Statistics, University of British Columbia, Vancouver, BC, Canada V6T 1Z2*

Correspondence should be addressed to James V. Zidek, jim@stat.ubc.ca

Academic Editor: Tareq Hussein

This paper develops and empirically compares two Bayesian and empirical Bayes space-time approaches for forecasting next-day hourly ground-level ozone concentrations. The comparison involves the Chicago area in the summer of 2000 and measurements from fourteen monitors as reported in the EPA's AQS database. One of these approaches adapts a multivariate method originally designed for spatial prediction. The second is based on a state-space modeling approach originally developed and used in a case study involving one week in Mexico City with ten monitoring sites. The first method proves superior to the second in the Chicago Case Study, judged by several criteria, notably root mean square predictive accuracy, computing times, and calibration of 95% predictive intervals.

1. Introduction

This paper compares two methods for temporally forecasting next-day hourly ground-level ozone concentrations over spatial regions. Software for implementing both methods along with demo files can be downloaded from http://enviro .stat.ubc.ca/. The paper focuses on a case study involving Chicago during the summer of 2000. The methods can be used to forecast the maximum eight-hour average ozone concentration, which is reported for many urban areas. For example, on June 27, 2009 the AIRNow website forecasts a maximum for Chicago of between 0 and 50 ppb, rating that as "Good." In contrast, for that day in one part of Los Angeles, the rating was "Unhealthy for sensitive groups," meaning a forecast maximum of between 101 and 150 ppb.

These forecasts are needed to forewarn susceptible groups of high ozone concentrations that are associated with acute health effects. Such effects are well documented in the air quality criterion document (Ozone [1]. See http://oaspub .epa.gov/eims/eimsapi.dispdetail?deid=149923), the basis of the recommendations made in 2007 to the US Environment Protection Agency by its Clean Air Scientific Advisory Committee for Ozone, in which the third author served. In fact,

the accumulated body of evidence was so strong that the committee recommended strengthening the air quality standards for this criterion pollutant to meet the requirements of the US Clean Air Act. In particular, the evidence pointed to a strong association between morbidity as well as reduced lung function and high levels of ground-level ozone concentrations. This points to a need for enhanced near-term forecasting methods.

One general method for making such forecasts relies on the fusion of measured hourly ozone concentration values and simulated values obtained from chemical transport models (CTMs) such as CMAQ. Two papers [2, 3] develop methods for doing this, albeit by different approaches unrelated to those in this paper. In future work, all these methods should be compared in domains where CTM data are available. Even without that hypothetical comparison, those in this paper have the advantage of being available in domains where they are not. These Bayes-empirical Bayes methods offer the flexibility needed to characterize environmental space-time processes, while fully representing the various kinds of uncertainty involved in their construction. Both have been developed for and successfully used to model hourly ozone air pollution concentrations in other contexts.

The first method in this paper denoted by M1 adapts a multivariate method developed for modeling space-time fields [4–7]. A univariate version of that method for hourly ozone concentrations is the subject of a companionable paper [8], which compares it with a state space model but for spatial prediction, not temporal forecasting, and it does so in a different geographical region. The goal there is mapping the ozone field for another requirement of the US Clean Air Act of 1970, namely, the protection of human welfare including such things as crop yields. M1 needs some new theory, which is presented in the sequel along with a demonstration on how it may be applied.

The second method denoted by M2 uses a method originally developed for modeling hourly ozone concentrations in Mexico City [9]. That method and the models on which it is based along with the computational algorithms used to implement it seem to have been quite successful in that application. Moreover, even though it was developed for use in Mexico City for one specific week, a strong prima facie case can be made for its applicability in other weeks and jurisdictions and that is why we assess the performance of that method here. Much recent work has been done in modeling random space-time pollution fields [10–12]. As one of the photo-oxidants, ozone is produced in the same way in all temporal and spatial domains by a complex interaction of oxides of nitrogen (NO_x) with volatile organic compounds (VOCs) in the presence of heat and sunlight. In most modern urban environments such as Chicago and Mexico City, vehicle emissions are a prime source of the NO_x and VOCs [1]. Furthermore the prima facie case is supported by an exploratory data analysis that shows very similar daily cycles in our domain as those observed in the Mexico City application. This led us to match, to the greatest possible extent [8], the method used there in our adaptation of it, and no originality is claimed for it.

The main finding in this paper is that in the case study M1 outperforms M2 in a number of ways. First is its computational efficiency. To run the M2 approach, it often took about a week or so to get the results, while M1 only took about ten to twelve hours at the same Linux server. Thus, M2 would not be suitable for making 24 ahead forecasts, while M1 running on a faster processor could be used for that purpose. We also found that M1 produced more accurate forecasts than M2, as measured by their root-mean-squared-prediction errors. Moreover, M1's predictive error bands proved to be better calibrated. In other space-time domains, a similar assessment would have to be made to select a forecasting procedure, and M2 may be superior in some. Overall, we believe that the value of this paper lies in the guidance on how that assessment could be made and the source of software that can be used for it.

The layout of this paper will now be described. Section 2 presents both approaches to forecasting hourly ozone concentrations. Section 2.1 introduces M1. Its forecasting (posterior) distribution is developed, and the corresponding pointwise predictive intervals at each gauged site constructed. Section 2.1 also extends the results to forecasting r-step-ahead responses for any $r \in \mathcal{N}$. Section 2.2 reviews M2. Section 3 implements these two methods in a case study involving Chicago and data from the US EPA's AQS air qua-lity database. Section 3.1 presents and compares the results for one-day-ahead predictions. Finally, Section 4 summarizes these results and gives our conclusions.

2. Methodology

This section presents our two temporal forecasting approaches. Although both are general and can be used in other contexts, we develop them as methods for forecasting an hourly response tomorrow given data up to today. The measured value of the response is available and serves as a "test value" at each of g monitoring sites for a comparative assessment of the two forecasted methods.

2.1. Method M1

2.1.1. Basic Theory. The general approach [7], on which M1 is based, assumes a multivariate space-time field of p-dimensional random response vectors indexed by their locations, a finite number of sites in a specified geographical domain. These sites need not lie on a lattice.

The general theory involves a geographical region that includes g monitored (i.e., gauged) sites and u unmonitored (i.e., ungauged) sites. Although this paper requires only the part of that theory for those that are monitored, we state it in generalable to link this paper to its companion publication [8]. The p dimensional response vectors at these sites for times $t = 1, \ldots, n$ are combined in the response matrix $\mathbf{Y}$: $n \times (u + g)p$. This matrix can be partitioned as $(\mathbf{Y}^{[u]}, \mathbf{Y}^{[g]})$, where $\mathbf{Y}^{[u]}$: $n \times up$ contains the response vectors at the ungauged sites and $\mathbf{Y}^{[g]}$: $n \times gp$, those at the gauged sites. The theory posits ([7], p. 145-146) that

$$\mathbf{Y} \mid \beta, \mathbf{\Sigma} \sim N(\mathbf{Z}\beta, \mathbf{I}_n \otimes \mathbf{\Sigma}),$$

$$\beta \mid \mathbf{\Sigma}, \beta_0, \mathbf{F} \sim N(\beta_0, \mathbf{F}^{-1} \otimes \mathbf{\Sigma}), \tag{1}$$

$$\mathbf{\Sigma} \sim \text{GIW}(\mathbf{\Psi}, \delta),$$

where $\mathbf{Z}$: $n \times l$ denotes the non-site-specific covariates (their site-specific counterparts could be included in the response vectors); β : $l \times (u + g)p$ denotes the matrix of site-specific random covariate coefficients; $\mathbf{\Sigma}$: $(u+g)p \times (u+g)p$ denotes the covariance matrix among the responses at any given time point. The hyperparameters for this hierarchical Bayes model are $\beta_0, \mathbf{F}, \delta$, and $\mathbf{\Psi}$, with $\mathbf{F}$ representing the variance component of β between its l rows. Here, GIW denotes the Generalized Inverted Wishart distribution, a conjugate prior for normal matrix distributions. Separability of the hyper-covariant matrices for both the response and random coefficient matrices is assumed for computational simplicity. Invoking Box's celebrated dictum that all models are wrong, we defend this assumption by the good performance of the resulting method as seen in the empirical assessment provided in the sequel.

Validating the modeling assumptions above will usually require a transformation of the random responses, a square root transformation in this paper's case study. Then systematic components such as the temporal trend over the whole regions will need to be removed. These can be accurately inferred from the typically large dataset formed by aggregating the data over all sites and times. Finally, something needs to be done to eliminate autocorrelation in the temporal sequence of responses. For example, the temporal series can often be filtered using a regional time series model without site-specific parameters. However, our relative abundance of data leads us here to a different approach described in detail in the next subsection, that splits the transformed, detrended residuals into separate, disjoint subsequences of responses, which are separated widely enough in time as to be uncorrelated and hence independent under our Gaussian sampling model. In our experience, the residuals obtained after these steps have been taken usually satisfy the model assumptions above, and these comprise the response vectors in M1. So the model above can then be applied to each subsequence and type II maximum likelihood estimators found for the hyperparameters. These can subsequently be averaged across the subsubsequences to get an overall estimate. While this approach would be less efficient than a full-data approach under a correctly specified model, it avoids the risk of model misspecification in complex situations like that of the case study. The forecasting model developed below can then be applied, and the preliminary steps above reversed, to get the forecasts back on the scale of the raw data.

To elaborate on our distributional assumptions, the GIW prior for

$$\mathbf{\Sigma} = \begin{pmatrix} \mathbf{\Sigma}^{[g,g]} & \mathbf{\Sigma}^{[g,u]} \\ \mathbf{\Sigma}^{[u,g]} & \mathbf{\Sigma}^{[u,u]} \end{pmatrix} : \begin{pmatrix} gp \times gp & gp \times up \\ up \times gp & up \times up \end{pmatrix} \quad (2)$$

can be defined, through the Bartlett decomposition, as follows:

$$\mathbf{\Sigma}^{[g,g]} \sim \mathrm{IW}(\mathbf{\Lambda}_1, \delta_1),$$

$$\mathbf{\Gamma}^{[u]} \sim \mathrm{IW}(\mathbf{\Lambda}_0, \delta_0), \quad (3)$$

$$\boldsymbol{\tau}^{[u]} \mid \mathbf{\Gamma}^{[u]} \sim N\left(\boldsymbol{\tau}_{00}, \mathbf{H}_0 \otimes \mathbf{\Gamma}^{[u]}\right),$$

where $\mathbf{\Gamma}^{[u]} = \mathbf{\Sigma}^{[u|g]} = \mathbf{\Sigma}^{[u,u]} - \mathbf{\Sigma}^{[u|g]}(\mathbf{\Sigma}^{[g,g]})^{-1}\mathbf{\Sigma}^{[g|u]}$, and $\boldsymbol{\tau}^{[u]} = (\mathbf{\Sigma}^{[g,g]})^{-1}\mathbf{\Sigma}^{[g,u]}$. Note that $\mathbf{\Sigma}^{[g,g]}$ has an inverted Wishart distribution with hyperparameters $(\mathbf{\Lambda}_1, \delta_1)$; the matrix $\boldsymbol{\tau}_{00}$ is the hypermean of $\boldsymbol{\tau}^{[u]}$, and the matrix $\mathbf{H}_0$ gives the covariance between the rows of $\boldsymbol{\tau}^{[u]}$. Denote the set of hyperparameters as $\mathcal{H} = \{\mathbf{F}, \boldsymbol{\beta}_0, \mathbf{\Omega}, \mathbf{\Lambda}_1, \delta_1, \mathbf{\Lambda}_0, \delta_0, \boldsymbol{\tau}_{00}, \mathbf{H}_0\}$.

Given the observations at the gauged sites (i.e., $\mathbf{Y}^{[g]}$), the predictive distribution of $\mathbf{Y}^{[u]}$ is completely determined, it is the distribution required in our companionable paper on spatial prediction [8]. Furthermore given partially observed responses at gauged sites, the predictive distribution of the missing responses at gauged sites given these observations can be derived after the hyperparameters have been estimated by an empirical Bayes approach ([7], p. 300–303), and that is how the theory is used in this paper.

Deriving that forecasting model requires a general result that concerns a sequence of n response vectors of which $n - w$ are observed, and w lies in the future. Then with the superscripts "m" and "o," respectively, standing for "missing" and "observed," we may further partition the random response

matrix already partitioned above, as $\mathbf{Y}^{[g]} = (\mathbf{Y}^{[g^m]\prime}, \mathbf{Y}^{[g^o]\prime})'$: $n \times gp$ with $\mathbf{Y}^{[g^m]}$: $w \times gp$ and $\mathbf{Y}^{[g^o]}$: $(n - w) \times gp$. Forecasting requires the predictive posterior distribution of $(\mathbf{Y}^{[g^m]} \mid \mathbf{Y}^{[g^o]}, \mathcal{H})$ given in the following result ([7], p. 160-161).

Theorem 1. *Let*

$$\mathbf{Z}\boldsymbol{\beta}_0^{[g]} = \begin{pmatrix} \mu_{(1)} \\ \mu_{(2)} \end{pmatrix} : \begin{pmatrix} w \times gp \\ (n - w) \times gp \end{pmatrix},$$

$$\mathbf{I}_n + \mathbf{Z}\mathbf{F}^{-1}\mathbf{Z}'$$

$$= \begin{pmatrix} \mathbf{A}_{11} & \mathbf{A}_{12} \\ \mathbf{A}_{21} & \mathbf{A}_{22} \end{pmatrix} : \begin{pmatrix} w \times w & w \times (n - w) \\ (n - w) \times w & (n - w) \times (n - w) \end{pmatrix}. \quad (4)$$

Conditional on the hyperparameters $\mathcal{H}$, the marginal posterior distribution is a matrix t distribution:

$$\mathbf{Y}^{[g^m]} \mid \mathbf{Y}^{[g^o]}, \mathcal{H} \sim t_{w \times gp}\left(\mu_{(u|g)}, \mathbf{\Phi}_{(u|g)} \otimes \mathbf{\Psi}_{(u|g)}, \delta_{(u|g)}\right), \quad (5)$$

where

$$\mu_{(u|g)} = \mu_{(1)} + \mathbf{A}_{12}\mathbf{A}_{22}^{-1}\left(\mathbf{Y}^{[g^o]} - \mu_{(2)}\right) : w \times gp,$$

$$\mathbf{\Phi}_{(u|g)} = \frac{\delta_1 - gp + 1}{\delta_1 - gp + n - w + 1}\mathbf{A}_{11\circ2} : w \times w,$$

$$\mathbf{\Psi}_{(u|g)} = \frac{1}{\delta_1 - gp + 1}\left\{\mathbf{\Psi}_{gg} + \left(\mathbf{Y}^{[g^o]} - \mu_{(2)}\right)' \right.$$

$$\left. \mathbf{A}_{22}^{-1}\left(\mathbf{Y}^{[g^o]} - \mu_{(2)}\right)\right\} : gp \times gp,$$

$$\delta_{(u|g)} = \delta_1 - gp + n - w + 1,$$

where $\mathbf{A}_{11\circ2} = \mathbf{A}_{11} - \mathbf{A}_{12}\mathbf{A}_{22}^{-1}\mathbf{A}_{21}$, and $\mathbf{\Psi}_{gg}$ is assumed to be $\mathbf{\Lambda}_1 \otimes \mathbf{\Omega}$, denoting spatial and between hour correlations, respectively.

Remark 2. This theorem gives the joint predictive distribution for w future response vectors given $(n - w)$ observed responses. As coordinates of those future vector are observed, this distribution yields in turn the conditional predictive distribution for the unobserved coordinate responses, That is how the theorem is applied in the next section.

2.1.2. One-Day-Ahead Forecasts. For expository simplicity, we describe the general method M1 in terms of the goal of forecasting ozone concentrations at a specific hour on Day 121 and each of $g = 14$ monitoring sites based on data collected at those sites during the preceding days, that being the objective in the case study described in the next section. However, we emphasize that M1 is generally applicable to other days and geographical domains with appropriate modifications. In fact, the approach can be adapted for use with other environmental processes. With that caveat, we now describe M1 in this subsection. Note that in our description, hour 1 refers to the period between midnight and 1 AM and so on.

To begin, we follow the standard practice of transforming the hourly data by taking their square roots to achieve a more nearly Gaussian data distribution [13]. Hereafter these transformed values will be our "responses." M1 then partitions the sequence of responses into blocs of $p = 24$ hours each. For $k < 24$ (hereafter referred to as Case 2), the last bloc spans the 24 hour period from hour $k + 1$ on day 120 to hour k on Day 121, and the first from hour $k + 1$ of Day 1 to hour k of day 2. However $k = 24$ (Case 1) is different, and there the blocs correspond to the days from 1 to 121. In either case, each bloc yields a 24-dimensional multivariate response vector with an unspecified covariance structure. Reformulating our model in this way gives us the advantage of avoiding the challenging task of specifying the complex short-term autocovariance structure, which varies over the day. But it does require the multivariate theory described in the previous subsection.

The next step in developing the forecast model would generally require the removal of any systematic, regional components in the series. In particular, it is necessary to learn which covariates/predictors to include in the design matrix $\mathbf{Z}$ so that in application of the method, site-specific coefficients can be fitted allowing deviations from the regional baselines established for them at the preliminary stage. Note that the EnviRo.stat software (see [14]) referred to in Section 1 automatically estimates the those baseline coefficients as prior hypermeans, using a maximum likelihood-based approach, the key preliminary step here is actually the identification of $\mathbf{Z}$. Note also that in our application the covariates need to be adapted in form to conform to the temporal span of the response vectors (the so-called "blocs" introduced below), once hour k has been specified.

Finally, we need to address the autocovariance structure. While the responses are primarily an AR(2) time series after removing their diurnal pattern [13], as we do in the case study, we need to allow for a lag 1 autocorrelation in series of the response vectors to capture small but potentially significant longer-term dependence [15]. To eliminate it, we use the approach described in the previous subsection and split the observed response vectors into two groups: those with odd numbers and those with even numbers. For each of the resulting subseries, the model assumptions will hold approximately, and the hyperparameters can be estimated as described in the previous subsection. More detail is given below where we turn to a more precise description of M1.

In Case 2 where $k < 24$, the 24-hour bloc containing hour k includes measured responses for the hours $k + 1, \ldots, 24$ on Day 120. Thus we may first apply Theorem 1 with $w = 1$ designating the last bloc in our construction above, to get a joint predictive distribution for its associated "future" response vector. Then we compute the marginal conditional predictive distribution for hour k implied by that joint distribution, given these data from Day 120 to finish the construction.

However, Case 1 is more difficult and the one we treat first. Among options considered by the authors for this case, was changing the bloc length to say $p = 25$ so that the last bloc could reach back to hour k on Day 120, which contains a measured response. That approach was discarded since the sequence of 7 bloc sequences would not synchronize with weeks. The latter are important structural features of the process that reflect the changing daily traffic patterns and give us the single covariate we have in the case study. Instead we accomplish the same thing, another way, which does preserve the week, namely, we apply Theorem 1 with $w = 2$ to get the initial joint predictive distribution. The required marginal conditional distribution is then obtained by conditioning on all the data from Day 120.

Note that in both cases, the future response or responses in Theorem 1 depend on the bloc immediately preceding them. To avoid the need to model in that dependence, we simply eliminate that data vector. We did explore the result of keeping it in and ignoring that dependence and found virtually no difference in the Case Study. We now turn to a more precise description of M1.

For that we need some notation. Let $Y_{t,i}^{[g_j^m]}$ denote the ith coordinate of the unobserved response vector for blocs t and gauged site j, while $Y_{t',i}^{[g_j^o]}$ denotes the observed response for bloc t', $t, t' = 1, \ldots, n$, $i = 1, \ldots, p$, $j = 1, \ldots, g$. For bloc t, Sites $g_1, \ldots, g_{14}$, and hours between i and j inclusive, that is, $i : j$, let $\mathbf{Y}_{t,i:j}^{[g_l]}, l = 1, \ldots, 14$ denote the random response. Throughout $p = 24$.

The two cases referred to above are as follows.

Case 1 (predict the response for the last hour $k = 24$ day 121). Hyperparameter estimates for its predictive distribution conditional on observed data are found first for the odd blocs $\mathbf{U}_t = (Y_{2t-1,1}^{[g_1^o]}, \ldots, Y_{2t-1,p}^{[g_g^o]}) : gp \times 1, t = 1, \ldots, 60$ as described in the previous subsection, using $\mathbf{Y}^{[g^o]'} = (\mathbf{U}_1, \ldots, \mathbf{U}_{60})$. Repeat this procedure for the even blocs, letting $\mathbf{V}_t = (Y_{2t,1}^{[g_1^o]}, \ldots, Y_{2t,p}^{[g_g^o]})$, $t = 1, \ldots, 59$. Finally average the resulting pairs of hyperparameter estimates to get overall estimates.

Construction of the predictive distribution of the response at hour $k = 24$, begins with the corresponding matrix-variate observed responses, $\mathbf{Y}^{[g^o]}$ made by extracting the corresponding responses from Day 1 to Day 118 according to the above constructed $\mathbf{U}_t$ or $\mathbf{V}_t$. Now suppose hypothetically that Day 120's responses have not been observed and suppose the corresponding matrix-variate missing responses, $\mathbf{Y}^{[g^m]}$ are constructed as follows:

$$\mathbf{Y}^{[g^m]} = \begin{bmatrix} Y_{121,1}^{[g_1^m]} & \cdots & Y_{121,p}^{[g_g^m]} \\ Y_{120,1}^{[g_1^o]} & \cdots & Y_{120,p}^{[g_g^o]} \end{bmatrix} = \begin{pmatrix} \mathbf{Y}_{121,1:p}^{[g_{1:g}^m]} \\ \mathbf{Y}_{120,1:p}^{[g_{1:g}^o]} \end{pmatrix} : 2 \times gp, \quad (7)$$

where $\mathbf{Y}_{121,1:p}^{[g_{1:g}^m]} : 1 \times gp$ is the unobserved future response vectors for days 121 and $\mathbf{Y}_{120,1:p}^{[g_{1:g}^o]} : 1 \times gp$, the observed response vector of Day 120. Hence we have $w = 2$ and $n = 121$ in Theorem 1. Thus the predictive posterior distribution of $\mathbf{Y}^{[g^m]}$ can be obtained by applying (6). To obtain that of $\mathbf{Y}_{121,1:p}^{[g_{1:g}^m]}$ given $\mathbf{Y}_{1:120,1:p}^{[g_{1:g}^o]}$, since in reality the latter are observed, one can decompose $\mu_{(u|g)}$ and $\mathbf{\Phi}_{(u|g)}$ as follows:

$$\mu_{(u|g)} = \begin{pmatrix} \mu_{1r} \\ \mu_{2r} \end{pmatrix}, \qquad \delta_{(u|g)}\mathbf{\Phi}_{(u|g)} = \begin{pmatrix} B_{11} & B_{12} \\ B_{21} & B_{22} \end{pmatrix}, \quad (8)$$

where $\mu_{ir} : 1 \times gp$ and $B_{ij} : 1 \times 1$ for $i, j = 1, 2$. Hence, the predictive posterior distribution of $\mathbf{Y}_{121,1:p}^{[g_{1:g}^m]}$ is given by

$$
\mathbf{Y}_{121,1:p}^{[g_{1:g}^m]} \mid \mathbf{Y}_{120,1:p}^{[g_{1:g}^o]}, \mathbf{Y}_{1:119,1:p}^{[g_{1:g}^o]}, \mathcal{H}
$$

$$
\sim t_{1 \times gp}\left(\mu_{1r} + B_{12} B_{22}^{-1}\left(\mathbf{Y}_{120,1:p}^{[g_{1:g}^o]} - \mu_{2r}\right), \frac{B_{11 \circ 2}}{\delta_{(u|g)} + 1}\right.
$$

$$
\otimes \mathbf{\Psi}_{(u|g)}\left(\mathbf{I}_{gp} + \mathbf{\Psi}_{(u|g)}^{-1}\left(\mathbf{Y}_{120}^{[g_{1:g}^o]} - \mu_{2r}\right)' B_{22}^{-1}\right.
$$

$$
\left.\left.\times \left(\mathbf{Y}_{120}^{[g_{1:g}^o]} - \mu_{2r}\right)\right), \delta_{(u|g)} + 1\right).
$$

(9)

To get the joint predictive distribution of responses at these gauged sites for Day 121's last hour, let $\mathbf{e}_k^l : k \times 1$ be such that $e_{kl} = 1$ and $e_{kj} = 0$ for $j \neq l, j = 1, \ldots, k$. Let $\mathbf{E}_1 = \text{blocs-diag-matrix } \{\mathbf{e}_p^p\} : gp \times g$. The joint predictive distribution of the pth unobserved response $\mathbf{Y}_{121,p}^{[g_{1:g}^m]}$, that is, $\mathbf{Y}_{121,1:p}^{[g_{1:g}^m]}\mathbf{E}_1$, is also a multivariate t-distribution:

$$
\mathbf{Y}_{121,p}^{[g_{1:g}^m]} \mid \mathbf{Y}_{120,1:p}^{[g_{1:g}^o]}, \mathbf{Y}_{1:119,1:p}^{[g_{1:g}^o]}, \mathcal{H}
$$

$$
\sim t_{1 \times g}\left(\mu^* \mathbf{E}_1, \phi^* \mathbf{E}_1' \mathbf{\Psi}^* \mathbf{E}_1, \delta_{(u|g)} + 1\right),
$$

(10)

where $\mu^* = \mu_{1r} + B_{12} B_{22}^{-1}(\mathbf{Y}_{120,1:p}^{[g_{1:g}^o]} - \mu_{2r})$, $\phi^* = B_{11 \circ 2}/(\delta_{(u|g)} + 1)$ and $\mathbf{\Psi}^* = \mathbf{\Psi}_{(u|g)}(\mathbf{I}_{gp} + \mathbf{\Psi}_{(u|g)}^{-1}(\mathbf{Y}_{120,1:p}^{[g_{1:g}^o]} - \mu_{2r})' B_{22}^{-1}(\mathbf{Y}_{120,1:p}^{[g_{1:g}^o]} - \mu_{2r}))$, that completes the description of M1 in this case.

Case 2 (predict the response for hour $k < 24$ on day 121). For bloc t, Sites $g_1, \ldots, g_{14}$, and hours between i and j inclusive, that is, $i : j$, let $\mathbf{Y}_{t,i:j}^{[g_l]}$, $l = 1, \ldots, 14$ denote the random response. Now let $\mathbf{Y}^{[g^m]}$ consist of k unobserved responses and $p - k$ observed ones at each of the gauged sites. What we need to do now at each site is to use the hours leading up to the first hour without data on Day 121, that is, hour k for which the forecast is needed. To do this, we can use any data from hours $1 : k$ on Day 121 that may be available, supplemented by the data from the $p - k$ preceding hours on Day 120, $(k + 1) : p = 21 : 24$. Thus we create a $p = 24$ dimensional response vector, $\mathbf{Y}^{[g^m]} = (Y_{120,(k+1):p}^{[g_1]}, Y_{121,1:k}^{[g_1]}, \ldots, Y_{120,(k+1):p}^{[g_{14}]}, Y_{121,1:k}^{[g_{14}]})$. The same routine as in Case 1 using odd and even blocs, is then used with the remaining data to obtain parameter estimates, albeit with these shifted 24 dimensional hourly response vectors.

For hour $k < 24$, $\mathbf{Y}^{[g^o]}$ holds the observed responses from Day 1 to the one ending on day 119. To predict the responses one-day-ahead at gauged sites in this field, we have $w = 1$ and $n = 120$ in Theorem 1 with $\mathbf{Y}^{[g^m]}$ rearranged so that all missing responses are at the beginning of the response vector. Specifically, let $\mathbf{e}_1^{i,j} : gp \times 1$, be 1 at the $(pj - i + 1)$th element and 0 otherwise, for $i = 1, \ldots, k$ and $j = 1, \ldots, g$; $\mathbf{e}_2^{i,j} : gp \times 1$, be 1 at the $(p(j - 1) + i)$th element and 0

otherwise, for $i = 1, \ldots, p - k$ and $j = 1, \ldots, g$. Let $\mathbf{E}_2 : gp \times gp = (\mathbf{e}_1^{1:k,1}, \ldots, \mathbf{e}_1^{1:k,g}, \mathbf{e}_2^{1:(p-k),1}, \ldots, \mathbf{e}_2^{1:(p-k),g})$. Applying Theorem 1 yields

$$
\mathbf{Y}^{[g^m]}\mathbf{E}_2 \mid \mathbf{Y}^{[g^o]}, \mathcal{H}
$$

$$
\sim t_{1 \times gp}\left(\mu_{(u|g)}\mathbf{E}_2, \Phi_{(u|g)} \otimes \mathbf{E}_2' \mathbf{\Psi}_{(u|g)}\mathbf{E}_2, \delta_{(u|g)}\right).
$$

(11)

Notice that $\mathbf{Y}^{[g^m]}\mathbf{E}_2$ is $(Y_{n,k:1}^{[g_1^m]}, \ldots, Y_{n,k:1}^{[g_{14}^m]}, Y_{n-1,1:(p-k)}^{[g_1^o]}, \ldots, Y_{n-1,(p-k)}^{[g_{14}^o]})$. To obtain the predictive distribution for the unobserved responses, we first need to decompose $\mathbf{Y}^{[g^m]}\mathbf{E}_2$, $\mu_{u|g}$ and $\mathbf{E}_2' \mathbf{\Psi}_{(u|g)}\mathbf{E}_2$ as follows: $\mathbf{Y}^{[g^m]}\mathbf{E}_2 = (\mathbf{T}_{1c}, \mathbf{T}_{2c}) : (1 \times gk, 1 \times g(p - k))$, where $\mu_{(u|g)} = (\mu_{1c}, \mu_{2c}) : (1 \times gk, 1 \times g(p - k))$, and

$$
\mathbf{E}_2' \mathbf{\Psi}_{(u|g)}\mathbf{E}_2
$$

$$
= \begin{pmatrix} \mathbf{C}_{11} & \mathbf{C}_{12} \\ \mathbf{C}_{21} & \mathbf{C}_{22} \end{pmatrix} : \begin{pmatrix} gk \times gk & gk \times g(p - k) \\ g(p - k) \times gk & g(p - k) \times g(p - k) \end{pmatrix}.
$$

(12)

Applying standard theory, for the multivariate t distribution yields the predictive distribution of the unobserved response $\mathbf{T}_{1c}$ given data $\mathbf{T}_{2c}$ as a t-distribution given by

$$
\mathbf{T}_{1c} \mid \mathbf{T}_{2c}, \mathbf{Y}^{[g^o]}, \mathcal{H}
$$

$$
\sim t_{1 \times gk}\left(\mu_{1c} + (\mathbf{T}_{2c} - \mu_{2c})\mathbf{C}_{22}^{-1}\mathbf{C}_{21},\right.
$$

$$
\frac{\delta_{(u|g)}}{\delta_{(u|g)} + g(p - k) + 1}\Phi_{(u|g)}
$$

$$
\times \left\{1 + \left(\delta_{(u|g)}\Phi_{(u|g)}\right)^{-1}(\mathbf{T}_{2c} - \mu_{2c})\mathbf{C}_{22}^{-1}\right.
$$

$$
\left.\left.\times (\mathbf{T}_{2c} - \mu_{2c})'\right\} \otimes \mathbf{C}_{11 \circ 2},\right.
$$

$$
\left. \delta_{(u|g)} + g(p - k) + 1\right).
$$

(13)

This completes the description of M1 in this case.

2.1.3. Multi-Day Ahead Forecasts. This subsection generalizes M1 to get a method that provides an r-step-ahead forecast, $(r \in \mathcal{N})$. Let N be the total number of days of observed responses. As before, we consider the multivariate setting, $p = 24$ being the total number of response coordinates and g, the total number of gauged sites. As well, we generalize the forecasting problem in two ways, namely, to forecasting the response on the last hour of the $(N + r)$th day; hour $k < 24$ on the $(N + r)$th day.

Case 1 (predict the final response for the last hour on day $N + r$). Here the odd bloc responses are $\mathbf{U}_t^{(1)} = (Y_{2t-1,1}^{[g_1^o]}, \ldots, Y_{2t-1,p}^{[g_g^o]})$. Note that $t^O = K$ if $N = 2K$ or $N = 2K - 1$ for some $K \in \mathcal{N}$. The even bloc responses are $\mathbf{V}_t^{(1)} = (Y_{2t,1}^{[g_1^o]}, \ldots, Y_{2t,p}^{[g_g^o]})$. Note that $t^E = K - 1$ if $N = 2K$ and $t^E = K$ if $N = 2K - 1$ for some $K \in \mathcal{N}$.

Remark 3. Notice that the total number of observations in data submatrices can be different for each N, being an odd or even number. Denote by t^N, the total number of observed responses. Then $t^N = t^O$ for the odd-day-responses and t^E, for the even-day-responses.

As in Section 2.1, we obtain the estimates of hyperparameters by averaging those two sets of estimates given odd or even bloc responses, respectively. Given these final estimates, we now obtain the predictive posterior distributions given all hourly observations up to Day N in Theorem A.1. They are also multivariate t-distributions. The proof and details can be seen in Appendix A.1.

Case 2 (predict the response for hour $k < 24$ on day $N + r$). Here the odds block responses are $\mathbf{U}_t^{(k)} = (Y_{2t-1,k+1}^{[g_1^o]}, \ldots, Y_{2t-1,p}^{[g_1^o]}, Y_{2t,1}^{[g_1^o]}, \ldots, Y_{2t,k}^{[g_1^o]}, \ldots, Y_{2t-1,k+1}^{[g_g^o]}, \ldots, Y_{2t-1,p}^{[g_g^o]}, Y_{2t,1}^{[g_g^o]}, \ldots, Y_{2t,k}^{[g_g^o]}) : 1 \times gp$. Note that $t^O = K \in \mathcal{N}$ if $N = 2K$ and $t^O = K - 1$ if $N = 2K - 1$ for some $K \in \mathcal{N}$. The even block responses are $\mathbf{V}_t^{(k)} = (Y_{2t,k+1}^{[g_1^o]}, \ldots, Y_{2t,p}^{[g_1^o]}, Y_{2t+1,1}^{[g_1^o]}, \ldots, Y_{2t+1,k}^{[g_1^o]}, \ldots, Y_{2t,k+1}^{[g_g^o]}, \ldots, Y_{2t,p}^{[g_g^o]}, Y_{2t+1,1}^{[g_g^o]}, \ldots, Y_{2t+1,k}^{[g_g^o]}) : 1 \times gp$. Note that $t^E = K - 1$ if $N = 2K$ or $N = 2K - 1$ for some $K \in \mathcal{N}$. We do the same thing here to obtain the "final" estimates for hyperparameters. Thus we are able to get the predictive posterior distribution given all observations up to bloc N and estimates for hyperparameters.

Remark 4. We also let t^N be the total number of observed response variables. So t^N is t^O for the odds-day-response blocs and t^E, for the even-day-response blocs.

Given the final estimates in Section 2.1, the predictive posterior distributions given all hourly observations up to N days are also multivariate t-distribution. The proof and details can be seen in Appendix A.2.

Remark 5. From Theorem A.2, the predictive distribution for the unobserved response variables from day $N+1$ to $N+r$ is the product of a sequence of matrix-t and t distributions. This implies no analytic form can be found for the response variable at the $(k-1)$th (for $k = 2, \ldots, p$) hour of the $(N+r)$th day at gauged site j ($j = 1, \ldots, g$).

2.2. Method M2.

An alternative approach to M1, through dynamic linear modeling, can also be used for forecasting and would seem an obvious choice, being an amalgamation of state-space time series models. Let $\mathbf{Y}_t$ be the response vector across all sites at hour t. As in the previous subsections, responses are square root transformed hourly ozone concentrations. Exploratory analysis of these transformed data, found 24- and 12-hour diurnal cycles, pointing to the approach in Huerta et al. [9] for ozone that is based on the same patterns.

Let β_t be the common temporal trend coefficient across the spatial sites at hour t. Furthermore, α_{1t} and α_{2t} denotes the coefficients of the periodic components with respect to 24- and 12-hour diurnal cycles, respectively. These two components are $S_{jt}(a_j) = \cos(\pi jt/12) + a_1 \sin(\pi jt/12)$ for $j = 1, 2$. Let $\mathbf{x}_t = (\beta_t, \alpha'_{1t}, \alpha'_{2t})' : (2n + 1) \times 1$ be the state vector at hour t and $\mathbf{F}_t : n \times (2n + 1) = [\mathbf{1}'_n, \text{diag}(S_{1t}(a_1)), \text{diag}(S_{2t}(a_2))]$. Furthermore, $\mathbf{V} : n \times n$ denotes the Euclidean intersite distance matrix for all site pairs. In this model, λ is the range parameter, σ^2, the variance parameter, and a_1, a_2, the phase parameters. Assume a constant covariance matrix for the state parameter vector $\mathbf{x}_t$ and denote it by $\mathbf{W}$. That covariance, which accounts for the spatial correlation between sites, is given by $\mathbf{W} = \text{diag}(\tau_y^2, \tau_1^2 \exp(-\mathbf{V}/\lambda_1), \tau_2^2 \exp(-\mathbf{V}/\lambda_2))$. The vector of hyperparameters, $\gamma = (\tau_y^2, \tau_1^2, \lambda_1, \tau_2^2, \lambda_2)$, is identical to that in Huerta et al. [9] based on assessments made for Dou et al. [8].

Thus, the measurement and state equations of the DLM are given by

$$\mathbf{Y}_t = \mathbf{F}'_t \mathbf{x}_t + \nu_t, \quad \nu_t \sim N\left(\mathbf{0}, \sigma^2 \exp\left(-\frac{\mathbf{V}}{\lambda}\right)\right),$$
$$\mathbf{x}_t = \mathbf{x}_{t-1} + \omega_t, \quad \omega_t \sim N(\mathbf{0}, \sigma^2 \mathbf{W}), \tag{14}$$

with initial information: $\mathbf{x}_0 \mid D_0 \sim N(\mathbf{m}_0, \sigma_0^2 \mathbf{C}_0)$. The hyperparameters $\mathbf{m}_0, \sigma_0^2$, and $\mathbf{C}_0$ are also identical to those in Huerta et al. [9]. One can obtain the posterior distribution of the state parameters at the last known time point, n, that is, $\mathbf{x}_n \mid \mathbf{y}_{1:n}, \theta \sim N(\mathbf{m}_n, \sigma^2 \mathbf{C}_n)$, using the Kalman filter, a smoothing method and the Metropolis-within-Gibbs sampling algorithm [8, 9, 16–18]. We omit details on updating and forecasting the state parameters given the model parameters and observations up to current time point.

Given the distribution of the state parameters at the last time point, n, the observed responses until time n, $\mathbf{y}_{1:n}$, and the model parameters, $\theta = \{\lambda, \sigma^2, a_1, a_2\}$, the r-step-ahead prediction is given by

$$\mathbf{y}_{n+r} \mid \mathbf{y}_{1:n}, \theta$$
$$\sim N\left(\mathbf{F}'_{t+r}\mathbf{m}_n, \sigma^2\left\{\mathbf{F}'_{t+r}(\mathbf{C}_n + r\mathbf{W})\mathbf{F}_{t+r} + \exp\left(-\frac{\mathbf{V}}{\lambda}\right)\right\}\right), \tag{15}$$

for $r \in \mathcal{N}$. Note that $\mathbf{F}_{t+r}$, $\mathbf{m}_n$ and $\mathbf{C}_n$ can be obtained by application of a standard method [8, 9, 17]. Here $n = 2880$ and $r = 1, \ldots, 24$ for the one-day-ahead prediction in the Case Study. For any fixed r, the predictive response, $\mathbf{y}_{n+r}$, can also be obtained by the MCMC (Markov Chain Monte Carlo) method. More specifically, at iteration j, suppose we have updated the vector of model parameters: $\theta^{(j)} = (\lambda^{(j)}, \sigma^{2(j)}, a_1^{(j)}, a_2^{(j)})$ using the FFBS (forward-filtering-backward-sampling) algorithm [8, 9, 16, 18]. That is, one has

$$\mathbf{x}_n \mid \mathbf{y}_{1:n}, \theta^{(j)} \sim N\left(\mathbf{m}_n^{(j)}, \sigma^{2(j)}\mathbf{C}_n^{(j)}\right). \tag{16}$$

Then, the predictive response at iteration j, $\mathbf{y}_{n+r}^{(j)}$ can be drawn from (15), that is,

$$
\mathbf{y}_{n+r} \mid \mathbf{y}_{1:n}, \theta^{(j)}
$$

$$
\sim N\left(\mathbf{F}_{t+r}^{(j)'}\mathbf{m}_n^{(j)}, \sigma^{2(j)}\right.
$$

$$
\left\{\mathbf{F}_{t+r}^{(j)'}\left(\mathbf{C}_n^{(j)} + r\mathbf{W}^{(j)}\right)\mathbf{F}_{t+r}^{(j)}\right. \tag{17}
$$

$$
\left.\left.+ \exp\left(-\frac{\mathbf{V}}{\lambda^{(j)}}\right)\right\}\right).
$$

Consequently, the predictive responses are obtained by the sample means of $\{\mathbf{y}_{n+r}^{(j)} : j = 1,\ldots,J\}$ ($J = 500$, where J denotes the total number of iteration after burn-in period; $r = 1,\ldots,24$). The empirical predictive intervals at the 95% nominal level can be obtained as the corresponding sample quantiles.

3. Case Study

This section implements the forecasting methods in the last section for one Chicago summer (from May 1 to August 31) using data for that urban area taken from the EPA's AQS database (2000). These extracted data come from fourteen irregularly distributed monitoring stations measuring hourly ozone concentrations in parts per billion (ppb), which, to assure the validity of our Gaussian model assumptions, are square-root-transformed as noted in the last section. Each has few missing values under the EPA 1997 Standard in 1997 (i.e., 80 parts per billion for the eight-hour ground level ozone concentrations) during the overall time span across all available sites in this region.

To assess the model's performance for temporal forecasting, 14 sites are selected as "gauged" sites (i.e., $g = 14$) and their observed responses on Day 121 are set aside as test values. Figure 1 shows the geographical locations of these fourteen gauged sites.

To explore these data further, weekday and hourly effects were computed for each site by averaging the transformed hour values over each of the seven weekdays over the whole summer. We found these effects to be very similar from one gauged site to the next. Thus, since "bloc" is the unit of time t, our approach puts the appropriate zero-one elements into the $\mathbf{Z}$ to mark off the progression of blocs as the t progressed. Baseline hourly effects are represented in the hypermean function and are automatically fitted by software, EnviroStat.1.0.1. This then represents the overall diurnal pattern, while allowing site-specific deviations within the model.

3.1. Results and Comparisons. The two methods considered in this paper were applied at all fourteen gauged sites (GSs for short) to predict the twenty-four left out (test) and square-root-transformed hourly observations on Day 121. For all sites, plots showing the observations during the six days leading up to test Day 121, as well as the twenty-four forecasts by both methods for that day, may be seen in Dou et al. [19].

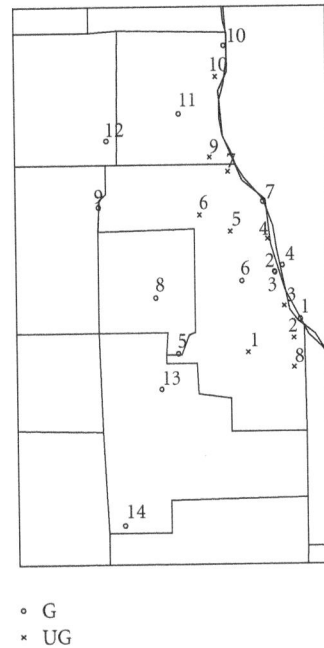

FIGURE 1: Geographical locations for the Chicago AQS database (2000), where the latitude and longitude are measured in degrees. (G: gauged sites; UG: ungauged sites).

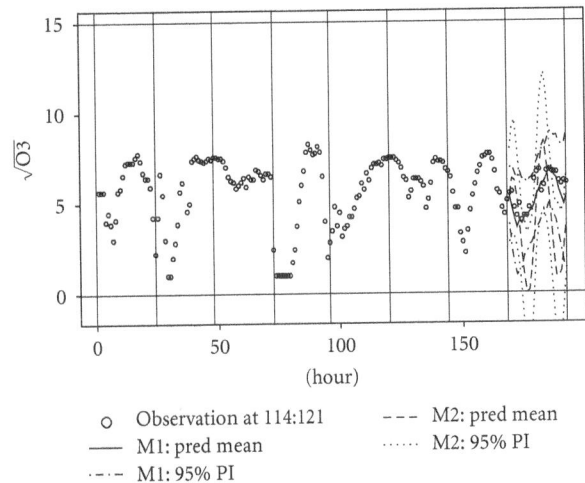

FIGURE 2: The observed square-root of ozone concentrations ($\sqrt{\text{ppb}}$) from Day 114 to Day 121, the predicted values using M1 and M2, and their 95% pointwise predictive intervals at GS 1.

These plots also include the 95% pointwise predictive intervals for that day for M1 and M2. For brevity, we include only figures that present some noteworthy features of these predictors.

To begin, Figure 2 for GS 1 shows a flat ozone field at this location on Day 119 and strongly varying one on Day 120 with two peaks. M1 does better than M2 in following not only the overall trend but tracking the turns quite well. All twenty-four of the test values lie within the 95% predictive credibility band although that for hour 12 overlaps the upper boundary. M2 follows ozone's peaks fairly well, but it forecasts valleys that do not turn up—the strong structure provided by the sines and cosines reduces this model's flexibility and capacity to track the series in this case. Those harmonics point to two twelve-hour cycles despite the general lack of same in the observed series during the days before.

In contrast to the case of spatial prediction [8] where generally four peaks are seen in the sinusoidal curves bounding the DLM's 95% predictive credibility bands, here there are just two. In fact, four peaks would be expected since the mean model's components of variation has sines and cosines that are squared when they enter into the posterior variance. Thus their valleys turn into peaks, one every six hours since both twelve- and twenty-four hour cycles are present in the mean. So why only two?

The answer seems to lie in the fact that the random coefficients for the twelve hour components of variance in the forecast at any monitoring site, is not as uncertain in forecasting at that site than they are in spatial prediction at other sites, which may be a substantial distance from the monitoring sites. Consequently these components, although small, would have large posterior variances in the predictor than these components of the forecaster at one of the monitoring sites. In other words, much more information is available in the data leading up to the last day for the forecasts for the test values at that site than is available at a remote and unmonitored sites.

Incidentally, the lower bounds for M2 forecasts can go below zero so in practice would need to be truncated. Moreover, few of the test values lie within the 95% credibility band.

Huerta et al. [9] give similar plots for three stations in their case study for Mexico City using the method that led to our method M2. These plots differ from ours in that their plotted ordinates represent ozone while ours represent square-root-transformed ozone. We elected to keep the latter since that is the scale on which our analysis was done and hence the one that provides maximal diagnostic benefit. Furthermore our square root scale does not risk exaggerating the observed differences between both methods being considered. At the same time, we recognize the practical importance of publishing forecasts on the ozone rather than square-root ozone scale, and hence Table 1 presents comparisons on the former scale.

Huerta et al. [9] focus on just one week, so their forecasts are for their last Day 7, based on the six preceding days (whose observed hourly concentrations are plotted). In contrast, since our forecasts are based on the entire summer, our last day is Day 121 and we plot the observed values for this day (as well as the preceding seven). Although the daily amplitudes of the sinusoids in Huerta et al. [9] vary from day to day, the periodicity is very consistent over those days, with two fairly distinct peaks each day. In contrast, the ozone patterns over the seven days preceding our forecast Day 121, differ markedly from one day to another. Thus while the data series for Day 120 shows two very distinct peaks, that for Day 116 is nearly flat, and that for Day 118 (essentially) shows only one peak, after a monotone increasing trend rising to the end of that day. Finally Day 7 in their plot shows

TABLE 1: The root-mean-square-predictive error (RMSPE) of the one-day-ahead prediction at fourteen gauged sites by using M1 and M2. M1 dominates M2 in all but 1 case.

Gauged site	RMSPE (M1)	RMSPE (M2)
1	0.71	3.06
2	0.63	2.72
3	0.63	2.16
4	0.70	2.06
5	**1.73**	**1.61**
6	0.86	1.52
7	0.47	2.05
8	1.00	2.73
9	0.77	2.65
10	0.71	2.35
11	0.70	3.04
12	0.67	1.85
13	0.85	3.50
14	1.04	2.37

good agreement at all three of their stations, between their "predictive median" and the test data values. In contrast, our averages of 500 MCMC generated predicted responses for M2 disagree markedly with many of the hourly test values on Day 121. We would conjecture that these discrepancies occur because the data series on Day 121 tends to be much flatter than on all the preceding days except Day 116. (It may also derive from the additional (temperature) data Huerta et al. [9] had to enhance their forecasts that we did not have).

We do not see in any of the plots in Huerta et al. [9], the four peaks seen in some of our 95% credibility bands, suggesting that the uncertainty in the coefficients of the twelve-hour cyclical components is quite well resolved by the six days of data preceding their test day. The credibility bands for their version of M2 contain all the test values for all three of their sites, as do M2's bands in Figure 2. (However, that is not the case for all fourteen of our stations as noted below. Moreover, their retrospective 95% credibility bands for all three sites are too narrow and fail to contain a large fraction of the observed values on a number of their days, e.g., on Sep 10 and Sep 12 at the Xalostoc site).

A summary very similar to that above for GS 1 also applies to the omitted figures for GS's 2–4, 7–13, although for GS 11 M1 forecasts diverge from the test values over the final four hours, and for GS 13 M1 underestimates those test values in the middle of the day.

Figure 3 shows that GS 5 is different. Although M2's forecast series has two peaks of moderate height, the bounds for the credibility bands have four of them consistent with two twelve-hour cycles. None of the forecast series tracks the series of test values well.

Summaries similar to that for GS 5 applies to GS 6 with the exception that M1 forecasts track the test value series quite well unlike M2 and to GS 14, where both methods underestimate the test values, M1 being closer overall than M2.

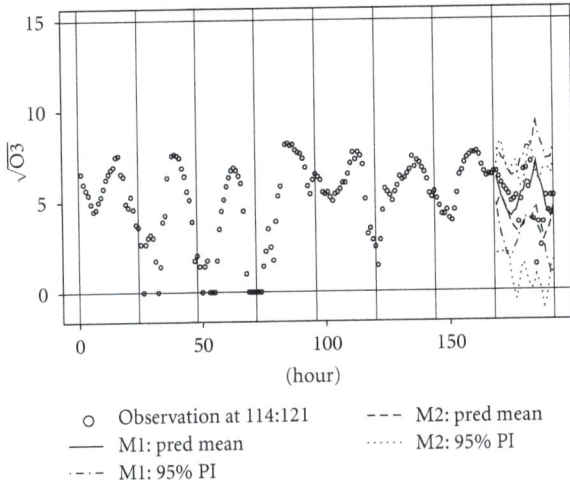

FIGURE 3: The observed square-root of ozone concentrations ($\sqrt{\text{ppb}}$) from Day 114 to Day 121, the predicted values using M1 and M2, and the 95% pointwise predictive intervals using M1 and M2 at GS 5.

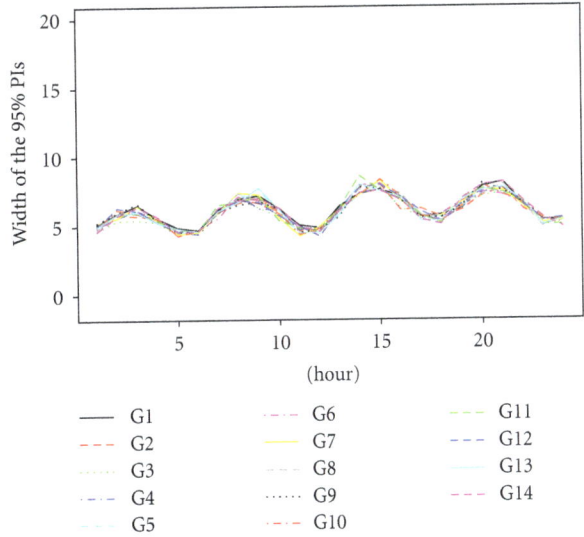

FIGURE 5: The width of the 95% pointwise predictive intervals (PIs) of the one-day-ahead prediction at 14 gauged sites using M2.

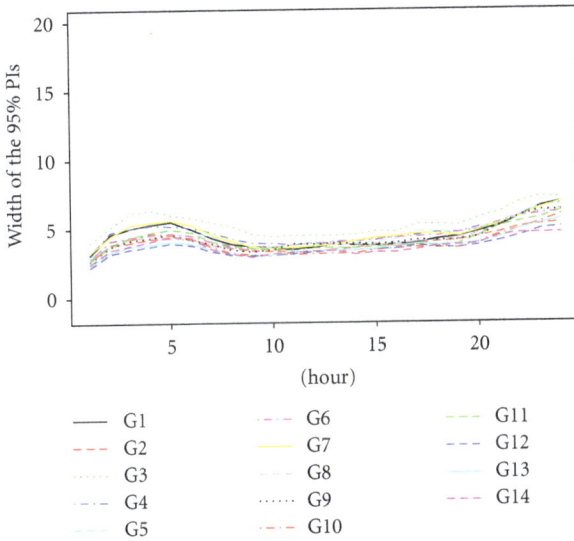

FIGURE 4: The width of the 95% pointwise predictive intervals of the one-day-ahead prediction at 14 gauged sites using M1.

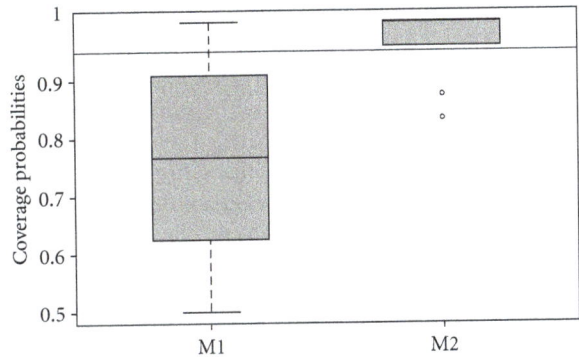

FIGURE 6: Boxplots of the coverage probabilities using M1 and M2 at the 95% nominal level.

Figure 4 plots the width of the 95% pointwise predictive credibility bands generated by the BSP at each of the twenty-four hours of Day 121. Starting from around 9 AM, these bands tend to increase and continue to do so until the last hour at 11 PM, reflecting the increasing uncertainty about the forecasts since increasingly fewer responses are observed as time increasing.

Figure 5 is a similar plot to that above, but this one for M2 instead of M1. These lengths are close to each other for various sites, exhibiting a wiggly periodic behavior across all gauges sites, a characteristic previously observed in Dou et al. [8, 18]. Although these lengths are very close to each other, M2 actually underestimates the predictive varian-ces at gauged sites as seen in Figure 6, which shows the

coverage probabilities of M1 and M2, and also shows a slightly overestimated predictive variance for M1, at the 95% nominal level.

Table 1 presents the root-mean-square-predictive error (RMSPE) of the predictive responses on the 121st day at each one of fourteen gauged sites using these two approaches. At GS j, the RMSPE of the prediction at hour h can be com-puted by

$$\text{RMSPE}^j = \sqrt{\sum_{h=1}^{24} \left(\text{PRED}_h^j - \text{OBS}_h^j \right)^2}, \qquad (18)$$

where PRED_h^j is the predictive response at hour h of Day 121 and OBS_h^j, the corresponding observed response at the same hour, same day, and same site. M2 has a larger RMSPE over all the gauged sites compared with M1. M1 has the smallest RMSPE across most gauged sites.

4. Discussion and Conclusions

For forecasting ground-level hourly ozone concentrations in a Chicago summer, M1 seems better than M2. It seems more accurate, and its 95% predictive credibility interval is better calibrated. However, in any practical application M1 and M2 would need to be assessed in the same manner as in this paper before making a final selection. It should be noted that a new model also based on the dynamic linear approach has been proposed by Sahu et al. [11] for ozone modeling. It would be interesting to compare this new approach with the methods in this paper.

Those methods M1 and M2 are two quite different approaches to modeling space-time process and comparing and contrasting them at a more fundamental level seems worthwhile. To begin, both are quasi-Bayesian models in that they rely on some preliminary data analyses. Thus the diurnal cycles are identified for the M2 mean function, while regional non-site-specific weekday effects are found for M1. Both methods can then incorporate predictors or covariates in their parametric mean functions with random coefficients as well as reflect diurnal patterns of variation. M1 proceeds with this in two steps. First, regional time-dependent covariates or predictors are identified for the construction of the design matrix $\mathbf{Z}$, in his case day-of-the-week bloc effects. Secondly, it estimates hypermeans for this predictor's coefficients as well as for the multivariate bloc vector's responses, in this case the hourly effects. At the same time, it allows site-specific deviations from these baseline estimates through the random mean coefficients. The case study suggests that these random coefficients capture site-specific hourly effects quite well. In contrast, M2 builds regional features and daily variation into its mean response function through the incorporation of mean trends and periodic components before implementation. Thus, its prescribed mean is fairly structured with Fourier components to describe daily 12- and 24-hour cycles. In contrast, M1 incorporates all general trends and diurnal patterns in the hypermean for its random coefficients and then allows site-specific deviations from this hypermean at all sites. The former is more flexible than the latter, in allowing the coefficients to change over time, but the second is more flexible than the first in allowing an arbitrary shape for the daily pattern of variation and allowing site-specific trends.

Both approaches put spatial covariance structures on their mean models as well as on the residuals. In contrast to M2, M1 does not require a nonstationary spatial covariance structure, and the form of the spatial covariance matrix is completely unspecified at level one of the Bayesian hierarchy. This is not important for the Chicago analysis where the spatial ozone field is quite flat, but we believe it would be an important difference between the models in say Los Angeles or Seattle, where M1 would be favored. M2 prescribes its temporal correlation structure through the structure of its mean function, notably a random walk model for its model coefficient vector. In contrast, M1's 24-hour bloc covariance matrix is unspecified at level one of the hierarchical model, leaving the data a big role in determining its form. However, this feature comes at the price of an assumption that the 24 autocovariance matrix is separable from the spatial covariance. Moreover, the covariance is constant over time. Both of these assumptions are limitations of M1.

Both M1 and M2 rely on both autocorrelation as well as temporal correlation for forecasting next day ozone levels. We believe responses will be somewhat autocorrelated from day to day and that feature can be exploited to enhance the forecasting performance. As formulated, M2 does borrow that additional strength, where M1 loses in the way we have implemented is parent, the BSP, by dropping a day to avoid having to formulate a multivariate time series model for the vectors of daily bloc responses. However, this is not strictly necessary. The more general version of the BSP approach does allow for that correlation, and in principle we would have estimated the hyperparameters that approach, suffered the consequences of possible misspecification and increased the computational burden of implementing M1. Thus M1 was formulated under the assumption of uncorrelated responses between days, unlike M2 which makes no such assumption, with the goal of ensuring timely 24 ahead ozone forecasts.

M2 has a much more general parent, in the dynamic linear model (DLM) and undoubtedly other implementations of the DLM could be made that retained its positive features while overcoming some of the limitations of M2 noted above. For example, a nonstationary spatial covariance could undoubtedly be used. As well the random walk model which has serious limitations could be replaced by say a more reasonable model like an AR(1), albeit with an added parameter burden. That would in turn further restrict the number of monitoring sites it could realistically handle in an urban area. As it stands, M1 computational efficiency enables it to handle a much larger number of sites than M2 in an urban area such as the greater Los Angeles area, which has 30 sites well beyond the reach of M2.

Although any ozone forecast for hourly concentrations 24 hours in advance cannot be much better than the baseline estimate, we have included Case 1 for completeness. Its Equation (9) is the basis of the forecast for that case. That equation actually gives a predictive distribution for all the hourly concentrations in Day 121 and could be used to forecast them all. However, the forecasts would not be very good compared to those given by the method in Case 2. The reason is that the latter exploits the strong AR(2) structure in any consecutive sequence of 24 hourly responses, unlike the former which assumes the daily vectors of responses are conditionally independent as an approximation made mainly for computational expediency. Thus for all other hours the forecaster in Case 2 should be used when the data in Day 120 are available. Note that within the Bayesian framework, the unconditional distribution of 24 dimensional response vectors are not independent, a feature that Case 1 exploits.

We have not considered the realistic case where only a limited number of hours of Day 120 data are available. That is because this case would be just a formalistic extension of methods M1 and M2.

Finally, we would emphasize that the results in Section 2.1 have been generalized in that section, another way in which M1 and M2 go beyond the limited application in the Case Study. Moreover, our approach for turning BSP into the temporal forecasting tool in M1 could well be used for any univariate time series. The approach would avoid the need to capture autocorrelation at fine temporal scales, something that can be difficult to do as in the ozone case, where the AR structure varies over the day.

Overall, we have found that for forecasting Chicago's next day ozone concentration levels, M1 would be more practical and more accurate than M2. With its well-calibrated forecast intervals, it seems a promising methodology for practical application.

Appendix

A. Supplementary Results

A.1. Theorem A.1 and Its Proof

Theorem A.1. *Let* $\mathbf{Y}^{[g^m]} = ((\mathbf{Y}^{[g^m_{1:g}]}_{N+r,1:p})', \dots, (\mathbf{Y}^{[g^m_{1:g}]}_{N+1,1:p})',$
$(\mathbf{Y}^{[g^o_{1:g}]}_{N,1:p})')' : (r+1) \times gp$ *and* $\mathbf{Y}^{[g^o]} = \mathbf{Y}^{[g^o_{1:g}]}_{1:(N-1),1:p} : (N-1) \times gp$.
Then, one has the following predictive distributions:

(i) $(\mathbf{Y}^{[g^m]} \mid \mathbf{Y}^{[g^o]}, \mathcal{H}) \sim t_{(r+1) \times gp}(\breve{\mu}_{(u|g)}, \breve{\Phi}_{(u|g)} \otimes$
$\breve{\Psi}_{(u|g)}, \breve{\delta}_{(u|g)})$, *where*

$$\breve{\mu}_{(u|g)} = \mu_{(1)} + \mathbf{A}_{12}\mathbf{A}_{22}^{-1}\left(\mathbf{Y}^{[g^o]} - \mu_{(2)}\right),$$

$$\breve{\Phi}_{(u|g)} = \frac{\delta_1 - gp + 1}{\delta_1 - gp + N + 1}\mathbf{A}_{11 \circ 2},$$

$$\breve{\Psi}_{(u|g)} = \frac{1}{\delta_1 - gp + 1}$$
$$\times \left\{ \Lambda_1 \otimes \Omega + \left(\mathbf{Y}^{[g^o]} - \mu_{(2)}\right)' \right. \tag{A.1}$$
$$\left. \times \mathbf{A}_{22}^{-1}\left(\mathbf{Y}^{[g^o]} - \mu_{(2)}\right)\right\},$$

$$\breve{\delta}_{(u|g)} = \delta_1 - gp + N + 1.$$

(ii) *The predictive distribution of* $Y^{[g^m_j]}_{N+r,p}$, *the pth unobserved response on the* $(N+r)$*th day at gauged site* j, *is t-distributed:*

$$Y^{[g^m_j]}_{N+r,p} \mid \mathbf{Y}^{[g^o_{1:g}]}_{1:N,1:p}, \mathcal{H}$$
$$\sim t_{\breve{\delta}}\left(\left(\mathbf{e}^j_r\right)'\breve{\mu}\mathbf{e}^j_g, \frac{\breve{\delta}}{\breve{\delta}-2}\left(\mathbf{e}^j_r\right)'\breve{\Phi}\mathbf{e}^j_r\left(\mathbf{e}^j_g\right)'\breve{\Psi}\mathbf{e}^j_g\right), \tag{A.2}$$

where

$$\breve{\mu} = \breve{\mu}_{1r} + \breve{\mathbf{B}}_{12}\breve{\mathbf{B}}_{22}^{-1}\left(\mathbf{Y}^{[g^o_{1:g}]}_{N,1:p} - \breve{\mu}_{2r}\right),$$

$$\breve{\Phi} = \frac{1}{\breve{\delta}_{(u|g)}+1}\breve{\mathbf{B}}_{11 \circ 2},$$

$$\breve{\Psi} = \breve{\Psi}_{(u|g)}\left[\mathbf{I}_{gp} + \breve{\Psi}_{(u|g)}^{-1}\left(\mathbf{Y}^{[g^o_{1:g}]}_{N,1:p} - \breve{\mu}_{2r}\right)'\right.$$
$$\left. \times \breve{\mathbf{B}}_{22}^{-1}\left(\mathbf{Y}^{[g^o_{1:g}]}_{N,1:p} - \breve{\mu}_{2r}\right)\right],$$

$$\breve{\delta} = \breve{\delta}_{(u|g)} + 1. \tag{A.3}$$

Proof. The result is straightforward by Theorem 1, where $w = r+1$ and $n = N+r$; decompose $\breve{\mu}_{(u|g)}$ and $\breve{\delta}_{(u|g)}\breve{\Phi}_{(u|g)}$ as follows

$$\breve{\mu}_{(u|g)} = \begin{pmatrix} \breve{\mu}_{1r} \\ \breve{\mu}_{2r} \end{pmatrix} : \begin{pmatrix} r \times gp \\ 1 \times gp \end{pmatrix},$$
$$\breve{\delta}_{(u|g)}\breve{\Phi}_{(u|g)} = \begin{pmatrix} \breve{\mathbf{B}}_{11} & \breve{\mathbf{B}}_{12} \\ \breve{\mathbf{B}}_{21} & \breve{\mathbf{B}}_{22} \end{pmatrix} : \begin{pmatrix} r \times r & r \times 1 \\ 1 \times r & 1 \times 1 \end{pmatrix}. \tag{A.4}$$

Hence, we have

$$\mathbf{Y}^{[g^m_{1:g}]}_{(N+1):(N+r),1:p} \mid \mathbf{Y}^{[g^o_{1:g}]}_{N,1:p}, \mathbf{Y}^{[g^o_{1:g}]}_{1:(N-1),1:p}, \mathcal{H} \sim t_{r \times gp}\left(\breve{\mu}, \breve{\Phi} \otimes \breve{\Psi}, \breve{\delta}\right), \tag{A.5}$$

where $\breve{\mu}, \breve{\Phi}, \breve{\Psi}$, and $\breve{\delta}$ are given in Theorem A.1.

We have $(\mathbf{e}^j_r)'\mathbf{Y}^{[g^m_{1:g}]}_{(N+1):(N+r),1:p}\mathbf{E}_1\mathbf{e}^j_g = Y^{[g^m]}_{N+r,p}$, that is, the unobserved response of the last hour of the $(N+r)$th day at Gauged Site j $(j = 1,\dots,g)$. Hence, we have

$$Y^{[g^m_j]}_{N+r,p} \sim t_{1\times 1}\left(\left(\mathbf{e}^j_r\right)'\breve{\mu}\mathbf{e}^j_g, \left(\mathbf{e}^j_r\right)'\breve{\Phi}\mathbf{e}^j_r \otimes \left(\mathbf{e}^j_g\right)'\breve{\Psi}\mathbf{e}^j_g, \breve{\delta}\right), \tag{A.6}$$

that is, $t_{\breve{\delta}}((\mathbf{e}^j_r)'\breve{\mu}\mathbf{e}^j_g, (\breve{\delta}/(\breve{\delta}-2))(\mathbf{e}^j_r)'\breve{\Phi}\mathbf{e}^j_r(\mathbf{e}^j_g)'\breve{\Psi}\mathbf{e}^j_g).$ $\square$

A.2. Theorem A.2 and Its Proof

Theorem A.2. *Let* $\mathbf{Y}^{[g^m]} = ((\mathbf{Y}^{[g^m_{1:g}]}_{N+r-1})', \dots, (\mathbf{Y}^{[g^m_{1:g}]}_{N+1})',$
$(\mathbf{Y}^{*[g^m_{1:g}]}_N)')':r \times gp$, *where* $\mathbf{W}^{[g^m_j]}_i = (Y^{[g^m_j]}_{i,k+1}, \dots, Y^{[g^m_j]}_{i,p}, Y^{[g^m_j]}_{i+1,1}, \dots,$
$Y^{[g^m_j]}_{i+1,k}) : 1 \times gp$, *and* $\mathbf{Y}^{*[g^m_j]}_N = (Y^{[g^m_j]}_{N,k+1}, \dots, Y^{[g^m_j]}_{N,p}, Y^{[g^m_j]}_{N+1,1}, \dots,$
$Y^{[g^m_j]}_{N+1,k}) : 1 \times gp$, *for* $i = N+1, \dots, N+r-1$ *and* $j = 1, \dots, g$.
We also let $\mathbf{Y}^{[g^o]} = \mathbf{Y}^{*[g^o_{1:g}]}_{1:(N-1)} : (N-1) \times gp$, *where*
$\mathbf{Y}^{*[g^o]}_i = (Y^{[g^o_j]}_{i,k+1}, \dots, Y^{[g^o_j]}_{i,p}, Y^{[g^o_j]}_{i+1,1}, \dots, Y^{[g^o_j]}_{i+1,k}) : 1 \times gp$. *One then has the following predictive distributions:*

(i)

$$\mathbf{Y}^{[g^m]} \mid \mathbf{Y}^{[g^o]}, \mathcal{H} \sim t_{r \times gp}\left(\tilde{\mu}_{(u|g)}, \tilde{\Phi}_{(u|g)} \otimes \tilde{\Psi}_{(u|g)}, \tilde{\delta}_{(u|g)}\right), \tag{A.7}$$

where

$$\widetilde{\mu}_{(u|g)} = \mu_{(1)} + \mathbf{A}_{12}\mathbf{A}_{22}^{-1}\left(\mathbf{Y}^{[g^o]} - \mu_{(2)}\right) : r \times gp,$$

$$\widetilde{\mathbf{\Phi}}_{(u|g)} = \frac{\delta_1 - gp + 1}{\delta_1 - gp + N}\left(\mathbf{A}_{11} - \mathbf{A}_{12}\mathbf{A}_{22}^{-1}\mathbf{A}_{21}\right),$$

$$\widetilde{\mathbf{\Psi}}_{(u|g)} = \frac{1}{\delta_1 - gp + 1}$$

$$\times \left\{ \mathbf{\Lambda}_1 \otimes \mathbf{\Omega} + \left(\mathbf{Y}^{[g^o]} - \mu_{(2)}\right)' \right. \tag{A.8}$$

$$\left. \times \mathbf{A}_{22}^{-1}\left(\mathbf{Y}^{[g^o]} - \mu_{(2)}\right)\right\},$$

$$\widetilde{\delta}_{(u|g)} = \delta_1 - gp + N.$$

(ii)

$$\left(\mathbf{Y}^{[g^m_{1:g}]}_{(N+1):(N+r-1),1:p}, \mathbf{Y}^{[g^m_{1:g}]}_{N+r,1:k} \mid \mathbf{Y}^{[g^o_{1:g}]}_{1:N,1:p}, \mathcal{H}\right)$$

$$\propto \prod_{j=1}^{g} p\left(\mathbf{T}^r_{1j} \mid \mathbf{Y}^{[g^m_j]}_{N+1,1:k}, \mathbf{Y}^{[g^o_j]}_{1:N,1:p}, \mathcal{H}\right)$$

$$\times \left(\mathbf{Y}^{[g^m_j]}_{N+1,1:k} \mid \mathbf{Y}^{[g^o_{1:g}]}_{1:N,1:p}, \mathcal{H}\right) \tag{A.9}$$

$$\sim \prod_{j=1}^{g} t_{(r-1)\times p}\left(\widetilde{\mu}^*_{1j}, \widetilde{\mathbf{\Phi}}^* \otimes \widetilde{\mathbf{\Psi}}^*_j, \widetilde{\delta}_{(u|g)} + 1\right)$$

$$\times t_{1\times k}\left(\widetilde{\mu}^*_{2j}, \widetilde{\mathbf{\Phi}}^*_{2j} \otimes \widetilde{\mathbf{\Psi}}^*_{2j}, \widetilde{\delta}_{(u|g)} + p - k\right),$$

where $\widetilde{\mu}^*_{1j}$, $\widetilde{\mathbf{\Phi}}^*$, and $\widetilde{\mathbf{\Psi}}^*_j$ are given in (A.16) and $\widetilde{\mu}^*_{2j}$, $\widetilde{\mathbf{\Phi}}^*_{2j}$, and $\widetilde{\mathbf{\Psi}}^*_{2j}$, in (A.19).

Proof. (i) The result is straightforward by Theorem 1 where $w = r$ and $n = N + r - 1$;

(ii) denote $\mathbf{E}_{2j} = (\mathbf{e}^{(j-1)p+1}_{gp}, \ldots, \mathbf{e}^{jp}_{gp}) : gp \times p$ for $j = 1, \ldots, g$. And let $\mathbf{E}_3 = (\mathbf{e}^p_p, \ldots, \mathbf{e}^1_p) : p \times p$. We will have the following results (details can be referred to in [19]):

$$\widetilde{\mathbf{Y}}^{[g^m_j]}$$

$$= \mathbf{Y}^{[g^m]}\mathbf{E}_{2j}\mathbf{E}_3$$

$$= \begin{pmatrix} Y^{[g^m]}_{N+r,k} & \cdots & Y^{[g^m]}_{N+r,1} & Y^{[g^m]}_{N+r-1,1} & \cdots & Y^{[g^m]}_{N+r-1,k+1} \\ \vdots & & \vdots & \vdots & & \vdots \\ Y^{[g^m]}_{N+1,k} & \cdots & Y^{[g^m]}_{N+1,1} & Y^{[g^o]}_{N,1} & \cdots & Y^{[g^o]}_{N,k+1} \end{pmatrix} : r \times p, \tag{A.10}$$

for $j = 1, \ldots, g$. From (i) in Theorem A.2, we have

$$\widetilde{\mathbf{Y}}^{[g^m_j]} \mid \mathbf{Y}^{[g^o]}, \mathcal{H} \sim t_{r\times p}\left(\widetilde{\mu}_j, \widetilde{\mathbf{\Phi}}_{(u|g)} \otimes \widetilde{\mathbf{\Psi}}_j, \widetilde{\delta}_{(u|g)}\right), \tag{A.11}$$

where

$$\widetilde{\mu}_j = \widetilde{\mu}_{(u|g)}\mathbf{E}_{2j}\mathbf{E}_3,$$

$$\widetilde{\mathbf{\Psi}}_j = \mathbf{E}'_3\mathbf{E}'_{2j}\widetilde{\mathbf{\Psi}}_{(u|g)}\mathbf{E}_{2j}\mathbf{E}_3. \tag{A.12}$$

We first decompose $\widetilde{\mathbf{Y}}^{[g^m_j]}$, $\widetilde{\mu}_j$ and $\widetilde{\delta}_{(u|g)}\widetilde{\mathbf{\Phi}}_{(u|g)}$ as follows:

$$\widetilde{\mathbf{Y}}^{[g^m_j]} = \begin{pmatrix} \mathbf{T}^r_{1j} \\ \mathbf{T}^r_{2j} \end{pmatrix} : \begin{pmatrix} (r-1) \times p \\ 1 \times p \end{pmatrix},$$

$$\widetilde{\mu}_j = \begin{pmatrix} \widetilde{\mu}_{1j} \\ \widetilde{\mu}_{2j} \end{pmatrix} : \begin{pmatrix} (r-1) \times p \\ 1 \times p \end{pmatrix},$$

$$\widetilde{\delta}_{(u|g)}\widetilde{\mathbf{\Phi}}_{(u|g)}$$

$$= \begin{pmatrix} \widetilde{\mathbf{\Phi}}_{11} & \widetilde{\mathbf{\Phi}}_{12} \\ \widetilde{\mathbf{\Phi}}_{21} & \widetilde{\mathbf{\Phi}}_{22} \end{pmatrix} : \begin{pmatrix} (r-1)\times(r-1) & (r-1)\times 1 \\ 1\times(r-1) & 1\times 1 \end{pmatrix}. \tag{A.13}$$

Consequently, we have

(a)

$$\mathbf{T}^r_{2j} \mid \mathbf{Y}^{[g^o]}, \mathcal{H} \sim t_{1\times p}\left(\widetilde{\mu}_{2j}, \widetilde{\mathbf{\Phi}}_{22} \otimes \widetilde{\mathbf{\Psi}}_j, \widetilde{\delta}_{(u|g)}\right), \tag{A.14}$$

(b)

$$\mathbf{T}^r_{1j} \mid \mathbf{T}^r_{2j}, \mathbf{Y}^{[g^o]}, \mathcal{H} \sim t_{(r-1)\times p}\left(\widetilde{\mu}^*_{1j}, \widetilde{\mathbf{\Phi}}^* \otimes \widetilde{\mathbf{\Psi}}^*_j, \widetilde{\delta}_{(u|g)} + 1\right), \tag{A.15}$$

where

$$\widetilde{\mu}^*_{1j} = \widetilde{\mu}_{1j} + \widetilde{\mathbf{\Phi}}_{12}\widetilde{\mathbf{\Phi}}_{22}^{-1}\left(\mathbf{T}^r_{2j} - \widetilde{\mu}_2\right),$$

$$\widetilde{\mathbf{\Phi}}^* = \frac{\widetilde{\delta}_{(u|g)}}{\widetilde{\delta}_{(u|g)} + 1}\left(\widetilde{\mathbf{\Phi}}_{11} - \widetilde{\mathbf{\Phi}}_{12}\widetilde{\mathbf{\Phi}}_{22}^{-1}\widetilde{\mathbf{\Phi}}_{21}\right), \tag{A.16}$$

$$\widetilde{\mathbf{\Psi}}^*_j = \widetilde{\mathbf{\Psi}}_j\left(\mathbf{I}_p + \widetilde{\mathbf{\Psi}}_j^{-1}\left(\mathbf{T}^r_{2j} - \widetilde{\mu}_{2j}\right)'\widetilde{\mathbf{\Phi}}_{22}^{-1}\left(\mathbf{T}^r_{2j} - \widetilde{\mu}_{2j}\right)\right).$$

We then decompose $\mathbf{T}^r_{2j}$, $\widetilde{\mu}^r_{2j}$, and $\widetilde{\mathbf{\Psi}}_j$ as follows:

$$\mathbf{T}^r_{2j} = \begin{pmatrix} \mathbf{T}^j_{21} & \mathbf{T}^j_{22} \end{pmatrix} : \begin{pmatrix} 1\times k & 1\times(p-k) \end{pmatrix},$$

$$\widetilde{\mu}^r_{2j} = \begin{pmatrix} \mu^j_{21} & \mu^j_{22} \end{pmatrix} : \begin{pmatrix} 1\times k & 1\times(p-k) \end{pmatrix},$$

$$\widetilde{\mathbf{\Psi}}_j = \begin{pmatrix} \widetilde{\mathbf{\Psi}}^j_{11} & \widetilde{\mathbf{\Psi}}^j_{12} \\ \widetilde{\mathbf{\Psi}}^j_{21} & \widetilde{\mathbf{\Psi}}^j_{22} \end{pmatrix} : \begin{pmatrix} k\times k & k\times(p-k) \\ (p-k)\times k & (p-k)\times(p-k) \end{pmatrix}. \tag{A.17}$$

Hence the predictive distribution of $\mathbf{T}^j_{21}$, that is, $\mathbf{Y}^{[g^m_j]}_{N+1,1:k}$ is given by

$$\mathbf{Y}^{[g^m_j]}_{N+1,1:k} \mid \mathbf{Y}^{[g^o]}_{1:N,1:p}, \mathcal{H} \sim t_{1\times k}\left(\widetilde{\mu}^*_{2j}, \widetilde{\mathbf{\Phi}}^*_{2j} \otimes \widetilde{\mathbf{\Psi}}^*_{2j}, \widetilde{\delta}_{(u|g)} + p - k\right), \tag{A.18}$$

where

$$\tilde{\mu}_{2j}^* = \mu_{21}^j + \left(\mathbf{T}_{22}^j - \mu_{22}^j\right)\left(\tilde{\mathbf{\Psi}}_{22}^j\right)^{-1}\tilde{\mathbf{\Psi}}_{21}^j,$$

$$\tilde{\mathbf{\Phi}}_{2j}^* = \frac{\tilde{\delta}_{(u\,|\,g)}}{\tilde{\delta}_{(u\,|\,g)} + p - k}\tilde{\mathbf{\Phi}}_{22}^j$$

$$\times \left(1 + \left(\tilde{\delta}_{(u\,|\,g)}\tilde{\mathbf{\Phi}}_{22}^j\right)^{-1}\left(\mathbf{T}_{22}^j - \mu_{22}^j\right)\right. \quad (A.19)$$

$$\left.\times\left(\tilde{\mathbf{\Psi}}_{22}^j\right)^{-1}\left(\mathbf{T}_{22}^j - \mu_{22}^j\right)'\right),$$

$$\tilde{\mathbf{\Psi}}_2^* = \tilde{\mathbf{\Psi}}_{11}^j - \tilde{\mathbf{\Psi}}_{12}^j\left(\tilde{\mathbf{\Psi}}_{22}^j\right)^{-1}\tilde{\mathbf{\Psi}}_{21}^j.$$

Therefore, we have

$$p\left(\mathbf{Y}_{(N+1):(N+r-1),1:p}^{[g_{1:g}^m]}, \mathbf{Y}_{N+r,1:k}^{[g_{j}^m]} \mid \mathbf{Y}_{1:N,1:p}^{[g_{1:g}^o]}, \mathcal{H}\right)$$

$$\propto \prod_{j=1}^{g} p\left(\mathbf{T}_{1j}^r \mid \mathbf{Y}_{N+1,1:k}^{[g_{j}^m]}, \mathbf{Y}_{1:N,1:p}^{[g_{1:g}^o]}, \mathcal{H}\right) \quad (A.20)$$

$$\times p\left(\mathbf{Y}_{N+1,1:k}^{[g_{j}^m]} \mid \mathbf{Y}_{1:N,1:p}^{[g_{1:g}^o]}, \mathcal{H}\right).$$

The predictive distribution for $\mathbf{Y}_{N+r,k}^{[g_{1:g}^m]}$ has no analytic form. $\quad\square$

Acknowledgments

This work was partially supported by funding from the Pacific Institute of the Mathematical Sciences as well as the Natural Science and Engineering Research Council of Canada.

References

[1] Ozone, *Air Quality Criteria for Ozone and Related Photochemical Oxidants*, US Environmental Protection Agency, 2005.

[2] S. K. Sahu, S. Yip, and D. M. Holland, "A fast Bayesian method for updating and forecasting hourly ozone levels," *Environmental and Ecological Statistics*, vol. 18, no. 1, pp. 185–207, 2011.

[3] J. V. Zidek, N. D. Le, and Z. Liu, "Combining data and simulated data for space-time fields: application to ozone," *Environmental and Ecological Statistics*, vol. 19, pp. 37–56, 2012.

[4] N. D. Le and J. V. Zidek, "Interpolation with uncertain spatial covariances: a Bayesian alternative to Kriging," *Journal of Multivariate Analysis*, vol. 43, no. 2, pp. 351–374, 1992.

[5] N. D. Le, W. Sun, and J. V. Zidek, "Bayesian multivariate spatial interpolation with data missing by design," *Journal of the Royal Statistical Society B*, vol. 59, no. 2, pp. 501–510, 1997.

[6] J. Zidek, L. Sun, N. Le, and H. Özkaynak, "Contending with space-time interaction in the spatial prediction of pollution: vancouver's hourly ambient PM10 field," *Environmetrics*, vol. 13, no. 5-6, pp. 595–613, 2002.

[7] N. D. Le and J. V. Zidek, *Statistical Analysis of Environmental Space-Time Processes*, Springer, New York, NY, USA, 2006.

[8] Y. P. Dou, N. D. Le, and J. V. Zidek, "Modeling hourly ozone concentration fields," *Annals of Applied Statistics*, vol. 4, Article ID 11831213, 2010.

[9] G. Huerta, B. Sansó, and J. R. Stroud, "A spatiotemporal model for Mexico City ozone levels," *Journal of the Royal Statistical Society C*, vol. 53, no. 2, pp. 231–248, 2004.

[10] V. J. Berrocal, A. E. Gelfand, and D. M. Holland, "A bivariate space-time downscaler under space and time misalignment," *Annals of Applied Statistics*, vol. 4, pp. 1942–1975, 2010.

[11] S. K. Sahu, A. E. Gelfand, and D. M. Holland, "High-resolution space-time ozone modeling for assessing trends," *Journal of the American Statistical Association*, vol. 102, no. 480, pp. 1221–1234, 2007.

[12] D. Wraith, C. Alston, K. Mengersen, and T. Hussein, "Bayesian mixture model estimation of aerosol particle size distributions," *Environmetrics*, vol. 22, no. 1, pp. 23–34, 2011.

[13] W. Meiring, P. Guttorp, and P. D. Sampson, "Space-time estimation of grid-cell hourly ozone levels for assessment of a deterministic model," *Environmental and Ecological Statistics*, vol. 5, no. 3, pp. 197–222, 1998.

[14] N. D. Le and J. V. Zidek, EnviRo.stat. A software for "Statistical Analysis of Environmental Space-Time Processes", 2006, http://enviro.stat.ubc.ca/statisticalanalysis/.

[15] C. S. Hirtzel and J. E. Quon, "Statistical analysis of continuous ozone measurements," *Atmospheric Environment A*, vol. 15, no. 6, pp. 1025–1034, 1981.

[16] C. K. Carter and R. Kohn, "On gibbs sampling for state space models," *Biometrika*, vol. 81, pp. 541–553, 1994.

[17] W. West and J. Harrison, *Bayesian Forecasting and Dynamic Models*, Springer, 1997.

[18] Y. P. Dou, N. D. Le, and J. V. Zidek, "A dynamic linear model for hourly ozone concentrations," Tech. Rep. 228, Department of Statistics, University of British Columbia, 2007.

[19] Y. P. Dou, N. D. Le, and J. V. Zidek, "Temporal prediction with a Bayesian spatial predictor: an application to ozone fields," Tech. Rep. 249, Department of Statistics University of British Columbia, 2009.

7

Wind Velocity Vertical Extrapolation by Extended Power Law

Zekai Şen, Abdüsselam Altunkaynak, and Tarkan Erdik

Hydraulics Division, Civil Engineering Faculty, Istanbul Technical University, Maslak, 34469 Istanbul, Turkey

Correspondence should be addressed to Tarkan Erdik, tarkanerdik@hotmail.com

Academic Editor: Harry D. Kambezidis

Wind energy gains more attention day by day as one of the clean renewable energy resources. We predicted wind speed vertical extrapolation by using extended power law. In this study, an extended vertical wind velocity extrapolation formulation is derived on the basis of perturbation theory by considering power law and Weibull wind speed probability distribution function. In the proposed methodology not only the mean values of the wind speeds at different elevations but also their standard deviations and the cross-correlation coefficient between different elevations are taken into consideration. The application of the presented methodology is performed for wind speed measurements at Karaburun/Istanbul, Turkey. At this location, hourly wind speed measurements are available for three different heights above the earth surface.

1. Introduction

Wind energy, as one of the main renewable energy sources in the world, attracts attention in many countries as the efficient turbine technology develops. Wind speed extrapolation might be regarded as one of the most critical uncertainty factor affecting the wind power assessment, when considering the increasing size of modern multi-MW wind turbines. If the wind speed measurements at heights relevant to wind energy exploitation lacks, it is often necessary to extrapolate observed wind speeds from the available heights to turbine hub height [1], which causes some critical errors between estimated and actual energy output, if the wind shear coefficient, n, cannot be determined correctly. The difference between the predicted and observed wind energy production might be up to 40%, due to turbulence effects, time interval of wind data measurement, and the extrapolation of the data from reference height to hub heights [2].

In the literature, the wind shear coefficient is generally approximated between 0.14 and 0.2. However, in real situations, a wind shear coefficient is not constant and depends on numerous factors, including atmospheric conditions, temperature, pressure, humidity, time of day, seasons of the year, the mean wind speed, direction, and nature of terrain [3–6]. Table 1 demonstrates the various wind shear

coefficients for different types of topography and geography [3].

According to the calculations of wind resource analysis program (WRAP) report, in 39 different regions, out of 7082 different wind shear coefficients, 7.3% are distributed between 0 and 0.14 and 91.9% above 0.14, while 0.8% are calculated as negative [7], due to the measurements error.

Different methods have been developed to analyze wind speed profiles, such as power, logarithmic, and loglinear laws [8]. Besides, in the literature various studies are conducted in order to estimate wind shear coefficient in the power law only if surface data is available at hand [9–11].

The wind speed undergoes repeated changes, as a result of which the roughness and friction coefficients also change depending on landscape features, the time of the day, the temperature, height and wind direction. The uncertainty is inhereted in the wind speed data and its extrapolation to the hub height should be considered carefully and preciously [12]. Moreover, this uncertainty is exacerbated in the offshore environment by the inclusion of the dynamic surface [13]. Therefore, the mean wind speed profile of the logarithmic type is developed by applying a stability correction for offshore sites [14].

It is crucial point for energy investors to accurately predict the average wind speed at different wind turbine

TABLE 1: Wind shear coefficient of various terrains [3].

TABLE 1: Wind shear coefficient of various terrains [3].

Terrain type	
Lake, ocean, and smooth-hard ground	0.1
Foot-high grass on level ground	0.15
Tall crops, hedges, and shrubs	0.2
Wooded country with many trees	0.25
Small town with some trees and shrubs	0.3
City area with tall buildings	0.4

hub heights and make realistic feasibility projects for these heights. In this study, a simple but effective methodology on the basis of the perturbation theory is presented in order to derive an extended power law for the vertical wind speed extrapolation and then the Weibull probability distribution function (pdf) parameters. It is observed that on the contrary to the classical approach not only the means of wind speeds are at different elevations, but also the standard deviations and the cross-correlation coefficient should be taken into consideration, if the wind speeds at different elevations are not independent from each other.

2. Power Law

This law is the simplest way for estimating the wind speed at a wind generator hub elevation from measurements at a reference level. In general, the power law expression is given as,

$$\left(\frac{Z_1}{Z_2}\right)^n = \left(\frac{V_1}{V_2}\right), \tag{1}$$

where terms in the brackets are the velocity and elevation ratios, $V_2 > V_1$ and Z_2/Z_1, respectively. Furthermore, $V_2 > V_1$ and $Z_2 > Z_1$; and n is the exponent of the power law, which is a complex function of the local climatology, topography, surface roughness, environmental conditions, meteorological lapse rate, and weather stability. It is clear that the effects of all these factors are embedded in the wind velocity time records, and consequently, their total reflections are also expected in the value of the exponent, n. Therefore, one tends to think whether there is a way of obtaining the estimation of this exponent from the wind speed time series. Power laws are used almost exclusively without any generally accepted methodology. Most often, only the arithmetic averages of the wind speed at two elevations are considered in the numerical calculation of the exponent. Logically, other than the mean values, standard deviations and cross-correlation coefficient should enter the calculations, because these additional parameters arise as a result of instability, roughness, and so forth. Provided that there are wind speed records at two or more elevations, the following approach provides an objective solution.

3. Extended Power Law

In (1), the only random variable that represents the weather situation is the wind speeds, which can be written in terms of the averages and perturbation terms about their averages that render (1) to

$$\left(\frac{Z_1}{Z_2}\right)^n = \left(\frac{\overline{V_1} + V_1'}{\overline{V_2} + V_2'}\right), \tag{2}$$

where V_1' and V_2' are the perturbation terms with averages equal to zero, $\overline{V_1'} = \overline{V_2'} = 0$. This last expression can be rewritten simply as

$$\left(\frac{Z_1}{Z_2}\right)^n = \left(\frac{\overline{V_1}}{\overline{V_2}}\right)\left(1 + \frac{V_1'}{\overline{V_1}}\right)\left(1 + \frac{V_2'}{\overline{V_2}}\right)^{-1}. \tag{3}$$

This expression is referred to as the extended power law in this paper. The third bracket on the right-hand side corresponds to a geometric series, which can be expressed by the Binomial expansion as

$$\left(\frac{Z_1}{Z_2}\right)^n = \left(\frac{\overline{V_1}}{\overline{V_2}}\right)\left(1 + \frac{V_1'}{\overline{V_1}}\right)\left[1 - \left(\frac{V_2'}{\overline{V_2}}\right) + \left(\frac{V_2'}{\overline{V_2}}\right)^2 \right.$$
$$\left. - \left(\frac{V_2'}{\overline{V_2}}\right)^3 + \left(\frac{V_2'}{\overline{V_2}}\right)^4 - \cdots \right]. \tag{4}$$

This expression can still be simplified after the expansion of the second and third brackets on the right-hand side and then by considering the second-order term approximately yields

$$\left(\frac{Z_1}{Z_2}\right)^n = \left(\frac{\overline{V_1}}{\overline{V_2}}\right)\left[1 - \left(\frac{V_2'}{\overline{V_2}}\right) + \left(\frac{V_2'}{\overline{V_2}}\right)^2 + \left(\frac{V_1'}{\overline{V_1}}\right) \right.$$
$$\left. - \left(\frac{V_1'}{\overline{V_1}}\right)\left(\frac{V_2'}{\overline{V_2}}\right) + \frac{V_1'}{\overline{V_1}}\left(\frac{V_2'}{\overline{V_2}}\right)^2 \right]. \tag{5}$$

After taking the arithmetic averages of both sides and then considering that the odd order power term averages are equal to zero, (5) yields to

$$\left(\frac{Z_1}{Z_2}\right)^n = \left(\frac{\overline{V_1}}{\overline{V_2}}\right)\left[1 - \frac{V_1'}{\overline{V_1}}\frac{V_2'}{\overline{V_2}} + \left(\frac{V_2'}{\overline{V_2}}\right)^2\right]. \tag{6}$$

By definition $\overline{V_1'} = \overline{V_2'} = 0$, exactly. In fact, for symmetrical (i.e., Gaussian) perturbation terms, the odd number arithmetic averages such as $\overline{V_1'V_2'^2}$ are also equal to zero approximately by definition.

In (6), the common arithmetic average of the perturbation multiplication, $\overline{V_1'V_2'}$, at two different elevations is equal to the covariance of the perturbations. This can be written in terms of the standard deviations S_{V_1} and S_{V_2} and cross-correlation, r_{12}, multiplication as $\overline{V_1'V_2'} = r_{12}S_{V_1}S_{V_2}$. The second-order perturbation term average, $\overline{V_2'^2}$, is equivalent to the variance of the perturbation term as $\overline{V_2'^2} = S_{V_2}^2$. The substitution of these last two expressions into (6) leads to

$$\left(\frac{Z_1}{Z_2}\right)^n = \left(\frac{\overline{V_1}}{\overline{V_2}}\right)\left(1 - \frac{S_{V_1}S_{V_2}}{\overline{V_1}\overline{V_2}}r_{12} + \frac{S_{V_1}^2}{\overline{V_2}^2}\right), \tag{7}$$

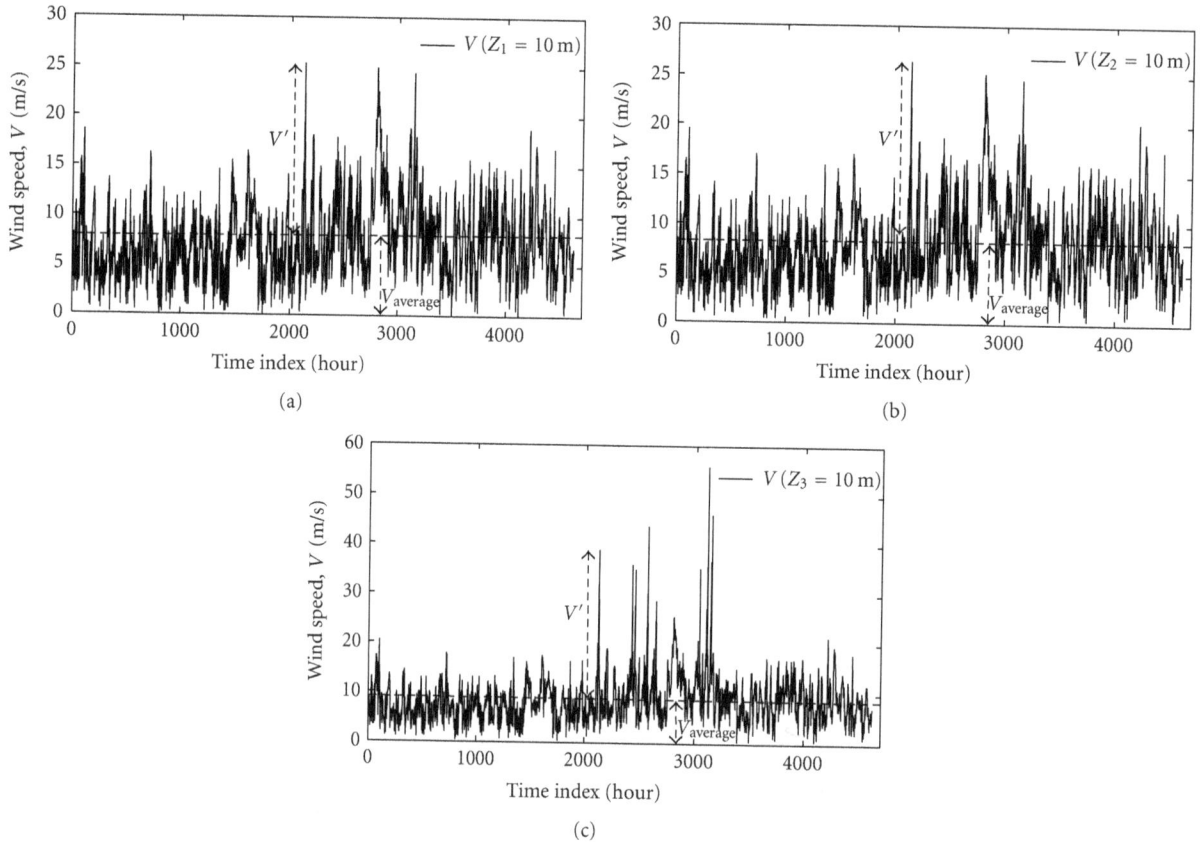

FIGURE 1: Karaburun wind speed records at different levels.

where S_{V_1} and S_{V_2} are the standard deviations and r_{12} is the cross-correlation coefficient between the wind speed time series at two elevations. In practice, most often the cross-correlation term in (7) is overlooked by assuming that there are no random fluctuations around the mean speed values which bring the implication that the standard deviations are equal to zero. These assumptions are not valid because in an actual weather, there are always fluctuations in the wind speed records as in Figure 1. By definition in statistics, the ratio of standard deviation to the arithmetic mean is the coefficient of variation, and hence (7) can be rewritten in parameterized form as,

$$\left(\frac{Z_1}{Z_2}\right)^n = \left(\frac{\overline{V_1}}{\overline{V_2}}\right)\left(1 - C_{V_1}C_{V_2}r_{12} + C_{V_2}^2\right). \quad (8)$$

Herein, C_{V_1} and C_{V_2} are the coefficients of variation for the wind speed records, V_1 and V_2, at two different elevations, respectively.

4. Weibull Distribution Parameter Extrapolation

Extrapolation of wind speed data to standard elevations poses a rather subjective approach based on the mean wind velocity only. Unreliability in such extrapolations is reflected

in the subsequent wind energy, E, calculations through the classical formulation

$$E = \frac{1}{2}\rho V^3, \quad (9)$$

where ρ is the standard atmosphere air density which is equal to $1.226 \, \text{gr/cm}^3$ at $25°C$, and V is the wind speed. Most often the wind speed at a meteorology station is measured along a tower at different elevations, and it is desired to be able to find the wind profile at this station for further wind loadings or energy calculations. Some researchers have employed the Weibull pdf for empirical wind speed relative frequency distribution (histogram), and a set of formulas are derived for the extrapolation of the Weibull pdf parameters [15–18]. In general, two-parameter Weibull pdf of wind speed, $P(V)$, is given as,

$$P(V) = \left(\frac{k}{c}\right)\left(\frac{V}{c}\right)^{k-1} \exp\left[-\left(\frac{V}{c}\right)^k\right], \quad (10)$$

where k is a dimensionless shape parameter, and c is a scale parameter with the speed dimension. In many applications,

FIGURE 2: The location of the Karaburun wind station.

the basic Weibull pdf statistical properties are the expectation, $E(V)$, variance, $\text{Var}(V)$, and the mth order moment $E(V^m)$ around the origin which are explicitly available as [19]

$$E(V) = c\Gamma\left(1 + \frac{1}{k}\right), \tag{11}$$

$$\text{Var}(V) = c^2\left[\Gamma\left(1 + \frac{2}{k}\right) - \Gamma^2\left(1 + \frac{1}{k}\right)\right], \tag{12}$$

$$E(V^m) = c^m\Gamma\left(1 + \frac{m}{k}\right), \tag{13}$$

respectively.

It is the purpose of this paper to present detailed extrapolation formulations for the Weibull pdf parameters on the basis of perturbation approach and power law of vertical wind velocity variation.

The two-parameter Weibull pdf has the average and standard deviation as in (11) and (12); and by definition their ratio gives the coefficient of variation as

$$C_V = \sqrt{\frac{\Gamma(1 + (2/k))}{\Gamma^2(1 + (1/k))} - 1}. \tag{14}$$

The k value can best be estimated by using the approximate relationship for (14) as given by Justus and Mikhail [17], that is:

$$k = \frac{1}{C_v^{1.086}} \tag{15}$$

and (11) yields the scale parameter as

$$c = \frac{\overline{V}}{\Gamma(1 + (1/k))}. \tag{16}$$

The substitution of these last two expressions for two elevations with labels 1 and 2 into (8) leads after some algebra to

$$\left(\frac{Z_1}{Z_2}\right)^n = \left(\frac{c_1}{c_2}\right)\left[1 - r_{12}(k_1 k_2)^{-0.921} + k_2^{-1.841}\right]. \tag{17}$$

By taking the logarithms of both sides, give

$$n = \frac{\text{Ln}(c_1/c_2) + \text{Ln}\left[1 - r_{12}(k_1 k_2)^{-0.921} + k_2^{-1.841}\right]}{\text{Ln}(Z_1/Z_2)}. \tag{18}$$

Hence, once the Weibull pdf parameters, c and k, are determined the power exponent can be calculated provided that the cross-correlation coefficient, r_{12} is found from the available wind speed time series data. It must be noticed that this last expression reduces to the classical counterpart in (1) after the substitution of $k_2 = 0$. Therefore, this expression can be written as

$$n = \frac{\text{Ln}(c_1/c_2)}{\text{Ln}(Z_1/Z_2)}. \tag{19}$$

5. Application

In this paper, wind speed data from the Karaburun wind station in Istanbul, Turkey are used, and this station is located at latitude 41.338′ N and longitude 28.677′ E (Figure 2). At this location hourly, wind speed measurements are available at three different heights (10 m, 20 m, and 30 m) above the earth surface. The average wind speed, the standard deviation, and the coefficient of variation for each height are given in Table 2. Wind speed data measurement empirical relative frequency distribution functions (histograms) at 10 m, 20 m, and 30 m are given together with the theoretically fitted Weibull pdf's in Figure 3 for each height. A good fit between the empirical and theoretical counterparts at different heights are obtained through the Kolmogorov-Smirnov test at significance level of 5%. Table 1 presents the Weibull pdf parameters, c (scale) and k (shape). The scale parameters are 8.32, 8.77, and 9.54 for heights of 10 m, 20 m, and 30 m, respectively. The shape parameters are determined as 2, 2.11, and 2.06, respectively for the same heights.

It is clear from this table that as the height increases, the mean speed and standard deviation increase as expected. Coefficient of variation varies with different heights as in Table 2. This shows also that the closer the height to the earth surface, the greater is the instability of the air. The power law exponent, n, calculation between any two heights are found from (18) and classically from (19); and they are presented in Table 3.

For both classical and Weibull pdf approaches the greatest value lies between 20 m and 30 m, whereas the lowest value is between 10 m and 20 m. For all levels, the average values are 0.1360 and 0.1238 for classical and extended power laws, respectively.

6. Conclusions

A simple methodology on the basis of the perturbation theory is presented in order to derive an extended power law for the vertical wind speed extrapolation and then the Weibull probability distribution parameters. It is observed that on the contrary to the classical approach not only the means of wind speeds are at different elevations, but also the standard deviations and the cross-correlation coefficient should be taken into consideration, if the wind speeds at different elevations are not independent from each other. Otherwise, consideration of the classical power law in the calculations embodies the assumption that there are no fluctuations in the wind speed time series around their

--- $V\,(Z_1 = 10\,\text{m})$ data
—— Weibull-pdf

(a)

--- $V\,(Z_2 = 20\,\text{m})$ data
—— Weibull-pdf

(b)

--- $V\,(Z_3 = 30\,\text{m})$ data
—— Weibull-pdf

(c)

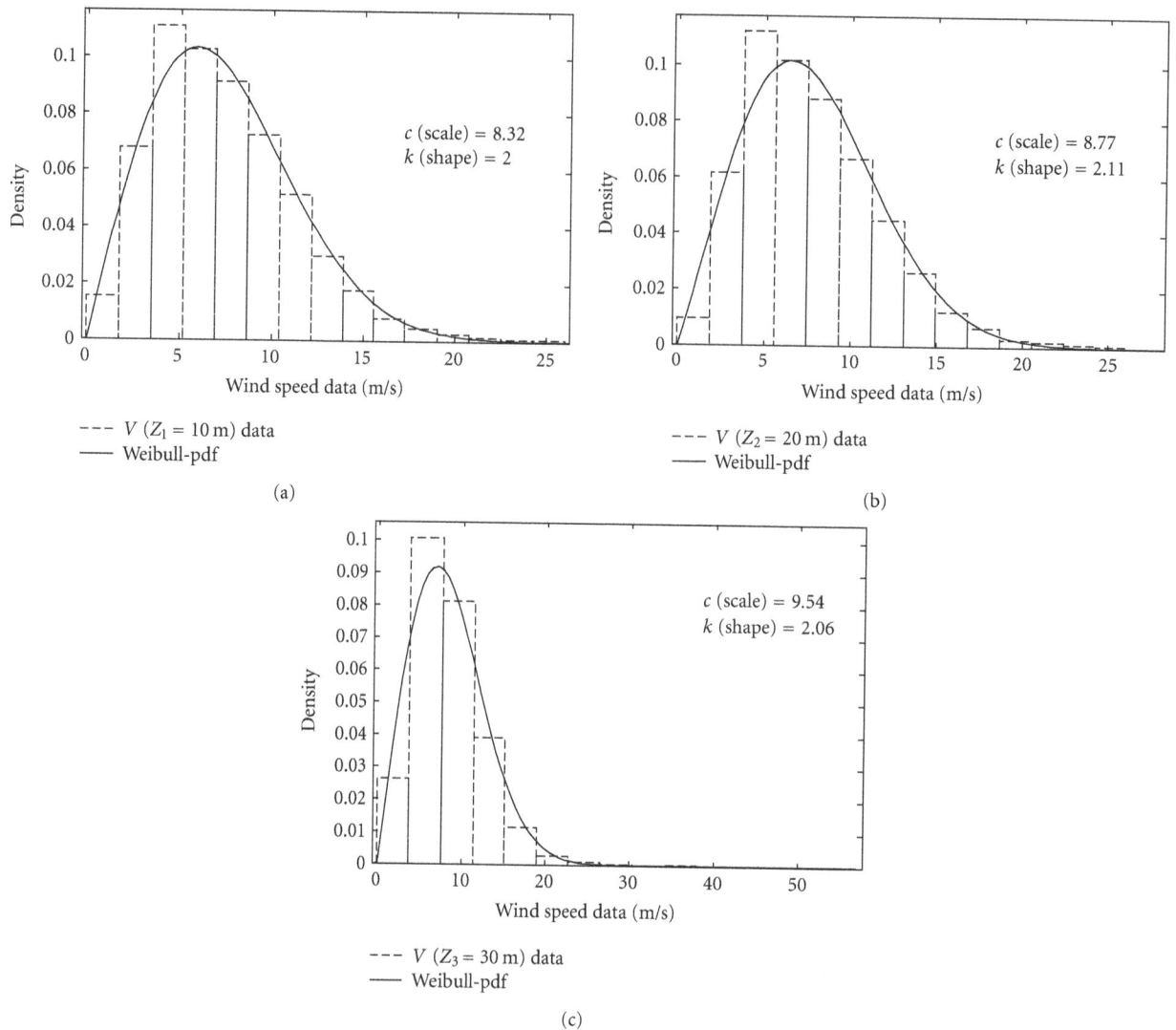

FIGURE 3: Karaburun Weibull pdf's at different levels.

TABLE 2: Wind speed summary statistics of Karaburun for three different heights.

Height (m)	Mean speed (m/s)	Standard deviation (m/s)	Coefficient of variation	Weibull pdf parameters	
				c	k
10	7.37	3.86	0.52	8.32	2.0
20	7.75	3.88	0.50	8.77	2.11
30	8.44	4.30	0.51	9.54	2.06

TABLE 3: Exponent calculations both classical and extended power law.

Height (m)	Classical (19)	Extended (18)
10–20	0.0760	0.0935
20–30	0.2076	0.1608
10–30	0.1245	0.1172
Average	0.1360	0.1238

respective mean values. The necessary formulations for the Weibull distribution function wind speed parameter extrapolations are presented in this paper. The application of the developed methodology is presented for Karaburun, Istanbul, near the Black Sea coast wind speed measurement station data at three different levels.

References

[1] M. Motta, R. J. Barthelmie, and P. Vølund, "The influence of non-logarithmic wind speed profiles on potential power output at danish offshore sites," *Wind Energy*, vol. 8, no. 2, pp. 219–236, 2005.

[2] A. Tindal, K. Harman, C. Johnson, A. Schwarz, A. Garrad, and G. Hassan, "Validation of GH energy and uncertainty predictions by comparison to actual production," in *Proceedings of the AWEA Wind Resource and Project Energy Assessment Workshop*, Portland, Ore, USA, September 2007.

[3] M. R. Patel, *Wind and Solar Power Systems*, CRC Press, 1999.

[4] M. R. Elkinton, A. L. Rogers, and J. G. McGowan, "An investigation of wind-shear models and experimental data trends for different terrains," *Wind Engineering*, vol. 30, no. 4, pp. 341–350, 2006.

[5] R. H. Kirchhoff and F. C. Kaminsky, "Wind shear measurements and synoptic weather categories for siting large wind turbines," *Journal of Wind Engineering and Industrial Aerodynamics*, vol. 15, no. 1–3, pp. 287–297, 1983.

[6] B. Turner and R. Istchenko, "Extrapolation of wind profiles using indirect measures of stability," *Wind Engineering*, vol. 32, no. 5, pp. 433–438, 2008.

[7] Minnesota Department of Commerce, "Wind resource analysis program (WRAP)," Minnesota Department of Commerce, St. Paul, Minn, USA, October 2002.

[8] G. Gualtieri and S. Secci, "Comparing methods to calculate atmospheric stability-dependent wind speed profiles: a case study on coastal location," *Renewable Energy*, vol. 36, no. 8, pp. 2189–2204, 2011.

[9] M. Hussain, "Dependence of power law index on surface wind speed," *Energy Conversion and Management*, vol. 43, no. 4, pp. 467–472, 2002.

[10] D. A. Spera and T. R. Richards, "Modified power law equations for vertical wind profiles," in *Proceedings of the Conference and Workshop on Wind Energy Characteristics and Wind Energy Siting*, Portland, Ore, USA, June 1979.

[11] A. S. Smedman-Högström and U. Högström, "A practical method for determining wind frequency distributions for the lowest 200 m from routine meteorological data," *Journal of Applied Meteorology*, vol. 17, no. 7, pp. 942–954, 1978.

[12] F. Bañuelos-Ruedas, C. Angeles-Camacho, and S. Rios-Marcuello, "Analysis and validation of the methodology used in the extrapolation of wind speed data at different heights," *Renewable and Sustainable Energy Reviews*, vol. 14, no. 7, pp. 2383–2391, 2010.

[13] R. J. Barthelmie, "Evaluating the impact of wind induced roughness change and tidal range on extrapolation of offshore vertical wind speed profiles," *Wind Energ*, vol. 4, pp. 99–105, 2001.

[14] M. Motta, R. J. Barthelmie, and P. Vølund, "The influence of non-logarithmic wind speed profiles on potential power output at danish offshore sites," *Wind Energy*, vol. 8, no. 2, pp. 219–236, 2005.

[15] C. G. Justus, W. R. Hargraves, and A. Yalcin, "Nationwide assessment of potential output from wind powered generators," *Journal of Applied Meteorology*, vol. 15, no. 7, pp. 673–678, 1976.

[16] C. G. Justus, W. R. Hargraves, and A. Mikhail, "Reference wind speed distributions and height profiles for wind turbine design and performance evaluation applications," ERDA ORO/5107-76/4, 1976.

[17] C. G. Justus and A. Mikhail, "Height variation of wind speed and wind distribution statistics," *Geophysical Research Letters*, vol. 3, pp. 261–264, 1967.

[18] A. Altunkaynak, T. Erdik, I. Dabanlı, and Z. Sen, "Theoretical derivation of wind power probability distribution function and applications," *Applied Energy*, vol. 92, pp. 809–814, 2012.

[19] K. Conradsen, L. B. Nielsen, and L. P. Prahm, "Review of Weibull statistics for estimation of wind speed distributions," *Journal of Climate & Applied Meteorology*, vol. 23, no. 8, pp. 1173–1183, 1984.

8

Self-Organized Criticality of Rainfall in Central China

Zhiliang Wang and Chunyan Huang

College of Mathematics and Informatics, North China University of Water Conservancy and Hydroelectric Power, 36 Beihuan Road, Henan, Zhengzhou 450011, China

Correspondence should be addressed to Zhiliang Wang, wzl@ncwu.edu.cn

Academic Editor: Harry D. Kambezidis

Rainfall is a complexity dynamics process. In this paper, our objective is to find the evidence of self-organized criticality (SOC) for rain datasets in China by employing the theory and method of SOC. For this reason, we analyzed the long-term rain records of five meteorological stations in Henan, a central province of China. Three concepts, that is, rain duration, drought duration, accumulated rain amount, are proposed to characterize these rain events processes. We investigate their dynamics property by using scale invariant and found that the long-term rain processes in central China indeed exhibit the feature of self-organized criticality. The proposed theory and method may be suitable to analyze other datasets from different climate zones in China.

1. Introduction

China is not only a big country for its population but also a big agriculture one. Rain is the main source of irrigation water, and it plays a key role in the crop growing period. No rain will cause drought while storm may cause flood. To keep sufficient agriculture production sustainable, it is necessary to identify the role of the rain clearly and to understand the characteristics of the rain deeply. In particular, analyzing the rain in central China is more important because this region is the main crop source and the population density is very high.

Rain is liquid precipitation, as opposed to nonliquid kinds of precipitation such as snow and hail and so on. Rainfall is the result of the atmosphere movement, which is influenced by sun radiation, sea water evaporation, and earth rotation. In the fact, the long-term rain record is a time series which can be regarded as a random process. The rainfall process is actually a complexity system because there are too many influencing factors.

In previous studies, many mathematical methods have been applied to find the rainfall pattern, such as periodic, trend, change point, and fractal. Based on the last 1033 years historic data set, Jiang analyzed the temporal and spatial climate variability by using a "Mexican hat" wavelet transform [1]. Bordi used Standardized Precipitation Index (SPI) to assess the climatic condition of this region and applied principal component to capture the pattern of covariability of the index at different gauge stations [2]. The results suggest that the northern part of east-central China is experiencing dry conditions more frequently from the 1970s onwards indicated by a negative trend in the SPI time series. Applying the binary cubic interpolation and optimal fitting method, Wang et al. set up a statistical model [3] and Yu used the application of gray and fuzzy methods [4], to make the rain forecast. Appling chaos dynamics theory on rainfall, Rodriguez-Iturbe et al. [5] and Wang et al. [6, 7] found that both the characteristics of the correlation integral and the Lyapunov exponents of the historical data give preliminary support to the presence of chaotic dynamics with a strange attractor. Using the correlation dimension method, the inverse approach of the nonlinear prediction method, and the method of surrogate data, Siva Kumar found that the rainfall data exhibit nonlinear behavior and possibly low-dimensional chaos, which imply that short-term prediction based on nonlinear dynamics might be possible [8, 9]. Ramirez used a feedforward neural network and resilient propagation learning algorithm to analyze the relation between the rain data and potential temperature, vertical component of the wind, specific humidity, air temperature, precipitable water, relative vorticity, and moisture divergence flux [10].

All the previous works are based on an assumption that the long-term rain process is stochastic or chaotic one. Nevertheless, the rain process is located at the brink between the chaos and determination. It is necessary to develop the new theory and methodology to address this context.

In recent years, a new perspective, which is called self-organized criticality (SOC), attracted applied mathematicians, meteorologists, climatologists, and environmentalists. Self-organized criticality is proposed by Bak et al. [11–15]. The term self-organized criticality refers to the tendency of many systems driven by an energy input at a slow and constant rate to enter states characterized by scale-free behavior. The statistics of the systems then resemble those of equilibrium systems near the critical point of a phase transition. Self-organized criticality is one of a number of important discoveries made in statistical physics and related fields over the latter half of the 20th century, discoveries which relate particularly to the study of complexity in nature. The most classical instances of SOC include the common natural phenomena, such as earthquakes and avalanches [16–19]. A rainfall event can be considered as an earthquake-like or an avalanche-like event [20, 21]. Further more, a long-term rain event series can be also seen as a similar event which is a complexity dynamics process and exhibits the feature of self-organized criticality.

Andrade analyzed long-term daily rain records of weather stations around the world with a special emphasis on the semiarid regions and found that there existed some evidences of SOC with these data [22]. Peters et al. investigated the European rain and fund it exhibits the feature of SOC [23–25]. However, up to now, we have not seen any report which related the China rain to SOC in the literature. In this work, we chose five meteorological stations in Henan, a central province of China, to try to find out the SOC evidence.

First of all we assume that the rainfall events that occurred in this region follow the power law distribution. And then based on the theory and method, we look for the SOC evidence through our calculating and analyzing. It is our aim for us to confirm the existence of SOC.

Henan is a leading province in grain, wheat, and oil seed output, and it is also an important producer of beef, cotton, pork, animal oil, and corn. With a population of approximately 93.6 million, Henan is the second most populous Chinese province after Guangdong. In this sense Henan is the big agriculture and population province in China. Precipitation, especially rain, has a dramatic effect on agriculture. All plants need at least some water to survive; therefore rain is important to agriculture. A regular rain pattern is usually vital to healthy plants; too much or too little rainfall can be harmful, even devastating to crops. Drought can kill crops and increase erosion while overly wet weather can cause harmful fungus growth. So studying the characteristics of the rainfall event is important to understand the dry and wet spell. It is consequently helpful to local flood and drought management.

2. Data Sets

We downloaded the data sets from China Meteorological Data Sharing Service System. These stations' geographic positions are displayed in Figure 1 in which Xinyang and Zhumadian are in the northern of Henan, Anyang and Zhengzhou in the northern part, Lushi in the western region. From the point of view of meteorological classification, the former two stations belong to humid subtropical climate zone while the later three stations are in the temperate climate zone.

The site name, site number, operation period and location are listed in Table 1. Station number is the general international code which is the WMO number. Obliviously the operation period is not equal in different stations. The format of degree, minute, and second is used to represent the location of the gauge station.

The original data sets should have contained the daily rain, but in some cases missing data may occur. Therefore these need to be pretreated by the method of interpolation before further analysis. After data pretreatment, we plot the five series in the Figure 2 whose horizontal axis represents the time of the rainfall day and vertical axis displays the daily rain amount (0.1 mm).

3. Method of Analysis

3.1. Scale Invariant. One of the great successes of physics in the last decades has been in the understanding of phenomena with fluctuations over many scales. In high-energy physics, critical phenomena and hydrodynamics it is often possible to establish the existence of a scaling or scale invariant regime in which the fluctuation (Δx) in the field of interest (x) at small scale Δt and at large scale $\lambda \Delta t$ ($\lambda > 1$) is amplified by the factor $\lambda^{-\tau}$, where τ is the scaling parameter. This may be written more concisely as

$$\Delta x(\lambda \Delta t) = \lambda^{-\tau} \Delta x(\Delta t), \qquad (1)$$

where $\Delta t = t_1 - t_0$, $\Delta x(\Delta t) = x(t_1) - x(t_0)$, $t_2 = t_0 + \lambda(t_1 - t_0)$, $\Delta x(\lambda \Delta t) = x(t_2) - x(t_0)$, and equality is understood in the sense of probability distributions, that is, $F(X) = F(Y)$ if $\Pr(X > c) = \Pr(Y > c)$ for all c, Pr means Probability.

In Section 1, we said that our work objective is to find the evidence of the SOC of rain process in central china. To look it out, we need construct mathematical models to mine the relations between the SOC and the long-term rain records. In the literatures on the SOC, power law distribution (scaling or scale invariance) is often used as a tool to prove a complexity dynamic process with the feature of SOC. To reveal the existing of SOC in the rain process, in the beginning we define a rain event as a sequence of consecutive nonzero measurements of the rain rate. Consequently, we use 3 variables, which contain rain duration, drought durations and accumulated rain amount to describe the feature of rain event in the region.

The probability density function of a physical variable X (such as rain duration, drought duration and accumulated rain amount in a rainfall event) is denoted by $n(x)$, which

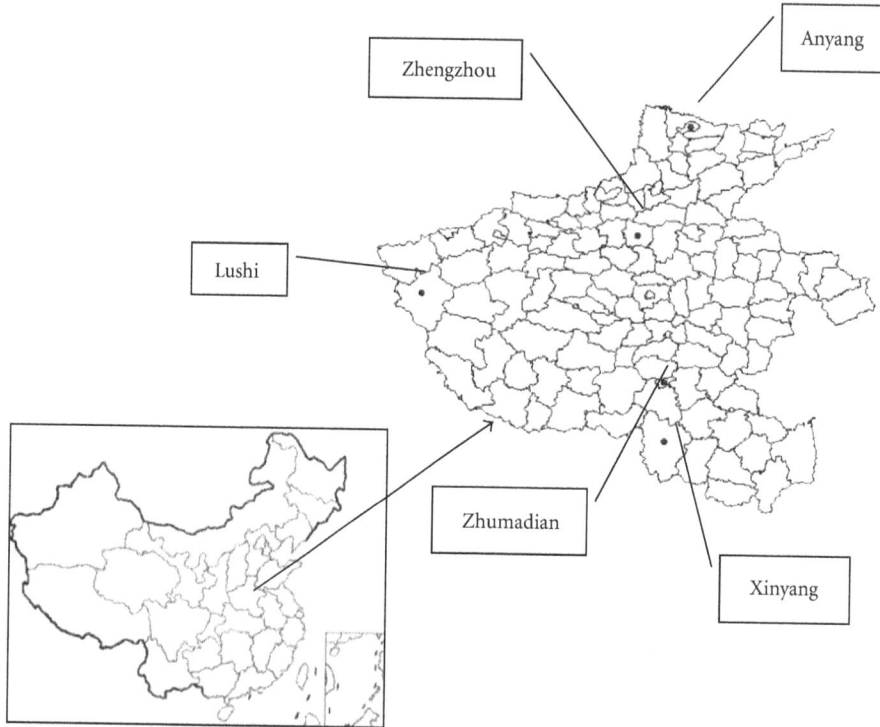

FIGURE 1: Distribution of stations, named Anyang, Zhengzhou, Lushi, Zhumadian, and Xinyang.

TABLE 1: Observation sites with station number, corresponding time periods estimated, and location.

Site name	Station number	Operation period	Location	
Zhengzhou	57083	195101–200712	34° 43′N	113° 39′E
Anyang	53898	195102–200712	36° 07′N	114° 22′E
Zhumadian	57290	195801–200712	33° 00′N	114° 01′E
Xinyang	57297	195101–200712	32° 08′N	114° 03′E
Lushi	57067	195207–200712	34° 03′N	111° 02′E

Longitude and latitude of the location is represented as degree and minute.

accumulated probability distribution (PDF) is $\tilde{N}(x)$. PDF can be defined as

$$\tilde{N}(x) = \int_x^{x_M} n(x)dx, \qquad (2)$$

where x_M is the maximum value in the data set. By using the integrated description instead of histograms we avoid data fluctuations in the low (high) value regime induced by the choice of logarithmic. If $n(x) \propto x^{-\tau}$, $x_M \to \infty$ and $\tau \to 1$, then $\tilde{N}(x) \propto x^{-\tau+1}$. Our rainfall data are generally confined to ranges $1 < I < 200$ days and $0.1\,\text{mm} < x < 400\,\text{mm}$, while higher values are observed only in extreme situations. Therefore, we cannot replace M by ∞ in (2) and obtain

$$N(x) = \frac{\tilde{N}(x)}{x} \propto \frac{1}{x^\tau}\left[1 - \left(\frac{x}{x_M}\right)^{\tau-1}\right]. \qquad (3)$$

Thus, the log-log plot of $N(x)$ versus x definitely departs from a straight line as $x \to x_M$.

3.2. *Algorithm.* Based on the principle of scale invariant, we develop 2 algorithms corresponding to rain duration, drought duration and accumulated rain amount, in which first algorithm is about the former two items and the second is about the accumulated rain amount.

3.2.1. *Procedure for Rain Duration and Drought Duration.* Rain duration is the life time of a successive rainfall, and drought duration is the waiting time between two rain events. Let RD and DD stand for rain duration and drought, respectively. Their distribution densities are discrete which are listed as shown in Table 2.

To test the scaling of rain duration and drought duration in central China, analysis can be conducted by the following procedure.

Algorithm 1 (Accumulated probability distribution computation and plot of rain duration and drought duration). We have the following:

Step 1. Wash and treat the original data.

TABLE 2: Distribution model of rain duration and drought duration.

RD/DD	x_1	x_2	$\cdots$	x_M
$n(x)$	$n(x_1)$	$n(x_2)$	$\cdots$	$n(x_M)$
$\tilde{N}(x)$	$\sum_{i=1}^{M} n(x_i)$	$\sum_{i=2}^{M} n(x_i)$	$\cdots$	$\sum_{i=M}^{M} n(x_i)$
$N(x)$	$\left(\sum_{i=1}^{M} n(x_i)\right)/x_1$	$\left(\sum_{i=2}^{M} n(x_i)\right)/x_2$	$\cdots$	$\sum_{i=M}^{M} n(x_i)/x_M$

x_i stands for the length of every rain or drought event. M is the longest time period.

Step 2. Plot the daily rain size on the rain date (see Figure 2).

Step 3. Count the numbers of every kind of rain events and assign $n(x)$ the number listed in the second row of Table 2.

Step 4. Calculate the accumulated number $\tilde{N}(x) = \sum_{i=j}^{M} n(x_i)$, $j = 1, 2, \ldots, M$ and list it in the third row.

Step 5. Calculate the number density $N(x) = (\sum_{i=j}^{M} n(x_j))/x_j$ of rain event corresponding to the fourth row and list it in the fourth row.

Step 6. According to formula (3), we take the logarithm of the first and fourth row. By the linear regression technology, we will obtain τ.

Step 7. Make the double-log distribution plot.

3.2.2. Procedure for Accumulated Rain Amount. In the present paper, the accumulated rain amount, which is symbolized as ARA, represents the total rain amount in a single rain event. Then, based on Step 1 and Step 2 in Algorithm 1, the procedure of ARA analysis can be followed by Algorithm 2 below.

Algorithm 2 (Accumulated probability distribution computation and plot of ARA). We have the following:

Step 1. By calculating the ARA we get a new ARA series $y_1, y_2, \ldots, y_N$, where N stands for the number of the rainfall event in the long-term record.

Step 2. Let $y_{\min} = \min(y_1, y_2, \ldots, y_N)$, $y_{\max} = \max(y_1, y_2, \ldots y_N)$, and partition the closed interval $[y_{\min}, y_{\max}]$ into subinterval $[y, y + \Delta y]$. Here, Δy can be computed by using the following equation:

$$\frac{y + \Delta y}{y} = 10^{1/5}. \qquad (4)$$

Step 3. If S is the number of all the subintervals, we calculate the times of the occurrences of $y_1, y_2, \ldots y_N$ in every interval and note them $n(s), s = 1, 2, \ldots, S$.

Step 4. Calculate the accumulated number $\tilde{N}(s) = \sum_{s=j}^{S} n(s)$, $j = 1, 2, \ldots S$.

Step 5. Calculate the number density $N(s) = (\sum_{s=j}^{S} n(s))/\Delta s_j$.

Step 6. Take the logarithm of $N(s)$ and left end point of $[y, y + \Delta y]$. By the linear regression technology, we will obtain τ in terms of Formula (3).

Step 7. Make the double-log distribution plot.

4. Result and Discussion

According to the steps of Algorithm 1, the long-term rains are shown in Figure 2. Xinyang station and Zhumadian station, which are located in the southern part of Henan Province, are plotted above. Contrast Anyang station, Zhenzhou station, and Lushi station, which are located in the northern part of this province.

4.1. Rain Duration and Drought Duration. Corresponding to Table 2, the calculation results of rain duration are listed in Table 3. For the dataset is too large, here we only list out Zhenzhou station's data.

The computation results about rain duration and drought duration show that the rainfall times and the rain duration, the drought days and the drought duration exhibit the power law relation. Taking the logarithm value of number density as the vertical axis and the logarithm value of the rain duration as horizontal axis, we find that the distribution of the rain duration number in Xinyang station, Zhumadian station, Zhengzhou station, and Lushi station is similar with their function profile (see Figure 3) while Anyang's number density of rain duration exhibits the different feature. The scale-free region ranges from 1 to 10 in Anyang and from 1 to 11 in other places. Here, $\tau_{anyang} = -1.35$, other τs fluctuate around -1.68. Influenced by the monsoon climate, the rain season begins in the early June and ends in the lately September. In fact, the Anyang is this region where the rain duration is the shortest in Henan province.

Looking at Figure 3 and Figure 4, it seems that the number density of drought duration displays the same distribution characteristics with the rain durations. Actually, the number density of rain duration is different in some aspects, such as the maximum time period and scaleless region. Furthermore, attention should be paid to the number density of drought duration in Zhengzhou and Anyang (see Figure 5). The scale region of Anyang station ranges from 1 day to 100 days, and Zhengzhou's scaling left end point closes to 100 but Xingyang's and Zhumadian's are about 50.

Here, $\tau_{xinyang}$, $\tau_{xinyang} \approx -1.71$, and other τ is about -1.5.

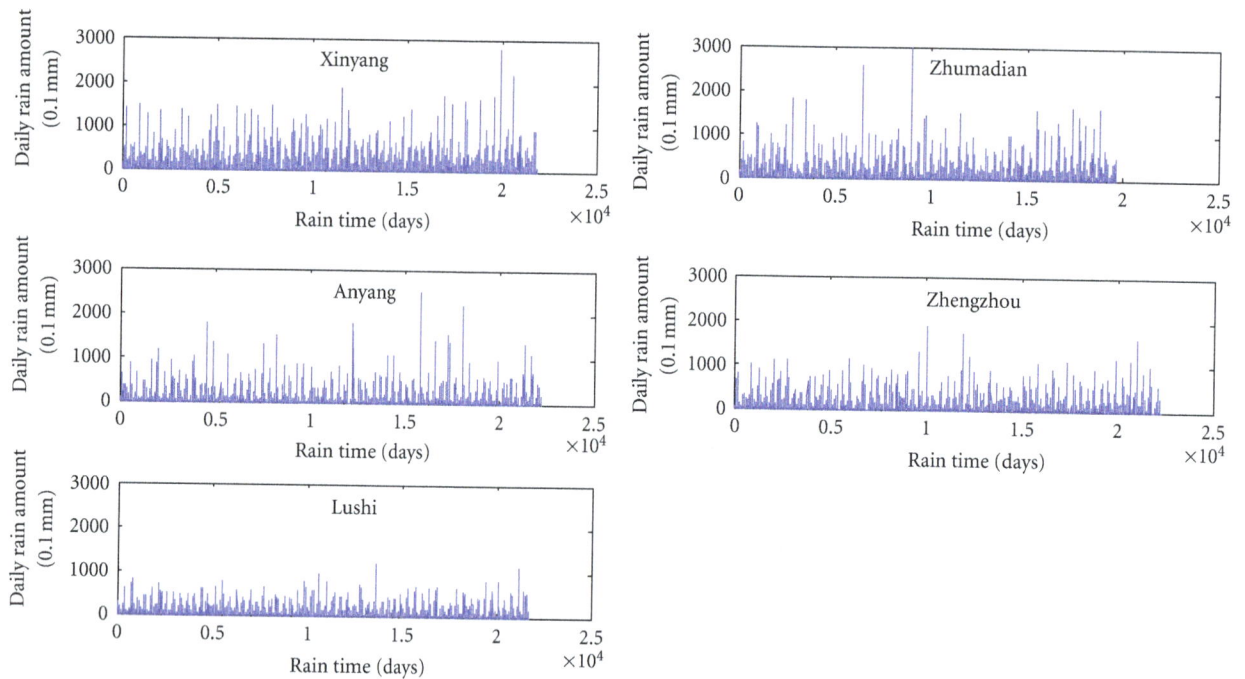

FIGURE 2: Plot of five rain series in Henan province.

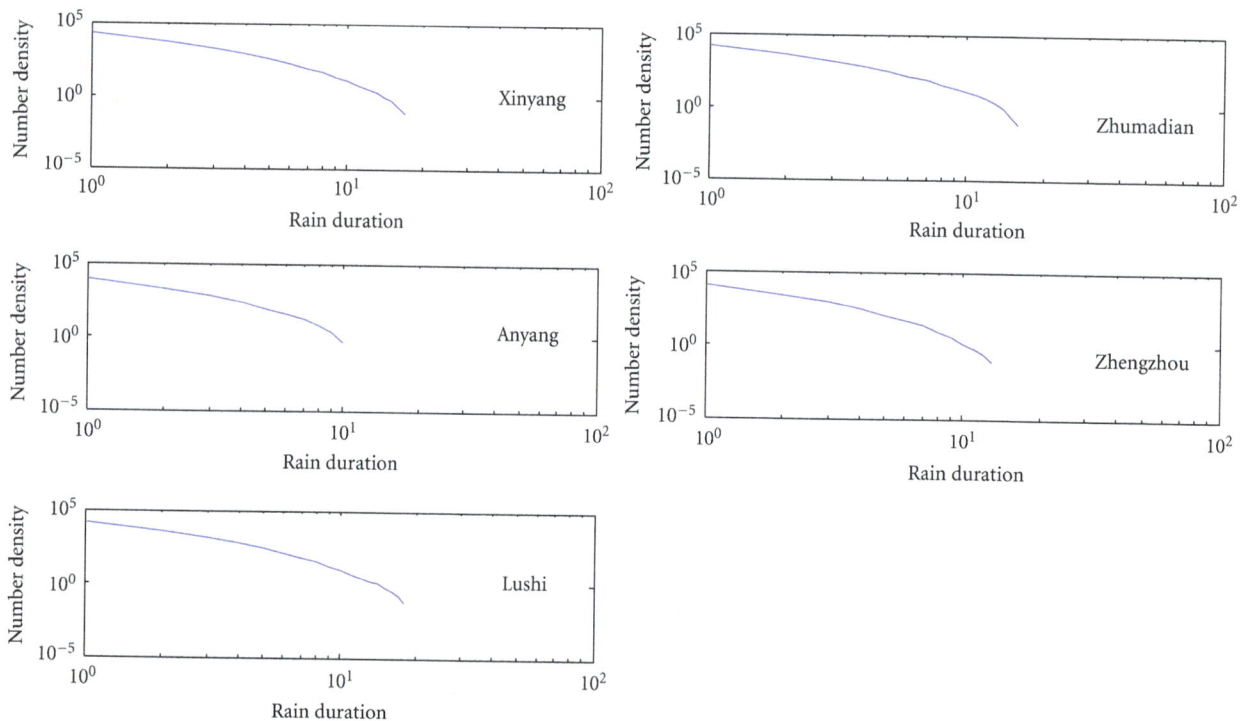

FIGURE 3: Log-log plot of the probability number density for rain duration.

4.2. Accumulated Rain Amount. Comparing with the analysis on rain duration and drought duration, it is more difficult to understand the behavior characteristics of long-term ARA dynamics process. To test if the ARA is scale invariant, we calculated the ARA number density in the partitioned intervals in terms of Algorithm 2.

The results show that all five series follow the power law relation in the scale range from 0.1 mm to 60 mm. Here $\tau_{xingyang} = -1.61$, $\tau_{zhumadian} = -1.63$, $\tau_{anyang} = -1.72$, $\tau_{zhengzhou} = -1.52$, $\tau_{lushi} = -1.55$. The ARA number density distributions of Xinyang and Zhumadian displays the similar profile since they are in the same

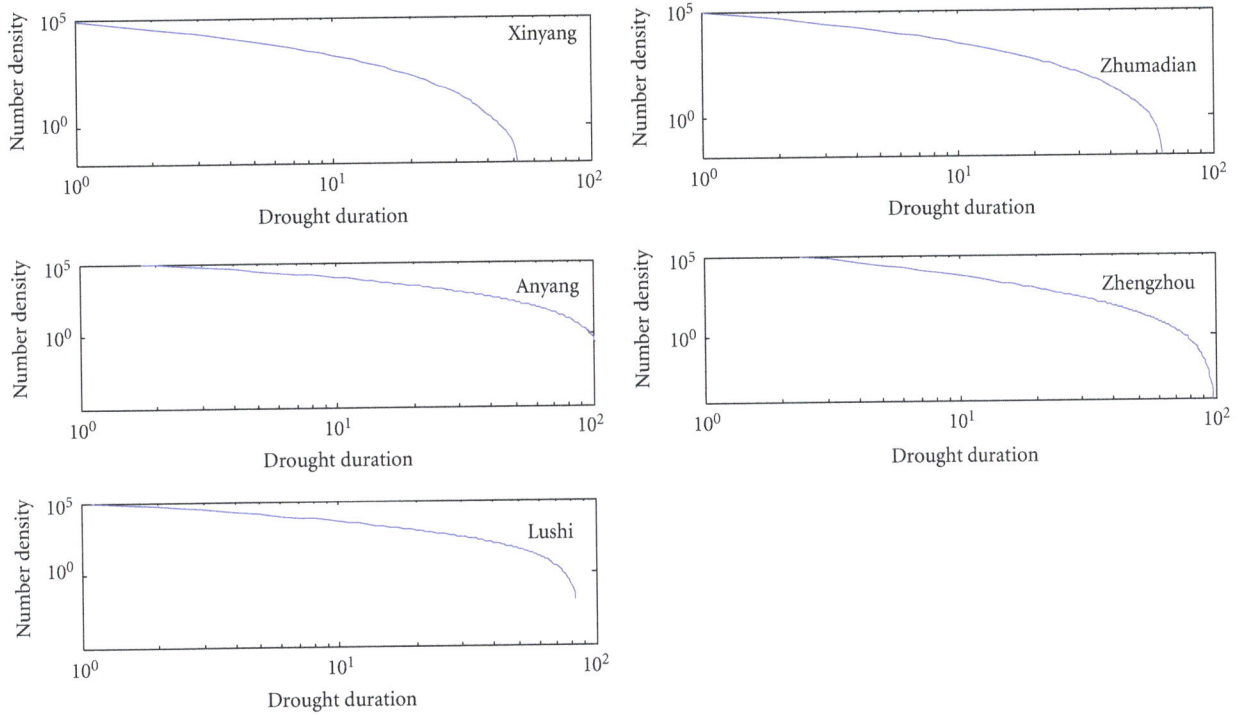

FIGURE 4: Log-log plot of the probability number density for drought duration.

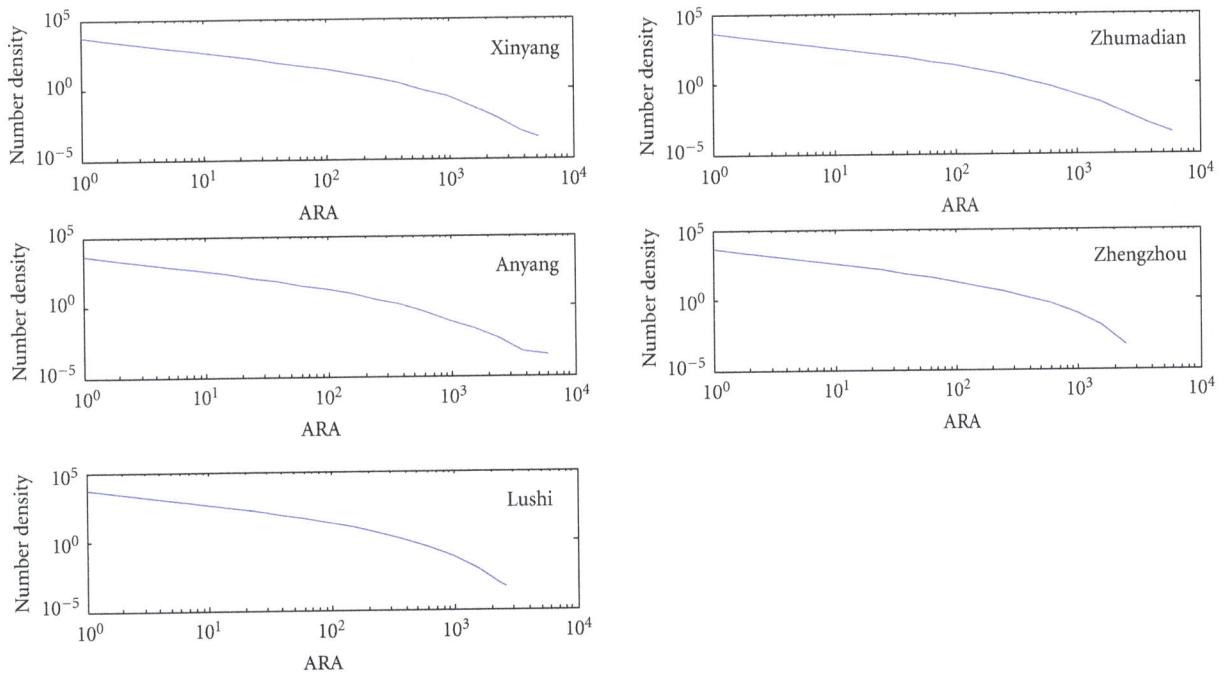

FIGURE 5: Log-log plot of the probability number density for accumulated rain amount.

climate classification region, the subtropical zone. Like Section 4.1, the feature of ARA in Anyang station is also different from the other stations. Although Lushi station is located in the western mountain region, however, the ARA behavior characteristic in Zhengzhou station is similar to it.

4.3. *Discussion.* In Sections 4.1 and 4.2 we have investigated the long-term series of precipitation records and succeeded in finding the evidence for power law distribution even though the five stations are located in different position in central China. This implies that our guess (i.e., the rainfall in central China takes on the feature of SOC) is right. This also

TABLE 3: Days of RD and DD; ARA (in 0.1 mm) per rain event.

RD	N (RD)	DD	N (DD)	DD	N (DD)	S.N	ARA	S.N	ARA	S.N	ARA
1	4309	1	14586	37	44	1	48	37	92	...	...
2	2427	2	11564	38	35	2	92	38	42	2501	15
3	1418	3	9553	39	27	3	7	39	18	2502	60
4	812	4	7615	40	21	4	61	40	55	2503	1
5	463	5	6205	41	17	5	39	41	14	2504	212
6	257	6	5084	42	13	6	8	42	20	2505	26
7	141	7	4179	43	11	7	6	43	88	2506	80
8	78	8	3448	44	9	8	3	44	1	2507	9
9	42	9	2868	45	8	9	42	45	320	2508	66
10	26	10	2425	46	7	10	35	46	95	2509	297
11	17	11	2072	47	6	11	75	47	56	2510	60
12	10	12	1783	48	5	12	58	48	296	2511	8
13	6	13	1542	49	4	13	780	49	68	2512	10
14	4	14	1333	50	3	14	85	50	38	2513	18
15	2	15	1161	51	2	15	7	51	93	2514	53
16	1	16	1105	52	1	16	278	52	16	2515	51
17	0	17	879	53	0	17	44	53	25	2516	223
18	0	18	762	54	0	18	468	54	121	2517	299
19	0	19	660	55	0	19	18	55	1	2518	66
20	0	20	566	56	0	20	555	56	653	2519	437
21	0	21	486	57	0	21	50	57	43	2520	213
22	0	22	421	58	0	22	59	58	30	2521	104
23	0	23	365	59	0	23	34	59	29	2522	11
24	0	24	317	60	0	24	21	60	79	2523	605
25	0	25	278	61	0	25	213	61	52	2524	381
26	0	26	241	62	0	26	225	62	237	2525	1561
27	0	27	211	63	0	27	13	63	1	2526	237
28	0	28	187	64	0	28	96	64	21	2527	355
29	0	29	164	65	0	29	11	65	8	2528	125
30	0	30	143	66	0	30	2	66	384	2529	267
31	0	31	125	67	0	31	812	67	47	2530	398
32	0	32	109	68	0	32	198	68	7	2531	422
33	0	33	95	69	0	33	7	69	15	2532	25
34	0	34	81	70	0	34	146	70	111	2533	0
35	0	35	68	71	0	35	189	71	71	2534	0
36	0	36	56	72	0	36	8	72	76	2535	0

supports the view that atmospheric dynamics is governed, at least in part, by SOC.

As described in the introduction, the concept of SOC refers to the state of nonequilibrium systems driven by slow constant energy input to organize themselves into critical systems. The intermediately stored energy is eventually released in sudden bursts with no typical scale. From this point of view, rainfall events which occur in the central China are not very different from those ones in the other place. Like earthquakes, a rainfall event is driven by a slow and constant energy input from the sun and water is evaporated from the Western Pacific Ocean. The energy is stored in the form of water vapor in the atmosphere. It is then suddenly released in bursts when the vapors condense to water drops. The

power-law distribution of the number density of rain events is equivalent to the Gutenberg-Richter law for earthquakes.

Although there are different geographic coordinates at the five stations, these τs show little change because the five stations are neighborhoods in one province after all. Except for the topography and landform the driving factors of rainfall, such as the sun radiation and the pattern of atmospheric circulation are nearly the same.

5. Conclusion

Testing the hypothesis, the existence of self-organized criticality in the long rain process in central China is our objective in this work. By calculating the number density of the rain duration, the drought duration, and the accumulated rain amount, we found that the relationship between the number density functions and rain duration, drought duration and accumulated amount exhibits the feature of power law. In other words, we have looked out the evidence of self-organized criticality in the long-term rain processes in central region of China. It is turned out that the long dry and wet process is indeed the complexity dynamics process.

Henan province is in the climate transition zone where the climate changes from the southern humid subtropical monsoon region to the northern semiarid temperature monsoon. It is also the topography transition zone. The western region is mountain and the eastern is the great plain. Because of these reasons, the weather and the climate change drastically in a year and interyears. This point has also been turned out from our results and plots. The feature of wet and dry spell is significantly different between Anyang station and Xinyang station.

In this work we only studied five meteorological stations due to the limitation of datasets. And the minimum time interval is a day which appears to be not small enough for the analysis of rainfall event. We will therefore collect more data sets in central China and other regions in the future work. The self-organized criticality needs to be tested in a small temporal interval and in a larger spatial scale, such as in the level of hour or second and in the level of the whole country. Furthermore, Algorithm 2 proposed should be improved on the number density computation of accumulated rain amount. The concept of number density of duration and intension should be redefined over and over again in the future work.

Acknowledgments

This work was founded by the Technologies R & D Program of Zhengzhou (Grant no. 0910SGYG21201-6, 43204-522) and Educational Commission of Henan Province of China (4113-521). The authors would like to thank Dr. Yanling Li for the helpful assistance. They thank their anonymous referees for their valuable comments and suggestions.

References

[1] J. Jiang, D. Zhang, and K. Fraedrich, "HIstoric climate variability of wetness in east China (960–1992): a wavelet analysis," *International Journal of Climatology*, vol. 17, no. 9, pp. 969–981, 1997.

[2] I. Bordi, K. Fraedrich, J. M. Jiang, and A. Sutera, "Spatio-temporal variability of dry and wet periods in eastern China," *Theoretical and Applied Climatology*, vol. 79, no. 1-2, pp. 81–91, 2004.

[3] Q. Wang, X. Liu, and A. Fang, "Mathematical model of rain fall forecast," in *Proceedings of the ETP International Conference on Future Computer and Communication (FCC '09)*, pp. 112–115, 2009.

[4] P.-S. Yu, C.-J. Chen, and S.-J. Chen, "Application of gray and fuzzy methods for rainfall forecasting," *Journal of Hydrologic Engineering*, vol. 5, no. 4, pp. 339–345, 2000.

[5] I. Rodriguez-Iturbe, B. Febres De Power, M. B. Sharifi, and K. P. Georgakakos, "Chaos in rainfall," *Water Resources Research*, vol. 25, no. 7, pp. 1667–1675, 1989.

[6] Z. Wang and W. Li, "Prediction of monthly precipitation in Kunming based on the chaotic time series analysis," in *Proceedings of the 4th Annual Meeting of Risk Analysis Council of China Association for Disaster Prevention*, Atlantis Press, 2010.

[7] Z. Wang and Y. Zhang, "Chaos analysis of time series of kunming annual precipitation," *Journal of North China Institute of Water Conservancy and Hydroelectric Power*, vol. 32, 2, pp. 8–10, 2011.

[8] S. Bellie, S.-Y. Liong, and C.-Y. Liaw, "Evidence of chaotic behavior in Singapore rainfall," *Journal of the American Water Resources Association*, vol. 34, 2, pp. 301–310, 1998.

[9] B. Sivakumar, S.-Y. Liong, C.-Y. Liaw, and K.-K. Phoon, "Singapore rainfall behavior: Chaotic?" vol. 4, no. 1, pp. 38–48, 1999.

[10] M. C. Valverde Ramírez, H. F. De Campos Velho, and N. J. Ferreira, "Artificial neural network technique for rainfall forecasting applied to the São Paulo region," *Journal of Hydrology*, vol. 301, no. 1–4, pp. 146–162, 2005.

[11] P. Bak, *How Nature Works: The Science of Self-Organized Criticality*, Springer, New York, NY, USA, 1996.

[12] P. Bak and S. Boettcher, "Self-organized criticality and punctuated equilibria," *Physica D*, vol. 107, no. 2-4, pp. 143–150, 1997.

[13] P. Bak, C. Tang, and K. Wiesenfeld, "Self-organized criticality: An explanation of the 1/f noise," *Physical Review Letters*, vol. 59, no. 4, pp. 381–384, 1987.

[14] C. Tang and P. Bak, "Critical exponents and scaling relations for self-organized critical phenomena," *Physical Review Letters*, vol. 60, no. 23, pp. 2347–2350, 1988.

[15] K. Christensen, Z. Olami, and P. Bak, "Deterministic 1/f noise in nonconserative models of self-organized criticality," *Physical Review Letters*, vol. 68, no. 16, pp. 2417–2420, 1992.

[16] S. C. Manrubia and R. V. Solé, "Self-organized criticality in rainforest dynamics," *Chaos, Solitons and Fractals*, vol. 7, no. 4, pp. 523–541, 1996.

[17] A. Sarkar and P. Barat, "Analysis of rainfall records in India: self-organized criticality and scaling," *Fractals*, vol. 14, no. 4, pp. 289–293, 2006.

[18] P. Lehmann and D. Or, "Concepts of Self-Organized Criticality for modeling triggering of shallow landslides," *Geophysical Research Abstracts*, vol. 10, no. 2, 2008.

[19] M. J. Van De Wiel and T. J. Coulthard, "Self-organized criticality in river basins: challenging sedimentary records of environmental change," *Geology*, vol. 38, no. 1, pp. 87–90, 2010.

[20] O. Peters and K. Christensen, "Rain viewed as relaxational events," *Journal of Hydrology*, vol. 328, no. 1-2, pp. 46–55, 2006.

[21] O. Peters and J. D. Neelin, "Critical phenomena in atmospheric precipitation," *Nature Physics*, vol. 2, no. 6, pp. 393–396, 2006.

[22] R. F. S. Andrade, H. J. Schellnhuber, and M. Claussen, "Analysis of rainfall records: possible relation to self-organized criticality," *Physica A*, vol. 254, no. 3-4, pp. 557–568, 1998.

[23] O. Peters and K. Christensen, "Rain: relaxations in the sky," *Physical Review E*, vol. 66, no. 3, Article ID 036120, 9 pages, 2002.

[24] O. Peters, C. Hertlein, and K. Christensen, "A complexity view of rainfall," *Physical Review Letters*, vol. 88, no. 1, Article ID 018701, 4 pages, 2002.

[25] G. Pruessner and O. Peters, "Self-organized criticality and absorbing states: lessons from the Ising model," *Physical Review E*, vol. 73, no. 2, Article ID 025106, 4 pages, 2006.

9

Ocean Cooling Pattern at the Last Glacial Maximum

Kelin Zhuang[1] and John R. Giardino[2]

[1] *University of Arizona, Tucson, Arizona, AZ 85721, USA*
[2] *Texas A&M University, College Station, Texas, TX 77843, USA*

Correspondence should be addressed to Kelin Zhuang, klzhuang@hotmail.com

Academic Editor: Youmin Tang

Ocean temperature and ocean heat content change are analyzed based on four PMIP3 model results at the Last Glacial Maximum relative to the prehistorical run. Ocean cooling mostly occurs in the upper 1000 m depth and varies spatially in the tropical and temperate zones. The Atlantic Ocean experiences greater cooling than the rest of the ocean basins. Ocean cooling is closely related to the weakening of meridional overturning circulation and enhanced intrusion of Antarctic Bottom Water into the North Atlantic.

1. Introduction

The cooling at the Last Glacial Maximum has been extensively studied geologically and numerically (e.g., [1–5]) where proxy data and numerical modeling were both employed to explore the climate sensitivity and mechanism. The Paleoclimate Modeling Intercomparison Project (PMIP2) presented large-scale features (e.g., [6]). However, the spatial pattern of cooling at that period was little studied previously. In this paper we use the newly released PMIP3 data to study the cooling pattern at the Last Glacial Maximum.

2. Methods

We analyze ocean potential temperature anomaly, ocean heat content (OHC) change, and meridional overturning mass stream function based on four available PMIP3 models of IPSL-CM5A-LR, MIROC-ESM, MPI-ESM-P, and MRI-CGCM3 so far (http://cmip-pcmdi.llnl.gov/cmip5/), which follow the PMIP3 21ka experimental design (Table 1; http://pmip3.lsce.ipsl.fr/) and make a comparison between the last 50 years of Last Glacial Maximum (LGM) experiments relative to the base period of the last 50 years of the preindustrial control run.

We calculate the temperature anomaly, meridional overturning mass stream function anomaly, and ocean heat content change of each model by regridding all the temperature data into a common 1° × 1° grid.

3. Results

3.1. Geographical Distribution. Ocean cooling shows pronounced spatial variations (Figure 1), both horizontally and vertically. The Northern Hemisphere exhibits a stronger cooling than the Southern Hemisphere of IPSL-CM5A-LR, MIROC-ESM, and MPI-ESM-P except MRI-CGCM3 with stronger cooling in the Southern Ocean revealing a notable north-south asymmetry.

At the surface (Figures 1(a) and 1(b)), significant ocean cooling (<−5°C) occurs in the North Pacific and North Atlantic of the first three models. All these three models have shown the maximum cooling regions in the North Pacific around 40°N and in the Nordic Seas. But, MRI-CGCM3 demonstrates a significant cooling around 60°S in the Southern Ocean. It is of note that south of Greenland the model even shows a slight warming in shallow water.

At the subsurface 500 m layer (Figure 1(c)) cooling in the Pacific around 40°N and the North Atlantic still maintain

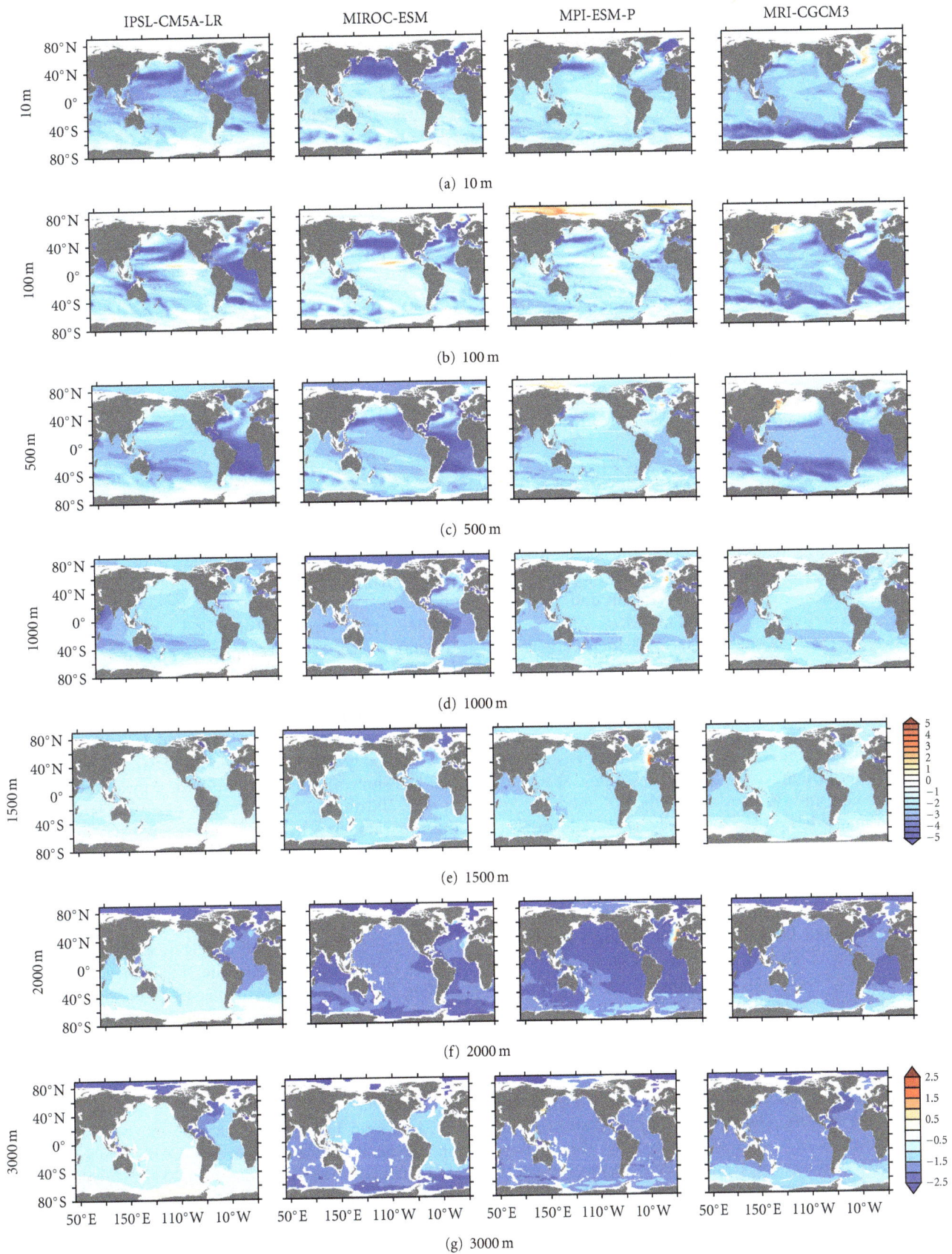

FIGURE 1: Geographical distribution of potential temperature anomaly in °C at different water depths. The abscissa is longitude in degrees and the ordinate is latitude in degrees. See text for detailed description.

TABLE 1: PMIP3 LGM models.

Model	Atmosphere	Ocean	Data output coverage
IPSL-CM5A-LR	$96 \times 95 \times L39$	$182 \times 149 \times L31$	4600–4699 (100 years)
MIROC-ESM	$128 \times 64 \times L80$	$256 \times 192 \times L44$	2501–2600 (100 years)
MPI-ESM-P	$196 \times 98 \times L47$	$256 \times 220 \times L40$	1850–1949 (100 years)
MRI-CGCM3	$320 \times 160 \times L48$	$364 \times 368 \times L51$	2501–2600 (100 years)

TABLE 2: Ocean heat content change ($\times 10^{24}$ J).

	Whole water depth				0–2000 m			
	IPSL	MIROC	MPI	MRI	IPSL	MIROC	MPI	MRI
Global	−7.39	−9.96	−10.70	−11.25	−5.43	−6.08	−5.78	−6.32
Atlantic	−1.88	−1.88	−1.86	−2.23	−1.31	−1.32	−0.94	−1.27
Arctic	−0.14	−0.13	−0.10	−0.08	−0.09	−0.09	−0.07	−0.04
Pacific	−2.78	−4.08	−4.41	−4.59	−2.04	−2.46	−2.30	−2.33
Indian	−1.08	−1.18	−1.45	−1.58	−0.82	−0.72	−0.82	−0.96
Southern	−1.31	−2.72	−2.69	−2.60	−0.98	−1.47	−1.46	−1.56

the surface cooling pattern. However, the South Atlantic demonstrates the strongest cooling in IPSL, MIROC, and MRI models. Strong cooling in the Indian and Southern Oceans is also presented in the MRI model.

The water body between 1000–2000 m experiences a moderate cooling of −2°C (Figures 1(d), 1(e) and 1(f)). All the four models have shown that north of the Gulf Stream region and South Atlantic experience more significant cooling than the rest of the regions. The Southern Ocean around 40°S has a larger amount of cooling as well. It is of note that the eastern boundary of the North Atlantic shows the slightest cooling and even warming in the MPI model.

The four models differ remarkably below 2000 m (Figures 1(f) and 1(g)). The IPSL model only shows a cooling of about −1.0°C in the large expanse with a maximum cooling of −2.5°C south of the Greenland whilst the Southern Ocean shows merely slight cooling. But, the other three models of MIROC, MPI, and MRI show a much stronger cooling of greater than −2°C. The MIROC shows the strongest cooling in the Southern Ocean. But, MRI shows strong cooling in the Indian and Southern Oceans.

3.2. Latitudinal Cross Sections and Depth Variations.
Figure 2 reveals that large cooling occurs in the upper 1000 m. The temperate and tropical zones between 40°S–40°N are marked by subsurface cooling. The northern high latitude north of 40°N experiences maximum surface cooling. It is of note that the deep Arctic Ocean experiences the largest cooling in MIROC and MRI whereas IPSL and MPI only show merely a subsurface cooling.

Ocean basins act differently. The cooling in the Atlantic (Figure 2(b)) penetrates to 3000 m with a strikingly subsurface cooling center between 20°S and 20°N. 40°N–70°N shows a surface cooling.

Compared to the Atlantic and Arctic, the Pacific, and Indian demonstrate a shallower cooling. All four models show a maximum cooling around 40°N. However, the cooling only penetrates to 600 m in IPSL and MPI, but the other two models have a deeper penetration depth.

Profiles of temperature anomaly in different ocean basins indicate that the largest surface cooling occurs in the North Atlantic (Figure 3). IPSL and MRI show a global average cooling of −2.5°C whilst MIROC and MPI have a cooling of −2°C. The North and South Atlantic in IPSL and MIROC show a similar pattern with a maximum of −5°C at 400–500 m. The Arctic cooling has a similar pattern in the four models where the coolest zone lies in the deep ocean. The other basins all exhibit a rapid subsurface cooling variation in the upper ocean.

3.3. Ocean Heat Content.
Ocean heat content (OHC) change reveals the integrated cooling. Figure 4 reveals the OHC change of the four models. The IPSL model shows most cooling happens in the Atlantic, parts of the Pacific and Indian. But MIROC, MPI, and MRI models demonstrate a much more significant cooling in the Pacific and Indian. Figure 5 lists the average and total OHC change. The North Pacific and Southern Ocean have the strongest change of average OHC.

More than half of OHC change occurs above 2000 m. Like temperature change in oceans, OHC change also takes place mainly in the upper ocean to 2000 m (Table 2). On a global scale, the upper 2000 m accounts for 73%, 61%, 54%, and 56% in IPSL, MIROC, MPI, and MRI, respectively.

OHC change has its individual feature in the five Ocean basins. The OHC change in Atlantic, Arctic, and Indian is almost the same for the four models but differs in the North Pacific and Southern Ocean, which also affect the total OHC

(a) Global

(b) Atlantic + Arctic

(c) Pacific + Indian

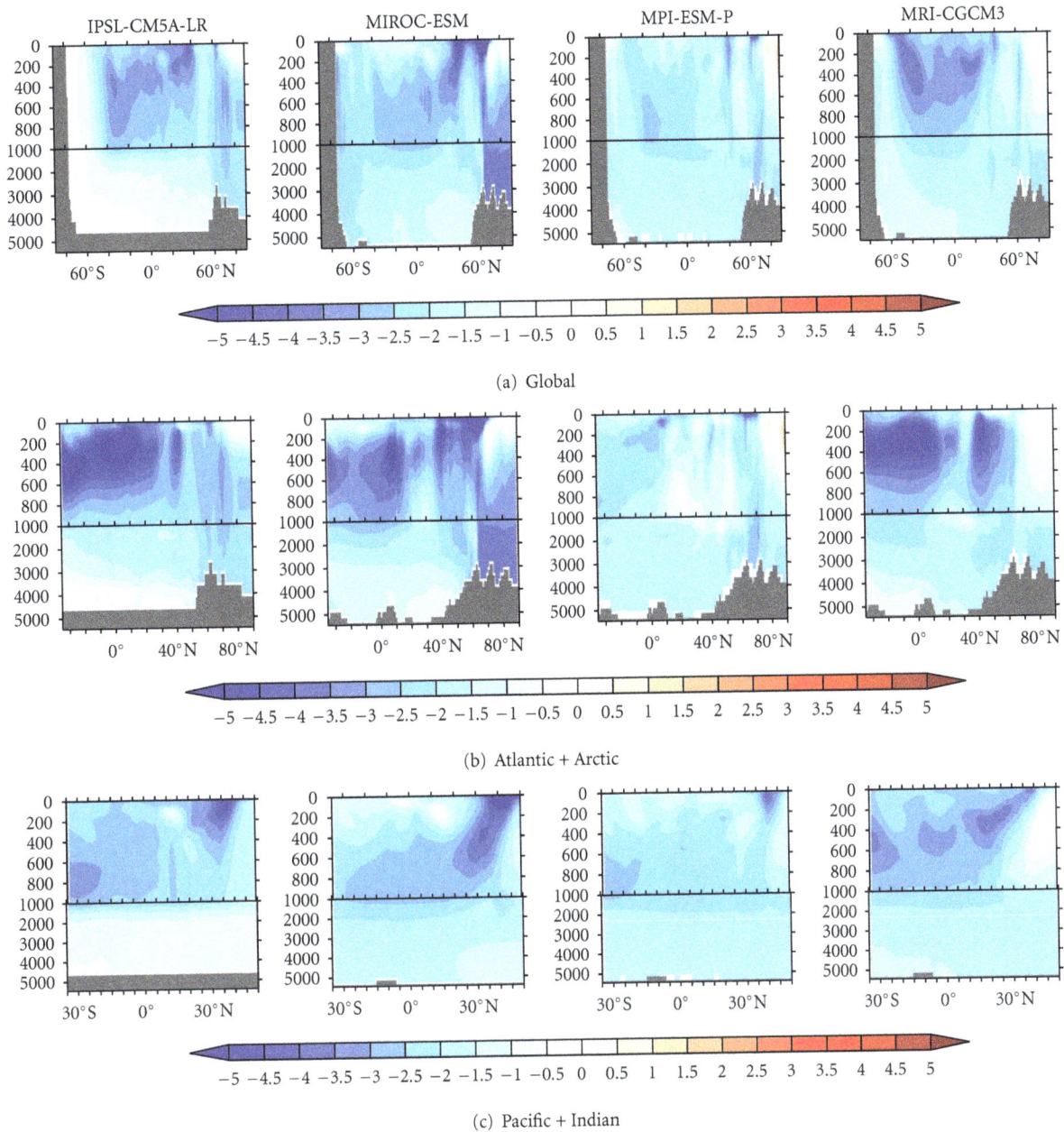

FIGURE 2: Latitudinal cross-sections of global mean (a) and ocean basins (b through c). The abscissa is latitude in degrees and the ordinate is water depth in meters. Notice that depth is unevenly distributed. The temperature anomaly is in °C. Cooling occurs mostly in the upper ocean. Cooling in the Atlantic and Arctic Oceans is deeper than in the Pacific and Indian Oceans.

change (Figure 5(b)). It is of note that the 2000 m OHC change is similar for all the four models and, therefore, the OHC change difference in the Pacific and Southern Ocean lies in the deep ocean.

4. Discussions

Unfortunately, not all PMIP3 models output ocean meridional overturning mass stream function. Among the four released model results, only MPI and MRI have meridional overturning variables. Meridional overturning circulation (MOC) changes are usually the responses to climate change. Here we explore the links between the MOC change and temperature anomaly on the global scale and regional scale at the Last Glacial Maximum.

Previous studies have revealed that cooling at the Last Glacial Maximum is closely related to meridional overturning circulation (MOC) [7–9]. They simulated a shallower and weaker North Atlantic Deep Water circulation and an enhanced intrusion of Antarctic Bottom Water (AABW) into

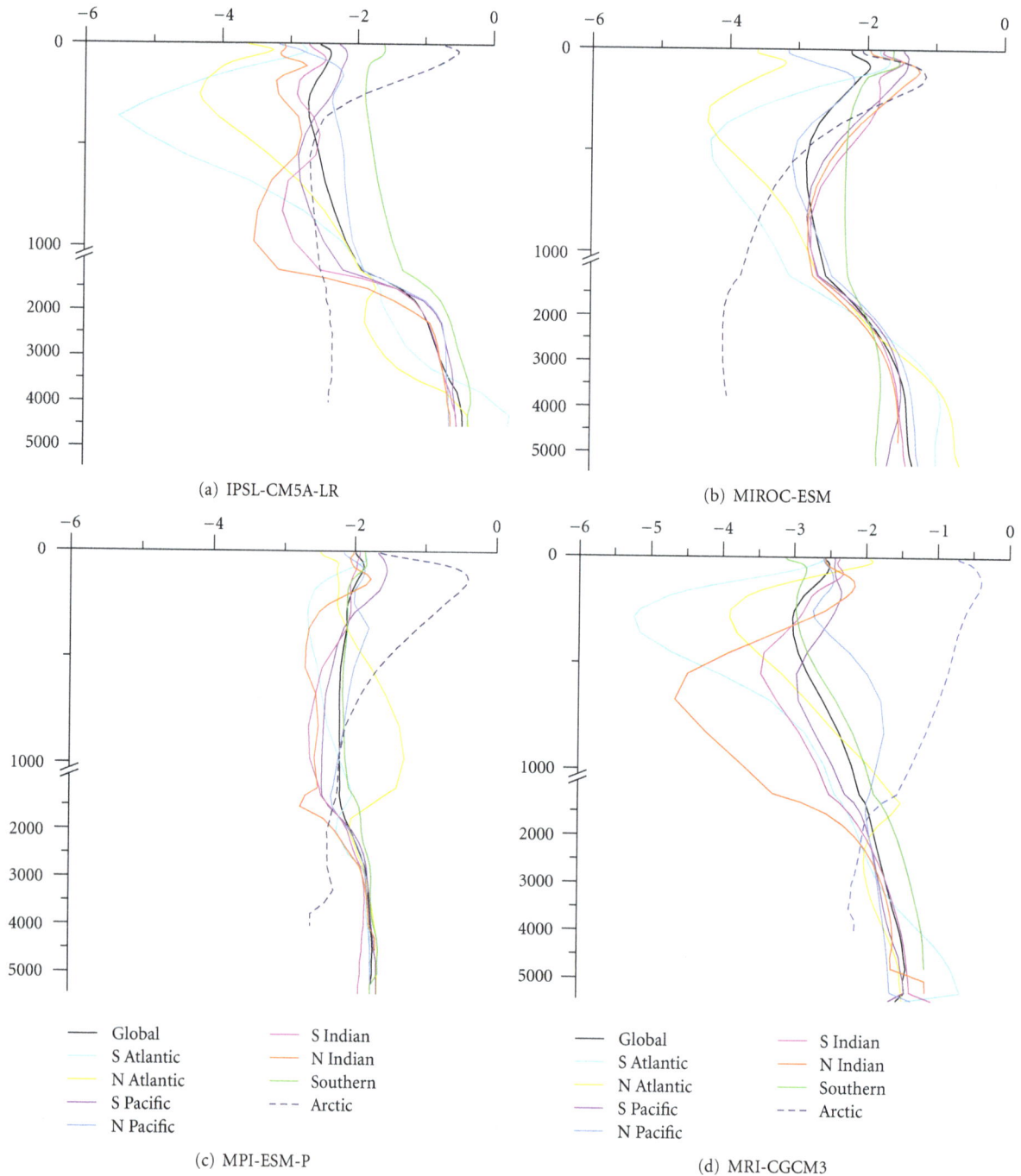

(a) IPSL-CM5A-LR

(b) MIROC-ESM

(c) MPI-ESM-P

(d) MRI-CGCM3

FIGURE 3: Ocean cooling profiles of global mean and ocean basins. The abscissa is temperature anomaly in °C and the ordinate is water depth in m. The maximum surface warming occurs in the North Atlantic. Ocean cooling mostly concentrates in the upper 1000 m with rapid variations in different basins.

the North Atlantic [10]. Ganopolski and Rahmstorf [8] proposed a cold mode MOC to explain the cooling mechanism at the Last Glacial Maximum.

On a global scale the upper 2000 m witnesses a weakening trend of the MOC although the magnitude varies in the two models (Figure 6(a)). MPI presents a negative MOC anomaly whilst MRI only shows a slight decrease of MOC compared to the preindustrial run. In the deep part we can

observe the enhancement of AABM. Both models exhibit the weakening of the Atlantic meridional overturning circulation (Figure 6(b)). Figures 1 and 2 have different temperature distributions with regard to MPI and MRI. Cooling in the upper Pacific and Indian Oceans simulated by MPI is not as strong as cooling by MRI, which is closely related to the enhanced MOC change in MPI upper ocean while MRI upper part shows a weakening MOC change.

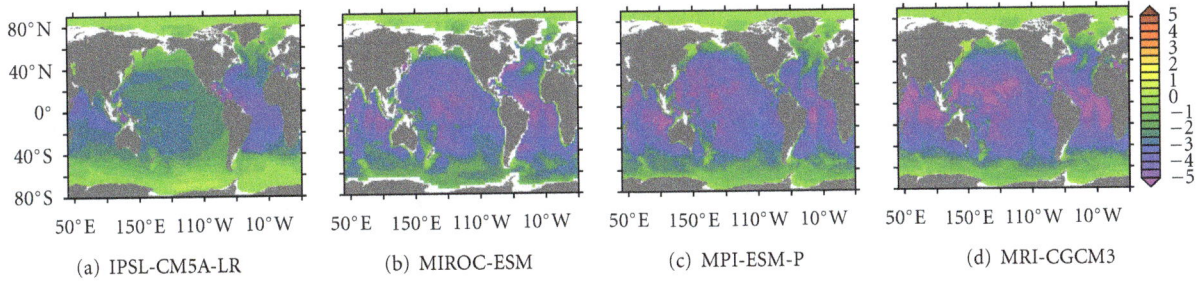

FIGURE 4: Ocean heat content change ($\times 10^{10}$ Jm^{-2}) at the Last Glacial Maximum. Most cooling occurs between 40°S and 40°N. All the four models show cooling in the Atlantic. But, cooling in the Pacific and Indian Oceans varies.

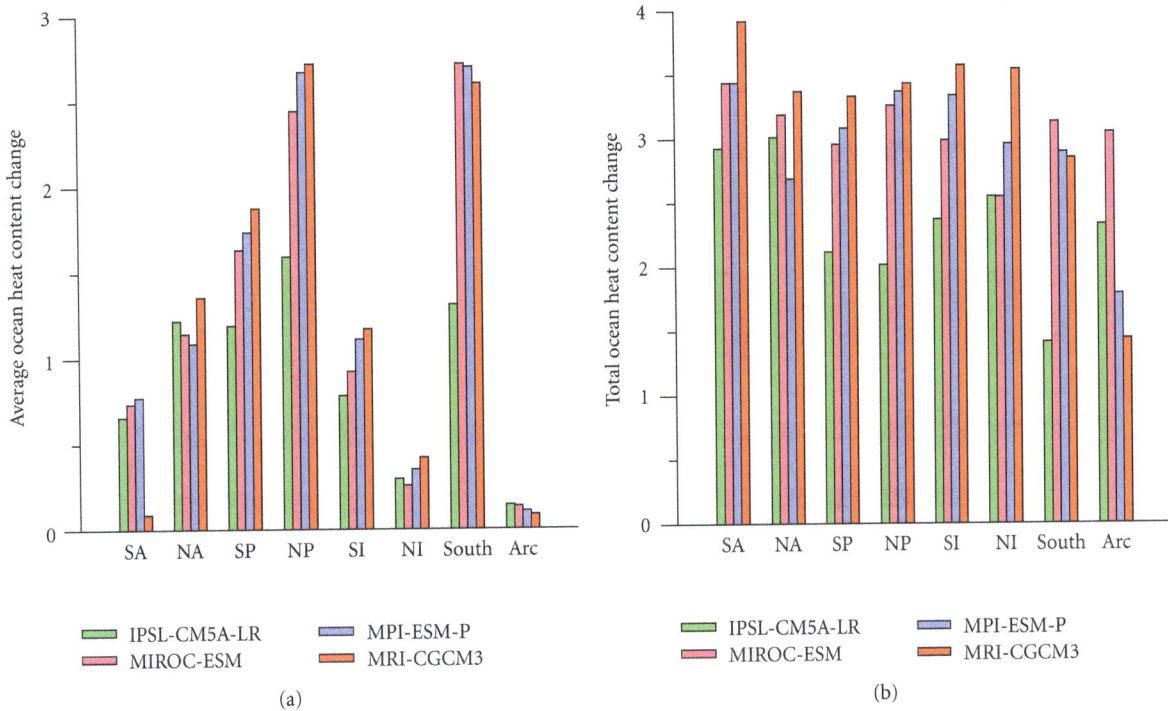

FIGURE 5: Ocean heat content change in ocean basins. (a) Average OHC change ($\times 10^{10}$ Jm^{-2}) and (b) total OHC change ($\times 10^{24}$ J). Cooling in the North Pacific and Southern Ocean differs in magnitude among the four models. SA—South Atlantic, NA—North Atlantic, SP—South Pacific, NP—North Pacific, SI—South Indian, NI—North Indian, South—Southern Ocean, Arc—Arctic.

Based on the MOC change and temperature anomaly in these two models, we can see that weakening of MOC in the upper ocean is closely related to upper cooling and enhancement of AABW in the deep part is associated with cooling in the deep ocean.

5. Summaries

Although the four models differ in cooling magnitude, we see that cooling at the Last Glacial Maximum varies both horizontally and vertically. Cooling mostly occurs between 40°S and 40°N in the upper 1000 m. More than half of OHC change happens in the upper 2000 m.

All the four models are in agreement that ocean basins, except the Arctic, are featured by surface to subsurface cooling. Cooling in the Atlantic and Arctic is much deeper than the Pacific and Indian. The Atlantic experiences the greatest cooling at the Last Glacial Maximum. Ocean cooling at the Last Glacial Maximum is closely related to the MOC change.

Acknowledgments

We acknowledge the modeling groups, the Program for Climate Model Diagnosis and Intercomparison (PCMDI) and the WCRP's Working Group on Coupled Modeling

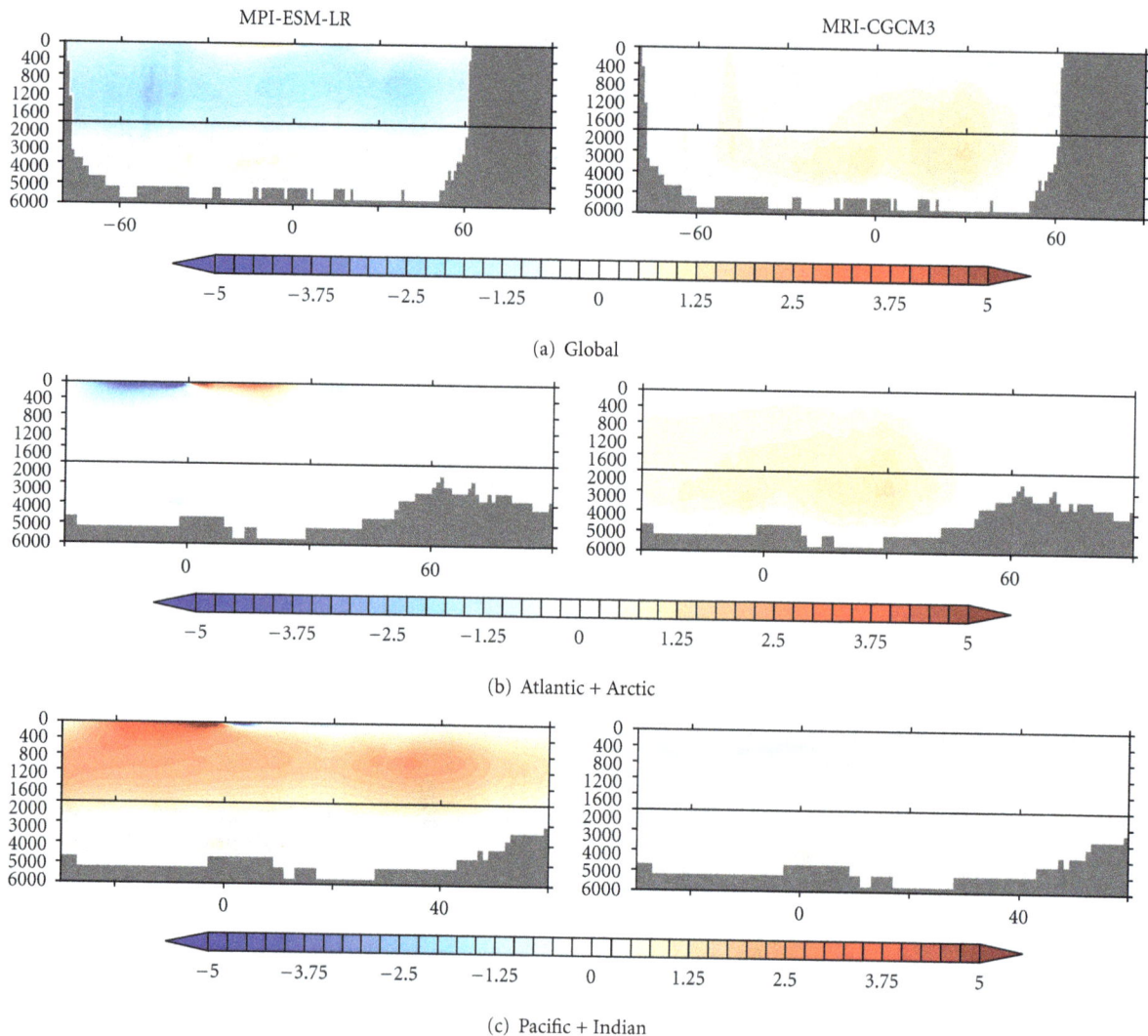

FIGURE 6: Ocean meridional overturning mass-stream function anomaly of global mean (a) and ocean basins (b through c). The abscissa is latitude in degrees and the ordinate is water depth in meters. Notice that depth is unevenly distributed. The overturning mass-stream function anomaly is in $\times 10^{10}$ kgs^{-1}. MOC weakens in the upper ocean whereas the AABW is enhanced. It is of note that MOC change in the Pacific and Indian Oceans are different in MPI and MRI.

(WGCM) for their roles in making available the WCRP CMIP5 multimodel dataset. Support of this dataset is provided by the Office of Science, US Department of Energy.

References

[1] A. Schmittner, N. M. Urban, J. D. Shakun et al., "Climate sensitivity estimated from temperature reconstructions of the Last Glacial Maximum," *Science*, vol. 334, pp. 1385–1388, 2011.

[2] Y. Okazaki, A. Timmermann, L. Menviel et al., "Deepwater formation in the North Pacific during the last glacial termination," *Science*, vol. 329, no. 5988, pp. 200–204, 2010.

[3] MARGO Project Members, "Constraints on the magnitude and patterns of ocean cooling at the Last Glacial Maximum," *Nature Geoscience*, vol. 2, pp. 127–132, 2009.

[4] M. Kageyama, A. Laîné, A. Abe-Ouchi et al., "Last Glacial Maximum temperatures over the North Atlantic, Europe and western Siberia: a comparison between PMIP models,

MARGO sea-surface temperatures and pollen-based reconstructions," *Quaternary Science Reviews*, vol. 25, no. 17-18, pp. 2082–2102, 2006.

[5] A. J. Weaver, M. Eby, A. F. Fanning, and E. C. Wiebe, "Simulated influence of carbon dioxide, orbital forcing and ice sheets on the climate of the Last Glacial Maximum," *Nature*, vol. 394, no. 6696, pp. 847–853, 1998.

[6] P. Braconnot, B. Otto-Bliesner, S. Harrison et al., "Results of PMIP2 coupled simulations of the Mid-Holocene and last glacial maximum—part 1: experiments and large-scale features," *Climate of the Past*, vol. 3, no. 2, pp. 261–277, 2007.

[7] A. Ganopolski, S. Rahmstorf, V. Petoukhov, and M. Claussen, "Simulation of modern and glacial climates with a coupled global model of intermediate complexity," *Nature*, vol. 391, no. 6665, pp. 351–356, 1998.

[8] A. Ganopolski and S. Rahmstorf, "Rapid changes of glacial climate simulated in a coupled climate model," *Nature*, vol. 409, no. 6817, pp. 153–158, 2001.

[9] A. F. Thompson and S. Rahmstorf, "Ocean circulation," in *Surface Ocean—Lower Atmosphere Processes*, C. L. Quere and E. S. Saltzman, Eds., Geographical Monograph 187, pp. 99–118, American Geophysical Union, 2009.

[10] S. I. Shin, Z. Liu, B. L. Otto-Bliesner, J. E. Kutzbach, and S. J. Vavrus, "Southern Ocean sea-ice control of the glacial North Atlantic thermohaline circulation," *Geophysical Research Letters*, vol. 30, no. 2, pp. 68–1, 2003.

10

Tracing Atlantic Water Signature in the Arctic Sea Ice Cover East of Svalbard

Vladimir V. Ivanov,[1, 2, 3] **Vladimir A. Alexeev,**[2] **Irina Repina,**[4]
Nikolay V. Koldunov,[5] **and Alexander Smirnov**[1]

[1] *Arctic and Antarctic Research Institute, St. Petersburg 199397, Russia*
[2] *International Arctic Research Centre, University of Alaska, Fairbanks, AK 99775, USA*
[3] *Scottish Marine Institute, Oban PA37 1 QA, UK*
[4] *A.M. Obukhov Institute of Atmospheric Physics of RAS, Moscow 119017, Russia*
[5] *Institute of Oceanography, University of Hamburg, 20146, Hamburg, Germany*

Correspondence should be addressed to Vladimir V. Ivanov, vladimir.ivanov@sams.ac.uk

Academic Editor: Igor N. Esau

We focus on the Arctic Ocean between Svalbard and Franz Joseph Land in order to elucidate the possible role of Atlantic water (AW) inflow in shaping ice conditions. Ice conditions substantially affect the temperature regime of the Spitsbergen archipelago, particularly in winter. We test the hypothesis that intensive vertical mixing at the upper AW boundary releases substantial heat upwards that eventually reaches the under-ice water layer, thinning the ice cover. We examine spatial and temporal variation of ice concentration against time series of wind, air temperature, and AW temperature. Analysis of 1979–2011 ice properties revealed a general tendency of decreasing ice concentration that commenced after the mid-1990s. AW temperature time series in Fram Strait feature a monotonic increase after the mid-1990s, consistent with shrinking ice cover. Ice thins due to increased sensible heat flux from AW; ice erosion from below allows wind and local currents to more effectively break ice. The winter spatial pattern of sea ice concentration is collocated with patterns of surface heat flux anomalies. Winter minimum sea ice thickness occurs in the ice pack interior above the AW path, clearly indicating AW influence on ice thickness. Our study indicates that in the AW inflow region heat flux from the ocean reduces the ice thickness.

1. Introduction

Steady reduction of the Arctic sea ice cover throughout 1990s has accelerated in the 2000s [1, 2]. As demonstrated in the recent studies, causative mechanisms for the extreme ice area/volume decay include an anomalous atmospheric circulation which forced ice out of the Canadian Basin towards Fram Strait [3, 4], the influence of warm inflow through Bering/Fram straits [5, 6], and the melting effect of warmed surface water [7]. However, long-term preconditioning occurred during three decades of steady ice thinning [8, 9]. This was largely a result of the fact that the Arctic has warmed up about twice as fast as lower latitudes due to the so-called polar amplification [10–12]. Years of reduced ice growth in winter and enhanced ice melt in summer led to the dominance of first-year ice over multiyear ice after 2004 [13].

Under conditions of enhanced seasonality, the influence of ocean heat on Arctic ice cover is expected to grow. The retreating summer ice edge increases the size of the marginal ice zones (MIZs)—the transient areas between open water and totally ice-covered ocean. For the Spitsbergen region, this process is particularly important due to the existence of an extended open water area (a quasi-steady-state polynya) bordered by an MIZ, the so-called Whalers Bay, close to the northern coast of the archipelago. In winter the presence of the large-scale open water zone substantially shapes local weather conditions, keeping air temperature well above average values for similar latitudes around the Arctic.

Large changes in the state of the ocean surface over the limited distance of MIZs build up high horizontal gradients of properties in the oceanic and atmospheric boundary layers below and above the MIZ. High gradients trigger horizontal motions in both media, providing favourable prerequisite conditions for intensive heat, moisture, and momentum exchange across the ocean-ice-air interface. This is true for the Pacific sector (circa 120°E–120°W), where the most dramatic ice edge retreat was reported in 2007 and 2011 (http://nsidc.org/arcticseaicenews/). On the opposite side of the Arctic Ocean the ice edge deviation from the climatic mean location was substantially smaller, for example, Figure 1 in [7]. Such anisotropy indicates that despite the fact that the strongest heat input to the high Arctic is associated with the eastward moving Atlantic cyclones and warm inflow of Atlantic-origin water through the Nordic Seas, the ice cover in the Atlantic sector seems to be rather insensitive to the increased heat impact from the lower latitudes [14]. In the present study, we use observational/reanalysis data and recent findings on the properties of the Atlantic water (AW) inflow to figure out whether this is actually the case.

We focus on the region of the Arctic Ocean between Svalbard and Severnaya Zemlya archipelagos, which is further referred to as the Western Nansen Basin (WNB: 15–60°E, 81–83°N). This is an area of complex ocean-ice-atmosphere interactions resulting in isolation of the inflowing AW from direct contact with ice and atmosphere. We base our study on the hypothesis that this isolation is primarily the consequence of intensive vertical mixing at the upper AW boundary. As a result, a substantial fraction of heat is released upwards contributing to the heat budget of the under-ice water layer and impacting the ice cover. Recent findings show conservation of a strong seasonal signal in the AW temperature at the location where the warm current encounters pack ice [15]. This conservation allows extensive penetration of warm "summer" AW into the subsurface layer (above 100 m) below the pack ice. We examine spatial and temporal variation of WNB ice concentration against relevant time series of wind, air temperature, and AW temperature. The main objective of this analysis is to separate the direct dynamic influence of wind from thermodynamic effects provided by heat fluxes at the ice-air and ice-water interfaces in order to assess the relative importance of the latter.

Following the paper objective, we introduce the physical concept of AW transformation into the intermediate water mass (Section 2), describe temporal and spatial variations of WNB ice concentration from 1979–2011 (Section 3), and discuss possible links between these variations and the most probable influential factors, including wind, air temperature, and AW temperature (Section 4). Discussion of the results and major conclusions are given in Section 5.

2. Atlantic Water Transformation East of Svalbard

A schematic of the two inflow branches is shown in Figure 1. The Barents sea branch of Atlantic water (BSBW) stays at the surface while in the Barents Sea. As a result, after this water finally reaches the Arctic Ocean interior it has substantially cooled and freshened. Contrary to BSBW, the Fram Strait branch of Atlantic water (FSBW) rapidly leaves the surface,

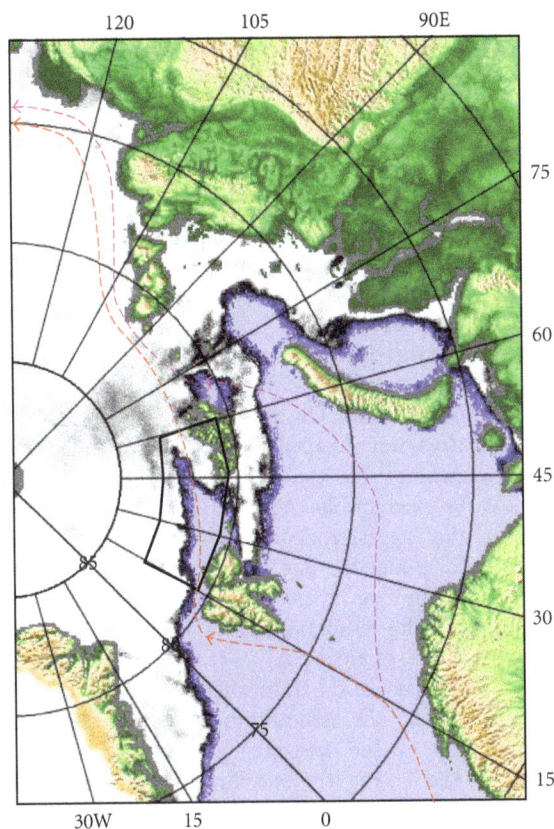

FIGURE 1: Schematic of two inflow branches (FSBW is shown in red and BSBW is shown in pink) of the AW against the special sensor microwave imager/Sounder (SSMIS) visual chart of ice concentration on February 4, 20112 (http://www.iup.uni-bremen.de:8084/ssmis/index.html). The Western Nansen Basin (WNB) area is marked by a black trapezium.

transforming into Arctic intermediate water (AIW). Due to this separation from the surface, AIW retains a large amount of its initial heat and salt, which are further transported around the deep ocean interior.

According to the Arctic Ocean climatology, north of Svalbard FSBW is capped by a cold and relatively fresh mixed water layer all year round [16]. The origin of this mixed layer is still under discussion. The traditional hypothesis states that this layer contains Arctic surface water which moves generally towards Fram Strait, opposing the FSBW inflow. Since this water is lighter than FSBW, it overlays the latter causing FSBW to sink beneath the surface mixed layer [17]. Westward flow in the surface mixed layer is consistent with the large-scale ice motion observed in the Arctic Ocean, for example Figure 1(e) in [18]. An alternative hypothesis suggests that the surface mixed layer originates directly in the upper part of inflowing AW, which cools down via heat loss to the atmosphere and freshens due to mixing with melted ice water [19]. This newly formed surface layer follows the warm bulk of AIW moving eastward along the continental slope, except for a very thin under-ice layer, which is deflected westwards by the drifting ice. Basically, both hypotheses agree that there must be some depth inside the water column at which the current turns around. The difference between these hypotheses is the depth at which this change of direction occurs. Observational data collected using traditional oceanographic methods, like occasional conductivity/temperature/depth (CTD) profiling at a limited number of transects, does not provide reliable justification for either of these hypotheses. For example, the typical inclination of isotherms between Svalbard and Franz Joseph Land (Figure 2) may be caused either by submerging of the AW water as it travels from west to east, or by cooling of the upper part of AW en route.

Recent measurements made at the autonomous moored station within the framework of the Nansen and Amundsen Basins Observational System (NABOS) project (http://www.iarc.uaf.edu/nabos.php) revealed strong seasonal variability in the inflowing FSBW (Table 1), which is conserved in the AIW far to the east of Fram Strait [20]. We suggest that this new knowledge provides some clue to understanding the FSBW transformation process. Applying the data from Table 1, we estimate that to increase the temperature of the $H = 217$ m thick layer over a unit square by $\Delta\overline{T} = 4.76$ K (the difference between the May and November vertically averaged temperature, under the assumption that the ocean surface is permanently at the freezing point), 4.2×10^9 J of heat is required (using the specific heat of sea water at constant pressure, $c_p = 4 \times 10^3$ J/kg/K and water density $\rho = 1.028 \times 10^3$ kg/m^3). Where does this heat come from? A crucial difference between the Arctic Ocean interior and the waters to the south is that summer warming due to the flux at the sea surface is tiny in the Arctic and is limited to a thin surface layer. A substantial part of the absorbed heat is spent on ice melting, thus it contributes little to the net water temperature increase. Therefore, the observed seasonal increase of heat content could only be caused by advection of warmer water, and not by the local ocean-air energy exchange. In contrast, seasonal cooling might be attributed to high negative local heat loss, which overcomes the positive advective influx.

The heat balance equation in the finite differences form may be written as follows:

$$c_p \rho H \frac{\Delta\overline{T}}{\Delta t} = Q - A, \qquad (1)$$

where Δt is the time interval, Q is the heat flux due to all nonadvective processes, and A is the heat flux caused by advection. Presuming that A is always negative, that is, the water coming from the west is warmer than the water at the position of the mooring, the change of sign on the left side of (1) is determined by Q. Positive $\Delta\overline{T}/\Delta t$ means that $Q - A > 0$. This is what happens in the warming season. During the cooling season, $Q - A < 0$. As shown in [15], the most probable reason for high negative Q in winter is thermal convection, which in the area of Whalers Bay (north of Svalbard) is able to reach deep into the water column [21].

Using the data from Table 1, and considering the water at the ocean surface to be permanently at the freezing point temperature, we can estimate heat loss from the upper 217 m, under the assumption of small seasonal

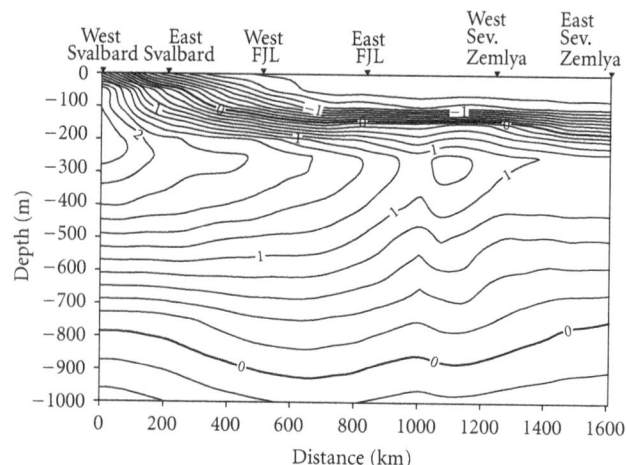

FIGURE 2: Temperature, °C section along the FSBW inflow in summer season (EWG, 1998) (FJL: Franz Joseph Land).

variation of advection [15]. Substituting vertically averaged temperature change between April-May and November in (1) and neglecting Q yields A equal to -235 W/m^2. Using this number for the cooling phase and taking into account the difference between the duration of warming and cooling seasons, we obtain an average heat loss of 560 W/m^2. This is a huge heat loss, having the same order of magnitude as is typically estimated for Arctic winter polynyas [22]. This number also matches well with the heat flux calculations done for the MIZ in the Barents Sea in winter [23]. Aagaard et al. [21] estimated winter heat loss in Whalers Bay from the 100–200 m layer to be 230 W/m^2. Applying the same equation (1) for the 113–217 m layer yields a similar result, 220 W/m^2. In the light of these numbers, the estimate done in [24], suggesting an increased surface heat loss of about 300 W/m^2 during pulses of anomalously warm AW inflow through Fram Strait, is also quite reasonable.

Progressive vector diagrams close to the core of the boundary current at 30°E calculated on the basis of 1-year-long continuous current meter measurements with 1-hour resolution show that the entire 70–217 m water layer is moving generally eastward with an average speed of 12–17 cm/s [15]. This indicates that the opposite motion (towards Fram Strait) may occur only in the subsurface layer above 70 m depth. The water at this level is apparently AW all year round despite the fact that its temperature may drop down to the freezing point. The AW "signature" is identified not by characteristic temperature and salinity values, but by the constant shape of the temperature-salinity (T-S) relationship [15]. Taking into account the steadiness of the flow direction and speed, we can argue that the actual cutoff depth of the reverse current is shallower than 70 m. Another argument in favor of this notion follows from the vertical distribution of temperature and salinity at cross-slope sections near the mooring position, see Figure 2 in [15]. In September 2006, the warm water core (over 5°C) resided at 50 m, while positive temperature water spread up to the ocean surface. A sharp salinity gradient at the 25–30 m depth

TABLE 1: Amplitude and phase of seasonal cycle in water temperature at 80°30′N and 31°E [15].

Depth (*m*)	Maximum, daily data		Minimum, daily data	
	Date	(*T*) °C	Date	(*T*) °C
70 ± 5	Nov 16 ± 10	5.12 ± 0.12	Apr 13	−1.77 ± 0.05
113 ± 5	Nov 16 ± 8	4.81 ± 0.17	Apr 23 ± 3	−0.40 ± 0.33
217 ± 5	Nov 24 ± 17	4.27 ± 0.29	May 13 ± 2	2.02 ± 0.06

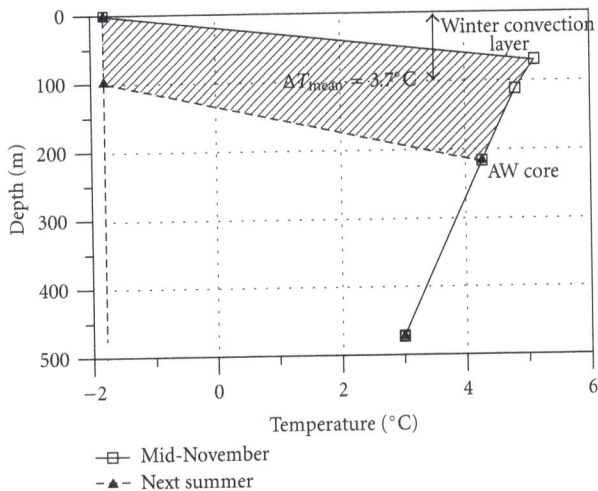

FIGURE 3: Sketch of the upper FSBW transformation during the winter season.

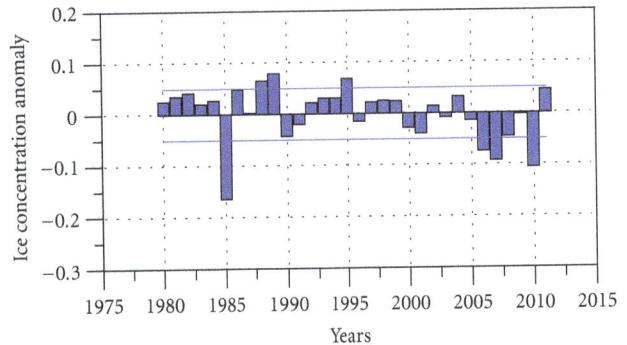

(a)

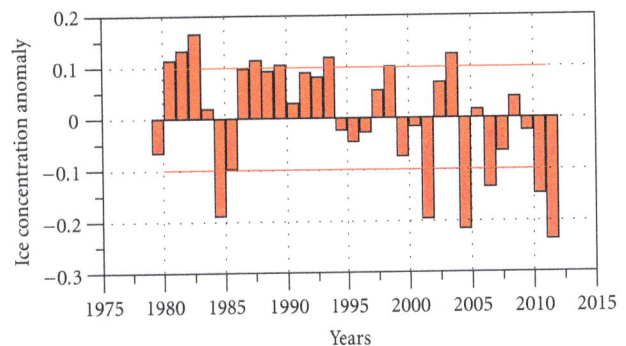

(b)

FIGURE 4: Time series of ice concentration (in parts of the unit) in the WNB (winter: blue; summer: red).

marks the location of the AW upper boundary in the vertical plane. The shape of the T-S relationship inside the 30–70 m layer also matches very well the typical AW T-S relationship, indicating that the water in this layer contains a considerable AW fraction.

The strong seasonal variation of heat content in the upper part of the water column at the mooring position in conjunction with the eastward direction of flow implies that a substantial amount of heat is advected by the boundary current. A persistent current moves warm water further to the east, bringing the upper part of this warm layer into close contact with pack ice drifting in the opposite direction. Taking into account that further to the east (in the Laptev Sea) the upper boundary of AW deepens to 150–200 m, we can anticipate that the heat stored above this depth is released en route, warming the under-ice layer. Applying simple theoretical considerations to the mid-November temperature profile from Table 1 enables us to estimate the order of magnitude of heat loss from the upper part of the water column. The evolution of temperature from mid-November to the following summer is sketched in Figure 3. Presuming that winter convection depth in the WNB is about 100 m [25], we calculate a mean temperature decrease above the AW core of 3.7°C. Applying this decrease of temperature to six winter-spring months yields a total heat loss of about 200 W/m². If only 7% of this heat goes upward [6], we calculate the value of heat flux from the ocean to the ice to be 7 times larger than the conventional mean

value [26]. This estimation points out that a considerable portion of seasonal heat input into the upper part of the WNB water column might be spent on ice melt and released to the atmosphere.

3. Structure and Variability of WNB Ice Conditions from 1979–2011

Data on ice concentration were taken from the Nimbus-7 Scanning Multichannel Microwave Radiometer (SMMR), and the Defense Meteorological Satellite Program (DMSP) Special Sensor Microwave/Imager (SSM/I), and Passive Microwave Data dataset [27, http://nsidc.org/data/nsidc-0051.html]. A brief description of these data is given in the Appendix. Time series of mean spatial winter and summer ice concentration (MSIC) in parts of the unit in the WNB are presented in Figure 4. Winter season was defined from November 1 to May 31 and summer season from June 1 to October 31. In winter, average MSIC is 0.88 ± 0.05, while in summer, it is 0.72 ± 0.10. Statistical properties are not uniform across time. Two distinct periods can be seen, with the border between them around 1995–1999. During the 1st time interval (1979–1995), positive MSIC anomalies prevail. There are 13 positive anomalies versus 3 negative in winter season and 12 positive versus 4 negative in summer. Values of anomalies lie within, or slightly exceed simple standard deviation (SSD) bounds. The only exception is the year 1985, which is characterized by large and coherent summer-winter negative anomalies. Anomalies exceed the SSD about twice in summer and more than 3 times in winter. During the 2nd time interval (1999–2011), the general pattern of anomalies is the opposite. In winter there are 8 negative anomalies versus 3 positive, while in summer there are 9 negative anomalies versus 4 positive. The negative anomalies often substantially exceed SSD, especially in summer. Although the time series are too short for performing robust correlation analysis, formal calculation shows that the correlation coefficient between the preceding summer and the next winter anomaly drops from 0.5 (in 1979–1995) to 0.1 (in 1999–2011). This suggests that the processes which control ice concentration during time intervals with high and low ice concentrations are not the same.

The difference between the average ice concentrations during two selected periods are plotted in Figure 5. In both seasons, ice concentration decreased during the 2nd time interval over the entire WNB area. In summer the maximum ice concentration decrease (−0.3) occurs within the WNB. Further to the north, the difference between the two time intervals is close to zero. The latter is explained by the fact that the region to the north of 83°N in the Atlantic sector of the Arctic Ocean is the "pack ice collector" for the new ice forming in Siberian shelf seas and being driven towards Fram Strait in the Transpolar Drift system [17]. In winter season the largest differences extend along two branches of AW (compare with Figure 1). During the 2nd time interval low-concentration "tongues" associated with two branches of AW merge, forming the large area with decreased ice concentration in the northern part of the Barents Sea and over the continental margin of the Nansen Basin (this plot is not shown).

In line with our basic hypothesis, which implies a substantial contribution of oceanic sensible heat in shaping the ice cover in the WNB (see Figure 1), we considered the unique ice thickness measurements made within the framework of the Ice, Cloud, and land Elevation Satellite (ICESat) campaigns [[28] http://rkwok.jpl.nasa.gov/icesat/download.html]. Two sequential maps showing ice thickness in October-November, 2007 (ON2007) and February-March, 2008 (FM2008) are presented in Figure 6. Our choice of 2007-2008 is dictated by the record summer minimum ice extent in the Arctic Ocean in 2007. Despite the record retreat of the ice edge in the Pacific sector, summer ice extent in the WNB was close to normal, which is reasonably explained by dynamics [29]. The FSBW pathway along the continental slope between Svalbard and Franz Joseph Land is within the seasonal summer MIZ; this does not provide any arguments

(a)

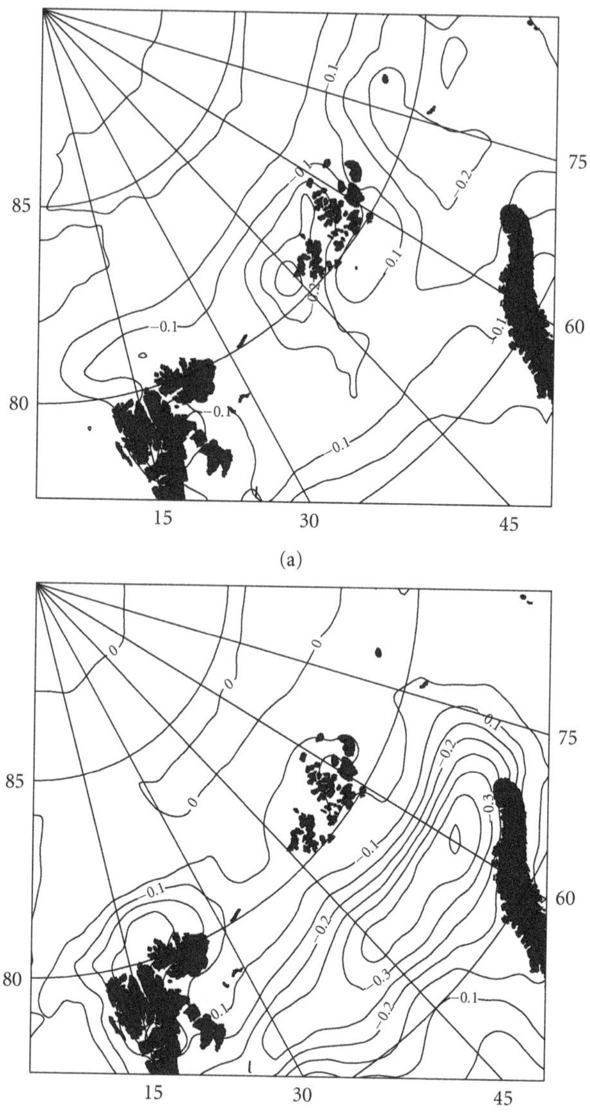

(b)

FIGURE 5: Difference between the average ice concentrations in the WNB in 1999–2011 versus in 1979–1995 in summer (a) and in winter (b).

for or against the hypothesis that AW heat impacts ice properties. A very different picture appears during the next winter survey (Figure 6(b)). The ice edge is shifted far to the south of the WNB. However, the local minimum ice thickness, surrounded by thicker ice, stretches from Svalbard to Severnaya Zemlya Archipelagos, visibly marking the FSBW inflow pathway.

Putting together the revealed features of mean seasonal distribution and interannual variability of ice concentration in the WNB and ice thickness data from 2007-2008, we suggest the following. The spatial patterns of ice distribution in winter and in summer are not the same. In winter, zones of decreased ice concentration extend along the branches of AW inflow, while in summer ice concentration decreases uniformly northward. Summer and winter MSIC

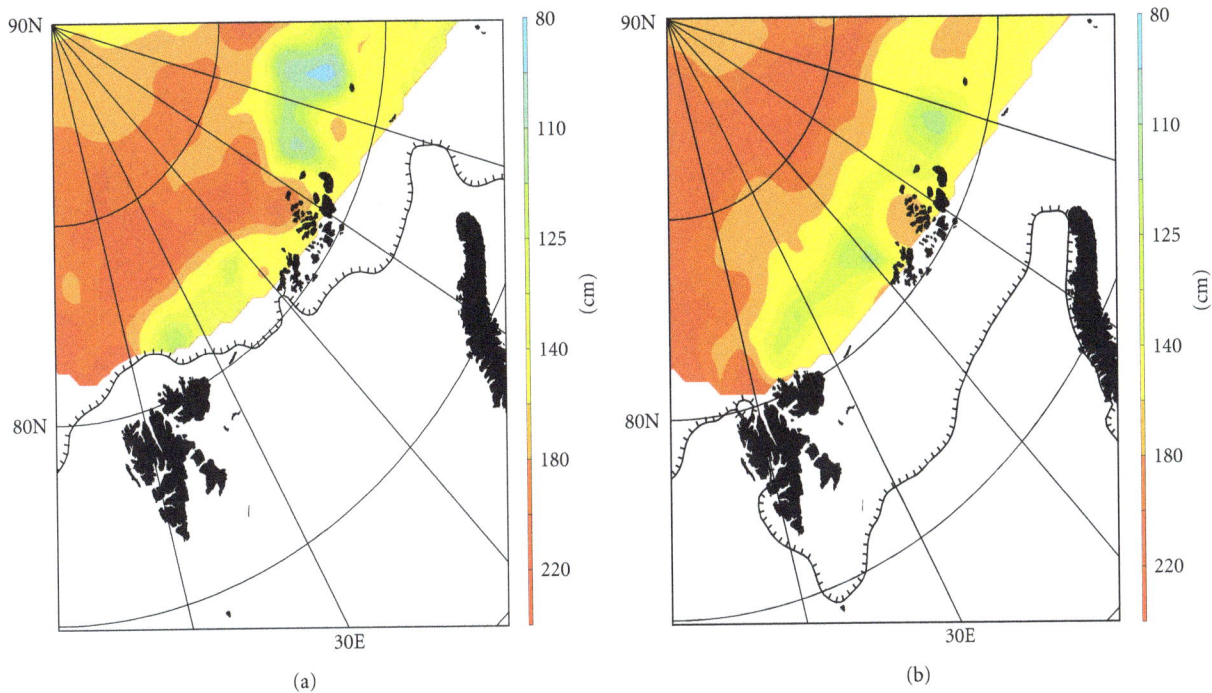

FIGURE 6: October-November, 2007 ice thickness (a); February-March, 2008 ice thickness (b). Black solid lines show the location of the ice edge, defined by 15% ice concentration.

time series are divided into two specific intervals: 1979–1995, with generally higher ice concentration and 1999–2011, with generally lower ice concentration. During the 1st time interval, the winter MSIC inherits features of the preceding season (summer-winter correlation coefficient is equal to 0.5). During the 2nd time interval, there is no link between summer and winter ice concentrations. Available ice thickness measurements support the notion that a substantial amount of heat reaches the under-ice surface and melts ice from below, rendering it thinner and more fragile. This process is especially noticeable in the winter season, when most of the Arctic Ocean is ice covered. In the next section, we test this observation-based hypothesis against available oceanographic and meteorological data and atmospheric reanalysis products in a search for the causative mechanisms that shape the sea ice cover in the WNB.

4. Causative Mechanisms

To elucidate possible mechanisms responsible for the observed features of the WNB ice conditions, we consider the following time series: (i) FSBW temperature at the WNB entry point; (ii) meridional wind component over the WNB; (iii) air temperature and surface-air temperature difference; (iv) sensible heat flux at the ocean surface. For this task, we used reanalysis products (http://www.ecmwf.int/research/era/do/get/era-interim). The features of this data set are briefly described in the Appendix.

FSBW interannual variability near the entry point. Annual time series of AW temperature were generated inside the cell bounded by 78–80°N and 5–10°E. This is the mean

climatic position of the West Spitsbergen Current main stream, see Figure 3 in [24]. The data from the Arctic and Antarctic Research Institute (AARI) collection were used [30]. We consider two water layers: 100–200 m and 200–300 m (Figure 7). This choice is explained by the features of AW transformation in the AIW discussed in Section 2. Provided that the concept of AW intensive mixing in the upper part is correct the layer shallower than ∼200 m initially contains the heat, which is totally released upwards and laterally during the FSBW transit from Fram Strait to the Laptev Sea. The layer below ∼200 m retains a large portion of its initial heat content up to where the AIW terminates its full circuit around the Arctic Ocean interior. Within the considered time interval (1979–2011) the temperature in both AW layers coherently increased. However, this increase was not monotonic, but rather cyclic with an 8–10-year period. The background positive trend is 1.5° per 32 years in the upper layer and 1.25° per 32 years in the lower layer. It is worth mentioning that after the most recent temperature increase, which culminated in 2007, the temperature did not drop back to the initial point as happened earlier, but remained about 1°C higher in both layers, leading to the conclusion that AW in the Arctic Ocean is shifting to a new warmer state [31, 32]. Hence, there is no doubt that the heat input to the Arctic Ocean interior has substantially increased since the end of the 1990s. The question is has this increased heat input provided the major forcing of the documented change in ice properties in the WNB? To answer this question let us examine the other potentially significant mechanisms.

Wind stress is usually considered to be the primary forcing which creates open water zones in the consolidated ice cover in winter [33]. Well-known quasi-steady-state

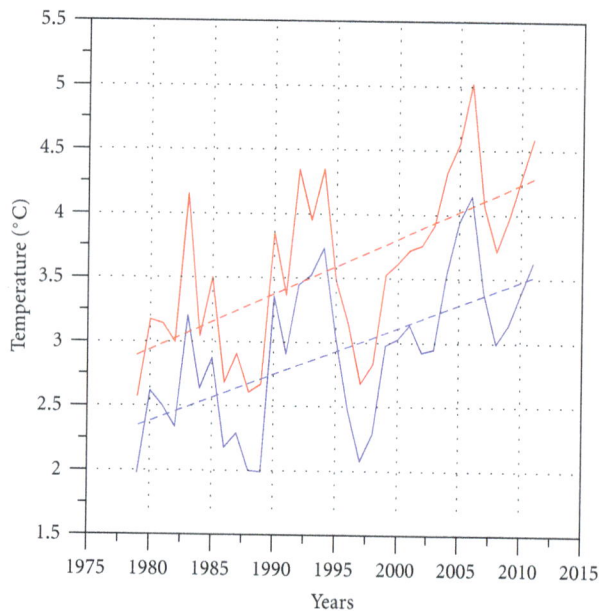

FIGURE 7: Water temperature in the core of the West Spitsbergen current at 79°N; red: 100–200 m layer; blue: 200–300 m layer.

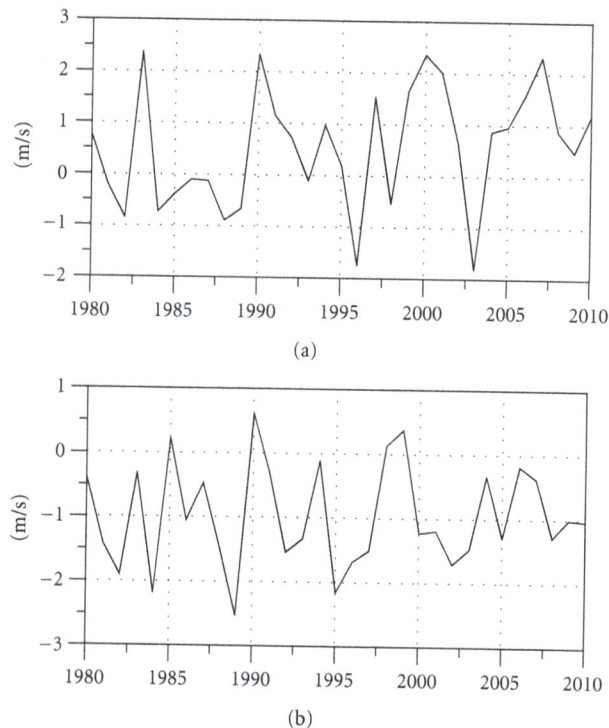

FIGURE 8: Meridional wind component in winter around Franz Joseph Land (a) and north of Svalbard (b).

polynyas along the Eurasian and American coastline are formed by katabatic wind and local currents which break up the fast ice and drive ice floes offshore [34]. In the conventional terminology such a polynya is known as a latent heat polynya (LHP) [35]. Another type of polynya, the sensible heat polynya (SHP), is thermodynamically driven and typically occurs when warm water upwelling keeps the surface water temperature above the freezing point. Favorable wind may also assist in maintaining an open SHP, but the major role is played by oceanic heat. To determine whether variation of the WNB open water area in winter is caused by wind stress, we plotted the time series of the meridional surface wind component in two localized regions where an increased area of open water was observed during the 2nd time interval (see Figure 8). No visible trends in the meridional wind component can be inferred from this plot or from the time series of wind speed (not shown), thus removing wind from the list of likely causative mechanisms.

Air temperature variation is another obvious candidate to be linked with the observed changes in the ice cover. In Figure 9, one can see that air temperature changed around the mid-1990s, at which time air temperature started to increase almost monotonically, which is in line with the ice concentration change. At about the same time the air-surface temperature difference decreased and then stayed at almost the same level up until present with a very weak upward tendency (Figure 10).

Wind speed and air-surface temperature difference determine sensible *heat flux* at the ocean ice/air interface [36]. In the absence of short-wave radiation (in winter season), sensible heat flux is the major contributor to the surface-air energy balance. The shape of the sensible heat flux curve in winter season (Figure 11) is very similar to the air-sea temperature difference curve, with a tipping point around

1995. However, it is important to note that the sensible heat flux shows more of a tendency to increase than does the air-surface temperature difference after 1995.

5. Discussion and Conclusions

Our analysis of WNB ice properties from 1979–2011 allowed us to detect a general tendency towards decreasing ice concentration that commenced after the mid-1990s. Combining the ice concentration data and the available ice thickness data allowed us to demonstrate that the location of local zones with thinner ice and lower ice concentration essentially mirror the pathway of the FSBW in the WNB. Time series of FSBW temperature in Fram Strait feature a monotonic increase after the mid-1990s, consistent with shrinking ice cover. This coincidence provides solid ground for the hypothesis that a substantial amount of the AW heat in the WNB is able to reach the under-ice layer and contribute to the ice melting from below. On the other hand, there is evidence in reanalysis products that WNB air temperature has exhibited trends consistent with those of ice concentration and FSBW temperature. Therefore, the question to pose is as follows. Are one or more of these correlated processes drive the observed changes, or is the driver of something else entirely? To answer this question let us briefly outline the physical background of ocean-ice-air interaction. In the ice-covered seas a key factor controlling the rate of energy exchange is the spatial irregularity of the ice cover. Irregularity primarily depends on the ice concentration and thickness [37]. Under similar

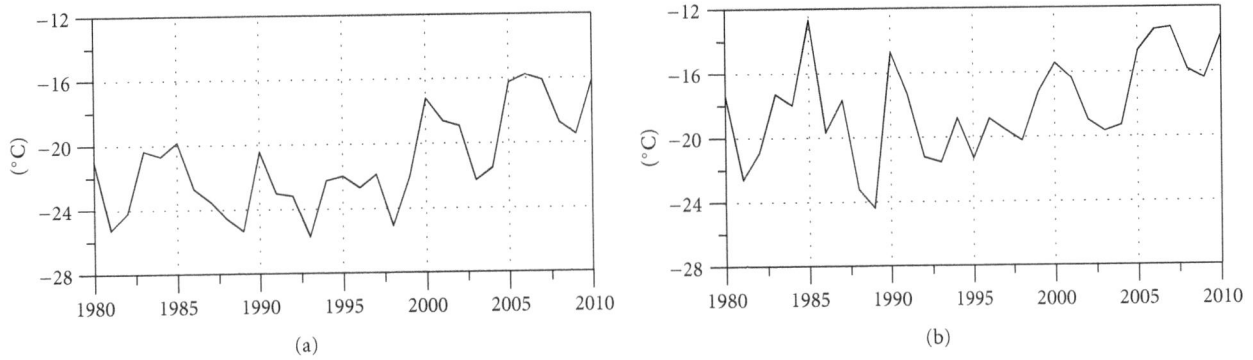

FIGURE 9: Air temperature at the surface in winter around Franz Joseph Land (a) and north of Svalbard (b).

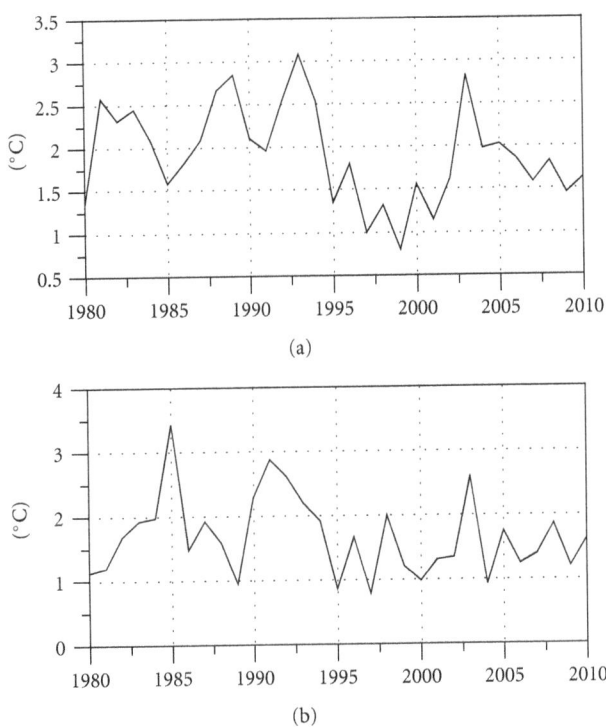

FIGURE 10: Surface-air temperature difference in winter around Franz Joseph Land (a) and north of Svalbard (b).

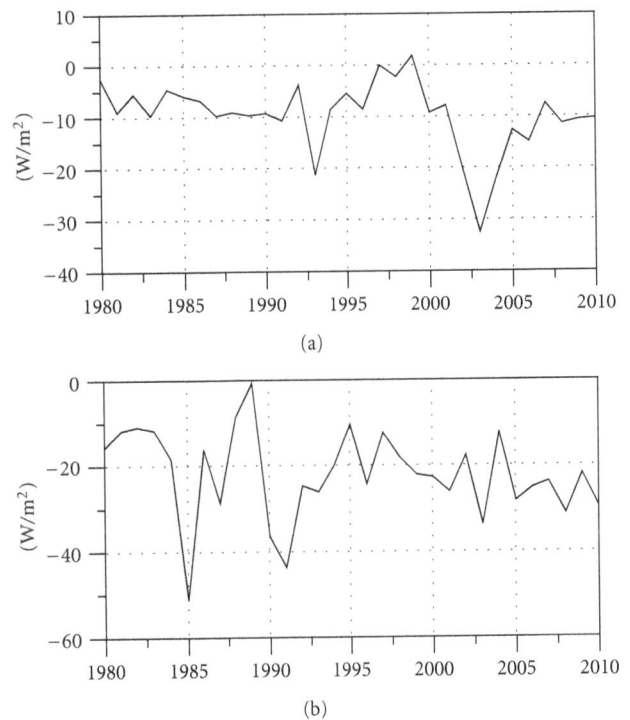

FIGURE 11: Sensible heat flux at the ocean ice/air interface in winter around Franz Joseph Land (a) and north of Svalbard (b).

meteorological conditions, heat exchange through the ice-free surface is two orders of magnitude greater than that through the surrounding pack ice [35]. Turning back to Figures 6, 7, and 8, we draw attention to the following facts. Air temperature permanently increased after the mid-1990s, as did negative sensible heat flux. Therefore, the amount of heat released from the ocean to the atmosphere increased. At the same time, the air-surface temperature difference did not show any significant trend. European reanalysis (ERA-Interim) surface heat fluxes from partially ice-covered ocean are obtained by averaging fluxes over open water and ice proportionally according to the concentration in a given model cell. Therefore, we argue that an increase of sensible heat flux from the ocean after the mid-1990s is a result of lower sea ice concentration in the area (perhaps due to an

increased number of small-scale leads [38]). Lower sea ice concentration is the consequence of ice thinning under the influence of an increased sensible heat flux from AW because ice erosion from below enables more effective breaking of the thinner ice by wind (which has not visibly changed) and by local currents. Lower sea ice concentrations are collocated with areas of anomalies in the surface heat fluxes (Figure 12). The fact that the minimum sea ice thickness lies in the interior of the ice pack right above the FSBW path (see Figure 6(b)) is a clear indication of the influence of heat flux from the FSBW on ice thickness. The physical processes that deliver warm water from the deep to the under-ice layer are yet to be studied in detail. We assume that in the study area the most likely candidate is winter convective mixing. In the WNB, in the absence of a cold halocline, convective mixing reaches the depth of ~100 m [25], entraining the upper part of AW, which is at its seasonal peak in late fall—early winter [15]. Additional forcing could be provided by upwelling events at the continental slope of Franz Joseph Land [39] and/or by Ekman pumping [14].

The possibility of AW heat impact on the Arctic Ocean ice cover has been debated since Fritjof Nansen's 1890s discovery of a warm water layer under the pack ice. The hypothesis that AW affects ice properties always had its supporters and opponents. However, the absence of robust observational data in specific locations and specific seasons prevented this dispute from growing beyond theoretical speculations and indirect estimations to justified statements, based on observations. In the present study we demonstrate that in the WNB region AW directly affects the ice thickness, providing an efficient thermodynamic mechanism by which ice volume is decreased. The significance of this influence in the pan-Arctic sea ice and fresh water budgets should be the subject of future studies.

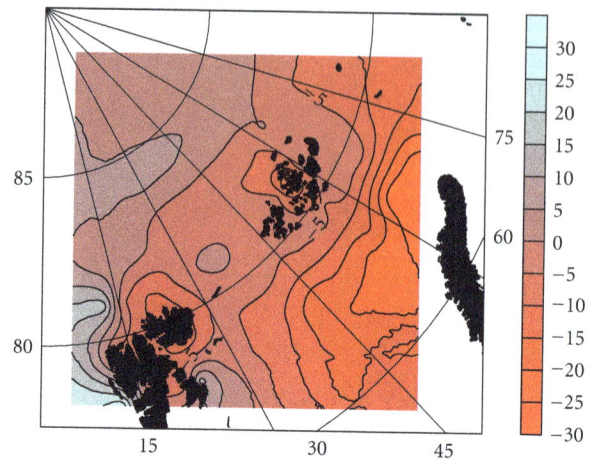

FIGURE 12: Averaged 1980–2010 winter distribution of sensible heat flux at the surface, W/m² (a); difference in heat flux between 1999–2010 and 1980–1995 (b).

Appendix

Data Sources and Uncertainties

For this study we used publicly available data sets. Originators of these data sets provide detailed description of metadata, including methods of measurements, accuracy, and so forth. Here, we briefly describe the sources of data and discuss possible uncertainties in the presented results.

CTD and mooring-based data on WNB temperature and salinity were taken from the NABOS data archive (http://nabos.iarc.uaf.edu/). These data had passed thorough quality control and have been used in multiple published studies, see [40]. The *AW temperature* in Fram Strait has been monitored for an extended period of time by the international scientific community, mainly by Russian, Norwegian, and German researchers [14, 24, 41]. The AW time series used in this study were generated from the oceanographic data base, collected, and routinely complemented by new data at the Arctic and Antarctic Research Institute (AARI). *Ice concentration* data were taken from the Nimbus-7 Scanning Multichannel Microwave Radiometer (SMMR) and the Defense Meteorological Satellite Program (DMSP) Special Sensor Microwave/Imager (SSM/I) Passive Microwave Data dataset [27, http://nsidc.org/data/nsidc-0051.html]. The spatial resolution of the regular grid is 25 km. For the year 2011, preliminary data from this dataset were used [42]. We calculated 10 day averages from the daily data and used the resulting products for the analysis. *Ice thickness* data were taken from the ICESat data collection [28] 2008; (http://rkwok.jpl.nasa.gov/icesat/download.html). *Meteorological* data and derived parameters (heat fluxes) were taken from the ERA-Interim reanalysis (http://www.ecmwf.int/research/era/do/get/era-interim). ERA-Interim is the latest ECMWF global atmospheric reanalysis covering the period from 1979 to present. It uses 4D variational data assimilation of a wide variety

of observations from surface based to aerological and satellite measurements. Available fields include all major meteorological variables plus important surface diagnostics including surface fluxes. We use the 1.5×1.5 degree resolution version of ERA Interim available from the ECMWF web site: http://www.ecmwf.int/.

Uncertainties are an inseparable component of any observation-based investigation, and our study is not free of them. We recognize two types of uncertainties: (i) uncertainties associated with the limitation of the data used and (ii) uncertainties caused by simplifications introduced during the analysis. Uncertainties in hydrographic data are usually caused by changing the location of sequential CTD casts and low horizontal resolution. Since the NABOS CTD data that we used were collected at the recurrent cross-slope section with ~5 km distance between stations over the steepest part of the slope, we do not expect a noticeable error in estimation of AW properties. Mooring-based data were collected close to the core of the AW flow, as was shown in [15], which also guarantees small errors of estimation based on these data. Uncertainties in ice concentration data were discussed in [43]. They conclude that according to data documentation in general, the accuracy of total sea ice concentrations is within ±5% of the actual sea ice concentration in winter, and ±15% in the Arctic during summer when melt ponds are present on the sea ice. Although these numbers are close to the MSIC variations presented in Figure 4, we would like to stress our use of ice concentration data averaged over time and space, which reduces the error proportional to the square root of the number of individual measurement points. Ice concentration data (as well as ice thickness data) have rather crude spatial resolution (25 km) and do not resolve small openings in the ice cover (cracks and flaw leads). However, for the purpose of the presented analysis these data fit reasonably well, since we discuss relatively large (~hundreds of kilometers) features. Uncertainty in the diagnostic fields provided by ERA-Interim, which are not constrained by any sort of data assimilation procedures, is always a big problem for any reanalysis products. For example, in [44] it is discussed that the overall performance of ERA-Interim products including fluxes over ocean. They found that the quality of many diagnosed variables (e.g., precipitation and surface fluxes) greatly improved in ERA-Interim in comparison with other products of this kind.

Uncertainties in the presented analysis may be associated with our failure to take ice drift into account. Drift provides key forcing of ice redistribution. We checked the pattern of ice thickness anomalies against the mean ice drift in February-March 2008 [18]. Within this time interval, mean ice drift in the WNB and around it was generally directed southward with speed not exceeding 5 km/day. Therefore, we conclude that the error of failing to take ice drift into account may not noticeably affect our results. Another possible source of error in ice thickness estimation is snow on ice, which is difficult to assess. Since large-scale snow anomalies are collocated with downwelling long wave (DLW) radiation, we estimated total accumulated snowfall from the ERA-Interim dataset by assuming snow density of 0.25, and that snowpack starts developing in September. No correlation between snowfall anomalies and ice thickness anomalies in 2008 was found. ERA-Interim uses satellite observed sea ice concentrations to calculate surface fluxes over the Arctic Ocean. Small-scale features like cracks and polynyas may not be well resolved. However, if sea ice concentration is less than 100% in a grid cell, open water and sea ice are calculated separately. The total heat flux over the grid cell is then calculated according to the proportion of open water and sea ice in that grid cell. Absolute values of air-surface temperature difference decreased according to our analysis. Since the wind change is relatively small, this could only lead to a decrease in the sensible heat flux if the surface properties stayed the same (e.g., ocean remained ice covered). Therefore, sensible heat flux can increase only at the expense of more open water in the area, because heat fluxes from open water are far greater than fluxes over ice-covered ocean. Note that we use the ERA-Interim sign convention: negative surface heat flux means the surface is heating the atmosphere.

We acknowledge that the presented analysis may be not perfect because the available data sets have limitations. However, we believe that the uncertainties of data and analytical method do not call into question the conclusions of this paper.

Acknowledgments

This study was supported by the following research grants/programs: EU FP7 "Arctic Climate Change Economy and Society" (ACCESS) project, NERSC-IAP 196174/S30: "The atmospheric boundary layer structure and surface-atmosphere exchange in the Svalbard area;" RFBR 11-05-12019-ofi-m-2011: "Modern polar climate change estimation on the base satellite microwave database GLOBAL-RT;" RFBR 11-05-01143: "Investigation of tide action on generation and dynamics of internal waves and their manifestation on the sea surface in Russian Arctic seas;" ONR-Global grant 62909-12-1-7013: "Decision Making Support System for Arctic Exploration, Monitoring and Governance;" European Commission 7th framework program through the MONARCH-A Collaborative Project, FP7-Space-2009-1 Contract no. 242446, NSF Grant ARC 0909525 and Japan Agency for Marine-Earth Science and Technology. ERA-Interim data used for the analysis were downloaded from the website of the European Centre for Medium Range Weather Forecasts (http://www.ecmwf.int/). The authors thank Anton Beljaars of ECMWF for providing very useful information on ERA-Interim data assimilation procedures.

References

[1] J. C. Comiso, C. L. Parkinson, R. Gersten, and L. Stock, "Accelerated decline in the Arctic sea ice cover," *Geophysical Research Letters*, vol. 35, no. 1, Article ID L01703, 2008.

[2] M. Wang and J. E. Overland, "A sea ice free summer Arctic within 30 years?" *Geophysical Research Letters*, vol. 36, no. 7, Article ID L07502, 2009.

[3] B. Y. Wu, J. Wang, and J. E. Walsh, "Dipole anomaly in the winter Arctic atmosphere and its association with sea ice motion," *Journal of Climate*, vol. 19, no. 1, pp. 210–225, 2006.

[4] V. A. Alexeev, I. N. Esau, I. V. Polyakov, S. J. Byam, and S. Sorokina, "Vertical structure of recent Arctic warming from observed data and reanalysis products," *Climatic Change*, vol. 111, no. 2, pp. 215–239, 2011.

[5] K. Shimada, T. Kamoshida, M. Itoh et al., "Pacific Ocean inflow: influence on catastrophic reduction of sea ice cover in the Arctic Ocean," *Geophysical Research Letters*, vol. 33, no. 8, Article ID L08605, 2006.

[6] I. V. Polyakov, L. A. Timokhov, V. A. Alexeev et al., "Arctic ocean warming contributes to reduced polar ice cap," *Journal of Physical Oceanography*, vol. 40, no. 12, pp. 2743–2756, 2010.

[7] D. K. Perovich, J. A. Richter-Menge, K. F. Jones, and B. Light, "Sunlight, water, and ice: extreme Arctic sea ice melt during the summer of 2007," *Geophysical Research Letters*, vol. 35, Article ID L11501, 4 pages, 2008.

[8] D. A. Rothrock, Y. Yu, and G. A. Maykut, "Thinning of the Arctic Sea-Ice cover," *Geophysical Research Letters*, vol. 26, no. 23, pp. 3469–3472, 1999.

[9] J. Rodrigues, "The rapid decline of the sea ice in the Russian Arctic," *Cold Regions Science and Technology*, vol. 54, no. 2, pp. 124–142, 2008.

[10] V. A. Alexeev, P. L. Langen, and J. R. Bates, "Polar amplification of surface warming on an aquaplanet in "ghost forcing" experiments without sea ice feedbacks," *Climate Dynamics*, vol. 24, no. 7-8, pp. 655–666, 2005.

[11] P. L. Langen and V. A. Alexeev, "Polar amplification as a preferred response in an idealized aquaplanet GCM," *Climate Dynamics*, vol. 29, no. 2-3, pp. 305–317, 2007.

[12] R. V. Bekryaev, I. V. Polyakov, and V. A. Alexeev, "Role of polar amplification in long-term surface air temperature variations and modern arctic warming," *Journal of Climate*, vol. 23, no. 14, pp. 3888–3906, 2010.

[13] R. Kwok, C. F. Cunningham, M. Wesnahan, I. Rigor, H. J. Zwally, and D. Yi, "Thinning and volume loss of the Arctic Ocean ice cover: 2003–2008," *Journal of Geophysical Research*, vol. 114, Article ID C07005, 16 pages, 2009.

[14] S. Lind and R. B. Ingvaldsen, "Variability and impacts of Atlantic water entering the Barents sea from the north," *Deep Sea Research I*, vol. 62, pp. 70–88, 2011.

[15] V. V. Ivanov, I. V. Polyakov, I. A. Dmitrenko et al., "Seasonal variability in Atlantic Water off Spitsbergen," *Deep-Sea Research I*, vol. 56, no. 1, pp. 1–14, 2009.

[16] Environmental Working Group (EWG), *1998, Joint U.S.-Russian Atlas of the Arctic Ocean (CD-ROM)*, National Snow and Ice Data Center, Boulder, Colo, USA, 1997.

[17] Y. G. Nikiforov and A. O. Shpaikher, *Features of the Formation of Hydrological Regime Large-Scale Variations in the Arctic Ocean*, Gydrometeoizdat, Leningrad, Russia, 1980.

[18] R. Kwok and J. Morrison, "Dynamic topography of the ice-covered Arctic ocean from ICESat," *Geophysical Research Letters*, vol. 38, Article ID L02501, 6 pages, 2011.

[19] B. Rudels, E. P. Jones, L. G. Anderson, and G. Kattner, "On the intermediate depth waters in the Arctic ocean," in *The Polar Oceans and Their Role in Shaping the Global Environment*, O. M. Johannessen, R. D. Muench, and J. E. Overland, Eds., pp. 33–46, AGU Geophysical Monograph 85, Washington, DC, USA, 1994.

[20] I. A. Dmitrenko, S. A. Kirillov, V. V. Ivanov et al., "Seasonal modification of the Arctic Ocean intermediate water layer off the eastern Laptev Sea continental shelf break," *Journal of Geophysical Research C*, vol. 114, no. 6, Article ID C06010, 2009.

[21] K. Aagaard, A. Foldvik, and S. R. Hillman, "The west spitsbergen current: disposition and water mass transformation," *Journal of Geophysical Research*, vol. 92, no. C4, pp. 3778–3784, 1987.

[22] M. A. Maqueda, A. J. Willmott, and N. R. T. Biggs, "Polynya dynamics: a review of observations and modeling," *Reviews of Geophysics*, vol. 42, no. 1, Article ID RG1004, 37 pages, 2004.

[23] S. Hakkinen and D. J. Cavalieri, "A study of oceanic surface heat fluxes in the Greenland, Norwegian, and Barents Seas," *Journal of Geophysical Research*, vol. 94, no. 5, pp. 6145–6157, 1989.

[24] U. Schauer, E. Fahrbach, S. Osterhus, and G. Rohardt, "Arctic warming through the fram strait: oceanic heat transport from 3 years of measurements," *Journal of Geophysical Research C*, vol. 109, no. 6, Article ID C06026, 14 pages, 2004.

[25] B. Rudels, E. P. Jones, U. Schauer, and P. Eriksson, "Atlantic sources of the Arctic Ocean surface and halocline waters," *Polar Research*, vol. 23, no. 2, pp. 181–208, 2004.

[26] N. Untersteiner, "On the mass and heat balance of Arctic sea ice," *Archives for Meteorology, Geophysics and Bioclimatology*, vol. 12, pp. 151–182, 1961.

[27] D. Cavalieri, C. Parkinson, P. Gloersen, and H. J. Zwally, *Updated Yearly. Sea Ice Concentrations from Nimbus-7 SMMR and DMSP SSM/I-SSMIS Passive Microwave Data, 1979–2010*, National Snow and Ice Data Center. Digital Media, Boulder, Colo, USA, 1996.

[28] R. Kwok and G. F. Cunningham, "ICESat over Arctic sea ice: estimation of snow depth and ice thickness," *Journal of Geophysical Research C*, vol. 113, no. 8, Article ID C08010, 2008.

[29] R. Kwok, "Outflow of Arctic ocean sea ice into the Greenland and Barent seas: 1979–2007," *Journal of Climate*, vol. 22, no. 9, pp. 2438–2457, 2009.

[30] S. Smirnov and A. Korablev, "A regional oceanographic database for the Nordic Seas: from observations to climatic dataset," in *Proceedings of the International Conference on Marine Data and Information Systems (IMDIS '10)*, p. 81, Paris, France, March 2010.

[31] I. V. Polyakov, A. Beszczynska, E. C. Carmack et al., "One more step toward a warmer Arctic," *Geophysical Research Letters*, vol. 32, no. 17, Article ID L17605, pp. 1–4, 2005.

[32] I. A. Dmitrenko, I. V. Polyakov, S. A. Kirillov et al., "Toward a warmer Arctic Ocean: spreading of the early 21st century Atlantic water warm anomaly along the Eurasian basin margins," *Journal of Geophysical Research C*, vol. 113, no. 5, Article ID C05023, 2008.

[33] S. Martin and D. J. Cavalieri, "Contribution of the Siberian shelf to the Arctic ocean intermediate and deep water," *Journal of Geophysical Research*, vol. 94, no. 12, pp. 12725–12738, 1989.

[34] V. F. Zakharov, *Sea Ice in the Climate System*, Hydrometeoizdat, St. Petersburg, Russia, 1996.

[35] S. D. Smith, R. D. Muench, and C. H. Pease, "Polynyas and leads: an overview of physical processes and environment," *Journal of Geophysical Research*, vol. 95, no. C6, pp. 9461–9479, 1990.

[36] T. Uttal, J. A. Curry, M. G. McPhee et al., "Surface heat budget of the arctic ocean," *Bulletin of the American Meteorological Society*, vol. 83, no. 2, pp. 255–275, 2002.

[37] I. A. Repina and A. S. Smirnov, "Heat and momentum exchange between the atmosphere and ice from the observational data obtained in the region of Franz Josef Land," *Atmospheric and Ocean Physics*, vol. 36, no. 5, pp. 618–626, 2000.

[38] S. Marcq and J. Weiss, "Influence of sea ice lead-width distribution on turbulent heat transfer between the ocean and the atmosphere," *The Cryosphere*, vol. 6, pp. 143–156, 2012.

[39] K. Aagaard, L. K. Coachman, and E. Carmack, "On the halocline of the Arctic Ocean," *Deep Sea Research A*, vol. 28, no. 6, pp. 529–545, 1981.

[40] I. V. Polyakov, V. A. Alexeev, I. M. Ashik et al., "Fate of early 2000's arctic warm water pulse," *Bulletin of the American Meteorological Society*, vol. 92, no. 5, pp. 561–566, 2011.

[41] I. V. Polyakov, G. V. Alekseev, L. A. Timokhov et al., "Variability of the intermediate Atlantic water of the Arctic ocean over the last 100 years," *Journal of Climate*, vol. 17, no. 23, pp. 4485–4497, 2004.

[42] W. Meier, F. Fetterer, K. Knowles, M. Savoie, and M. J. Brodzik, *Updated Quarterly. Sea Ice Concentrations from Nimbus-7 SMMR and DMSP SSM/I-SSMIS Passive Microwave Data, 2010*, National Snow and Ice Data Center. Digital Media, Boulder, Colo, USA, 2006.

[43] N. V. Koldunov, D. Stammer, and J. Marotzke, "Present-day arctic sea ice variability in the coupled ECHAM5/MPI-OM model," *Journal of Climate*, vol. 23, no. 10, pp. 2520–2543, 2010.

[44] D. P. Dee, S. M. Uppala, A. J. Simmons et al., "The ERA-Interim reanalysis: configuration and performance of the data assimilation system," *Quarterly Journal of the Royal Meteorological Society*, vol. 137, no. 656, pp. 553–597, 2011.

11

A Parameterized Method for Air-Quality Diagnosis and Its Applications

J. Z. Wang,[1] S. L. Gong,[1,2] X. Y. Zhang,[1] Y. Q. Yang,[1] Q. Hou,[1] C. H. Zhou,[1] and Y. Q. Wang[1]

[1] *Center for Atmospheric Composition Observing & Service, Chinese Academy of Meteorological Sciences, Beijing 100081, China*
[2] *Air Quality Research Division, Science & Technology Branch, Environment Canada, 4905 Dufferin Street, Toronto, ON, Canada M3H 5T4*

Correspondence should be addressed to X. Y. Zhang, xiaoye@cams.cma.gov.cn

Academic Editor: Zhanqing Li

A parameterized method is developed to diagnose the air quality in Beijing and other cities with an index termed (parameters linking air-quality to meteorological elements PLAM) derived from a correlation between PM_{10} and relevant weather elements based on the data between 2000 and 2007. Key weather factors for diagnosing the air pollution intensity are identified and included in PLAM that include atmospheric condensation of water vapour, wet potential equivalent temperature, and wind velocity. It is found that the poor air quality days with elevated PM_{10} are usually associated with higher PLAM values, featuring higher temperature, humidity, lower wind velocity, and higher stability compared to the averaged values in the same period. Both 24 h and 72 h forecasts provided useful services for the day of the opening ceremony of the Beijing Olympic Games and subsequent sport events. A correlation coefficient of 0.82 was achieved between the forecasts and (air pollution index API) and 0.59 between the forecasts and observed PM_{10}, all reaching the significant level of 0.001, for the summer period. A correction factor was also introduced to enable the PLAM to diagnose the observed PM_{10} concentrations all year round.

1. Introduction

Meteorological elements, including water vapor content, surface wind speed, visibility, and the diurnal temperature range are important parameters to effect air pollutant concentrations [1]. Studies using data from seven air quality stations during the period of 2000 to 2008 in Taiwan area showed that PM_{10} and $PM_{2.5}$ concentrations were significantly controlled by the weather elements, resulting in high concentrations in spring and winter and low concentrations in fall and summer [2]. Even in summer when relatively low pollutant concentrations were found, the weather elements still had a large impact on air quality as the high temperature and humidity tended to increase PM_{10} and $PM_{2.5}$ concentrations, due to the formation of secondary pollution [3, 4]. Recently, substantial advances have been made in the studies of the impact of human activities on air quality caused by the increasing pollutant concentrations, especially ozone and particulate matters (PM) [5–9]. The concentration variations of different aerosols in various regions of China, as well as

the impacts of aerosol concentration on sand/dust storms (SDS) and hazy weather in the Asia and North American were investigated [1, 10–14]. It is found that the variation of aerosol and gaseous concentrations is closely related to the changing weather elements [3, 8, 15–19]. The key weather elements influencing the air pollution include wind speed, wind direction, air pressure, temperature, humidity, precipitation, and atmosphere stability [20]. These factors control the dilution, transports, accumulation, and removal processes of air pollutants and hence dominate the air concentration of various pollutants under a given emission condition.

The relationship between the evolution of the general circulation and foggy/hazy weather systems was also examined to link the wind/cloud fields with low-visibility weather, indicating that atmospheric aerosols such as mist, fog, fume, smoke, smog, and haze can be formed from a condensation or nucleation process [21–25]. Studies also show that in the case of heavy pollution due to the increase of NO_2 and SO_2, significant differences in condensability were observed before and after a hazy weather process. Before the haze onset, with the increase of SO_2 and NO_2 concentrations, condensability was increasing quickly. On the contrary, during a haze event, both SO_2 and NO_2 concentrations were decreased [19]. This was related to the consumption and washout of the nucleus in the haze formation. Atmospheric condensation is a key factor for diagnosing and forecasting the air pollution intensity in relation to weather conditions.

The above studies indicated that meteorological elements have a significant control over the accumulation and dilution of air pollutants. One of the most important scientific questions is how to quantify the meteorological impacts and use them to diagnose and forecast the air quality under the assumption that the time scale of emission changes is much longer than the meteorology changes.

Numerical prediction of air quality by 3D chemical transport models (CTM) has been implemented in a number of weather services around the globe with various degree successes [26–28] to forecast air quality. Due to the emission inventories used for the forecasts usually lagging behind the current date, this method has the limitation for an accurate forecast if the emission inventory is not updated to the current time for a region with ever changing industries such as in China. Through an analysis of observed PM and meteorological parameters, a parameterized method was developed and defined as the PLAM index. PLAM links the air pollution variations to the meteorological conditions. During the period of 2008 Beijing Olympic Games, PLAM was used to forecast the air quality in Beijing and achieved reasonable results.

2. Data and Methods

2.1. Data. For the sake of data comparison, analysis, and operational run, both PM_{10} and weather observations were collected simultaneously at the National Climate Observatory in Beijing (NCOB). PM_{10} is measured by a TEOM (Series 1400a Ambient Particulate Monitor). The instrument measures the PM_{10} mass concentrations every 5 minutes automatically. The mass transducer minimum detection limit is $0.01\,\mu g$. The precision for 10-minute and 1-hour averaged data is $5.0\,\mu g\,m^{-3}$ and $1.5\,\mu g\,m^{-3}$, respectively [29]. The daily mean PM_{10} data and weather observations were obtained after careful QA/QC processes. NCOB is one of the national climate observatories providing data for international exchange under Word Meteorological Organization (WMO) through World Weather Watch (WWW). The NCOB (N 39.8, E116.5, 32 m) station is in the suburban/rural area of Beijing. Compared with meteorological variables, PM measurements are substantially influenced by the local environment.

2.2. Correlation Analysis. Previous study has indicated that a positive correlation of PM_{10} observations with maximum temperature (t_m) and relative humidity (rh) was found for the period from July to September in 2000–2007 at NCOB [30]. It was suggested that for a better signal analysis and forecasts, the peak PM_{10} values could be identified on a daily basis in the same month.

In this study, for the forecasting purpose, the observed PM_{10} (y) with preceding 24 h weather elements $x_i(p, t_m, w, rh, e,$ etc.) was used with daily average meteorological data for period from 1 July to 31 September 2000–2007 (total 736 days) managed by the National Climate Centre of China. The elements of p, t_m, w, rh, evaporability, water vapour, dew point, wind direction, wind speed, cloudiness, rainfall, and so forth were included in the dataset. Figure 1 presents the correlations of PM_{10} values measured at NCOB with the maximum temperature, relative humidity, water vapor observed 24 hours before and with the evaporability at the same time. The evaporability is to describe the evaporation capacity of the atmosphere with a unit of mm. The correlations of PM_{10} with t_m, rh, air pressure, water vapor amount and evaporability are given in Table 1. Positive correlations of PM_{10} observations were found with the maximum temperature, relative humidity, and water vapor with preceding 24 hours, and negative correlations were found with evaporability from July to September in 2000–2007 at NCOB. The correlation coefficients are 0.6039, 0.6246, 0.5458, and −0.2557, respectively, reaching a significant level over 0.001 [31].The correlation of PM_{10} with air pressure is also noted (Table 1).

Based on the analysis, a set of weather sensitive parameters signalling a forthcoming hazy event are identified, which can be observed *several hours before* the hazy weather occurs. Among these elements, high temperature, high humidity, moderate wind, and high stability will dominate the poor air-quality in the summer of Beijing. The effects of seasonal changes of meteorological elements on the PLAM will be discussed in next section. Therefore, the PLAM is established as a function of the following parameters:

$$PLAM\,(F) \in f\,(p, t, w, \text{rh}, e, s, c', \ldots), \qquad (1)$$

where p, t, w, rh, e, s, and c' represent air pressure, air temperature, winds, relative humidity, evaporability, stability, and effect parameter associated with the contribution of air pollution $\beta(c')$, respectively.

2.3. Parameterization Method for Air-Quality Diagnosis. Some studies [32–34] have showed that the microprocesses in cloud physics can be described in a parameterization scheme with large-scale observations. It was also found that the stabilized summer weather with high air temperature, high relative humidity, moderate winds, and stability might create a microenvironment for a high PM_{10} concentration in August over Beijing [30], corresponding to a static dynamic forcing (baroclinicity) and thermal forcing (equivalent potential temperature (θ_e) gradient) in the atmospheric moist adiabatic processes [35]. It is shown that the lowest diffusion efficiencies occur more than 90% in moderate winds and stable conditions during the time for the summer period of July and August. Part of the temperature profile ranges between the adiabatic lapse rate (neutral stability) and the isothermal lapse rate [36]. The condensation is also a key factor for diagnosing pollution intensity under given weather conditions in the atmospheric moist adiabatic processes. Apparently, the final PLAM can be attributed to two

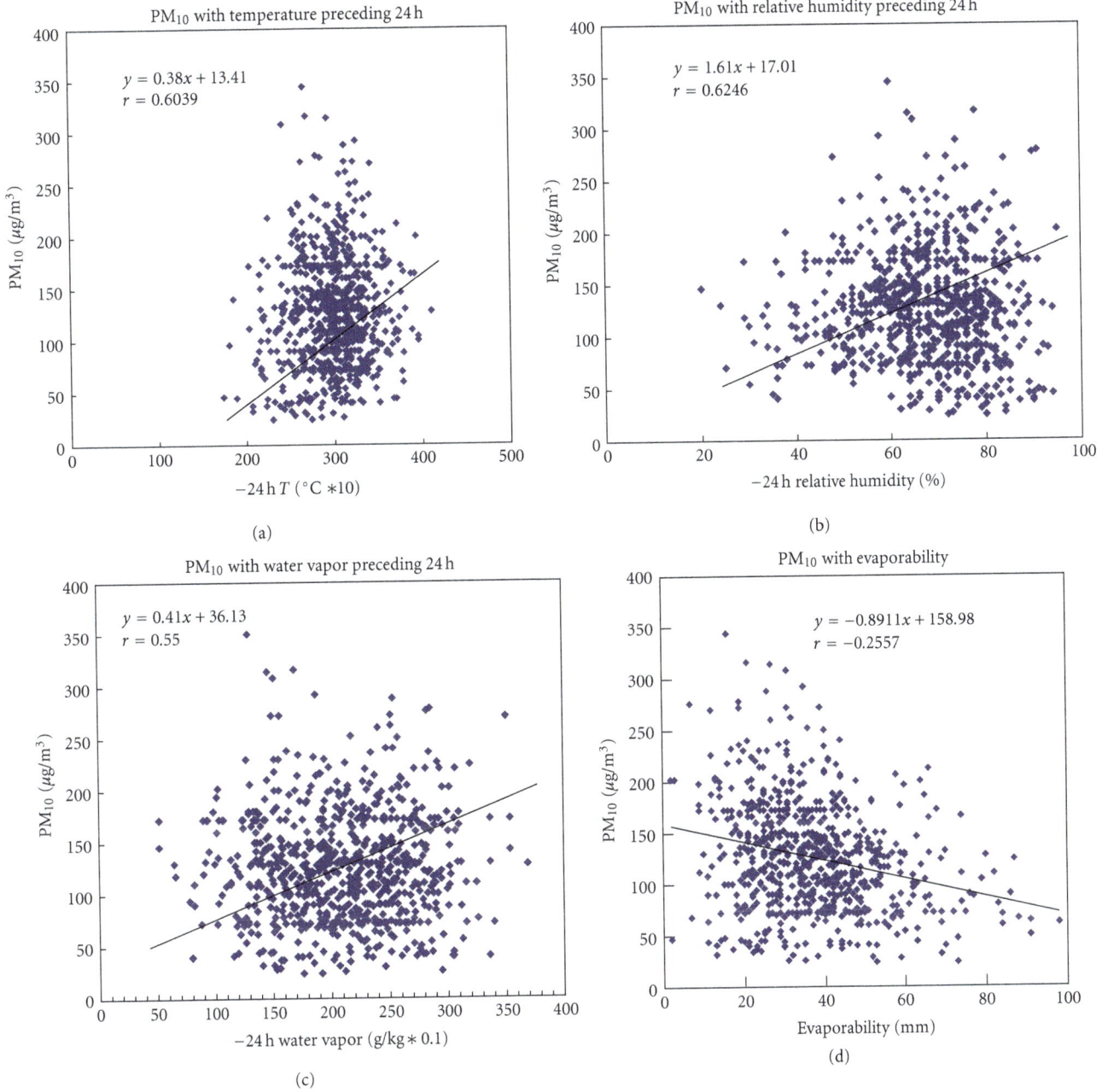

FIGURE 1: Correlation of PM_{10} with temperature (a), relative humidity (b), water vapor observed (c), and evaporability (d) preceding 24 h.

TABLE 1: Correlation analyses for PM_{10} with meteorological elements x_i.

Element	b_0 (Slop)	b_1	Correlation coefficient	Significant level	Period	N
t_m	0.38	13.41	0.6039	>0.001	pre 24 h	736
rh	1.61	17.01	0.6246	>0.001	pre 24 h	736
w (water vapor)	0.41	36.13	0.5458	>0.001	pre 24 h	736
p	−0.02	0.01	0.6481	>0.001	Synchronization	736
Evaporability	−0.89	158.98	−0.2557	>0.001	Synchronization	736

$Y_i = b_0 X_i + b_1$.

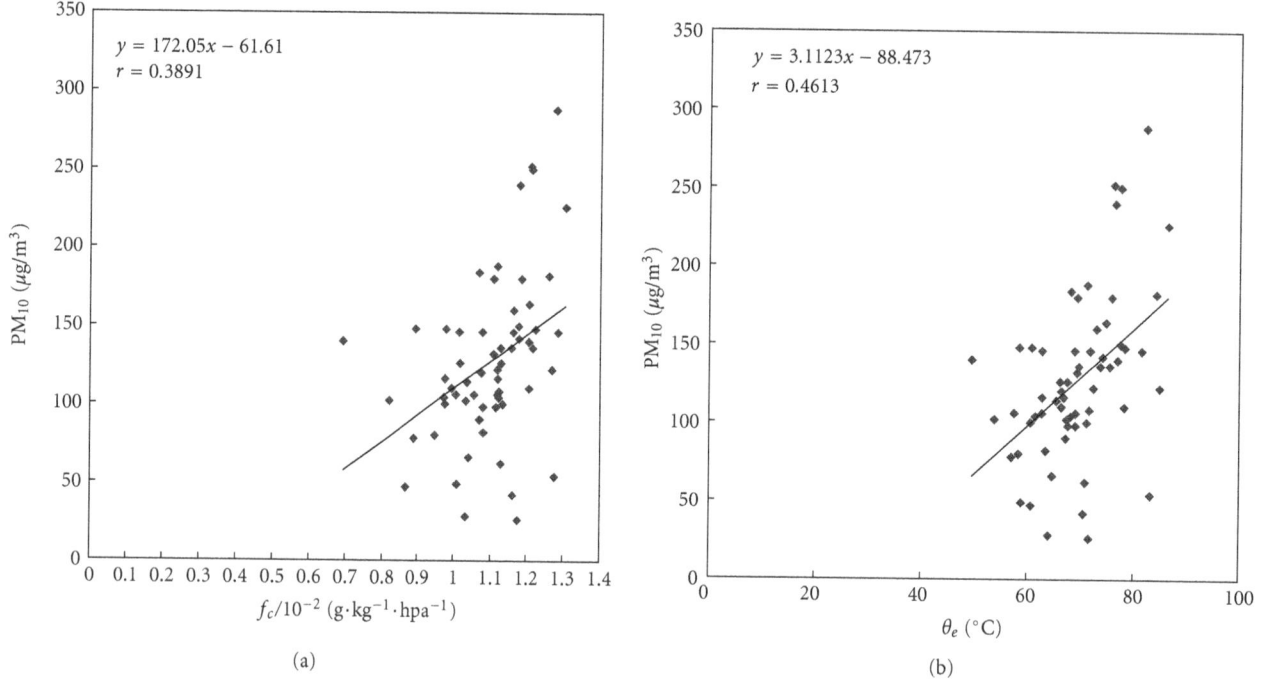

FIGURE 2: Correlation of PM_{10} with f_c (a) and θ_e (b) for July–August 2007.

separate factors: (1) initial meteorological conditions $\alpha(m)$ associated with the atmospheric condensation processes and (2) a dynamic effect parameter associated with the initial contribution of air pollution $\beta(c')$ such as

$$\text{PLAM} = \alpha(m) \times \beta'(c). \tag{2}$$

In the following sections, these two terms are derived and discussed in details.

2.3.1. Initial Meteorological Conditions $\alpha(m)$. According to Gao et al. [37], the meteorological dynamical identification for haze and humid weather in summer Beijing indicated that, in general, the obvious variation of wet-equivalent potential temperature (θ_e) may be observed, which is associated with the releasing of latent heat by adiabatic condensation processes.

Initial meteorological contribution can be expressed as the variation of wet-equivalent potential temperature (θ_e):

$$\alpha(m) = \frac{d\theta_e}{dt} = \theta_e \frac{f_c}{C_p T}, \tag{3}$$

where the condensation function f_c is described by [19, 38]

$$f_c = \frac{f_{cd}}{\left[\left(1 + \left(L/C_p\right)\left(\partial q_s/\partial T\right)\right)_p\right]}, \tag{4}$$

Where C_p is the heat capacity of air, L is the latent heat for condensation or evaporation of water vapor, and q_s is the

specific humidity. f_{cd} is the dry condensation function as defined below:

$$f_{cd} = \left[\left(\frac{\partial q_s}{\partial P}\right)_T + \gamma_p \left(\frac{\partial q_s}{\partial T}\right)_p\right]$$
$$\gamma_p = \frac{R_d}{C_p} \frac{T}{P}, \tag{5}$$

where R_d is the gas constant. In order to analyze the statistical relationship between air quality and meteorological parameters for the special period during the 2008 Beijing Summer Olympic Games (July to August, 2008), the data for July to August 2007 were used in calculating the correlation of PM_{10} with f_c and θ_e in (3). Figures 2(a) and 2(b) show the correlations of the condensation function f_c (4) and wet-equivalent potential temperature θ_e with the observed PM_{10} for July to August, 2007. It is seen that these two meteorological parameters are correlated with the PM_{10} and can be used for forecasting short term variations of PM due to changes in meteorology.

The f_c and θ_e only account for the meteorological contributions to the PM. In reality, the PM concentrations are effected by emissions, meteorology, and atmospheric processes such as nucleation, condensation, and dry/wet depositions.

2.3.2. Relative Dynamic Affect Parameter. In order to quantify the relative impact of weather conditions on air pollution and eliminate the effect of total aerosol concentration change, a ratio of the initial weather conditions to the

$y_{NDJF} = 9.5026x + 87.134$
$r^2 = 0.1547$

$y_{JAS} = 1.2134x + 36.22$
$r^2 = 0.0699$

$y_{MAM} = 3.0695x + 129.85$
$r^2 = 0.0772$

(a)

$y_{all} = 0.4612x + 95.681$

$r^2 = 0.211$

(b)

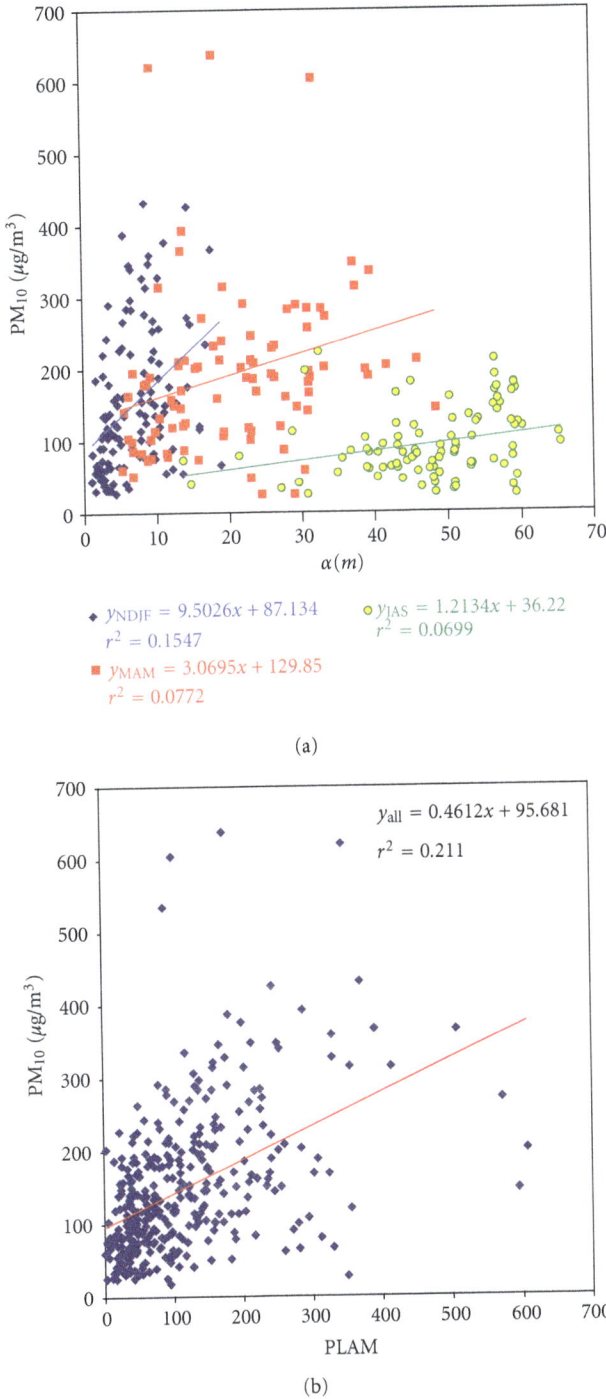

FIGURE 3: Correlations of $\alpha(m)$ and PLAM with PM_{10} observations for a year.

observed PM concentrations is introduced as the relative dynamic affect parameter μ:

$$\mu = \frac{\alpha(m)}{c'}, \qquad (6)$$

where μ denotes the meteorological contribution increment per unit PM concentrations. Initial contribution of air pollution (c') implicitly contains the information of all the

processes on PM before a forecast, including the emission contributions.

There exists a sharp seasonal variation in meteorological parameters. For example, the average temperature in Beijing can range from 28°C in summer to −8°C in winter, and relative humidity also varies a lot from summer to winter. The degree of the variation for $\alpha(m)$ with seasons is larger than that for PM_{10} observations. In order to derive a parameter applicable for a wider range of conditions, an adaptive function β', is introduced

$$\beta' = \frac{(1 - \mu)^{i-1}}{\mu}, \qquad (7)$$

which completes the definition of (2) for the PLAM.

For β' (μ, i), there exist several cases, depending on the size of the relative dynamic parameter value in this spectrum:

(1) $\mu \geq 1$, $i = 1$: favourable meteorological conditions for pollutants either to be diffused or to be maintained;

(2) $0.5 \leq \mu \leq 1$, $i = 2$: more adjustable weather conditions for pollutants to be accumulated;

(3) $\mu < 0.5$, $i = 3$: most favourable weather conditions for pollutants to be accumulated.

Figure 3 shows the correlations of $\alpha(m)$ and PLAM with PM_{10} observations for a year. The slope of the correlation and the magnitude of $\alpha(m)$ reflects the impact degree of meteorology on PM concentrations. In dry and cold season of Beijing and northern China area (NDJF), the observed concentration of PM_{10} can reach as high as $400 \sim 450\,\mu g\,m^{-3}$ and the initial meteorological influence parameter $\alpha(m)$ varies 0~20. Figure 3(a) also indicates that the air pollution situation in the winter time of Beijing is very heavy. However, the influences and contributions of meteorological conditions on the high PM_{10} were not high (shown in blue dot). In hot and humid season of Beijing and northern China area (JAS), the concentration of PM_{10} observations is less than $200\,\mu g\,m^{-3}$. However, due to the hot and humid stabilizing weather conditions, $\alpha(m)$ is larger than that in the winter time, ranging from 40 to 70, which is the highest among all seasons (shown in green dot). In the transition season, that is, spring, (MAM), the concentration of PM_{10} observations can also reach as high as $400 \sim 600\,\mu g\,m^{-3}$ with a medium contribution and influence of initial meteorological $\alpha(m)$ (in red dot). The correlation coefficient (r) of PLAM for all season with PM_{10} observations reaches 0.46 ($R^2 = 0.2116$, Figure 3(b)), indicating that PLAM is a good indicator for air quality diagnosis.

Because the levels of air pollutants are usually controlled by a combination of meteorological variables within the stable air masses, the impacts of individual meteorological variables on air pollution may not be significant and/or may be cancelled each other, as they do not account for the interrelation between variables. PLAM, as an integrated index, considered the interactions among the meteorological variables such as the individual wind direction, temperature and hence reflected the better performance than any individual variable.

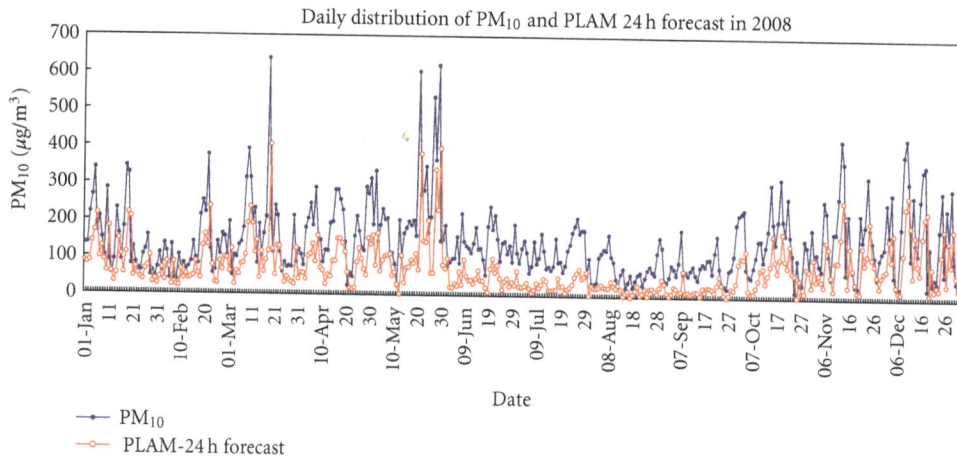

FIGURE 4: Daily PLAM distribution in 24 h forecasts, PM_{10} observed at NCOB from 1 January to 31 December 2008.

3. Annual Forecasts for 2008

To evaluate the performance of PLAM forecastability, an annual run was carried out to obtain the 24 hr forecasts of daily PLAM and compared with the observed PM_{10} at NCOB and the Air Pollution Index (API) defined by the Ministry of Environment of China taking into account of the air concentrations of NO_x, NO_2, SO_2, TSP, and PM_{10}. The API is calculated as follows:

$$I_i = \frac{\left(C_i - C_{ij}\right)}{\left(C_{ij+1} - C_{ij}\right)}\left(I_{ij+1} - I_{ij}\right) + I_{ij}, \qquad (8)$$

where I_i is the API index for species i, C_i is the observed daily averaged concentrations for species i. C_{ij}, C_{ij+1} and I_{ij}, I_{ij+1} are the two standard daily averaged concentrations and API indices that are adjacent to the observed C_i and are determined from the lookup table (Table 2). The largest index for species i is regarded as the API index of the day. In China, SO_2, TSP, and PM_{10} are usually the major air pollutants and correlated very well [23].

Figure 4 shows the daily PLAM distribution in 24 h forecasts from 1 January to 31 December 2008. A good agreement was found between the forecasted and observed, including the daily variations, peaks and lows. There were a number of heavy pollution days between March and May as well as in September and December with PM_{10} greater than $200 \, \mu g \, m^{-3}$, among which there existed several extreme high pollution episodes around March 18, May 20–22, and May 27-28 with $PM_{10} > 200$–637, 280–604, and 200–450 $\mu g \, m^{-3}$, respectively. These high episodes were all captured by the PLAM index. The features of comparison between the PLAM index forecast and PM_{10} observations at NCOB from 1 June to 31 August 2008 during the 2008 Olympic Games will be discussed in Section 4. The correlations between forecasted PLAM index and observed PM_{10} and API are shown in Figure 5. A correlation coefficient was achieved for 0.85 between the forecasts and API and 0.43 between the forecasts and observed PM_{10} for whole of the year 2008. All reached the significant level of 0.001. During the summer

season from mid-June to late September, that is, 2008 Beijing Olympic Games period, the air quality was relatively high with low PLAM indices forecasted. Detailed forecasts will be given Figure 6 in Section 4.

Comparing Figures 5(a) and 5(b), API is calculated from the average PM_{10} measurements at different sites in the Beijing metropolitan area, indicating that the PLAM has the better capability to capture the regional pollution rather than a single site.

4. Applications of PLAM

4.1. Olympic Applications from June to September in 2008. Figure 6 shows the time series of the PLAM index and PM_{10} observations at NCOB from 1 June to 31 August 2008. During this period of time, the averaged PLAM index was small (<50) with a declining rate of 35%, which signaled an improving meteorology for the period. However, a much larger declining rate of 83% for PM_{10} was observed. Under normal conditions, the variations of PLAM and PM_{10} should be rather consistent. The significant discrepancy of the variation trend should be caused by some nonmeteorological factors. It is well known that very active emission control measures were taken during the 2008 Olympic Games [3]. It is likely that the effective emission controls have contributed to the large declining rate for the PM_{10} in Beijing. Excluding the meteorological influence, that is, PLAM, the emission control may have contributed to a 43% reduction in PM_{10} from June 1 to August 31, 2008.

To evaluate the performance of the PLAM method, the correlative analysis has been made on the 24 h PLAM forecasts and observations (Figure 7). A correlation coefficient was achieved for 0.8209 between the forecasts and API and 0.5924 between the forecasts and observed PM_{10}. All reached the significant level of 0.001.

The good agreement of the forecasted PLAM index with the API and PM_{10} indicates the ability of PLAM as a criterion for diagnosing air quality problem. The high correlation illustrates that the meteorology has a dominant control over the variation of air quality in Beijing.

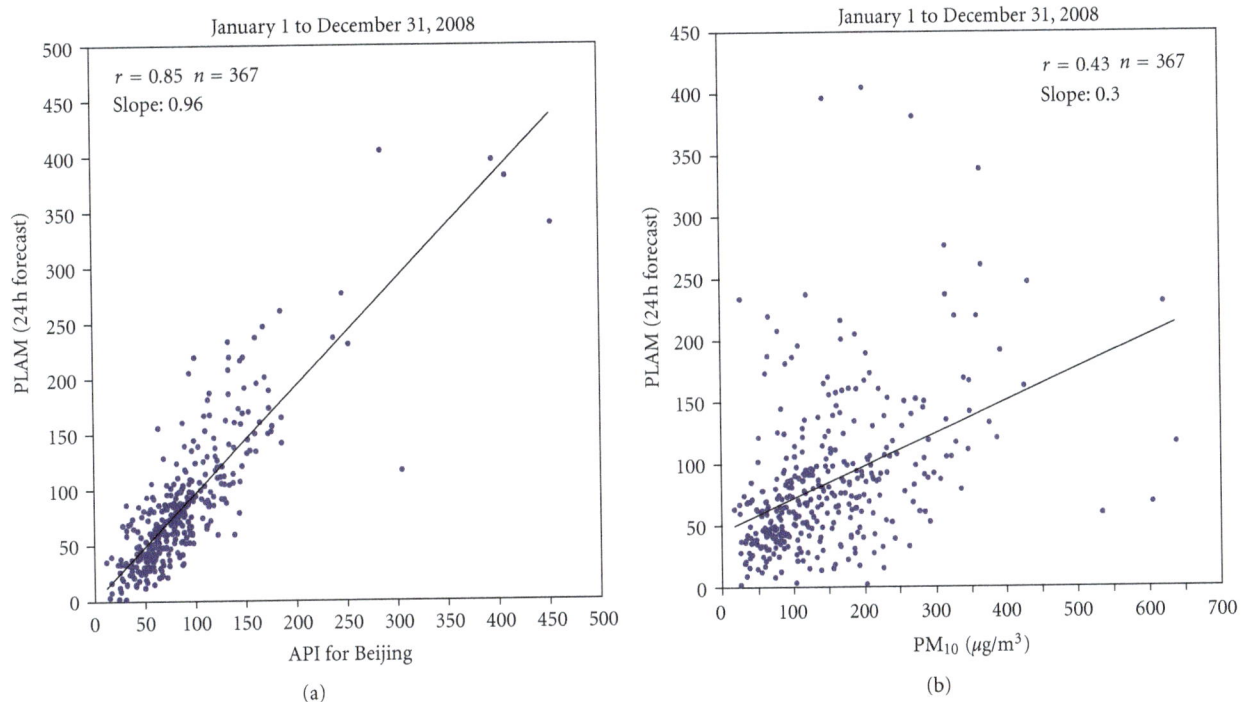

FIGURE 5: Correlations of 24 h PLAM forecast results with API of Beijing (a) and actual PM$_{10}$ in Beijing (b) for the period from 1 January to 31 December 2008.

TABLE 2: Standard air pollutant concentrations and indices.

API		500	400	300	200	100	50
Pollutant concentrations (mg/m³)	SO$_2$	2.62	2.100	1.60	0.80	0.15	0.05
	NO$_x$	0.94	0.750	0.565	0.15	0.10	0.05
	NO$_2$	0.94	0.750	0.565	0.28	0.12	0.08
	TSP	1.00	0.875	0.625	0.50	0.30	0.12
	PM$_{10}$	0.60	0.500	0.420	0.35	0.15	0.05

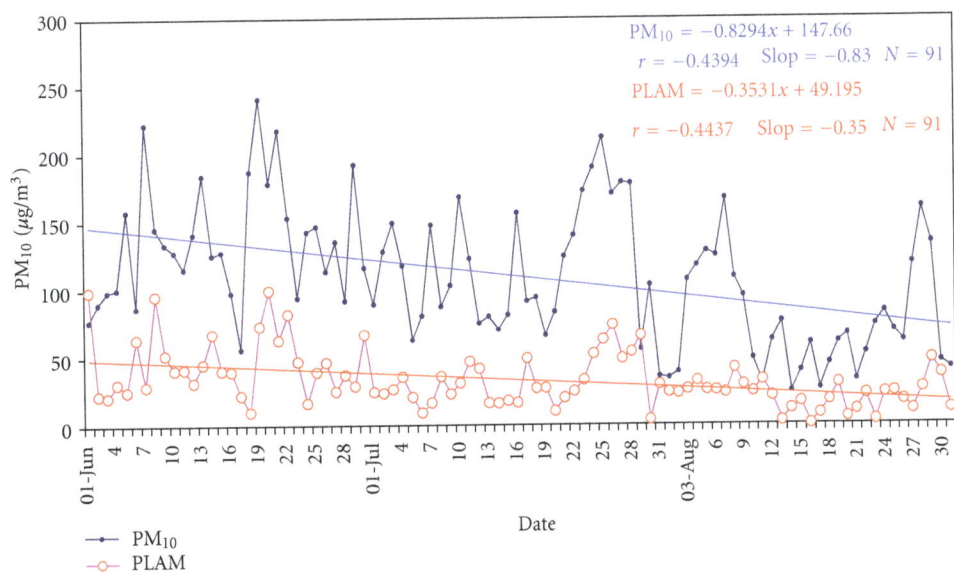

FIGURE 6: Daily PLAM distribution and PM$_{10}$ observed in NCOB in Beijing.

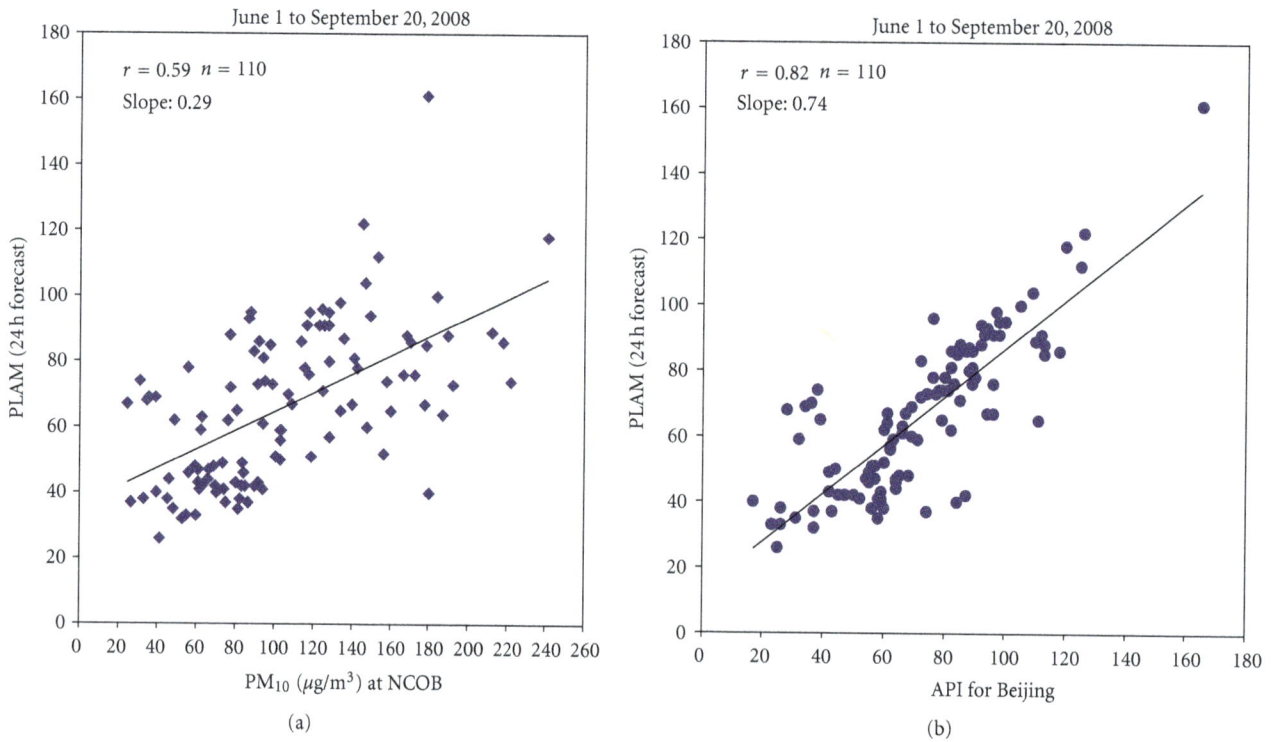

FIGURE 7: Correlations of 24 h PLAM forecast results with actual PM_{10} in Beijing Nanjiao (a) and API of Beijing (b) during the period of Beijing Olympic Game.

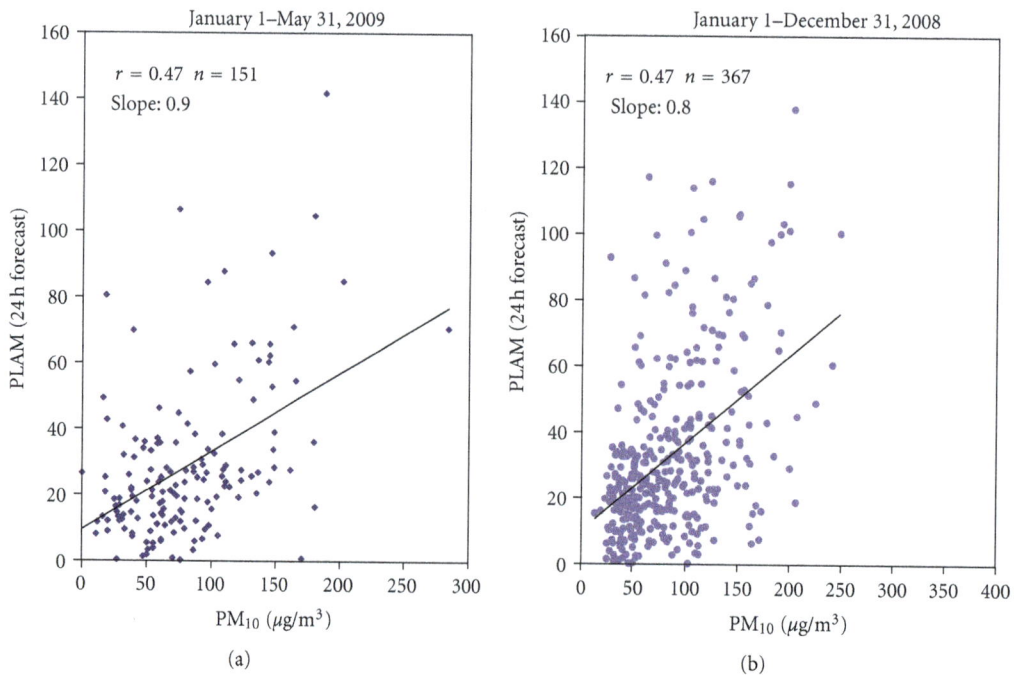

FIGURE 8: Correlation analysis between PLAM 24 h prediction and PM_{10} January 1–December 31, 2008 in Shanghai (a) and correlation analysis between PLAM 24 h prediction and of PM_{10} January 1–May 31, 2009 in Guangzhou (b).

4.2. PLAM Applications in Shanghai and Guangzhou. In order to check the applicability of the PLAM in various regions of China, two other cities were chosen, that is, Shanghai and Guangzhou, to evaluate the performance of the PLAM. Shanghai (31.1_N, 121.4_E, 8.2 m) is the largest city in Yangtze River Delta in East China with climatic characteristics different from North China and is located in the East Asian monsoon region with humid and rainy climate on the East coast of the Asian continent. Guangzhou (23.1,1_N 13.3_E, 7.3 m) located in the coast of South China has a subtropical climate features. Figure 8(a) shows correlation analysis between the PLAM 24 h prediction and PM_{10} during January 1–December 31 in 2008 in Shanghai. Figure 8(b) shows correlation analysis between the PLAM 24 h prediction and of PM_{10} during January 1–May 31 in 2009 in Guangzhou. From Figure 8 it can be seen that the PLAM also applies to eastern and southern areas of China for different climatic characteristics regions in China.

5. Conclusions and Discussions

A parameterized method has been developed to predict the air quality in BISA and used successfully during the 2008 Beijing Olympic Games. The following was found.

(1) Water vapor condensation (f_c), wet potential equivalent temperature (θ_e), air pressure, air temperature, relative humidity, and evaporability are found to be the key meteorological factors for diagnosing air pollution intensity.

(2) With the help of two independent data systems (PM_{10} values reported in API in Beijing and daily PM_{10} data collected by Beijing Nanjiao Observatory), the tests of PLAM performance showed that the forecasts have a cohesive correlation with them, reaching to a significant level.

(3) It is found that the correction factor β' enables the ability for PLAM to diagnose the observed PM_{10} concentrations all year round. The 24 ~ 72 h forecasts by PLAM provided valuable services for the day of opening ceremony and subsequent events during the whole Beijing Olympic Games. PLAM also was applied to eastern and southern areas of China for different climatic characteristics regions in China.

Acknowledgments

This research is supported by the National "973" Program of China under Grant no. 2011CB403404, no. 2011CB403401, and the Project (2009Y002).

References

[1] I. Sabbah, "Impact of aerosol on air temperature in Kuwait," *Atmospheric Research*, vol. 97, no. 3, pp. 303–314, 2010.

[2] G. C. Fang and S. C. Chang, "Atmospheric particulate (PM$_{10}$ and PM2.5) mass concentration and seasonal variation study in the Taiwan area during 2000–2008," *Atmospheric Research*, vol. 98, no. 2-4, pp. 368–377, 2010.

[3] X. Y. Zhang, Y. Q. Wang, W. L. Lin et al., "Changes of atmospheric composition and optical properties over beijing 2008 olympic monitoring campaign," *Bulletin of the American Meteorological Society*, vol. 90, no. 11, pp. 1633–1651, 2009.

[4] Y. Liu, W. Li, and X. Zhou, "Simulation of secondary aerosols over North China in summer," *Science in China D*, vol. 48, no. 2, pp. 185–195, 2005.

[5] G. Z. Zhang, X. D. Xu, J. Z. Wang, and Y. Q. Yang, "A study of characteristics and evolution of urban heat island over Beijing and its surrouding area," *Journal of Applied Meteorological Science*, vol. 13, p. 41, 2002.

[6] X. A. Xia, H. B. Chen, P. C. Wang, X. M. Zong, J. H. Qiu, and P. Gouloub, "Aerosol properties and their spatial and temporal variations over North China in spring 2001," *Tellus B*, vol. 57, no. 1, pp. 28–39, 2005.

[7] M. Xue, J. Ma, P. Yan, and X. Pan, "Impacts of pollution and dust aerosols on the atmospheric optical properties over a polluted rural area near Beijing city," *Atmospheric Research*, vol. 101, no. 4, pp. 835–843, 2011.

[8] J. Wang, Y. Yang, G. Zhang, and S. Yu, "Climatic trend of cloud amount related to the aerosol characteristics in Beijing during," *Acta Meteorologica Sinica*, vol. 24, no. 6, pp. 762–775, 2010.

[9] M. El-Metwally, S. C. Alfaro, M. M. Abdel Wahab, A. S. Zakey, and B. Chatenet, "Seasonal and inter-annual variability of the aerosol content in Cairo (Egypt) as deduced from the comparison of MODIS aerosol retrievals with direct AERONET measurements," *Atmospheric Research*, vol. 97, no. 1-2, pp. 14–25, 2010.

[10] C. H. Zhou, S. L. Gong, X. Y. Zhang et al., "Development and evaluation of an operational SDS forecasting system for East Asia: CUACE/Dust," *Atmospheric Chemistry and Physics*, vol. 8, no. 4, pp. 787–798, 2008.

[11] Y. Q. Wang, X. Y. Zhang, R. Arimoto, J. J. Cao, and Z. X. Shen, "Characteristics of carbonate content and carbon and oxygen isotopic composition of northern China soil and dust aerosol and its application to tracing dust sources," *Atmospheric Environment*, vol. 39, no. 14, pp. 2631–2642, 2005.

[12] S. L. Gong, L. A. Barrie, J. P. Blanchet et al., "Canadian aerosol module: a size-segregated simulation of atmospheric aerosol processes for climate and air quality models 1. Module development," *Journal of Geophysical Research D*, vol. 108, no. 1, pp. 3–16, 2003.

[13] Y. Q. Yang, J. Z. Wang, Q. Hou, Y. Li, and C. H. Zhou, "Discriminant Genetic Algorithm Extended (DGAE) model for seasonal sand and dust storm prediction," *Science China Earth Sciences*, vol. 54, no. 1, pp. 10–18, 2011.

[14] L. Han, G. Zhuang, Y. Sun, and Z. Wang, "Local and non-local sources of airborne particulate pollution at Beijing—the ratio of Mg/Al as an element tracer for estimating the contributions of mineral aerosols from outside Beijing," *Science in China B*, vol. 48, no. 3, pp. 253–264, 2005.

[15] K. K. Sui, Z. F. Wang, J. Yang, F. B. Xie, and Y. Zhao, "Beijing persistent PM$_{10}$ pollution and its relationship with general meteorological features," *Research of Environmental Sciences*, vol. 20, p. 77, 2007.

[16] Y. Ji, H. J. Fan, Q. F. Wang, and L. H. Nie, "Air particle concentration and meteorological factors," *Journal of Environmental Health*, vol. 25, p. 554, 2008.

[17] X. Xiao, L. Pengfei, G. Fuhai et al., "Comparison of black carbon aerosols in urban and suburban areas of Shanghai," *Journal of Applied Meteorological Science*, vol. 22, p. 158, 2011.

[18] S. Y. Yu, Z. Zhang, C. Q. Peng et al., "Effects of meteorological factors on SO_2 and other atmospheric pollutions in Shenzhen

China," *Journal of Environment and Health*, vol. 25, p. 483, 2008.

[19] G. Zhang, L. Bian, J. Wang, Y. Yang, W. Yao, and X. Xu, "The boundary layer characteristics in the heavy fog formation process over Beijing and its adjacent areas," *Science in China D*, vol. 48, no. 2, pp. 88–101, 2005.

[20] X. Pang, Y. Mu, X. Lee, Y. Zhang, and Z. Xu, "Influences of characteristic meteorological conditions on atmospheric carbonyls in Beijing, China," *Atmospheric Research*, vol. 93, no. 4, pp. 913–919, 2009.

[21] X. Y. Tang, Y. H. Zhang, and M. Shao, *Atmospheric Environment Chemistry*, Higher Education Press, Beijing, Japan, 2006.

[22] Y. Q. Yang, Q. Hou, C. H. Zhou, H. L. Liu, Y. Q. Wang, and T. Niu, "Sand/dust storm processes in Northeast Asia and associated large-scale circulations," *Atmospheric Chemistry and Physics*, vol. 8, no. 1, pp. 25–33, 2008.

[23] S. Wang and X. L. Zhang, "Meteorological features of PM_{10} pollution in Beijing," *Journal of Applied Meteorology*, vol. 13, p. 177, 2002.

[24] C. D. O'Dowd, J. A. Lowe, and M. H. Smith, "The effect of clouds on aerosol growth in the rural atmosphere," *Atmospheric Research*, vol. 54, no. 4, pp. 201–221, 2000.

[25] Y. Wang, J. Guo, T. Wang et al., "Influence of regional pollution and sandstorms on the chemical composition of cloud/fog at the summit of Mt. Taishan in northern China," *Atmospheric Research*, vol. 99, no. 3-4, pp. 434–442, 2011.

[26] C. Honoré, L. Rouïl, R. Vautard et al., "Predictability of European air quality: assessment of 3 years of operational forecasts and analyses by the PREV'AIR system," *Journal of Geophysical Research D*, vol. 113, no. 4, Article ID D04301, 2008.

[27] S. A. McKeen, S. H. Chung, J. Wilczak et al., "Evaluation of several PM2.5 forecast models using data collected during the ICARTT/NEAQS 2004 field study," *Journal of Geophysical Research D*, vol. 112, no. 10, Article ID D10S20, 2007.

[28] M. D. Moran et al., "Particulate-Matter Forecasting with GEM-MACH15, A New Canadian Air-Quality Forecast Model," in *Proceedings of the 30th NATO/SPS ITM on Air Pollution Modelling and Its Application*, San Francisco, Calif, USA, 2009.

[29] H. Che, G. Shi, A. Uchiyama et al., "Intercomparison between aerosol optical properties by a PREDE skyradiometer and CIMEL sunphotometer over Beijing, China," *Atmospheric Chemistry and Physics*, vol. 8, no. 12, pp. 3199–3214, 2008.

[30] Y. Q. Yang, J. Z. Wang, Q. Hou, and Y. Q. Wang, "A plam index for beijing stabilized weather forecast in summer over Beijing," *Journal of Applied Meteorological Science*, vol. 20, p. 649, 2009.

[31] F. Y. Wei, *Modern Diagnostic Techniques for Climatologically Statistics*, China Meteorology Press, 1999.

[32] H. L. Kuo, "Convective weather in conditionally unstable atmosphere," *Tellus*, vol. 13, p. 441, 1961.

[33] H. L. Kuo, "On formation and intensification of tropical cyclone through latent heat release in cumulus convection," *Journal of the Atmospheric Sciences*, vol. 22, p. 40, 1965.

[34] H. L. Kuo, "Further studies on the parameterization of the influence of cumulus convection in large-scale flows," *Journal of the Atmospheric Sciences*, vol. 31, p. 1232, 1974.

[35] S. Gao, X. Wang, and Y. Zhou, "Generation of generalized moist potential vorticity in a frictionless and moist adiabatic flow," *Geophysical Research Letters*, vol. 31, no. 12, p. L12113, 2004.

[36] A. B. Johnson and D. G. Baker, "Climatology of diffusion potential classes for Minneapolis-St. Paul," *Journal of Applied Meteorology*, vol. 36, no. 12, pp. 1620–1628, 1997.

[37] S. Gao, Y. Zhou, T. Lei, and J. Sun, "Analyses of hot and humid weather in Beijing city in summer and its dynamical identification," *Science in China, Series D: Earth Sciences*, vol. 48, no. 2, pp. 128–137, 2005.

[38] J. Z. Wang and Y. Q. Yang, *Contemporary Weather Engineering*, Meteorological Press, Beijing, Japan, 2000.

12

Modeling Effects of Climate Change on Air Quality and Population Exposure in Urban Planning Scenarios

Lars Gidhagen,[1] Magnuz Engardt,[1] Boel Lövenheim,[2] and Christer Johansson[2, 3]

[1] Swedish Meteorological and Hydrological Institute, 601 76 Norrköping, Sweden
[2] Environment and Health Administration, Box 8136, 104 20 Stockholm, Sweden
[3] Department of Applied Environmental Science, Stockholm University, 106 91 Stockholm, Sweden

Correspondence should be addressed to Magnuz Engardt, magnuz.engardt@smhi.se

Academic Editor: Eugene Rozanov

We employ a nested system of global and regional climate models, linked to regional and urban air quality chemical transport models utilizing detailed inventories of present and future emissions, to study the relative impact of climate change and changing air pollutant emissions on air quality and population exposure in Stockholm, Sweden. We show that climate change only marginally affects air quality over the 20-year period studied. An exposure assessment reveals that the population of Stockholm can expect considerably lower NO_2 exposure in the future, mainly due to reduced local NOx emissions. Ozone exposure will decrease only slightly, due to a combination of increased concentrations in the city centre and decreasing concentrations in the suburban areas. The increase in ozone concentration is a consequence of decreased local NOx emissions, which reduces the titration of the long-range transported ozone. Finally, we evaluate the consequences of a planned road transit project on future air quality in Stockholm. The construction of a very large bypass road (including one of the largest motorway road tunnels in Europe) will only marginally influence total population exposure, this since the improved air quality in the city centre will be complemented by deteriorated air quality in suburban, residential areas.

1. Introduction

Worldwide air pollution cause more than 2 million premature deaths annually [1]. A majority of the world's population today live in cities. One of the most challenging tasks of city planning is to cope with demands for efficient transport systems of the increasing urban population without risking adverse health impacts through poor air quality. The European air pollution directive states that if a proposed plan will lead to exceedances of air quality limit values, it should not be implemented. Estimates of future air quality are based on air quality dispersion models using assumptions of future air pollutant emissions. Future air quality is, however, not only a matter of emissions. City planners need also to consider climate change in order to fully assess the impacts on the environment (water and air quality), local climate (heat waves, storm water, and river flooding events), and the effects on population health. This requires much higher spatial resolution than available from global climate modeling [2]. Downscaling of global climate models to local scales has been done to assess effects on land use [3], agricultural potential [4], rain events, runoff, and local meteorology [5]. The effect of climate change on European air quality levels has been investigated in earlier studies, for example, [6, 7], but larger effects can be expected over urban areas and during pollution episodes [8]. Studies that address climate change effects on urban air quality and population exposure are only recently published. Mahmud et al. [9] assessed future exposure levels of particulate matter in California.

In this work we investigate the role of climate change on future urban air quality and compare it with the effects of changing emissions in Europe and locally in a specific urban area, namely, Stockholm, Sweden. We compare the future year 2030 pollution levels in the urban background air with current levels (ca. year 2010). We also assess the effects of different traffic solutions on future urban air quality. The impact on air quality and population exposure of a road transit scenario, in which a new bypass highway is constructed, is compared to that of a reference scenario

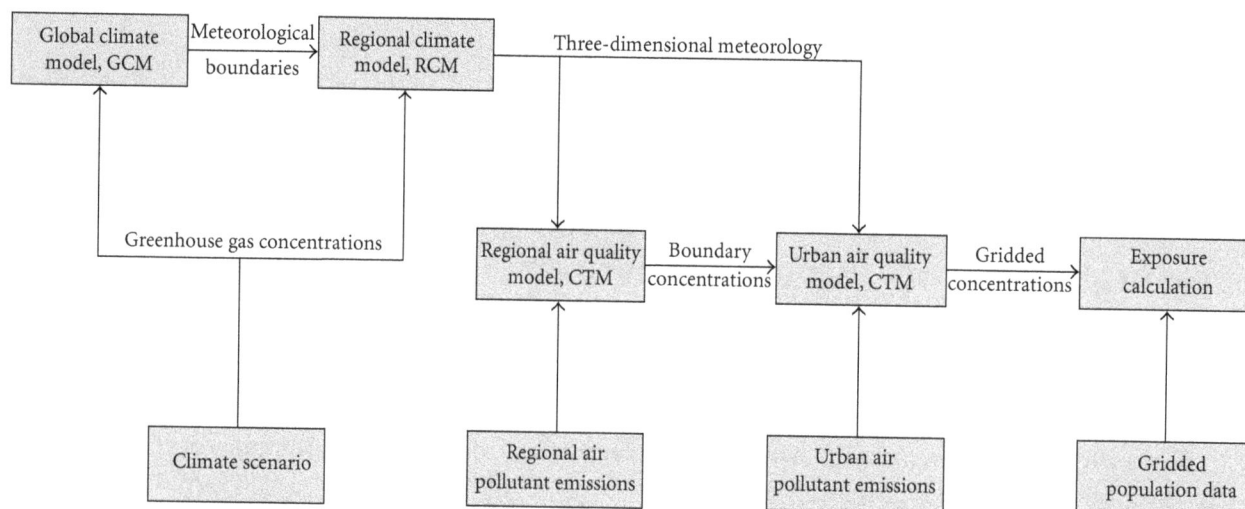

FIGURE 1: Conceptual overview of the model chain used in the present study. The global climate models (*ECHAM5 A1B-r3* and *HadCM3 A1B-ref*) utilizing the A1B climate scenario are being regionally downscaled by the Rossby centre regional climate model, RCA3. The regional and urban air quality models (MATCH) are driven by the same three-dimensional meteorology but forced by different emissions estimates, the urban air quality model takes boundary concentrations from the European-scale air quality model. The calculated near-surface concentrations from the urban scale air quality simulation are finally coupled to gridded population data to produce population exposure in the urban domain.

where no new roads are created. The comparison is made for year 2030 when the road project should be completed.

2. Methods

Figure 1 gives an overview on how different input and model results are coupled in the present study. Below we explicitly discuss some of the more important steps in our analysis of future air quality and population exposure in Stockholm.

2.1. Climate Scenarios. To assess the present and future climate in Stockholm we make use of European climate data provided by the Rossby Centre [5]. The Rossby Centre has completed a suite of regional climate simulations (climate downscaling), which provide more details over Europe than the global climate models (GCMs). The regional climate simulations address a range of uncertainties regarding projections of future climate in Europe through studying different climate scenarios (i.e., greenhouse gas concentrations), utilizing different global climate models as drivers of the regional climate model (RCM), taking into account different initial states and different climate sensitivities of the global climate models. The largest spread in the regional climate projections originates from varying the driving GCM. For our air quality simulations we will use climate downscaling of two different GCMs (*ECHAM5 A1B-r3* and *HadCM3 A1B-ref*, cf. Kjellström et al. [5]) to cover some of the inherent uncertainty in predicting the future climate. Three-dimensional regional climate data is available at model levels every 3 or 6 hours for use in the air quality models.

As an illustration of the expected climate change and intermodel differences, Figure 2 shows annual-average temperature and precipitation in Stockholm for the years used as present (2009–2011) and future (2029–2031) climate.

We note that both *ECHAM5 A1B-r3* and *HadCM3 A1B-ref* feature a clear increase in local temperature, while precipitation changes only little (*ECHAM5 A1B-r3*) or increases slightly (*HadCM3 A1B-ref*); both tendencies are in line with the long-term future trend towards a warmer and wetter climate in Scandinavia reported by Kjellström et al. [5]. The short averaging times—only three years—and the limited timespan of only 20 years, result in natural year-to-year variations obscuring the long-term trends in climate. The change in wind speed is not significant in any of the two climate downscaling.

2.2. Air Quality on the European Scale. To simulate present and future air quality over Europe, we use a regional chemistry transport model (CTM)—MATCH [10, 11] driven by meteorology (precipitation updated every 3 hours, all other parameters every 6 hours) from the regional climate model, time varying emissions of air pollutants from RCP4.5 [12], and with seasonally varying tracer concentrations at the lateral and top boundaries. The boundaries were identical during the present and the future period; for numerical values, see Andersson et al. [11]. The pan-European air quality simulations operate on the same horizontal grid as the regional climate model covering Europe with 50 km resolution but using 15 model levels up to ~6 km. The set-up was recently evaluated [13], and it was concluded that the CTM simulations using climate model output were able to capture the major features of the observed distribution of surface O_3 over Europe, although the spatial correlation is lower compared to results obtained using meteorological data constrained by observations. Further details of the set-up can be found in Andersson and Engardt [7]. The RCP4.5 emissions are available on a global $0.5° \times 0.5°$ grid every 10 years from 1960 to 2100. The RCP4.5 scenario assumes that

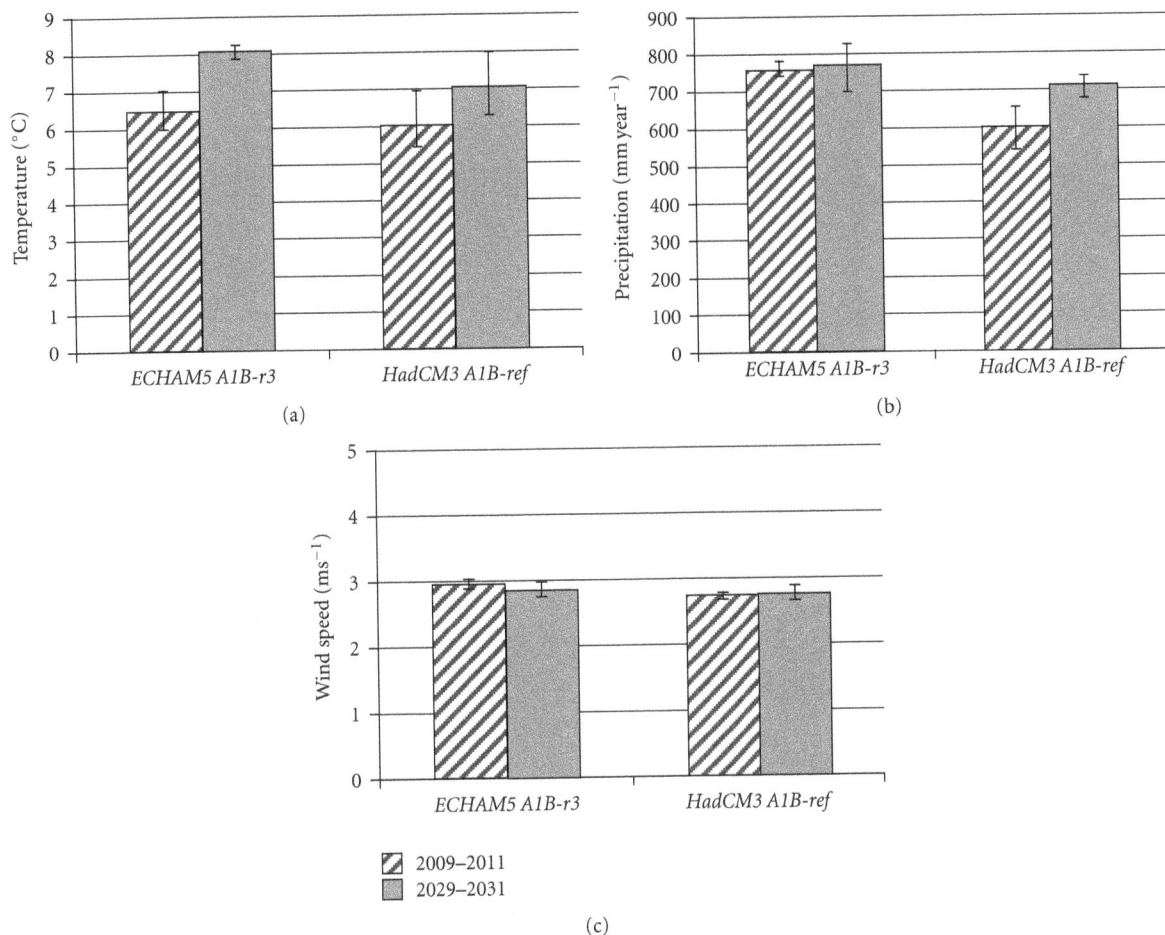

FIGURE 2: Three-year-averages of 2 m temperature (a), precipitation (b), and wind speed (c) at Torkel Knutssonsgatan in city center of Stockholm for the period 2009–2011 (striped) and 2029–2031 (grey). Error bars indicate maximum and minimum annual average for the 3 years.

TABLE 1: Local (2 × 2 km² grid over Stockholm) and European emissions for present situation (year 2010) and for two alternative future (year 2030) scenarios.

Substance	Present 2010 (tons/year)	Reference 2030 (compared to Present)	Road transit project 2030 (compared to Present)
Local NO_x	15 511	58.9%	59.1%
Local PM_{10}	3 879	113%	114%
Europe NO_x	26×10^6	81%	81%

all nations mitigate their emissions in response to a shared price system for greenhouse gases where different gases are priced according to their global warming potential [12]. With this driving mechanism, substantial reductions in European NOx emissions are projected between 2010 and 2030, from 26 Tg year⁻¹ in 2010 to 21 Tg year⁻¹ in 2030, cf. Table 1.

2.3. Air Quality Downscaling over Stockholm. The Stockholm air quality downscaling is performed by operating a high-resolution set-up of MATCH forced with interpolated meteorology from the regional climate model; the methodology follows Gidhagen et al. [14]. The urban air quality simulations take boundary concentrations—including top boundary—every three hour from the pan-European set-up of MATCH. For assessing future NO_2, and O_3 levels the urban downscaling was performed over a 102 × 102 km² region which includes the Stockholm Metropolitan area and also the city of Uppsala north of Stockholm (Figure 3). The horizontal resolution is 2 km, and the vertical resolution identical to the European set-up, with the lowest model layer being 60 m thick. The PM_{10}, NO_2 and O_3 exposure assessment was made through a downscaling over a smaller 36 × 30 km² domain with 1 km spatial resolution (domain also indicated in Figure 3). For PM_{10} we do not present total concentrations since the European scale model only simulates the tendencies for secondary inorganic aerosols and does not project what will happen in the future with important part of the PM mass such as organic aerosols and sea salt. However, since the downscaling over Stockholm involves primary PM and we have access to a high-quality emission inventories for both present and future periods, we can project the changes in PM exposure for different emission scenarios within Stockholm.

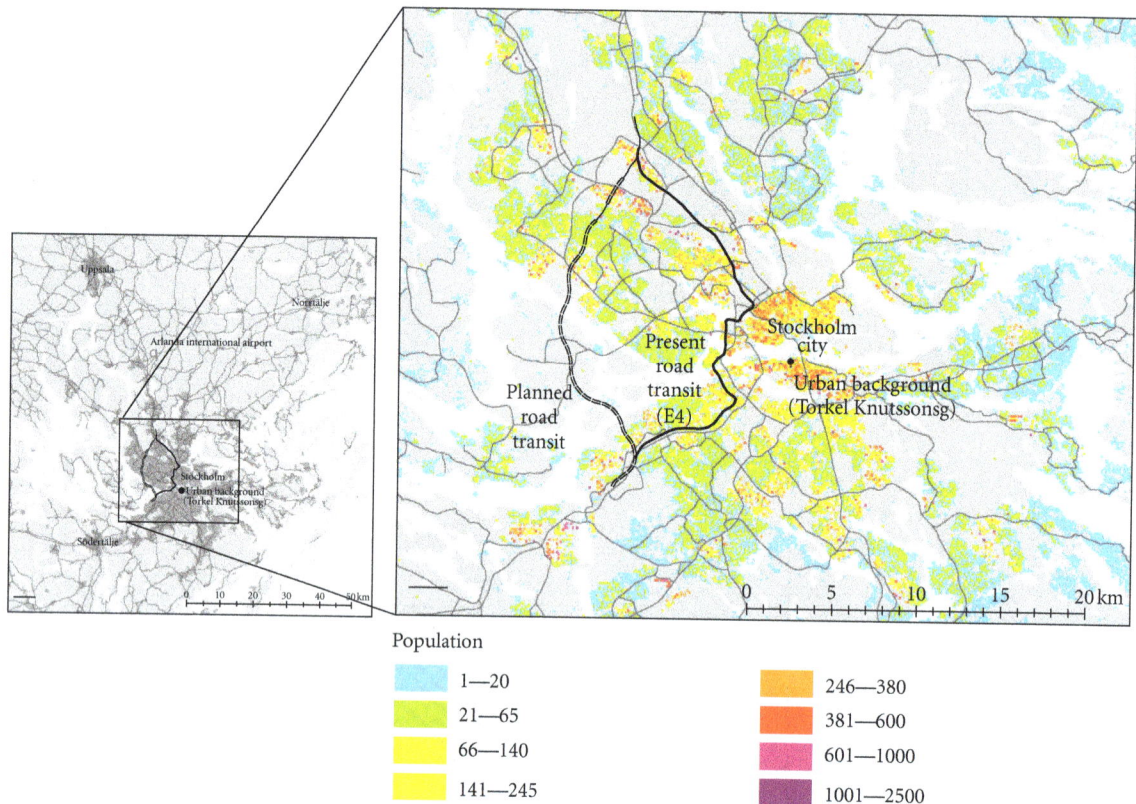

Population

🟦	1—20	🟧	246—380
🟩	21—65	🟥	381—600
🟨	66—140	🟪	601—1000
🟨	141—245	🟪	1001—2500

FIGURE 3: Map showing the air quality downscaling modeling domain (102×102 km^2, left) and the local domain (36×30 km^2, right) used for population exposure assessment. The locations of present and planned road transit are indicated with solid and broken lines, respectively. The location of the urban background monitoring station is indicated with a filled circle. The colors indicate population density (number of people in 100×100 m^2 squares). Thin black lines are roads. White and grey areas indicate water and land.

2.4. *Urban Emissions*. A detailed local emission database is administered and updated annually. It includes the two counties of Stockholm and Uppsala, covering an area with some 30 municipalities and ca. 2 million inhabitants [15]. The estimates of total traffic volumes are primarily based on *in situ* measurements. Such measurements are of different kinds: regular automatic traffic counting by the local traffic and street authorities within municipalities, automatic traffic counting on main roads by the Swedish National Road Administration, and manual surveys of traffic volumes. Variations of vehicle compositions and temporal variation of the traffic volumes are described for different road types. Present and future vehicle fleet composition and vehicle exhaust emission factors are based on the Swedish application of the ARTEMIS model [16]. In addition to the vehicle exhaust emissions there are large nontailpipe emissions of particulate matter due to wear of road surfaces, brakes, and tires. In Stockholm the nontailpipe emissions dominate and emission factors are estimated based on local measurements [17, 18].

The local air pollution emissions in the Stockholm subregions are described for three different situations: (a) current (2010) situation, (b) a future (2030) scenario with a new transit road, and (c) a future (2030) scenario without the bypass. The transit road is mainly an underground road tunnel (21 km). Emissions are described for all important

sectors but the difference in emissions in the three situations is only due to differences in road traffic emissions. Traffic prognoses for the future scenarios are obtained from a national traffic forecast model system called SAMPERS [19]; a travel demand forecasting tool. SAMPERS is based on travel enquiries and describes the transports using cars, public transport, cycling, and walking depending on the distances, destination, availability of different transportation systems and so forth. It also includes a model that considers peoples willingness to pay in order to account for taxes, for example, the congestion tax in central Stockholm [20]. Traffic forecasts are based on assumptions on future developments and involve many uncertainties. Likely the largest uncertainty is due to input data, such as assumptions on future economic development, salaries, car ownership, and so forth [21]. The uncertainties of the SAMPERS model itself (due to the model, not due to input data assumptions) have been evaluated using a bootstrap method [22]. Beser Hugosson [22] found a 95-percent confidence interval to be ±8% to ±11% of the total traffic volume on the road links studied. But the uncertainties in the difference in travel demand between two traffic scenarios have not been assessed. Some of the uncertainties in input data (e.g., economic development) are likely to be less important for the difference in traffic volumes in the two future scenarios, since such input will be the same in the two future scenarios.

The two scenarios with and without the transit road have the same land-use (e.g., with respect to locations of residential areas). With the bypass road the existing congestion tax zone is extended and includes a tax on the highway-ring around the inner city. This additional tax extension is not included in the scenario without the bypass road, the motivation being that there must be a way to bypass Stockholm without having to pay a tax. The decrease in local NOx emissions between the two periods, as estimated from the planned renewal of traffic fleet and stricter vehicle emission limits, is expected to be around 40% (cf. Table 1).

2.5. Population Exposure. While assessing the environmental consequences of future urban planning scenarios, population exposure can be used as a complement to a pure comparison of air pollution concentrations. In this study we produce exposures using a $100 \times 100\,m^2$ resolution population density grid (Figure 3), based on home addresses for the year of 2008 provided by Statistics Sweden. The exposure output is a population-weighted average exposure level for each scenario, together with tables indicating how many people that are exposed to a certain pollution level, and statistics on how many Stockholm citizens that will experience an improved or deteriorated air quality outside their residence as a consequence of a certain future scenario.

The 2008 population data was used to assess exposure both for present and future scenarios (future population projections with a comparable spatial precision do not exist). It is clear that the number of people exposed will be underestimated in future scenarios as population will grow. The merit of population weighted averages is that they take into account both the distribution of pollution levels and the location of densely populated areas. If population will grow without major redistribution between residential areas, then population weighted pollution averages will not change significantly with increasing population. The exposure calculation uses simulated concentrations on a $1 \times 1\,km^2$ grid representing outdoor urban background concentrations, that is, no correction is made for indoor concentration levels being different.

2.6. Measurement Methods and Sites. NO/NO$_2$ and O$_3$ were measured in the urban background site (Torkel Knutssonsgatan) in the city centre of Stockholm (Figure 3). Torkel Knutssonsgatan monitoring station is located at roof-top level (25 m, close to the middle of the lowest model layer of MATCH, 30 m) not directly affected by nearby emissions [14, 20, 23]. Continuous measurement of O$_3$ is based on its absorption of ultraviolet (UV) light, with an absorption maximum of 254 nm (Environment S A, Model 42M). NO and NO$_2$ was measured by chemiluminescence (Environment S.A., Model AC31M).

3. Results and Discussion

3.1. Present and Future Air Quality. Simulations of air quality were made for a $102 \times 102\,km^2$ domain that includes the Stockholm Metropolitan area as well as Uppsala, the fourth largest city in Sweden. The different combinations of meteorology, European emissions, and local emissions are summarized in Table 1. The assessment of how future air quality will evolve is focused on urban background levels in the central parts of Stockholm, as represented by Stockholm's main urban air quality monitoring station, Torkel Knutssonsgatan.

Figure 4 shows the spatial distributions of three-year average O$_3$ and NO$_2$ concentrations for the present conditions (i); for the future situation with current emissions but future climate (ii); the Reference scenario 2030 with future emissions both in Europe and in Stockholm, but without the transit road (iv). While climate change alone has a very small effect on NO$_2$ and O$_3$ concentrations, emission reductions significantly decrease future concentrations of both pollutants over the modeling domain. During both present and future periods NO$_2$ concentrations are highest in Stockholm, close to Arlanda international airport, and in the city of Uppsala; major roads and sea lanes are also visible. O$_3$ is anti-correlated with NO$_2$ and features the lowest concentrations in central Stockholm and the highest concentrations over the eastern, sea-dominated, part of the domain. The spatial pattern reflects relatively higher local traffic emissions of NO$_X$ in central Stockholm and the associated O$_3$-NO chemistry.

Figures 5 and 6 show three-year average concentrations at the urban background monitoring station (its location shown in Figure 3), where the simulated levels for present conditions are compared to measured concentrations. The figures also show the concentrations resulting from the two different climate scenarios as meteorological driver (*ECHAM5 A1B-r3* or *HadCM3 A1B-ref*), all other input kept the same. In total four different experiments were evaluated (cf. Table 2):

(i) The present situation (2009–2011) using European emissions valid for 2010 as well as local Stockholm emissions for 2010.

(ii) The climate change effect on air quality in Stockholm is assessed through retaining all emissions at their 2010 level, but using the climate around 2030 for both the regional and local air quality simulations.

(iii) The climate change effect together with the European emission reductions as given by the RCP4.5 scenario, but retaining local Stockholm emissions at the 2010 level. This experiment illustrate the expected evolution of pollution levels in the long-range incoming air.

(iv) Climate change, time-varying European emissions according to RCP4.5 and also a local Stockholm emission scenario projected for 2030.

Figure 5 shows that the simulated average concentrations of O$_3$ for the present situation (i) are within the variability of the measured levels both for the *ECHAM5 A1B-r3* and *HadCM3 A1B-ref* simulation, but *HadCM3 A1B-ref* predict 8% higher concentrations compared to *ECHAM5 A1B-r3*. The simulated 8-hour daily maximum concentrations are lower than observed concentrations, indicating that the

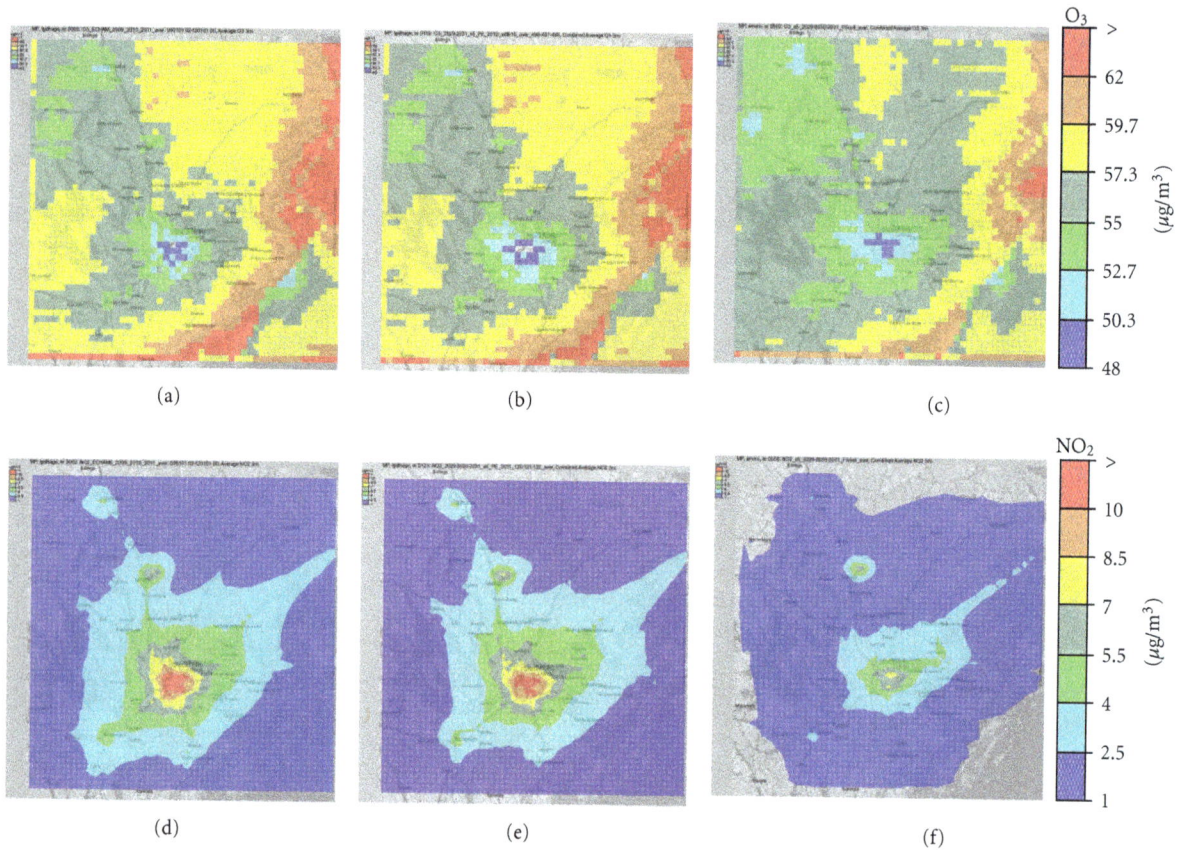

FIGURE 4: Simulated three-year-average O$_3$ (top) and NO$_2$ (bottom) concentrations in 2009–2011 (case i, left), 2029–2031 (case ii, middle), and 2029–2031 (case iv, right) in the downscaled area over the Stockholm region, based on *ECHAM5 A1B-r3* climate and RCP4.5 emissions in Europe.

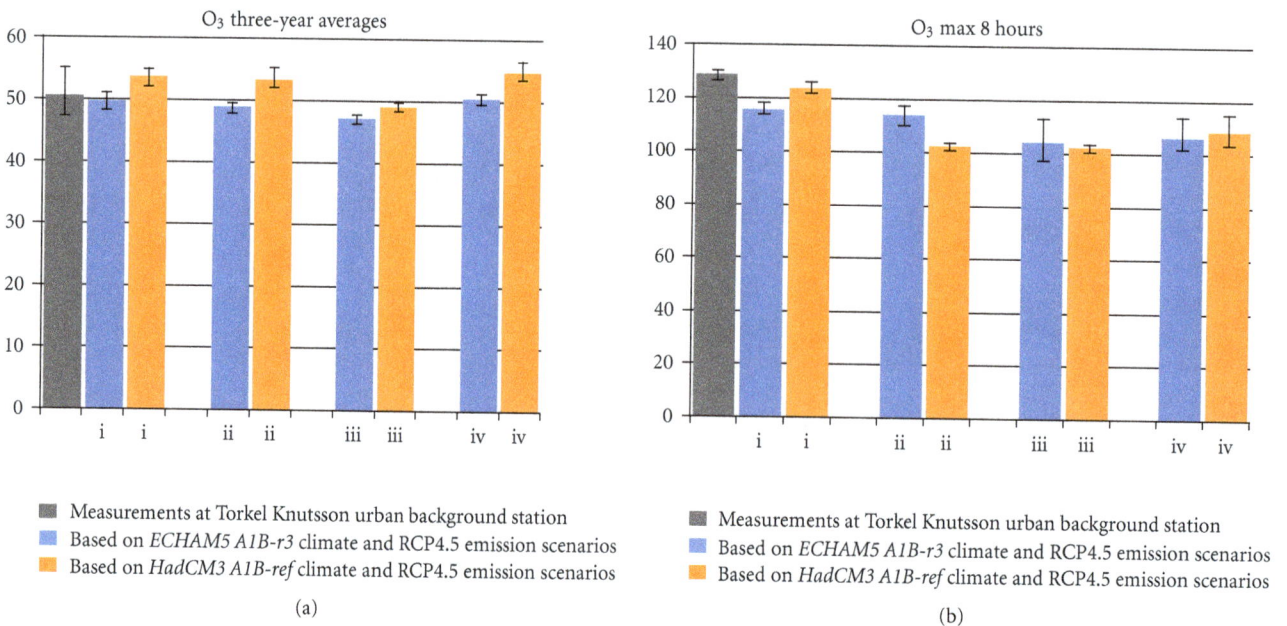

FIGURE 5: Observed and simulated levels of O$_3$ diurnal mean (a) and maximum 8-hour levels (b) at Torkel Knutssonsgatan (city centre of Stockholm), for present conditions (i) and simulated projections of future levels with only climate change effect (ii), including also emission changes in Europe (iii), and adding local emission changes in Stockholm (iv), see Table 2 for explanations. Error bars indicate lowest and highest annual average value of the three years. Unit: μ gm^{-3}. Simulated concentrations are bilinearly interpolated to the location of the monitor location.

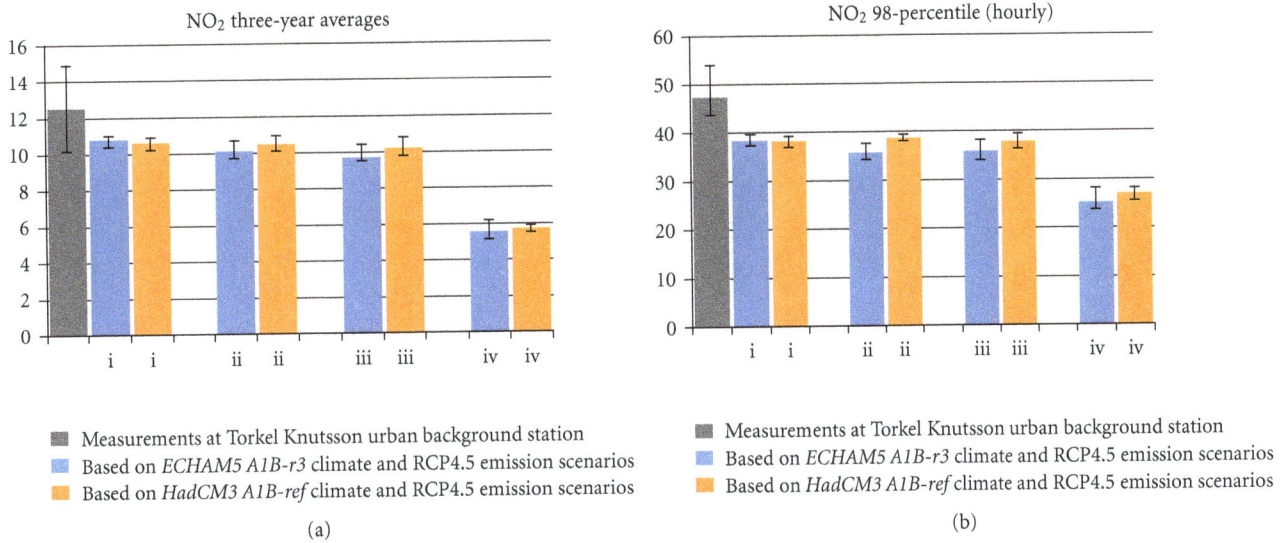

FIGURE 6: Observed and simulated levels of NO_2 diurnal mean (a) and 98-percentile of hourly levels (b) at Torkel Knutssonsgatan (city centre of Stockholm), for present conditions (i) and simulated projections of future levels with only climate change effect (ii), including also emission changes in Europe (iii) and adding local emission changes in Stockholm (iv), see Table 2 for explanations. Error bars indicate lowest and highest annual average value of the three years. Unit: $\mu\,gm^{-3}$. Simulated concentrations are bilinearly interpolated to the location of the monitor location.

TABLE 2: Model simulations performed for the large metropolitan area with $2 \times 2\,km^2$ spatial resolution. Present and future meteorology represent 3 full years centered around 2010 and 2030, receptively. Data generated by the RCA3 model forced with *ECHAM5 A1B-r3* and *HadCM3 A1B-ref* on the boundaries.

Case	Simulation scenario	European emissions	Local Stockholm emissions
(i)	Present meteorology Present European emissions Present Stockholm emissions	RCP4.5 at 2010	database 2010
(ii)	Future meteorology Present European emissions Present Stockholm emissions	RCP4.5 at 2010	database 2010
(iii)	Future meteorology Future European emissions Present Stockholm emissions	RCP4.5 at 2030	database 2010
(iv)	Future meteorology Future European emissions Future Stockholm emissions	RCP4.5 at 2030	Reference 2030

climate and air quality simulations representing present conditions underestimate the extreme O_3 concentrations. This tendency also holds for comparison of simulated and measured NO_2 concentrations as shown in Figure 6. The average simulated levels are within the inter-annual variability of the measured concentrations, but for the 98-percentile of the hourly values, the calculations underestimate the concentrations both in the *ECHAM5 A1B-r3* and *HadCM3 A1B-ref* simulations of the present situation. A multimodel study where the *ECHAM5 A1B-r3* climate scenario was used for an evaluation against measured ozone levels across Europe during 1997–2003 also showed that the MATCH model gives a good estimate of average ozone levels but an underestimation of summertime extreme values [24].

The simulation with changed climate but present (2010) European air pollution emissions (ii) demonstrate that the effect of climate change is minor on the average concentrations, this applies both to O_3 (Figure 5) and NO_2 (Figure 6). A larger impact is seen on the average 8-hour daily maximum O_3 concentrations, which reflects the potentially larger impact of climate change on the extreme values of O_3. A weak decreasing trend in background O_3 concentration in northern Europe from 2000–2009 to 2040–2049 was also detected in a recent multimodel study [24] on the impact of climate change on European surface O_3 concentrations.

The reduction of European NO_x emissions (as well as other O_3 precursors according to RCP4.5 [12]), case iii, results in slightly reduced average ozone concentrations. This

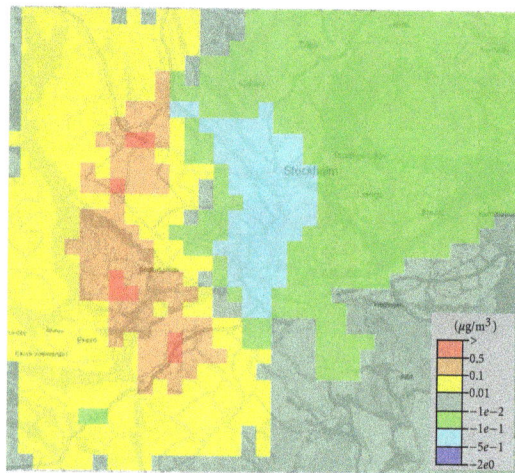

(a)

TABLE 3: Population-weighted average exposure given for a total population of 1 463 780 persons living inside the modeling domain. Model simulations performed for the smaller Stockholm domain with $1 \times 1\,km^2$ spatial resolution. Air quality simulations were made for the single year 2010 and 2030, with RCA3 downscaled *ECHAM5 A1B-r3* meteorology and time varying RCP4.5 emissions over Europe. Unit: $\mu\,gm^{-3}$.

Simulation scenario	Local Stockholm emissions	Population weighted exposure	
		NO_2	O_3
Present conditions	database 2010	6.85	54.13
Future: Reference	Reference 2030	3.62	53.05
Future: Road project	Road transit project 2030	3.60	53.06

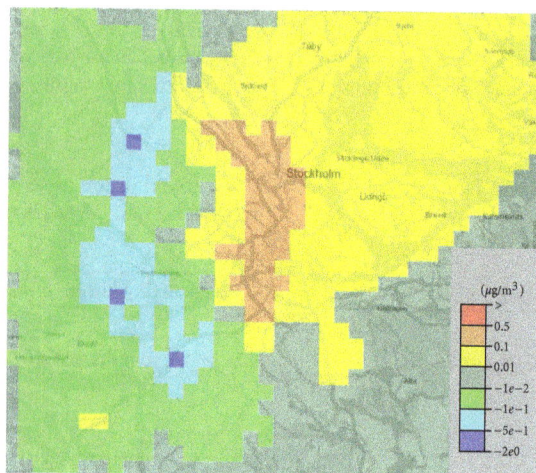

(b)

is valid for both climate scenarios, but with *HadCM3 A1B-ref* forcing there is a stronger reduction of O_3 in Stockholm during the studied period. For NO_2 average values the impact of reduced European NO_x emissions is small and comparable with the climate change-only effect. Extreme NO_2 values are also not significantly affected by the emission reduction in Europe; a consequence of the strong influence of local NO_x emissions, which in this case are kept constant at the year 2010 level.

The last simulation, case iv, (see also Figure 4) includes the combined effect of climate change and changes in European as well as local emissions. For NO_2 this results in a significant reduction (halving the average levels). This scenario, where local NO_x emissions are reduced with 40%, results in increased ozone levels in the city centre compared to present levels.

3.2. Population Exposure. Simulations for the exposure calculations were made with $1 \times 1\,km^2$ spatial resolution over the smaller domain covering $36 \times 30\,km^2$, which has a population of close to 1.5 million distributed as shown in Figure 3. This experiment utilized the *ECHAM5 A1B-r3* meteorology with European air pollutant emissions valid for either 2010 or 2030. The objective was to quantify and compare, on one hand, the present and the future air pollutions levels and, on the other hand, assess the expected differences in future air quality resulting from two different traffic solutions.

The general characteristics of how NO_2 and O_3 concentrations develop between present and future (Reference scenario without road project) were shown for the larger modeling domain in Figure 4. Figure 7 shows the difference in one-year average NO_2, O_3, and PM_{10} concentrations between the two future scenarios (with and without the bypass road). Even though the total local emissions of NO_x and PM_{10} are similar for the two local 2030-scenarios (see Table 1), there are considerable changes in emissions in certain areas. Since most of the new transit road will be constructed as an underground highway tunnel, the location

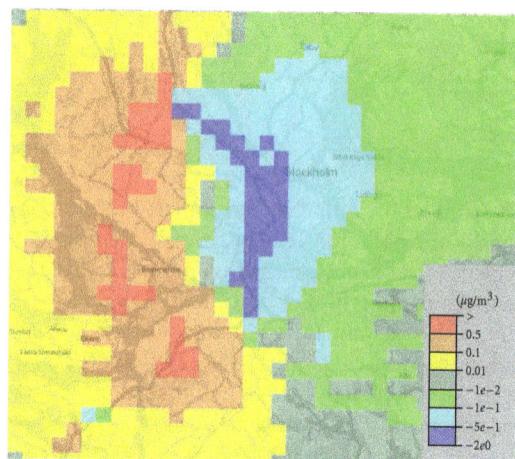

(c)

FIGURE 7: Differences in simulated NO_2 (a), O_3 (b), and PM_{10} (c) between the 2030 Road project and Reference average concentrations. Unit: $\mu\,gm^{-3}$.

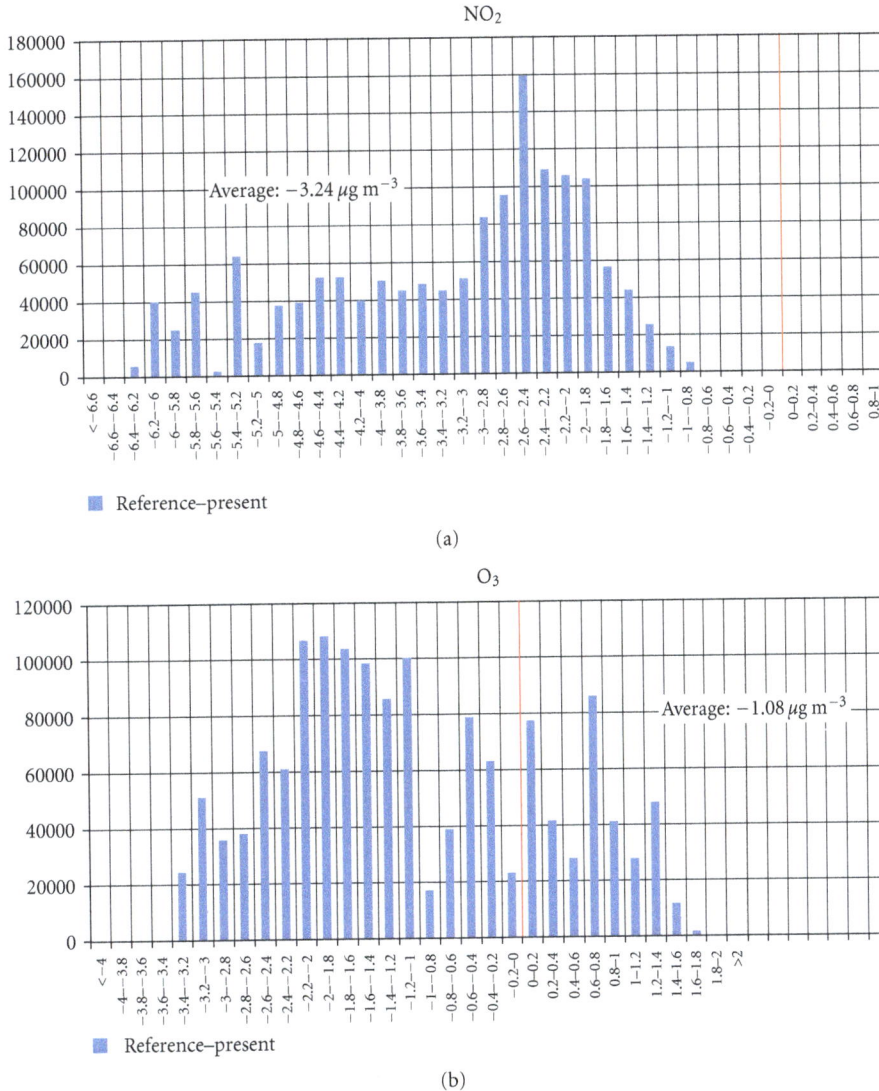

FIGURE 8: Differences in the number of inhabitants exposed to NO_2 (a) and O_3 (b) concentration levels "Reference" (2030)—"Present" (2010). The red vertical line indicates zero difference. All simulations made with *ECHAM5 A1B-r3* and RCP4.5 time varying emissions. Unit x axis: $\mu\,gm^{-3}$.

of the emissions will be very different compared to a situation with a highway on the ground. Most emissions from the tunnel will be ventilated in 10 to 20 meter high towers, with some emissions occurring at tunnel exits at ground surface level. The Figure 7 concentration differences indicate the locations of the pointwise emissions from the underground highway's ventilation towers.

The main difference between the two alternatives is that there will be much less traffic emissions close to the city centre with the new transit road, but this is accompanied with a negative impact on the air quality (NO_2 and PM_{10}) west of the city centre. Figure 7 also illustrate how ozone levels are anticorrelated with traffic-induced air pollution. It should be noted that concentration differences are small, for NO_2 and O_3 rarely exceeding $0.5\,\mu\,gm^{-3}$. For PM_{10} we can see a somewhat larger effect where large parts of the city centre concentrations are lowered 0.5–$2.0\,\mu\,gm^{-3}$.

Table 3 summarizes the population-weighted concentrations for NO_2 and O_3. For both pollutants population-weighted exposure will decrease from 2010 to 2030; for NO_2 the exposure reduction is almost 50%. The reason is, as discussed earlier, the expected strong reductions in local road traffic emissions. For ozone it is the lower concentrations in incoming, background, air that will result in a reduction of the average population weighted exposure level. Figure 8 shows that all residents can expect lower NO_2 concentrations outside their homes in 2030 compared to present situation. However, despite the overall reduced population-weighted ozone concentration, some 25% of the population will experience higher ozone levels in the future. The reason is the decreased amount of NO available in the city-centre for destroying O_3 in the future. Table 3 also shows that the average population-weighted exposure for the two future scenarios are only marginally different, although

NO_2

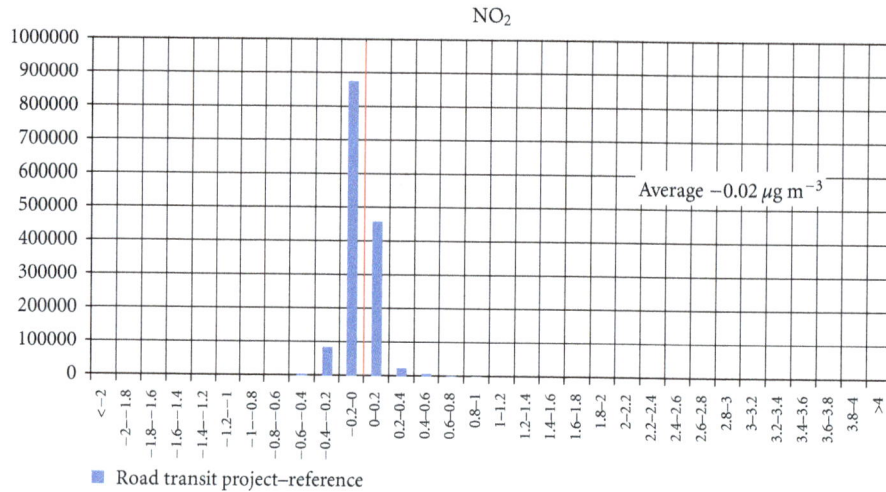

Road transit project–reference

(a)

O_3

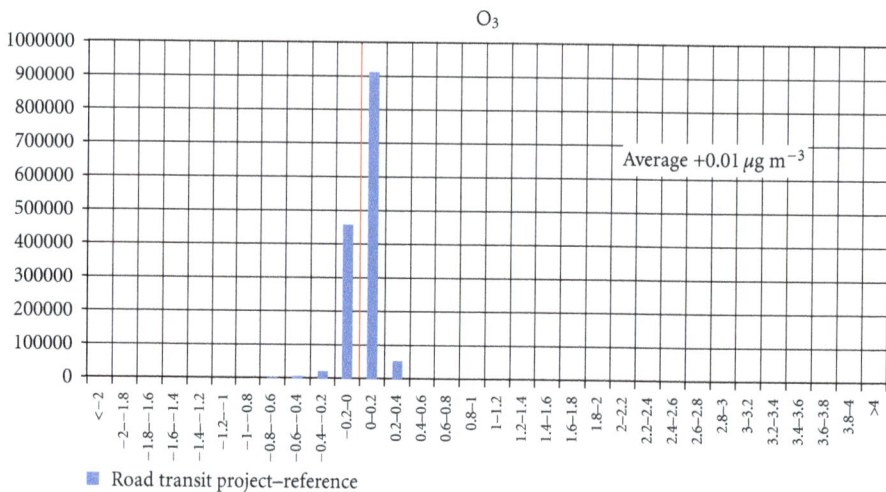

Road transit project–reference

(b)

PM_10

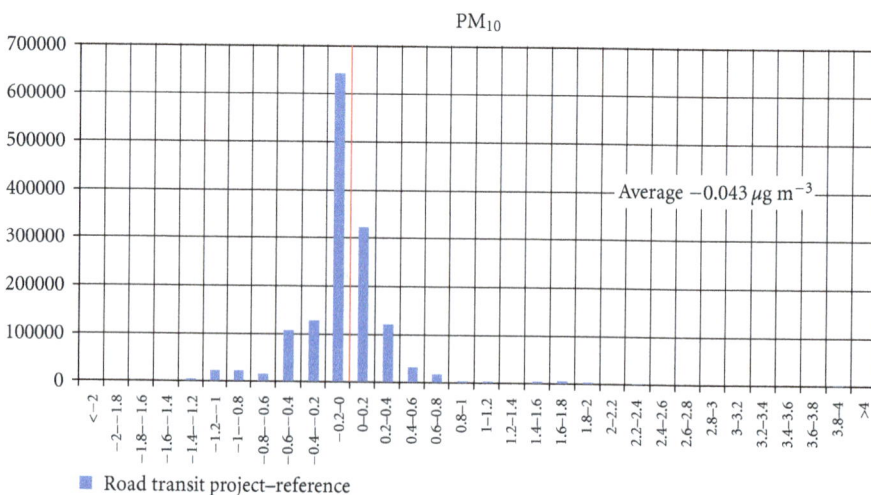

Road transit project–reference

(c)

FIGURE 9: Differences in the number of inhabitants exposed to NO_2 (a), O_3 (b), and PM_10, (c) concentration levels between "Road transit project"—"Reference", both scenarios valid for 2030. The red vertical line indicates zero difference. All simulations made with *ECHAM5 A1B-r3* and RCP4.5 time varying emissions. Unit *x* axis: μ g m^{-3}.

they for individuals may imply a considerable difference. Figure 9 shows differences in population exposure of NO_2, O_3, and PM_{10} between the two future scenarios, with and without the planned road transit. With the realization of the road transit project, a majority of people will experience a minor reduction in NO_2 concentrations compared with the situation with present roads, but at the same time a minor increase in ozone concentrations. The PM_{10} level differences are similar to those of NO_2 (the same traffic source), but the local impact is stronger since local emission are not expected to decrease. Figure 9 illustrates that the bypass is not an effective action to lower air pollution exposure in Stockholm. Even if there is a small average reduction of population-weighted PM_{10} exposure of $0.04\,\mu\,\mathrm{gm}^{-3}$, there will be a population share of more than 100 000 persons that will experience an exposure increase of 0.2–$0.4\,\mu\,\mathrm{gm}^{-3}$ and more than 10 000 persons will experience exposure increases $>1.5\,\mu\,\mathrm{gm}^{-3}$.

As described in Section 2, the simulation of future long-range concentrations of PM_{10} are not fully described in our modeling-system, which means that we can only discuss changes caused by local scenarios. A recent study from California [9] found that climate change will contribute to a small decrease in future PM_{10} levels, comparing the conditions year 2000 to those simulated for 2050. The net change over this period was however smaller than the interannual variations and mostly driven by an increase in average wind speed (more dilution of local sources). They could also see that higher temperatures will increase the formation of secondary inorganic aerosols. Climate change conditions in Stockholm are different from California, so a similar assessment in Stockholm is definitively of interest (Sweden is also much influenced by primary PM emissions in Europe, so that future levels will depend a lot on the rate of emission reductions).

Mahmud et al. [9] also reported more stagnation periods in the future that also contributed to higher extreme values in local (traffic and wood combustion) PM. From our assessment of NO_2 extreme values (Figure 6)—which should respond in a similar way as PM10 to stagnation conditions-responding to the sole effect of climate change, we cannot see any trend towards higher values in the future. Note, however, that we have only simulated three years, which is a too short period to conclude on extreme values.

4. Conclusions

In this study we have used regional downscaling of two different global climate models (*ECHAM5 A1B-r3* and *HadCM3 A1B-ref*) together with the RCP4.5 air pollution emission data to assess the importance of changing climate and changing emissions for the concentrations of O_3 and NO_2 in the urban area of Stockholm in 2030. Comparison of simulations for present situation (mean values for the period 2009–2011) shows acceptable agreement with measurements in the urban background for both climate realisations. The effect of a changing climate is small. Decreased future emissions of O_3 precursors in Europe will reduce O_3

production in Europe, but this effect is partly compensated for by increased hemispheric background concentrations of O_3 [13], leading to only modest changes in O_3 affecting the urban area of Stockholm.

The exposure assessment revealed that all residents can expect considerably lower NO_2 exposure in the future. Ozone exposure will change only marginally, partly due to decreased concentrations in the suburban areas and increased concentrations in the city centre.

We have also shown that a very large road transit project (involving the construction of one of the largest motorway road tunnels in Europe) will only marginally influence population exposure, since the improved air quality in the city centre will be complemented by deteriorated air quality in other residential areas.

Acknowledgments

This work has been cofunded by SUDPLAN: Sustainable Urban Development Planner for Climate Change Adaptation, European Framework Program 7, ICT-2009-6.4 ICT for Environmental Services and Climate Change Adaptation of the Information and Communication Technologies program, Project no. 247708.

References

[1] WHO. World Health Organisation (WHO) Air Quality Guidelines, *World Health Organisation (WHO) Regional Office For Europe*, Copenhagen, Denmark, 2006, Global Update 2005.

[2] C. M. Cooney, "Downscaling climate models: sharpening the focus on local-level changes," *Environ Health Perspect*, vol. 120, no. 1, pp. A24–A28, 2012.

[3] W. D. Solecki and C. Oliveri, "Downscaling climate change scenarios in an urban land use change model," *Journal of Environmental Management*, vol. 72, no. 1-2, pp. 105–115, 2004.

[4] M. A. Semenov and E. M. Barrow, "Use of a stochastic weather generator in the development of climate change scenarios," *Climatic Change*, vol. 35, no. 4, pp. 397–414, 1997.

[5] E. Kjellström, G. Nikulin, U. Hansson, G. Strandberg, and A. Ullerstig, "21st century changes in the European climate: uncertainties derived from an ensemble of regional climate model simulations," *Tellus A*, vol. 63, no. 1, pp. 24–40, 2011.

[6] J. Langner, R. Bergström, and V. Foltescu, "Impact of climate change on surface ozone and deposition of sulphur and nitrogen in Europe," *Atmospheric Environment*, vol. 39, no. 6, pp. 1129–1141, 2005.

[7] C. Andersson and M. Engardt, "European ozone in a future climate—the importance of changes in dry deposition and isoprene emissions," *Journal of Geophysical Research*, vol. 115, Article ID D02303, 13 pages, 2010.

[8] D. J. Jacob and D. A. Winner, "Effect of climate change on air quality," *Atmospheric Environment*, vol. 43, no. 1, pp. 51–63, 2009.

[9] A. Mahmud, M. Hixson, and M. J. Kleeman, "Quantifying population exposure to airborne particulate matter during extreme events in California due to climage change," *Atmospheric Chemistry and Physics Discussions*, vol. 12, pp. 5881–5901, 2012.

[10] L. Robertson, J. Langner, and M. Engardt, "An Eulerian limited-area atmospheric transport model," *Journal of Applied Meteorology*, vol. 38, no. 2, pp. 190–210, 1999.

[11] C. Andersson, J. Langner, and R. Bergström, "Interannual variation and trends in air pollution over Europe due to climate variability during 1958–2001 simulated with a regional CTM coupled to the ERA40 reanalysis," *Tellus B*, vol. 59, no. 1, pp. 77–98, 2007.

[12] A. M. Thomson, K. V. Calvin, S. J. Smith et al., "RCP4.5: a pathway for stabilization of radiative forcing by 2100," *Climatic Change*, vol. 109, no. 1-2, pp. 77–94, 2011.

[13] J. Langner, M. Engardt, and C. Andersson, "European summer surface ozone 1990–2100," *Atmospheric Chemistry and Physics Discussions*, vol. 12, pp. 4901–4939, 2012.

[14] L. Gidhagen, C. Johansson, J. Langner, and V. L. Foltescu, "Urban scale modeling of particle number concentration in Stockholm," *Atmospheric Environment*, vol. 39, no. 9, pp. 1711–1725, 2005.

[15] C. Johansson, A. Hadenius, Johansson, P. Å, and T. Jonson, *NO$_2$ and Particulate Matter in Stockholm—Concentrations and Population Exposure. The Stockholm Study on Health Effects of Air Pollution and Their Economic Consequences*, Swedish National Road Administration, Borlänge, Sweden, 1999.

[16] Å. Sjödin, M. Ekström, U. Hammarström et al., "Implementation and Evaluation of the ARTEMIS Road Model for Sweden's International Reporting Obligations on Air Emissions," in *Proceedings of the 2nd Conference Environment & Transport including 15th Conference Transport & Air Pollution*, vol. 1, no. 107, pp. 375–382, Reims, France, June 2006.

[17] G. Omstedt, B. Bringfelt, and C. Johansson, "A model for vehicle-induced non-tailpipe emissions of particles along Swedish roads," *Atmospheric Environment*, vol. 39, no. 33, pp. 6088–6097, 2005.

[18] M. Ketzel, G. Omstedt, C. Johansson et al., "Estimation and validation of PM2.5/PM10 exhaust and non-exhaust emission factors for practical street pollution modelling," *Atmospheric Environment*, vol. 41, no. 40, pp. 9370–9385, 2007.

[19] M. Beser and S. Algers, "SAMPERS—the new Swedish national travel demand forecasting tool," in *National Transport Models*, L. Lundqvist and L. G. Mattsson, Eds., pp. 101–118, Springer, Heidelberg, Germany, 2001.

[20] C. Johansson, B. Forsberg, and L. Burman, "The effects of congestions tax on air quality and health," *Atmospheric Environment*, vol. 43, no. 31, pp. 4843–4854, 2009.

[21] G. De Jong, M. Pieters, S. Miller et al., "Uncertainty in traffic forecasts. Literature review and new results for The Netherlands," RAND Europé WR-268-AVV, Leiden, The Netherlands.

[22] M. Beser Hugosson, "Quantifying uncertainties in a national forecasting model," *Transportation Research A*, vol. 39, no. 6, pp. 531–547, 2005.

[23] C. Johansson, M. Norman, and L. Gidhagen, "Spatial & temporal variations of PM10 and particle number concentrations in urban air," *Environmental Monitoring and Assessment*, vol. 127, no. 1–3, pp. 477–487, 2007.

[24] J. Langner, M. Engardt, A. Baklanov et al., "A multi-model study of impacts of climate change on surface ozone in Europe," *Atmospheric Chemistry and Physics Discussions*, vol. 12, pp. 4901–4939, 2012.

13

A Comparison of Two Dust Uplift Schemes within the Same General Circulation Model

Duncan Ackerley,[1,2] Manoj M. Joshi,[3] Eleanor J. Highwood,[1] Claire L. Ryder,[1] Mark A. J. Harrison,[4] David N. Walters,[4] Sean F. Milton,[4] and Jane Strachan[1]

[1] Department of Meteorology, University of Reading, Reading RG6 6BB, UK
[2] Monash Weather and Climate, Monash University, VIC, Clayton 3800, Australia
[3] National Centres for Atmospheric Science (Climate), University of Reading, Reading RG6 6BB, UK
[4] Met Office, Exeter EX1 3PB, UK

Correspondence should be addressed to Duncan Ackerley, duncan.ackerley@monash.edu

Academic Editor: Ralph A. Kahn

Aeolian dust modelling has improved significantly over the last ten years and many institutions now consistently model dust uplift, transport and deposition in general circulation models (GCMs). However, the representation of dust in GCMs is highly variable between modelling communities due to differences in the uplift schemes employed and the representation of the global circulation that subsequently leads to dust deflation. In this study two different uplift schemes are incorporated in the same GCM. This approach enables a clearer comparison of the dust uplift schemes themselves, without the added complexity of several different transport and deposition models. The global annual mean dust aerosol optical depths (at 550 nm) using two different dust uplift schemes were found to be 0.014 and 0.023—both lying within the estimates from the AeroCom project. However, the models also have appreciably different representations of the dust size distribution adjacent to the West African coast and very different deposition at various sites throughout the globe. The different dust uplift schemes were also capable of influencing the modelled circulation, surface air temperature, and precipitation despite the use of prescribed sea surface temperatures. This has important implications for the use of dust models in AMIP-style (Atmospheric Modelling Intercomparison Project) simulations and Earth-system modelling.

1. Introduction

Airborne mineral dust is important in all aspects of Earth-system modelling as it impacts on the Earth's radiation budget [1]; weather [2], and climate [3, 4] while also providing a source of nutrients to oceanic and land biota [5, 6].

Despite advances in dust modelling and the representation of dust uplift on a case by case basis (e.g., [7–9]), such studies have focussed on running a single dust scheme in a given general circulation model (GCM), regional climate model (RCM) or numerical weather prediction (NWP) model. However, the representations of dust uplift, transport and deposition in atmospheric models are still highly uncertain. While some of the uncertainty arises from the parameterization schemes used for representing the dust

cycle there are also influences from the driving model, which may also add to the uncertainty.

Modelling intercomparison studies such as the Atmospheric Model Intercomparison Project (AMIP [10]) are important as they aid in understanding model uncertainty and variability across a range of state-of-the art climate models. Similar model intercomparison studies, with a particular focus on mineral dust, can be found in [11] (for the Bodélé Depression in Chad) and [12] (for Asia under the framework of the Dust Model Intercomparison Project, DMIP). The studies by [11, 12] identified that dust simulations are particularly sensitive to:

(1) The dust-uplift parameterization scheme.

(2) The representation of surface soil characteristics, which vary between models and will impact on uplift.

(3) Surface wind speeds: these are typically small-scale processes that are not resolved explicitly given the size of model grid-boxes and therefore need to be parameterized.

Global model studies of aerosol processes [13] highlight the diversity in model output when research groups run their models in "standard" configurations. However, the schemes used by [13] not only contain differences from the parameterization of the dust cycle but also in the capabilities of those overriding atmospheric models to represent the processes that lead to dust uplift, transport and deposition. If the numerous dust schemes employed by the modelling community could be included in an ensemble of simulations that are driven by one GCM, for example, then differences in dust emission, transport, and deposition across the parameterization schemes could be understood better. Differences in modelled dust climatology would in this case be almost entirely governed by the dust cycle parameterization rather than the other parameterized processes in the driving GCMs. Such studies have been undertaken for either specific case studies [14, 15] or seasonal dust properties [16]; however, none of these studies [14–16] has looked at the effects of perturbing the uplift parameterization characteristics in multiple-year AMIP-type climatological simulations. Running a single driving GCM for a time period that is long enough to acquire long-term climatological averages is therefore key to this study as it allows us to identify robust differences in the climatological dust distribution that are not influenced by model-generated variability. The work presented in this paper therefore, attempts to make a first step towards understanding the climatological impact of two different dust uplift schemes in one GCM.

Descriptions of the models used are provided in Section 2 and an analysis of the modelled Aerosol Optical Depths (AODs), dust size distributions, deposition, and the impacts of the different dust schemes on the modelled climate are discussed in Sections 3–6. The discussions and conclusions are given in Section 7.

2. Model Setup and Experiments

2.1. GCM Description.
The HadGEM2-A (Hadley Centre Global Environmental Model version 2—Atmosphere-only) model is a state-of-the-art global general circulation model based on the HadGEM1 model [17, 18]. Some additional physics changes have been made to the model: these are described in [19, 20]. The model has 38 layers in the vertical reaching a height of approximately 40 km. The horizontal resolution in the present work is 3.75° longitude × 2.5° latitude (referred to as N48 as this is the maximum number of waves that can be represented in the zonal direction).

In addition to the above, a parameterization of land/sea breezes has been implemented in this version of HadGEM2-A in order to alleviate problems in the model associated with underestimation of wind speeds at coastal points (those points that are partially ocean and partially land). A scalar

TABLE 1: Particle size ranges for each size bin taken from HadGEM2-A during this study. Further discussion and the derivation of these values are given in [20].

Dust size bin	Particle radius range (μm)
1	0.0316–0.1
2	0.1–0.316
3	0.316–1.0
4	1.0–3.16
5	3.16–10.0
6	10.0–31.6
7	31.6–100.0
8	100.0–316.0
9	316.0–1000.0

term proportional to the cube root of the temperature difference between the land and the ocean fractions of the grid box is added to the calculation of the heat and moisture fluxes. Therefore, when the land-ocean temperature difference is 20 K, the effective wind used to calculate the scalar fluxes is doubled. The addition of the parameterisation results in a significant decrease in the dry rainfall bias that HadGEM1 exhibited over the Maritime Continent region [18].

2.2. Dust Uplift Scheme 1: CLIM.
The climatological dust uplift scheme used ordinarily in HadGEM2-A (denoted CLIM from now on) is based on the work by [21]. The version used in this study uses a 9-bin scheme (see Table 1, which contains the size ranges of each bin), where bins 7–9 represent larger particles than the original 6-bin scheme. Bins 7–9 are used in calculating the total horizontal flux from which the vertical dust flux is calculated (in a similar method to the original work by [21]). The version of CLIM used in this study is similar to the version of the Met Office dust scheme used by UK-HiGEM [22, 23] and the HadGEM2 model developers [20]. We have stated where the scheme in this study differs from [20].

To initiate dust uplift, a threshold wind velocity must be reached to overcome the cohesive forces between the dust particles and is known as the threshold friction velocity ($U^*_{t(\text{bins } 1-9)}$, m s^{-1}). The threshold friction velocity is calculated for each size bin (see Table 1 for particle sizes) and is very similar to the original setup in [21], given as:

$$U^*_{t(\text{bins } 1-9)} = A\left(r_{p(\text{bins } 1-9)}\right) + BW + C, \quad (1)$$

where $A(r_{p(\text{bins } 1-9)})$ is the dry threshold friction velocity (m s^{-1}) as a function of particle size (for each of the 9 size bins given in Table 1) and is calculated from the derivation given in [24], W is the soil moisture content of the top 10 cm of the soil (kg m^{-2}), and B and C are constants. B and C were originally derived empirically by [21] in HadCM3, however the values resulted in unrealistic dust uplift when applied to HadGEM2-A. The values were set to 0.10 and −0.08 in this study (following several iterative test experiments) to give the best representation of the aerosol optical depths over Africa.

Once the values of $U^*_{t(\text{bins } 1-9)}$ have been calculated, the horizontal flux of dust in each bin can be initiated once the

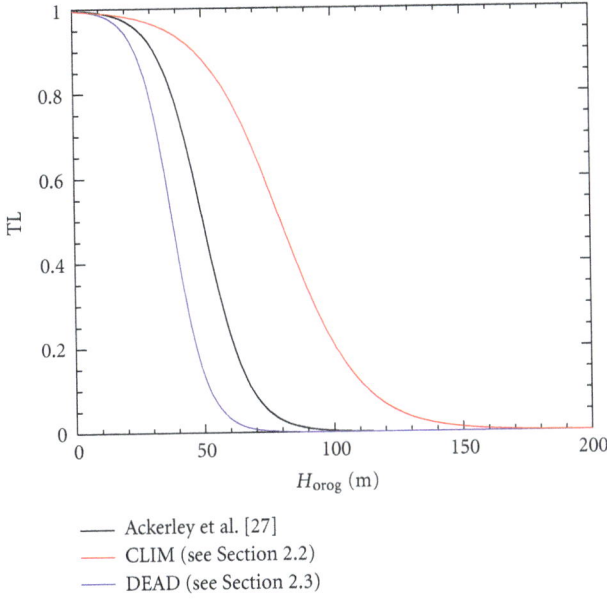

— Ackerley et al. [27]
— CLIM (see Section 2.2)
— DEAD (see Section 2.3)

FIGURE 1: A plot of the topographic "low" source term as given in (3) (red line), (5) (blue line) and the version used in [27] (black line).

friction velocity over bare soil at the surface (U^*, m s^{-1}) exceeds U_t^*. The horizontal flux (H(bins 1-9), kg m^{-2} s^{-1}) in each bin is calculated from the following:

$$H(\text{bins } 1\text{–}9) = F_{\text{soil}} C \rho_* U^{*^3} \left(1 + \left(\frac{U^*_{t(\text{bins } 1\text{–}9)}}{U^*}\right)\right),$$

$$\times \left(1 - \left(\frac{U^*_{t(\text{bins } 1\text{–}9)}}{U^*}\right)^2\right) \frac{M_{\text{rel}}}{G} D \text{ TL}, \quad (2)$$

where F_{soil} is the fraction of bare soil within a grid box, C is an empirically derived constant of proportionality (defined in [25]), ρ_* is the surface air density over land (kg m^{-3}). M_{rel} is the mass of dust in each bin relative to the total mass of dust at each grid point and is calculated from the silt, sand and clay fraction values taken from the International Geosphere-Biosphere Programme (IGBP) global soil data set [26], D is a globally uniform tuning parameter (set to 18 in these experiments following a series of test simulations) and TL is the "topographic low" source term (based on the work by [27]), which is calculated as:

$$\text{TL} = 0.5 \left(1 - \tanh\left(\frac{H_{\text{orog}} - 80}{30}\right)\right). \quad (3)$$

Where H_{orog} is half of the peak-to-trough height of the model surface elevation. The values in (3) (shown in Figure 1) were chosen following test experiments and are different to those used in [27] as a result of the lower resolution orography used in this study and the different dust uplift scheme (CLIM). The version of the horizontal flux used by [20] does not use the TL function used in this study.

The vertical dust flux (G (bins 1 to 6), kg m^{-2} s^{-1}) is then calculated for bins 1 to 6 from:

$$G(\text{bins 1 to 6}) = H(\text{bins 1 to 6}) * \left(1 + \frac{\sum H(\text{bins 7 to 9})}{\sum H(\text{bins 1 to 6})}\right)$$

$$* 10^{(13.4 F_c - 6.0)}. \quad (4)$$

As can be seen in (4), modelled dust particles in bins 7–9 are not subject to vertical transport as they are too large; however, the dust in these bins contributes to the saltation and sandblasting that occurs during horizontal dust transport, which is also responsible for liberating smaller particles. H (bins 1 to 6) is the horizontal dust mass flux for bins 1 to 6, ΣH(bins 7 to 9) is the total horizontal dust mass flux in bins 7 to 9, ΣH(bins 1 to 6) is the total horizontal dust mass flux in bins 1 to 6, and F_c is the clay fraction of the soil. The middle expression in the brackets increases the horizontal dust flux in each bin to account for the extra saltation and sand blasting from the large dust particles in bins 7 to 9 and is an extension to the relationship derived in [21]. The vertical flux term used is identical to that used in [20].

For more recent developments of the Met Office Unified Model's (MetUM) dust scheme, see [19, 20, 25].

2.3. Dust Uplift Scheme 2: DEAD. The dust uplift scheme from the freely available "mineral dust entrainment and deposition model" (DEAD, [28]) was downloaded and incorporated into the MetUM by [27]. Initial tests using the DEAD scheme from [27] resulted in excessive global emissions of dust in the N48 model (not shown). Therefore several adaptations were made to reduce the dust uplift in the model. The changes to the parameterizations given in [27] are:

(1) The factors in the topographic "low" source term (TL) were changed to the following:

$$\text{TL} = 0.5 \left(1 - \tanh\left(\frac{H_{\text{orog}} - 38}{13}\right)\right). \quad (5)$$

As a result of the lower resolution orography used in this study. Equation (5) (shown in Figure 1) is more stringent than the version in [27] (also shown in Figure 1), which is associated with the greater smoothing of surface orography at the N48 resolution used in this study. H_{orog} is half of the peak-to-trough height of the model surface elevation (m).

(2) Following several iterative test simulations, the global tuning parameter (GT) was increased to 0.016 from 0.014 used in [27].

(3) The gravimetric water content included an extra factor of the surface clay fraction as derived by [29] (see Equation (5) in [28]), which was not included in the [27] version. [28] multiplied the function by a model dependent "ad hoc" coefficient, which was included and set to 0.1 in this study (similar to GT) following the test runs.

Apart from the changes listed in this section, all other aspects of the dust uplift scheme (such as the horizontal and vertical flux calculations) are identical to those given in [27]. This includes the use of the same globally fixed size distribution for the emitted dust used by [27]. Also, the DEAD uplift scheme only used bins 1–6 for both the vertical and horizontal fluxes (see Table 1 for the size ranges) as in [27].

2.4. Experiments. Two GCM simulations were undertaken with one using the CLIM uplift scheme and the other using DEAD (the transport and deposition schemes are identical for CLIM and DEAD and are discussed in more detail in [21]). Both experiments used AMIP sea-surface temperatures (SSTs) as boundary conditions, and ran for 17 years from 1979–1995. The first year of both model integrations were regarded as spin-up and not included in this analysis.

3. Aerosol Optical Depth at 550 nm

The aerosol optical Depth (AOD) is a parameter that is derived in many global and regional atmospheric models including HadGEM2-A (see [23, 27]) and is also retrieved from numerous ground-based and satellite based instruments [30, 31]. The model derivation for calculating the AODs can be found in [32]. The values have also been derived in a selection of atmospheric models as part of the AeroCom project (see [13, 33]). [34] found that the global annual mean AOD values (all reference to AOD will be at 550 nm) ranged from 0.01–0.053 with a median value of 0.023 when simulating the year 2000. The simulated annual mean AODs at 550 nm over the full 1980–1995 period were 0.014 (CLIM) and 0.023 (DEAD), which both lie within the range given in [34] using AeroCom simulations. However, as the AeroCom simulations were only representative of the year 2000, the range given in [34] may not be representative of the long-term dust induced AOD unlike the 16-year averages for the CLIM and DEAD simulations.

The time series of the annual mean AOD has been plotted for both models in Figure 2 along with the full (dotted lines) and "80% of models" (dashed lines) range estimated in [34]. Nine of the sixteen years simulated using the CLIM scheme lie within the [34] range of which two lie within the "80% of models" range. All of the modelled annual mean AODs using the DEAD scheme lie within the [34] range and all but three lie within the range of the middle 80% estimated by the AeroCom models. While the description above is not a comparison between identical simulations, the values from the simulations using CLIM and DEAD lie within the same order of magnitude as the [34] simulations. Also, the majority of the simulated annual mean AOD estimates using CLIM and DEAD lie within the total range of modelled AODs in [34], which gives further confidence in each models' dust simulation.

Reference [23] discussed the global distribution of dust using maps of global AOD associated with dust and biomass aerosol. Similar maps are given in Figure 3 (for dust only)

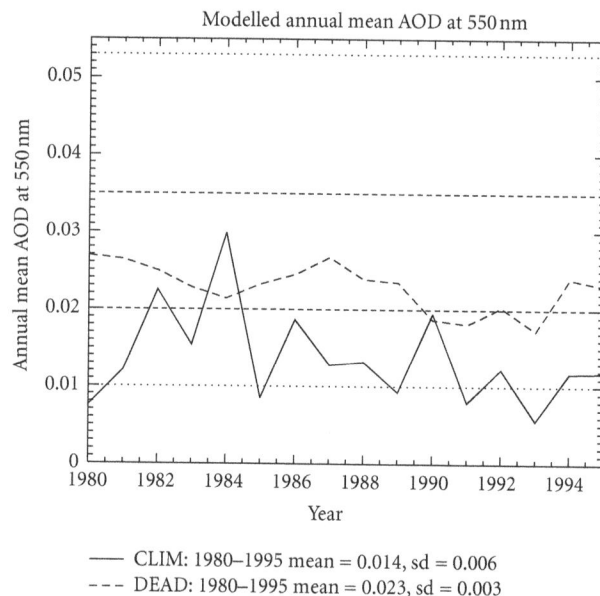

Modelled annual mean AOD at 550 nm

—— CLIM: 1980–1995 mean = 0.014, sd = 0.006
- - - DEAD: 1980–1995 mean = 0.023, sd = 0.003

FIGURE 2: Time series of global, annual mean AOD (550 nm) for the simulations with the CLIM (solid) and DEAD (dashed) dust uplift schemes. The full and "80% of models" ranges estimated by [34] are given by the dotted and dashed lines, respectively.

and show the distribution of dust resulting from using the CLIM and DEAD schemes. Both CLIM and DEAD have peak dust AODs over West Africa in the annual mean (see Figures 3(a) and 3(b)), which is similar to [23]. The values of AOD over West Africa are higher with the CLIM scheme than those of the DEAD scheme. The dust concentrations in the Northern Hemisphere (NH) are also larger than in the Southern Hemisphere (SH) (again similar to [23]) as there is a larger proportion of land in the NH and therefore more source regions (such as the Sahara Desert). The largest contribution in the SH dust load using both uplift schemes comes from Australia.

Despite the similarities between the two schemes there are some differences (compare Figures 3(a) and 3(b)). The AODs are much lower (less than 0.002) in CLIM north of 60°N and south of 30°S than in the simulation with DEAD, which may be associated with weaker transport to areas remote from dust sources (for CLIM). Conversely, the CLIM simulation has higher AODs closer to the dust source areas (such as West Africa) and may be associated with a higher vertical mass flux than for DEAD. Despite these differences the main global dust sources (Sahara, Asia, and Australia) that are apparent in both simulations and the global distributions of dust given in Figure 3 are similar to those of other studies [22, 23].

4. Dust Size Distributions

While there were many similarities between the global dust distributions given in Figure 3, the AODs of the two model simulations differed greatly in regions remote from the main dust sources (e.g., in the polar regions). As the

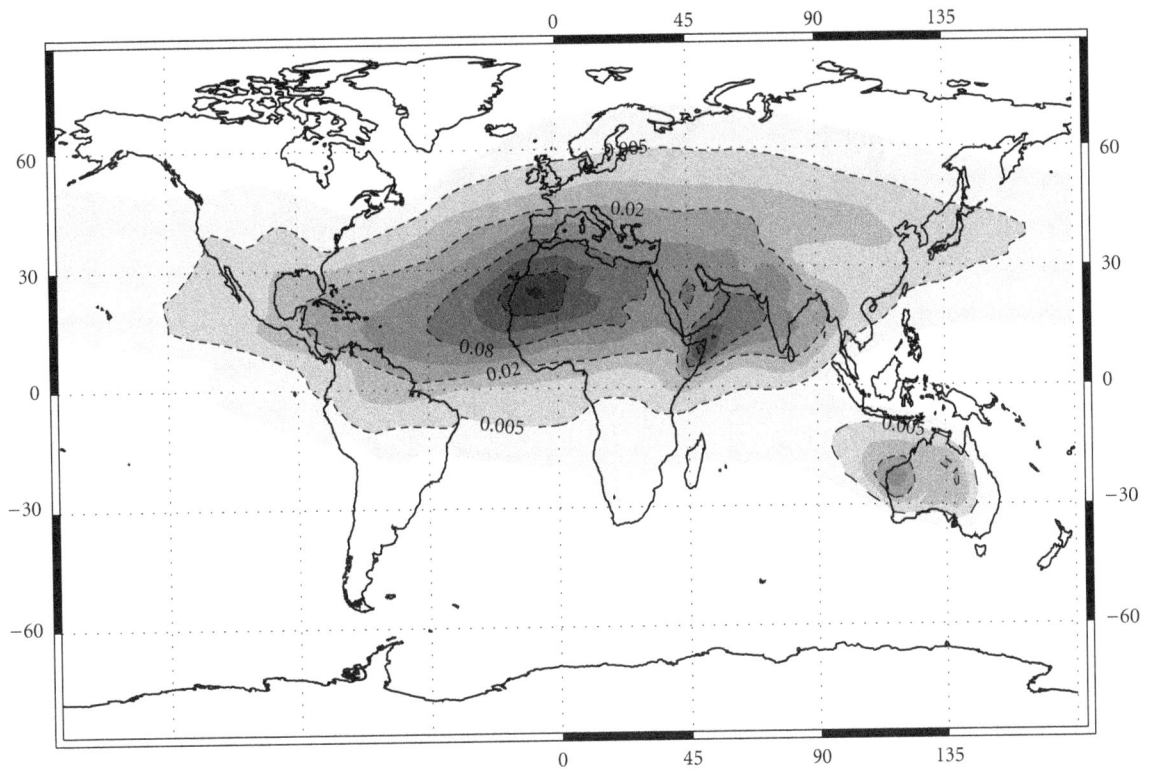

(a)

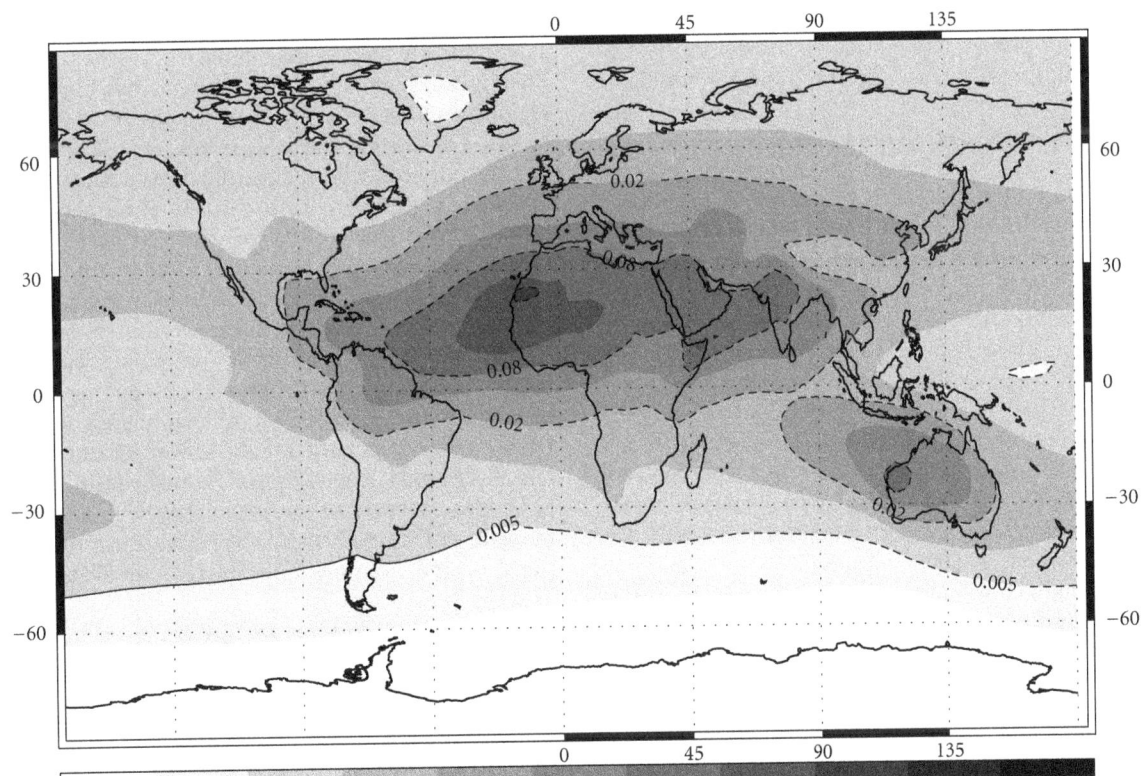

(b)

FIGURE 3: Annual mean AOD at 550 nm for (a) CLIM and (b) DEAD averaged over the last 16 years of each model simulation.

--- CLIM: model levels 1–16 (surface–5 km)
— DEAD: model levels 1–16 (surface–5 km)
△ DODO: values measured between 50 m–5 km

(a)

--- CLIM: model levels 1–16 (surface–5 km)
— DEAD: model levels 1–16 (surface–5 km)
△ DODO: values measured between 50 m–5 km

(b)

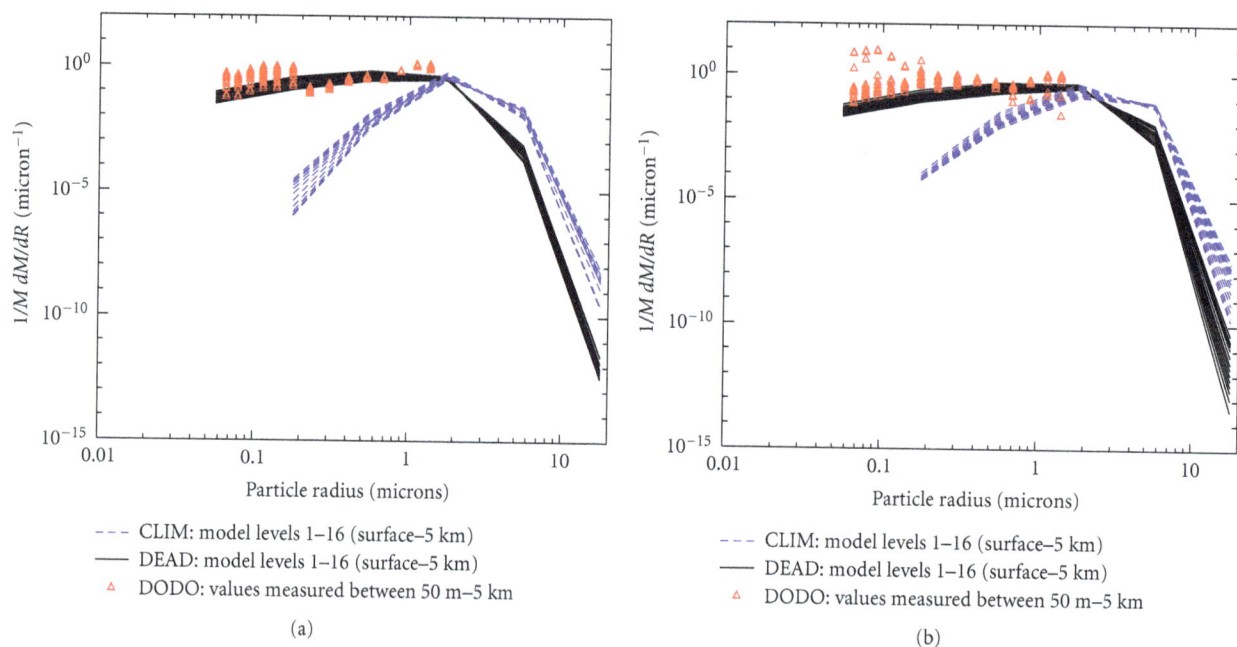

FIGURE 4: Normalised mass size distributions as a function of particle radius taken at 330°E and 17.5°N in CLIM (dashed blue lines) and DEAD (solid line) at each of the levels 1–16 (approximately from the surface to 5 km) for (a) DJF and (b) JJA. Corresponding values of the normalised mass size distribution from the DODO campaigns, between the surface and 5 km, (red triangles) are overlaid for (a) DJF and (b) JJA.

simulation using CLIM has lower AODs in remote regions than DEAD and as both schemes use identical dust transport and deposition schemes, it is likely that the size distribution of the dust transported throughout the model domain of the simulation using CLIM is different to the one simulated using DEAD.

To compare the size distributions in each model, we make use of airborne size distribution measurements collected during the Dust Outflow and Deposition to the Ocean experiment (DODO, see [35] for more details on the flight campaigns), which were subsequently used in the case study analysis by [27]. The flight campaign observations were taken during February and August 2006 in the vicinity of the West African coast at various heights above the surface. To compare with the DODO observations the December-January-February (DJF) and June-July-August (JJA) seasonal mean, mass weighted dust size distributions at 330°E and 17.5°N were taken from each model as being representative of dust transported off the African coast. Mass weighting the distributions allows a fairer comparison between the DODO observations (individual specific events) and the model-simulated output (climatological averages), as this will reduce any bias in the dust size distributions caused by individual dust events. The height of the observed data varied between approximately 50 m and 5 km and so the model output at levels 1 to 16 (surface to approximately 5 km above the surface) were taken to compare with the DODO observations.

The DJF dust size distributions for the models and observations are shown in Figure 4(a). The observed size distributions were taken from various heights and positions over West Africa. The observations were included this way

to give an idea of the range in possible size distributions taken during the DODO flights. Firstly, the DEAD scheme simulation contained more mass in bins 1, 2 and 3 (see Table 1 for size ranges) than in the CLIM scheme simulation. This indicates why the DEAD scheme has higher AODs in regions remote from the major global dust sources than CLIM. The smaller particles in the DEAD simulation can be transported further before deposition occurs (relative to CLIM) and therefore persist in the atmosphere longer to influence the model's radiation field northward (southward) of 60°N (30°S).

In comparison to the DODO observations, the DEAD scheme compares better with the DODO observations than CLIM, which contains very little dust in bin 1. However, there is no information from observations on the distribution of larger particles, which CLIM may represent better than DEAD. Although the size distribution measurements were taken over only a few days in February 2006, they appear to be representative of the seasonal average as several other more recent aircraft campaigns have measured a similar slope of the accumulation mode size distribution [36, 37].

The size distributions for JJA are given in Figure 4(b) and highlight again the larger proportion of smaller particles in DEAD compared to CLIM although DEAD has a better overall representation of the size distribution than CLIM. Again, the CLIM simulation contains almost no mass in bin 1, which does not agree well with the observations. Again, the main caveat is that the observed size distributions were taken in August and may not be representative of a seasonal mean; however, the biases appear to be systematic for both schemes over both seasons.

5. Deposition

The global transport and deposition of dust is important not just in terms of the local radiative forcing and the remote circulation anomalies that the presence of atmospheric dust generates, but also in terms of the whole Earth system [38, 39]. In some regions of the ocean, primary production by phytoplankton is limited by iron: dust that settles onto the sea surface containing iron can therefore fertilise ocean ecosystems and play an important role in regulating the entire carbon cycle. For this reason one important metric of the skill of a dust transport scheme in an Earth system context is the amount of dust transported to the oceans, and especially the southern oceans, where the effect on atmospheric CO_2 may be highest [39]. Also, Saharan dust transport and deposition to the Atlantic Ocean is particularly important in providing iron, potassium and other nutrients for ocean biological organisms.

To evaluate the model performance with respect to deposition, this study made use of data from the Dust Indicators and Records in Terrestrial and Marine Paleoenvironments (DIRTMAP) database [40–42]. Twelve stations were used and are given in Table 2 and their locations have also been plotted in Figure 5(a). Both marine sediment trap (with more than one thousand consecutive recorded days) and ice core data were used to provide the best estimate of the climatological mean dust deposition from the observations. The observed and modelled deposition values are plotted for both CLIM and DEAD in Figures 5(b) and 5(c). As the deposition schemes used in the CLIM and DEAD simulations are identical any differences in deposition between the two simulations can only be caused by the use of different uplift schemes. For the CLIM scheme deposition, eight of the twelve model points lie within a factor of ten of the observed deposition and for the DEAD scheme all twelve lie within a factor of ten of the observations. The range of values given in Figures 5(b) and 5(c) lie within the spread of model values given by [34].

The deposition in the DEAD and CLIM schemes are similar at 4 points (4, 5, 7, and 9, see Figure 5 and Table 2 for the locations), which are all in Greenland except for the marine sediment trap in the Sargasso Sea (9 in Table 2). This suggests that the westward and northward transport of dust over the Atlantic Ocean is similar in the two models. While the transport to the Sargasso Sea and eastern Greenland compare very well with the observations there is higher dust deposition than observed over central Greenland.

Both dust schemes simulate the deposition to Antarctica well (points 1, 3, and 8, see Table 2 and Figure 5 for the locations) with DEAD overestimating and CLIM only slightly underestimating the amount of dust deposited in Antarctica. The difference is likely to be due to CLIM having a smaller proportion of fine particles than DEAD (as discussed in Section 4), which implies that a higher percentage of the dust is deposited before reaching Antarctica.

Both CLIM and DEAD simulate too little dust deposition in north-western Greenland (2 in Table 2 and Figure 5), which may be associated with too much deposition over

TABLE 2: Names, data types, and locations of the DIRTMAP deposition data. The numbers on the left hand side of the first column correspond with those in Figure 5.

Site ID (numbers correspond with Figure 5).	Data source	Longitude (degrees)	Latitude (degrees)
(1) Byrd	Ice core	−120	−80
(2) Camp century	Ice core	−61	77
(3) Dome C	Ice core	124	−75
(4) Dye 3	Ice core	−44	65
(5) GRIP summit	Ice core	−38	73
(6) Huascarán	Ice core	−78	−9
(7) Renland	Ice core	−27	71
(8) Vostok	Ice core	107	−78
(9) Sargasso	Marine sediment trap	−64	32
(10) East	Marine sediment trap	69	15
(11) Cast	Marine sediment trap	65	14
(12) Wast	Marine sediment trap	60	16

the ice sheet as deposition was overestimated in central Greenland.

Finally, the CLIM scheme overestimates the deposition over western South America and the northern Indian Ocean (6, 10, 11, and 12, see Table 2 and Figure 5 for the locations), which is again likely to be due to the advection of the large particles downwind from the African and Asian dust sources. The deposition in the model using the DEAD scheme compares well with the observations in the northern Indian Ocean but slightly underestimates the deposition in western South America. The modelled deposition into the Atlantic and towards Antarctica compares well with the observations and so both regions are considered in the next sections.

5.1. North Atlantic.
The aim of DODO was to quantify the seasonal dust footprint from the Sahara to the Atlantic Ocean and subsequently estimate the iron deposition along with the variability in that flux. The dust deposition into the Atlantic needs to be compared to estimates in other work to identify whether the models are representing the real world well.

For this study, a region of the North Atlantic has been chosen in which to quantify the dust deposition to the surface and is shown in Figures 5(a) and 6. The region is similar in size to that used in the study by [31], which derives North Atlantic dust deposition from satellite data. The annual mean dust deposition and variability into the North Atlantic region can be seen in Table 3. The deposition from CLIM is much higher than for DEAD (almost a factor of 12) and CLIM also displays a much higher variability (more than 10 times) than DEAD. However, the coefficients of variation (standard deviation divided by the mean) are very similar for both simulations (Table 3), which imply that the apparent larger variability in CLIM is likely to be due to the larger mean deposition relative to DEAD.

(a)

(b)

(c)

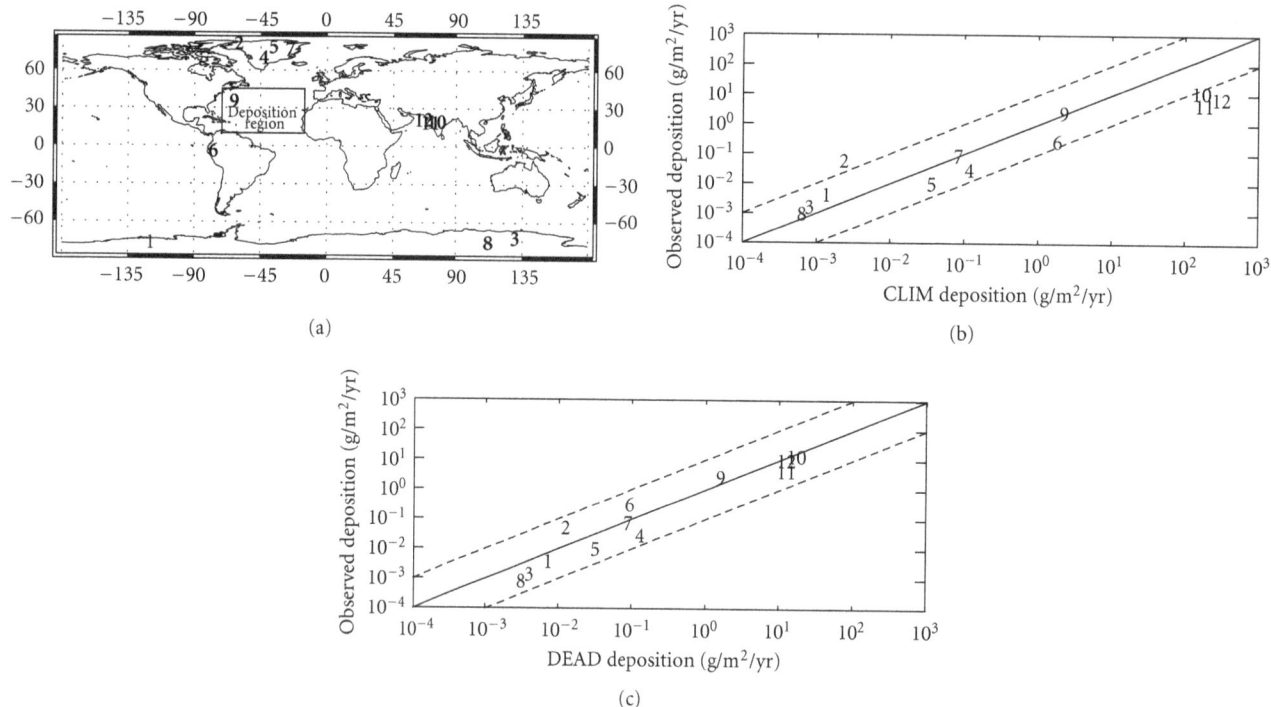

FIGURE 5: (a) The location of the DIRTMAP stations (1–9, see Table 2) used in this study and scatter plots of the DIRTMAP and nearest model grid box estimated deposition ($g\,m^{-2}\,yr^{-1}$) for (b) CLIM and (c) DEAD uplift schemes. Also shown in (a) is the area under consideration for the dust deposition values given in Table 3 and referred to in Section 5.1. Note in (c) that points 10, 11, and 12 lie close to the same point.

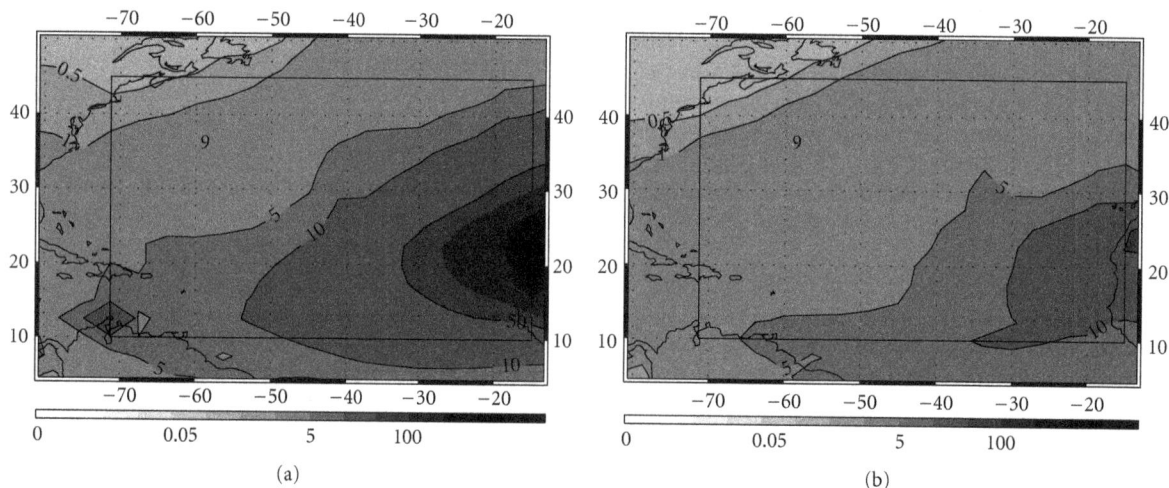

(a)

(b)

FIGURE 6: Deposition into the North Atlantic ($g\,m^{-2}yr^{-1}$) for (a) CLIM and (b) DEAD. The box and the "9" correspond to the box and DIRTMAP observation location given in Figure 5(a).

[31] suggest that the annual deposition from Africa to the Atlantic Ocean (in a similar region to Figure 6) is 140 ± 40 Tg dust, which agrees reasonably well with the value given in Table 3 for DEAD. The deposition into the region in Figure 6 for CLIM lies outside the error bounds for the [31] estimate and also lies outside the estimates for other studies (see Table 3 in [31]). However, in both simulations, the modelled deposition in the Sargasso Sea (point 9, Table 2 and Figure 5) compares very well with the DIRTMAP deposition

and suggests that further downstream from the Sahara the deposition is comparable between CLIM and DEAD, which can also be seen in Figure 6.

The higher deposition in CLIM is due to the size distribution of the transported dust (see Section 4) with large deposition rates adjacent to the West African coast, near the Saharan source (Figure 6(a)). By the time the prevailing easterlies have transported the dust to the West Atlantic, the majority of the large particles have been deposited and

TABLE 3: Annual (ANN) mean dust deposition into the North Atlantic for the region shown in Figures 5(a) and 6. Also included are the standard deviation, the coefficient of variation and the maximum and minimum deposition amounts of all modelled years.

	ANN mean, μ (Tg dust)	Standard deviation, σ (Tg dust)	Coefficient of variation σ/μ	Maximum/ minimum (Tg dust)
CLIM	1356	398	0.29	1826/691
DEAD	116	35	0.30	164/56

the deposition rates become comparable with DIRTMAP. Therefore, the high deposition amounts in CLIM cannot be ruled out as erroneous as deposition in other regions of the globe is comparable to DIRTMAP, although the high deposition values in the Indian Ocean (points 10, 11, and 12, Figures 5(b) and 5(c)) suggest there may be too much dust transported in CLIM.

5.2. Southern Ocean and Antarctica. Figure 7 shows that the simulation using the CLIM scheme results in more deposition of dust north of 47°S compared to the DEAD scheme. However, at all latitudes south of 47°S DEAD has higher deposition relative to CLIM. It is no coincidence that this latitude is the location of the Southern Ocean storm track, and the results of Figure 7 are consistent with the more numerous small particles in DEAD being advected into the southern polar region. In addition, the higher dust AODs exhibited by DEAD compared to CLIM south of 30°S in Figure 3 are consistent with observations of dust density in CLIM being too low in this region (see Figure 8 in [21] for King George's Island, Palmer Station, and Mawson station data). This can also be seen in the Antarctic (points 1, 3, and 8 in Figure 5) where CLIM underestimates and DEAD overestimates the dust deposition.

Observations of dust deposition are extremely sparse in the Southern Ocean region however. Although some measurements suggest that the reduction in deposition with latitude is perhaps more consistent with DEAD than CLIM [40], many more observations of dust deposition are needed before any conclusions can be drawn about the relative performance of DEAD and CLIM in depositing dust in the Southern Ocean.

6. Impacts on the Modelled Climate

We now discuss the effects of the differing spatial characteristics of the dust on the modelled climate. We confine ourselves to the JJA season, as this is the season with the largest differences between the CLIM and DEAD cases. Figure 8(a) shows the temperature difference at 1.5 m for DEAD relative to CLIM. Overall, the land surface is slightly cooler (particularly in Northern Africa and the Middle East), consistent with the higher dust AOD in DEAD, which intercepts incoming solar radiation over a greater area than CLIM. However, this signal is reversed in western Africa, where DEAD is warmer than CLIM and has lower dust AODs. The air temperatures at 200 hPa do not have the same

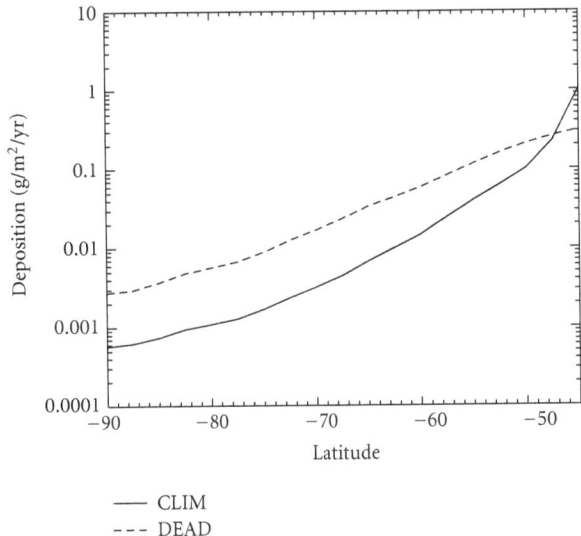

FIGURE 7: Zonal annual mean dust deposition ($g\,m^{-2}yr^{-1}$) for CLIM (solid line) and DEAD (dashed line) south of 45°S. Note: the y-axis is logarithmic.

dipole structure over western Africa (not shown) as those seen in the surface air temperatures, which indicates that the pattern in Figure 8(a) is a low-level structure.

The pattern evident in Figure 8(a) does have an effect on local circulation in western Africa. Figure 8(b) shows that a high-pressure anomaly exists in JJA in DEAD relative to CLIM; associated with this pressure anomaly is an anticyclonic circulation anomaly, as shown in Figure 9 at 850 hPa. The circulation anomaly has an effect on the west African monsoon circulation, as it transports dry air southwestwards toward the monsoonal region, the result is a small decrease in precipitation and relative humidity (RH) at 850 hPa over west Africa, as shown in Figures 10(a) and 10(b).

While the local effects are considerable, which is unsurprising given the large differences in dust AOD over Africa, some remote effects are evident from Figures 9 and 10. There is some evidence for an effect on the South Asian monsoon, with anomalous easterly winds off the coast of Somalia, and anomalous westerlies over southeastern Asia. Figure 10 also shows a reduction in precipitation over the equatorial Indian Ocean.

7. Discussion and Conclusions

The work presented in this study has attempted to identify the differences arising in the representation of dust in a GCM by using two separate uplift schemes. By incorporating two different schemes in one GCM, we have been able to understand how each uplift model responds to the simulated GCM climate and subsequently how the GCM responds to the differences in dust uplift from the two schemes.

Both simulations (CLIM and DEAD) had global annual mean AODs at 550 nm (averaged from 1980–1995) that lay within the range simulated by the AeroCom project [33].

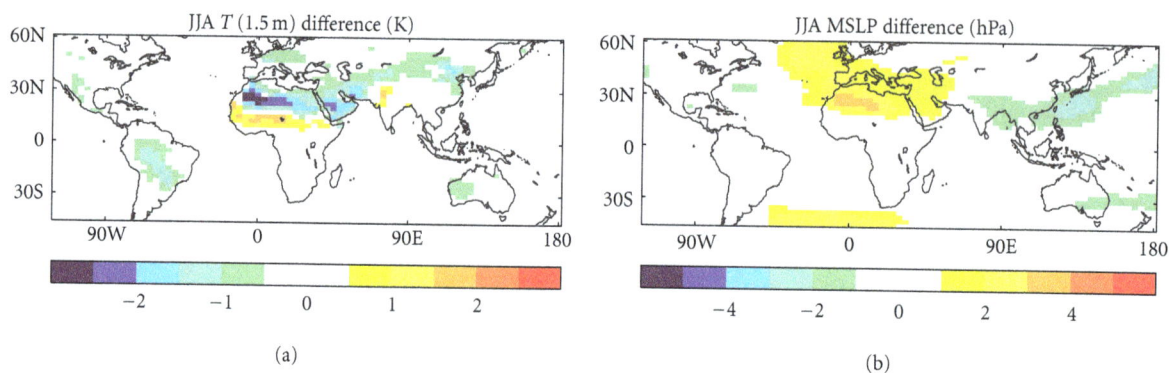

FIGURE 8: The difference in JJA (a) surface air temperature at 1.5 m (K) and (b) mean sea level pressure (MSLP, hPa)) for DEAD relative to CLIM.

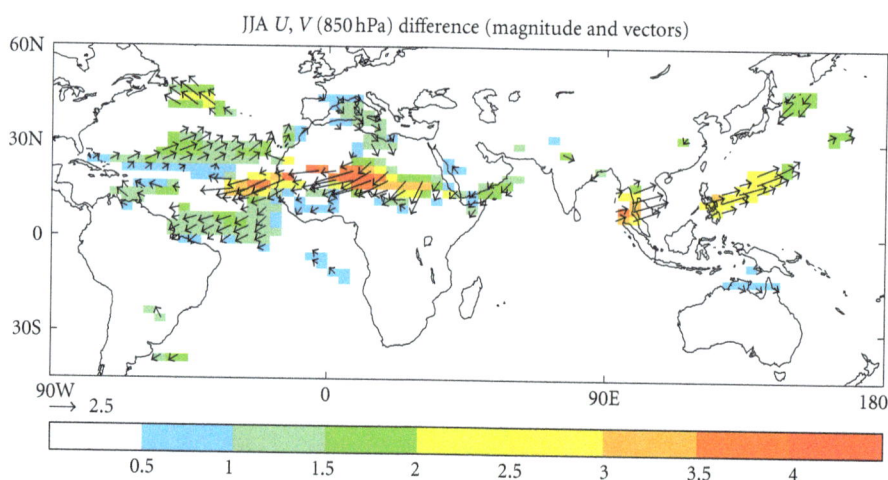

FIGURE 9: The changes in wind speed and direction at 850 hPa for DEAD relative to CLIM during JJA.

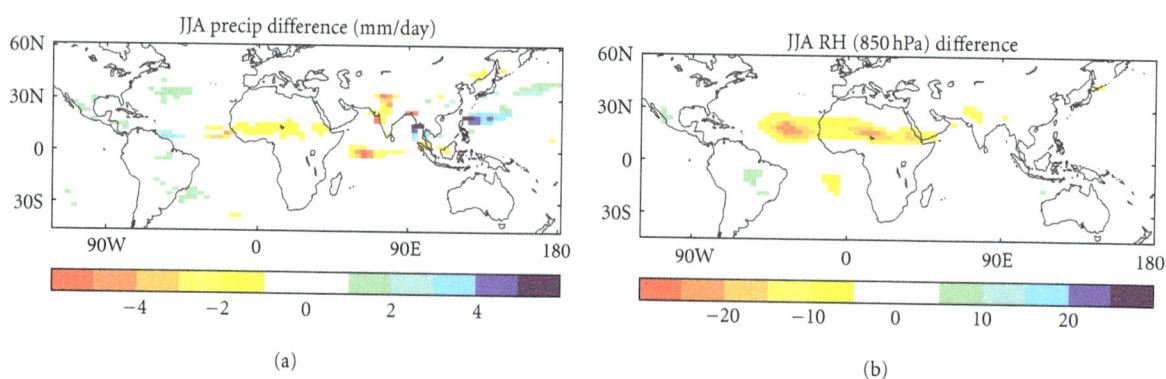

FIGURE 10: The difference in (a) precipitation (mm/day) and (b) relative humidity (RH, %) in JJA for DEAD relative to CLIM.

However, the AeroCom simulations only represented the year 2000, whereas the simulations undertaken in this study (using CLIM and DEAD) were run over sixteen years and may be more representative of the long-term climatological mean. Despite this difference, the majority of the annual mean AODs in both simulations, for any given individual year, lay within the AeroCom estimated range of global mean

AODs, which suggests that neither model lies outside the range of current dust modelling capabilities.

There were systematic differences between the two simulations however. The global annual mean AODs simulated using the CLIM scheme, were almost always consistently lower (fourteen of the sixteen simulated years) than those when using DEAD for any given year. Subsequently the climatological mean (1980–1995) dust AOD was lower in CLIM than for DEAD. The cause of these differences was due to the CLIM simulation containing a higher proportion of large particles, which could be deposited more rapidly from the atmosphere compared to those for DEAD, decreasing the dust load. Additionally DEAD contained more submicron particles than CLIM which enabled dust transport of particles with a high mass extinction efficiency [43] to greater distances in DEAD. Therefore the dust uplifted in the CLIM simulation had less time to interact effectively with radiation than in the DEAD simulation, resulting in lower AODs.

The dust size distribution in the CLIM scheme was also found to have too few small particles compared with the observational data compiled during the DODO field campaign. The size distribution of the dust uplifted using DEAD however was more comparable with the DODO observations, although there were still too few small particles. The close agreement between DEAD and DODO indicates that specifying the emitted dust size distribution based upon depositional data (as described in [28] and applied to the MetUM by [27]) may be more appropriate in GCM simulations. However, the size distribution of aeolian dust may vary regionally from the values calculated from the DODO campaign therefore specifying the modelled size distribution may not be appropriate for regional model simulations. A combination of the techniques used by CLIM and DEAD therefore may be appropriate for specifying the size distribution of aeolian dust.

Despite the differences in the modelled size distributions, the deposition rates in the CLIM and DEAD simulations compared well with the DIRTMAP observations. The CLIM scheme did less well than DEAD close to the major dust sources (such as the Sahara and Asia), which was due to the large proportion of coarse particles that could be deposited faster. Both CLIM and DEAD compared well with DIRTMAP in Antarctica except DEAD (CLIM) had slightly higher (lower) deposition rates than DIRTMAP, which was unsurprising given the higher proportion of fine (coarse) particles in DEAD (CLIM). The modelled deposition over eastern Greenland also compared well with DIRTMAP in both simulations although CLIM and DEAD had too little dust deposition in western Greenland.

While the differences in the simulated dust properties are important for understanding the uncertainty associated with specific uplift schemes, those differences also affected the modelled climate in each simulation. The higher global mean AODs with the DEAD scheme led to a large-scale cooling in JJA, particularly in North Africa and the Middle East, relative to CLIM (where the AODs reduced more rapidly away from the source). Conversely, within the source regions, the CLIM scheme had lower surface air temperatures, associated with the high AODs close to the source, than DEAD. The radiative impact of the dust also led to differences in the low-level circulation and precipitation rates close-to and remote from the Saharan dust source. These circulation changes still occurred despite the prescribed sea surface temperatures used in these simulations. This implies that the choice of dust uplift scheme has the potential to affect the tropical and subtropical circulation patterns even in AMIP-type simulations. Additionally, the differences in dust deposition in the Atlantic and Southern Ocean have important implications for ocean biogeochemistry and the carbon cycle. Therefore reducing the uncertainty associated with global dust modelling is important for all Earth-system modelling studies as it not only affects the global distribution of aeolian dust but also impacts significantly on the modelled climate.

Acknowledgments

DODO was funded by NERC via the SOLAS directed program (Grant NE/C517276/1) with the work undertaken by M. M. Joshi was funded by National Centres for Atmospheric Science (NCAS) Climate. The modelling work was carried out with the assistance of NCAS Computational Modelling Support (CMS). The aircraft data would not have been available without the enthusiastic work of the FAAM staff, DirectFlight, and Avalon Engineering. The authors would also like to thank Stephanie Woodward and Margaret Woodage for extremely helpful suggestions when reviewing earlier versions of this paper. Finally the authors would like to thank Charlie Zender for making the DEAD scheme code available to aid in generating the new uplift module in the MetUM.

References

[1] S. A. Christopher, P. Gupta, J. Haywood, and G. Greed, "Aerosol optical thicknesses over North Africa: 1. Development of a product for model validation using ozone monitoring instrument, multiangle imaging spectroradiometer, and aerosol robotic network," *Journal of Geophysical Research D*, vol. 113, no. 23, Article ID D00C04, 2008.

[2] P. Jiménez-Guerrero, C. Pérez, O. Jorba, and J. M. Baldasano, "Contribution of Saharan dust in an integrated air quality system and its on-line assessment," *Geophysical Research Letters*, vol. 35, no. 3, Article ID L03814, 2008.

[3] N. M. Mahowald, D. R. Muhs, S. Levis et al., "Change in atmospheric mineral aerosols in response to climate: last glacial period, preindustrial, modern, and doubled carbon dioxide climates," *Journal of Geophysical Research D*, vol. 111, no. 10, Article ID D10202, 2006.

[4] M. Yoshioka, N. M. Mahowald, A. J. Conley et al., "Impact of desert dust radiative forcing on sahel precipitation: relative importance of dust compared to sea surface temperature variations, vegetation changes, and greenhouse gas warming," *Journal of Climate*, vol. 20, no. 8, pp. 1445–1467, 2007.

[5] J. K. Moore, S. C. Doney, D. M. Glover, and I. Y. Fung, "Iron cycling and nutrient-limitation patterns in surface waters of the world ocean," *Deep-Sea Research II*, vol. 49, no. 1–3, pp. 463–507, 2002.

[6] I. Koren, Y. J. Kaufman, R. Washington et al., "The Bodélé depression: a single spot in the Sahara that provides most of the mineral dust to the Amazon forest," *Environmental Research Letters*, vol. 1, no. 1, Article ID 014005, 2006.

[7] C. Bouet, G. Cautenet, R. Washington et al., "Mesoscale modeling of aeolian dust emission during the BoDEx 2005 experiment," *Geophysical Research Letters*, vol. 34, no. 7, Article ID L07812, 2007.

[8] B. Heinold, J. Helmert, O. Hellmuth et al., "Regional modeling of Saharan dust events using LM-MUSCAT: model description

and case studies," *Journal of Geophysical Research D*, vol. 112, no. 11, Article ID D11204, 2007.

[9] G. Kallos, A. Papadopoulos, P. Katsafados, and S. Nickovic, "Transatlantic Saharan dust transport: model simulation and results," *Journal of Geophysical Research D*, vol. 111, no. 9, Article ID D09204, 2006.

[10] W. L. Gates, J. S. Boyle, C. Covey et al., "An overview of the results of the Atmospheric Model Intercomparison Project (AMIP I)," *Bulletin of the American Meteorological Society*, vol. 80, no. 1, pp. 29–55, 1999.

[11] M. C. Todd, D. Bou Karam, C. Cavazos et al., "Quantifying uncertainty in estimates of mineral dust flux: an intercomparison of model performance over the Bodélé depression, northern Chad," *Journal of Geophysical Research D*, vol. 113, no. 24, Article ID D24107, 2008.

[12] I. Uno, Z. Wang, M. Chiba et al., "Dust Model Intercomparison (DMIP) study over Asia: overview," *Journal of Geophysical Research D*, vol. 111, no. 12, Article ID D12213, 2006.

[13] S. Kinne, M. Schulz, C. Textor et al., "An AeroCom initial assessment—optical properties in aerosol component modules of global models," *Atmospheric Chemistry and Physics*, vol. 6, no. 7, pp. 1815–1834, 2006.

[14] P. R. Colarco, O. B. Toon, and B. N. Holben, "Saharan dust transport to the Caribbean during PRIDE: 1. Influence of dust sources and removal mechanisms on the timing and magnitude of downwind aerosol optical depth events from simulations of in situ and remote sensing observations," *Journal of Geophysical Research D*, vol. 108, article 8589, 20 pages, 2003.

[15] E. Nowottnick, P. Colarco, R. Ferrare et al., "Online simulations of mineral dust aerosol distributions: comparisons to namma observations and sensitivity to dust emission parameterization," *Journal of Geophysical Research D*, vol. 115, no. 3, Article ID D03202, 2010.

[16] C. Luo, N. Mahowald, and C. Jones, "Temporal variability of dust mobilization and concentration in source regions," *Journal of Geophysical Research D*, vol. 109, no. 20, Article ID D20202, 13 pages, 2004.

[17] T. C. Johns, C. F. Durman, H. T. Banks et al., "The new Hadley Centre Climate Model (HadGEM1): evaluation of coupled simulations," *Journal of Climate*, vol. 19, no. 7, pp. 1327–1353, 2006.

[18] G. M. Martin, M. A. Ringer, V. D. Pope, A. Jones, C. Dearden, and T. J. Hinton, "The physical properties of the atmosphere in the new Hadley Centre Global Environmental Model (HadGEM1). Part 1: model description and global climatology," *Journal of Climate*, vol. 19, no. 7, pp. 1274–1301, 2006.

[19] W. J. Collins, N. Belloiun, M. Doutriaux-Boucher et al., "Evaluation of HadGEM2 model," Hadley Centre Technical Note 74, 2008, http://www.metoffice.gov.uk/media/pdf/8/7/HCTN_74.pdf.

[20] The HadGEM2 Development Team, "The HadGEM2 family of Met Office Unified Model climate configurations," *Geoscientific Model Development*, vol. 4, pp. 723–757, 2011.

[21] S. Woodward, "Modeling the atmospheric life cycle and radiative impact of mineral dust in the Hadley centre climate model," *Journal of Geophysical Research D*, vol. 106, no. 16, pp. 18155–18166, 2001.

[22] L. C. Shaffrey and Coauthors, "U.K. HiGEM: the new U.K. high-resolution global environment model—model description and basic evaluation," *Journal of Climate*, vol. 22, pp. 1861–1896, 2011.

[23] M. J. Woodage, A. Slingo, S. Woodward, and R. E. Comer, "U.K. HiGEM: simulations of desert dust and biomass burning aerosols with a high-resolution atmospheric GCM," *Journal of Climate*, vol. 23, no. 7, pp. 1636–1659, 2010.

[24] R. A. Bagnold, *The Physics of Blown Sand and Desert Dunes*, Methuen, London, UK, 1941.

[25] S. Woodward, "Mineral dust in HadGEM2," Hadley Centre Technical Note 87, 2011, http://www.metoffice.gov.uk/media/pdf/l/p/HCTN_87.pdf.

[26] Global Soil Data Task Group, *Global Gridded Surfaces of Selected Soil Characteristics (IGBP-DIS)*, Oak Ridge National Laboratory Distributed Active Archive Center, Oak Ridge, Tenn, USA, 2000.

[27] D. Ackerley, E. J. Highwood, M. A. J. Harrison et al., "The development of a new dust uplift scheme in the Met Office Unified Model," *Meteorological Applications*, vol. 16, no. 4, pp. 445–460, 2009.

[28] C. S. Zender, H. Bian, and D. Newman, "Mineral Dust Entrainment and Deposition (DEAD) model: description and 1990s dust climatology," *Journal of Geophysical Research D*, vol. 108, no. 14, pp. 1–19, 2003.

[29] F. Fécan, B. Marticorena, and G. Bergametti, "Parametrization of the increase of the aeolian erosion threshold wind friction velocity due to soil moisture for arid and semi-arid areas," *Annales Geophysicae*, vol. 17, no. 1, pp. 149–157, 1999.

[30] B. N. Holben, T. F. Eck, I. Slutsker et al., "A federated instrument network and data archive for aerosol characterization," *Remote Sensing of Environment*, vol. 66, no. 1, pp. 1–16, 1998.

[31] Y. J. Kaufman, I. Koren, L. A. Remer, D. Tanré, P. Ginoux, and S. Fan, "Dust transport and deposition observed from the Terra-Moderate Resolution Imaging Spectroradiometer (MODIS) spacecraft over the Atlantic ocean," *Journal of Geophysical Research D*, vol. 110, no. 10, pp. 1–16, 2005.

[32] N. Bellouin, O. Boucher, J. Haywood et al., "Improved representation of aerosols for HadGEM2," Hadley Centre Technical Note 73, 2007, http://www.metoffice.gov.uk/media/pdf/8/f/HCTN_73.pdf.

[33] C. Textor, M. Schulz, S. Guibert et al., "Analysis and quantification of the diversities of aerosol life cycles within AeroCom," *Atmospheric Chemistry and Physics*, vol. 6, no. 7, pp. 1777–1813, 2006.

[34] N. Huneeus, M. Schulz, Y. Balkanski et al., "Global dust model intercomparison in AeroCom phase i," *Atmospheric Chemistry and Physics*, vol. 11, no. 15, pp. 7781–7816, 2011.

[35] C. L. McConnell, E. J. Highwood, H. Coe et al., "Seasonal variations of the physical and optical characteristics of Saharan dust: results from the Dust Outflow and Deposition to the Ocean (DODO) experiment," *Journal of Geophysical Research*, vol. 113, Article ID D14S05, 19 pages, 2008.

[36] B. T. Johnson and S. R. Osborne, "Physical and optical properties of mineral dust aerosol measured by aircraft during the GERBILS campaign," *Quarterly Journal of the Royal Meteorological Society*, vol. 137, no. 658, pp. 1117–1130, 2011.

[37] B. Weinzierl, A. Petzold, M. Esselborn et al., "Airborne measurements of dust layer properties, particle size distribution and mixing state of Saharan dust during SAMUM 2006," *Tellus B*, vol. 61, no. 1, pp. 96–117, 2009.

[38] A. J. Ridgwell, "Dust in the Earth system: the biogeochemical linking of land, air and sea," *Philosophical Transactions of the Royal Society A*, vol. 360, no. 1801, pp. 2905–2924, 2002.

[39] T. D. Jickells, Z. S. An, K. K. Andersen et al., "Global iron connections between desert dust, ocean biogeochemistry, and climate," *Science*, vol. 308, no. 5718, pp. 67–71, 2005.

[40] I. Tegen, S. P. Harrison, K. Kohfeld, I. C. Prentice, M. Coe, and M. Heimann, "Impact of vegetation and preferential source areas on global dust aerosol: results from a model study," *Journal of Geophysical Research D*, vol. 107, no. 21, article 4576, 2002.

[41] K. E. Kohfeld and S. P. Harrison, "DIRTMAP: the geological record of dust," *Earth-Science Reviews*, vol. 54, no. 1–3, pp. 81–114, 2001.

[42] K. E. Kohfeld, "DIRTMAP version 2. LGM and late holocene eolian fluxes from ice cores, marine sediment traps, marine sediments, and loess deposits," IGBP PAGES/World Data Center for Paleoclimatology Data Contribution Series #2002-045, NOAA/NGDC Paleoclimatology Program, Boulder, Colo, USA, 2002.

[43] J. H. Seinfeld and S. Pandis, *Atmospheric Chemistry and Physics*, John Wiley and Sons, New York, NY, USA, 1998.

14

Desert Dust Outbreaks over Mediterranean Basin: A Modeling, Observational, and Synoptic Analysis Approach

F. Calastrini,[1,2] F. Guarnieri,[1,2] S. Becagli,[3] C. Busillo,[1] M. Chiari,[4] U. Dayan,[5] F. Lucarelli,[4,6] S. Nava,[4] M. Pasqui,[2] R. Traversi,[3] R. Udisti,[3] and G. Zipoli[2]

[1] LaMMA Consortium, Laboratory of Monitoring and Environmental Modeling for the Sustainable Development, Via Madonna del Piano 10, 50019 Sesto Fiorentino, 50019 Sesto Fiorentino, Italy
[2] IBIMET, National Research Council, Via G. Caproni 8, 50145 Florence, Italy
[3] Department of Chemistry, University of Florence, Via della Lastruccia 3, 50019 Sesto Fiorentino, Italy
[4] INFN, National Institute of Nuclear Physics, Via G. Sansone 1, 50019 Sesto Fiorentino, Italy
[5] Department of Geography, The Hebrew University of Jerusalem, Jerusalem 91905, Israel
[6] Department of Physics and Astronomy, University of Florence, Via G. Sansone 1, 50019 Sesto Fiorentino, Italy

Correspondence should be addressed to F. Calastrini, calastrini@lamma.rete.toscana.it

Academic Editor: Pawan Gupta

Dust intrusions from African desert regions have an impact on the Mediterranean Basin (MB), as they cause an anomalous increase of aerosol concentrations in the tropospheric column and often an increase of particulate matter at the ground level. To estimate the Saharan dust contribution to PM_{10}, a significant dust intrusion event that occurred in June 2006 is investigated, joining numerical simulations and specific measurements. As a first step, a synoptic analysis of this episode is performed. Such analysis, based only on meteorological and aerosol optical thickness observations, does not allow the assessment of exhaustive informations. In fact, it is not possible to distinguish dust outbreaks transported above the boundary layer without any impact at the ground level from those causing deposition. The approach proposed in this work applies an ad hoc model chain to describe emission, transport and deposition dynamics. Furthermore, physical and chemical analyses (PIXE analysis and ion chromatography) were used to measure the concentration of all soil-related elements to quantify the contribution of dust particles to PM_{10}. The comparison between simulation results and in-situ measurements show a satisfying agreement, and supports the effectiveness of the model chain to estimate the Saharan dust contribution at ground level.

1. Introduction

Aerosols have direct and indirect effects on global climate [1], altering the radiative balance of the Earth-atmosphere system [2–4] and changing the microphysical and the radiative properties of clouds. In fact, aerosol particles can act as cloud condensation nuclei, modifying cloud lifetime and amount [3–5]. The mineral dust from desert areas, which represent a great source of aerosol injected into the atmosphere, suppresses precipitation in thin low-altitude clouds [6–8]. In addition, dust deposition can modify the ocean biogeochemical cycle, providing an important source of micronutrients [9]. It has also an impact on terrestrial ecosystems, providing nutrients as phosphorus to the soil [10]. Saharan desert is one of the most important sources of mineral dust [11], which has a considerable impact both at the global [12] and at the regional scale, as the Mediterranean Basin (MB) [13].

More specifically, Saharan dust outbreaks can cause an anomalous rise of PM_{10} concentrations over large parts of the MB, with impacts on Spain, Greece, and Italy [14–16]. The exceedance of daily thresholds established by the European Union (Directive 2008/50/EC) may depend not only on human emissions but also on the mineral dust contribution. The same directive includes the possibility of removing exceedances due to natural sources, such as atmospheric resuspension or the transport of natural particles from dry regions. Although the assessment of a recognised procedure to quantify and subtract the effective Saharan dust contribution to PM_{10} concentration levels is still under discussion, preliminary guidelines are proposed in a working paper of the European Commission (http://www.ec.europa.eu/environment/air/quality/legislation/pdf/sec_2011_0208.pdf), mainly following the approach reported in Escudero et al. [17].

The method proposed in the European guidelines suggests to combine the use of satellite retrievals, model systems, and PM_{10} measurements at the ground level. The meteorological analysis (ECMWF http://www.ecmwf.int/) checks the evolution of synoptic systems prone to the generation of strong gradient winds able to transport dust plumes from North Africa towards Europe. The satellite observations (http://oceancolor.gsfc.nasa.gov/SeaWiFS/HTML/dust.html) and the aerosol index (TOMS http://toms.gsfc.nasa.gov/ozone/ozone_v8.html) can be useful in the detection of the dust episode and in the identification of the area affected by the plumes. Daily results of numerical models (e.g., SKIRON http://forecast.uoa.gr/, BSC-DREAM http://www.bsc.es/projects/earthscience/BSC-DREAM/ and NAAPs http://www.nrlmry.navy.mil/aerosol/) can be checked to identify the occurrence and the duration of the dust episode. Using only satellite observations and model simulations of aerosol optical thickness, it is not possible to clearly distinguish episodes that involve dust transport above the boundary layer without any impact at the ground level from those that cause dust deposition. For this reason, the European guidelines also require an analysis of PM_{10} mass concentration, which is routinely measured at the ground level by air quality station networks. Distinguishing between the Saharan dust contribution is not trivial since the PM_{10} concentration may be contributed by many different sources [16]. In a near future, the results of this method will be provided by GMES-Atmosphere service (http://www.gmes-atmosphere.eu/).

Within this framework, a study of Saharan dust intrusions over the MB is proposed, focusing on the deposition mechanism to assess the dust impact at the ground level. For this reason, a comprehensive model chain is configured to reconstruct the dynamic evolution of the Saharan desert dust. Numerical models play an important role in the description of the process of dust emission, transport, and deposition from desert zones. It is possible to reconstruct the spatial distributions of the dust with a good description of the vertical concentration profile. To characterise the PM_{10} mass concentration and composition at the ground level, the model-based analysis is complemented by physical and chemical analyses of the PM_{10} sampled in different sites. In particular, the concentration of all soil-related elements is measured by PIXE (Particle-Induced X-ray Emission) to assess the mineral dust concentration. The ionic composition is also used as ancillary data to subtract the sea-salt contributions to Na and Mg. Because the impact of desert dust is well characterised by an increase of all soil-related elements (and by changes in elemental ratios), this approach is useful for assessing the real impact of dust episodes on PM_{10}. The measurement of all crustal elements may allow a quantitative assessment of the desert dust contribution, which is more accurate with respect to estimations that may be obtained simply by the analysis of PM_{10} mass concentration data. In addition to the specific quantitative information obtained by the PM_{10} measurements in the selected sampling sites, the integrated use of model simulations and in situ experimental data may provide an overall picture of the mineral dust distribution.

This approach, which joins numerical simulations and specific measurements, is employed to study the impact of Saharan dust event of June 2006, a month characterised by significant dust outbreaks over the MB [18, 19]. Since at the moment, the GMS-Atmosphere service is not operational, a preliminary analysis of this long spell is conducted using the available data as the synoptic meteorological data from NCEP, the satellite observation images (MODIS/AQUA, MODIS/TERRA http://modis.gsfc.nasa.gov/), and the aerosol index (TOMS). The global model GOCART (http://disc.sci.gsfc.nasa.gov/ges-News/gocart_data_V006), developed by the Georgia Institute of Technology-Goddard to simulate aerosol optical thickness [20, 21], is used to integrate the satellite data. The HYSPLIT model (Hybrid Single Particle Lagrangian Integrated Trajectory model, http://ready.arl.noaa.gov/HYSPLIT.php) is used to generate backward trajectories to trace back the sources of the air masses at different levels and for different hours of the day.

The results of model chain, configured to describe the spatiotemporal evolution of this specific dust outbreak, are compared with GOCART data. Physical and chemical analyses are conducted on PM_{10} samples collected from three sites in central Italy (Tuscany). Finally, the numerical simulation results are compared with these specific measurements to evaluate the effectiveness of the model chain in the quantitative estimation of the Saharan dust contribution at the ground level.

2. Instruments and Methods

2.1. The Model Chain. The natural phenomena involved in the dust cycle in the atmosphere consist of two major physical processes: a wind stress lifting mechanism enabling the dust particles to rise up from bare soil surfaces and then the transport and deposition of this mineral dust [22–24]. To provide regional characteristics of Saharan dust intrusion over the MB, an atmospheric emission and dispersion model chain is developed. The model chain is based on three different modules: the atmospheric, dust emission, and transport/deposition modules. The Regional Atmospheric Modeling System (RAMS) [19], used by CNR-IBIMET in operational mode [25–27], provides the input data for the other modules. The DUST Emission Model (DUSTEM), specifically developed for this aim, simulates the dust emission from the desert. The Comprehensive Air quality Model with extensions (CAMx) (http://www.camx.com/home.aspx) takes the meteorological inputs from RAMS and the emission rate from DUSTEM [28], providing the dynamical transport and deposition of the dust particles.

This section describes the model chain configuration used for this case study from 1 June to 5 July 2006.

TABLE 1: Features of the four typical dust particles.

Type	Name	Typical particle diameter (μm)	Typical particle radius (μm)	Particle density (g/cm^3)	Erodible fraction
Clay	CCR1	**01-02**	**0.73**	**2.50**	**0.08**
Silt, small	CCR2	**02-20**	**6.1**	**2.65**	**1.00**
Silt, large	CCR3	20-50	18	2.65	1.00
Sand	CCR4	50-100	38	2.65	0.12

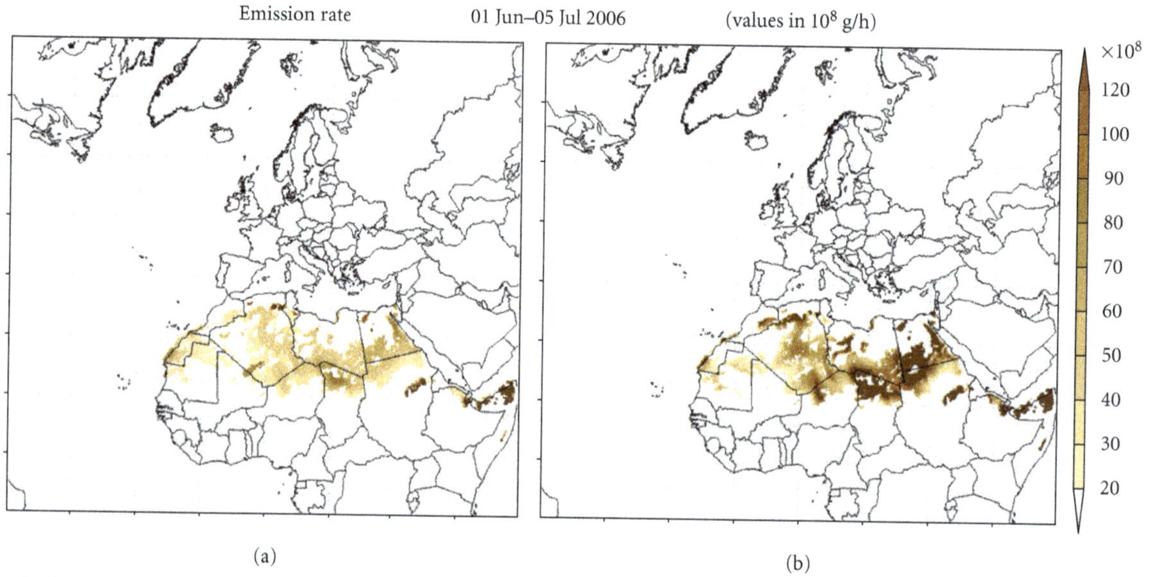

FIGURE 1: Emission rate, cumulated on the period 01–05 June 2006 for the first dimensional class (a) and for the second dimensional class (b).

2.2. The Regional Atmospheric Modeling System (RAMS). The RAMS 6.02 version is run over a domain including a large part of the Northern Hemisphere and performed by parallel computing. The initial and boundary atmospheric conditions need to be set as well as the forcing data during the period of simulation. To address these necessities, the Reanalysis2 dataset [29], with a 2.5-degree horizontal resolution, is employed. In particular, the geopotential height, temperature, relative humidity, and zonal and meridional wind component fields are used and forced as boundary conditions every 6 hours throughout the simulation period. The Kain-Fritsch convective scheme is also adopted. The configuration is set with the following features: the domain is centered on 40°N–5°E, with 200 × 80 grid points, 32 sigma vertical levels (with a stretching factor to obtain a greater resolution near the soil and a lesser one above 2000 m) and 11 soil levels. The horizontal resolution is 0.54 degrees, and the time step is 120 seconds, with a temporal output of 1 hour.

2.3. The Dust Emission Model (DUSTEM). Dust emissions are estimated by developing an ad hoc model called the DUST Emission Model (DUSTEM) [28]. This model estimates dust emission rates using empirical relationships based on soil texture and friction velocity [30]. DUSTEM can take into account four different soil types: clay, small silt, large silt, and sand [30, 31]. The main features of these classes are shown in Table 1. For the selected case study, only the first two dimensional classes, clay and small silt, are considered because they are the only ones involved in long-range transport [31]. DUSTEM takes the soil information from

the GLC2000 land cover [32] and FAO Textural Map [33] as input data to obtaining the bare soil map. The hourly meteorological fields (soil moisture and friction velocity) are provided by the RAMS model. The computational domain, in Polar Stereographic coordinates (with the pole at 40 degrees north and 5 degrees east), is formed by 380 × 340 cells with a 30 km resolution to provide the input data to CAMx with a temporal resolution of 1 hour. The maps of emission rates cumulated on the whole simulation period, relative to the first and second soil types, are shown in Figure 1.

2.4. The Comprehensive Air Quality Model (CAMx). The Comprehensive Air quality Model (CAMx), developed by ENVIRON International Corporation California, is an Eulerian photochemical dispersion model that allows for an integrated "one-atmosphere" assessment of gaseous and particulate air pollution over many scales, ranging from urban to superregional (http://www.camx.com/home.aspx). CAMx simulates the dispersion, chemical reactions, and removal of pollutants in the troposphere by solving the pollutant continuity equation for each species. The aerosol deposition is handled by adopting the algorithm of S. A. Slinn and W. G. N. Slinn [34] and Seinfeld and Pandis [35] for the dry and wet deposition, respectively. To simulate the transport and deposition of mineral dust, the chemical module is switched off in this study. The meteorological input data are provided by RAMS, and the emission rates of clay and small silt types are provided by DUSTEM (Table 1). The initial and boundary conditions are set to zero. To consider the atmospheric dust loading, the simulation starts on 1 June, some days in advance of the outbreak over Europe. The horizontal computational domain is the same as that used for DUSTEM. There are 18 vertical levels, from 10 m to 10,500 m, with a finer resolution near the ground. The concentration outputs for clay and small silt soil types are provided with a temporal resolution of 1 hour.

2.5. PM$_{10}$ Sampling and Analysis. To characterise the PM$_{10}$ mass concentration and composition at the ground level, the concentration of all soil-related elements is measured by PIXE to assess the mineral dust concentration [36–40]. The simultaneous detection, with high sensitivity, of all the elements that compose mineral dust makes PIXE highly effective for these investigations. A simultaneous increase in the concentration of all crustal elements is indeed a first indication of the occurrence of dust events; clearly, it is a stronger signature than the increase of the PM mass concentration, which may be due to many other sources, both natural and anthropogenic. However, an increase of crustal element concentrations may also be due to local soil dust resuspension: in this context, the analysis of changes in elemental ratios may help distinguish between these two kinds of events. Moreover, when the aerosol is sampled at more than one site, the observation of a simultaneous increase of soil-related elements at the different sites may also suggest the impact of long-range transported soil dust affecting a wide area, while the observation of different patterns from site to site clearly hints at the contribution of local sources.

2.6. PM$_{10}$ Sampling. Aerosol samples are collected on 47 mm Teflon filters on a daily basis (from midnight to midnight) using low-volume (2.3 m^3/h) sequential samplers (HYDRA Dual Sampler, FAI Instruments) equipped with a PM$_{10}$ inlet in accordance with the European rule EN 12341.

The PM$_{10}$ daily mass concentrations are obtained by weighing the filters with an analytical balance (sensitivity 1 μg) before and after the sampling, always after a storage period (48 hours) in a temperature and humidity controlled room with an ambient temperature $T = (20 \pm 1)°C$ and a relative humidity RH = $(50 \pm 5)\%$. Electrostatic effects are avoided by using a deionising gun.

In particular, a comprehensive dataset of PM$_{10}$ measured concentrations and composition at the ground level is obtained in the framework of the PATOS project, the first extensive field campaign for PM$_{10}$ characterisation in Central Italy (Tuscany), which was supported by the Regional Government (http://servizi.regione.toscana.it/aria/index.php?idDocumento=18348). PM$_{10}$ samples are collected from September 2005 to September 2006 at six sampling sites in Tuscany that are representative of different types of areas: Florence (urban background), Prato (urban traffic), Capannori-Lucca (urban background), Arezzo (urban traffic), Grosseto (urban background), and Livorno (suburban background). In this field campaign, the three available samplers are used to collect the PM$_{10}$ data in the six sites previously cited. These samplers are moved every 15 days from three sampling sites (Arezzo, Prato, and Livorno) to

the other three (Florence, Capannori-Lucca, and Grosseto). In this way, data relative to the long-range transport dust intrusion of June and the background level due to local dust resuspension is sampled.

2.7. PM$_{10}$ Compositional Analysis. The PM$_{10}$ samples are analysed by Particle-Induced X-ray Emission (PIXE) to determine the aerosol elemental composition and, in particular, the concentrations of all the main soil-related elements (Na, Mg, Al, Si, K, Ca, Ti, Mn, Fe, and Sr). The same samples are also analysed by ion chromatography (IC) to assess the soluble fraction of inorganic ions, specifically the Na$^+$ and Mg^{2+} concentrations. These values are used to determine the sea-salt contributions to Na and Mg to be subtracted for the calculation of the soil dust fraction of these elements.

The PIXE measurements are performed at the 3 MV Tandetron accelerator of the LABEC laboratory of INFN in Florence using the external beam setup dedicated to environmental applications [39, 41]. For this case study, each sample is irradiated for ~500 s with a 3.2 MeV proton beam with an intensity of ~5 nA over a spot of ~2 mm^2. During irradiation, the filter is moved in front of the beam so that most of the area of the deposit is analysed. The PIXE spectra are fitted using the GUPIX code [42] and elemental concentrations are obtained via a calibration curve from a set of thin standards (Micromatter Inc.). The minimum detection limits (MDLs) are ~10 ng/m^3 for low-Z elements and ~1 ng/m^3 for medium-high-Z elements. The elemental concentration uncertainty is determined by taking into account the independent uncertainties on the standard sample thickness ($\pm 5\%$), the aerosol deposition area ($\pm 2\%$), the airflow ($\pm 2\%$), and the X-ray counting statistics (from ± 2 to $\pm 20\%$ or higher when concentrations approach MDLs).

Ion analysis is performed on the water extract obtained from a quarter of each Teflon filter. Each sample is analysed for cations (Na$^+$, NH$_4^+$, K$^+$, Mg^{2+}, and Ca^{2+}), inorganic anions (F$^-$, Cl$^-$, NO$_3^-$, and SO$_4^{2-}$), and some organic anions (methanesulphonate (MSA), acetate, formate, glycolate, and oxalate) by 3 Dionex ion chromatographic systems operating in parallel under the working conditions summarised by Becagli et al. [43]. The detection limits are several orders of magnitude lower than the concentrations found in these samples. The uncertainty is mainly determined by the ion chromatography accuracy, which is typically $\pm 5\%$.

3. Results and Discussion

3.1. Synoptic Conditions Featuring the Dust Outbreak. The synoptic atmospheric conditions favourable for dust raising are those associated with air mass advection coming from Northern Europe or the Balkans regions and heading towards Algeria, Libya, and Egypt up to Chad [44]. These air masses are characterised by strong and constant wind in the lower layers. The associated thermal gradient has two important effects for dust emissions: the first is the increase of the soil friction in correspondence with the air mass front due to the acceleration of the thermal wind; the latter is the reduction of

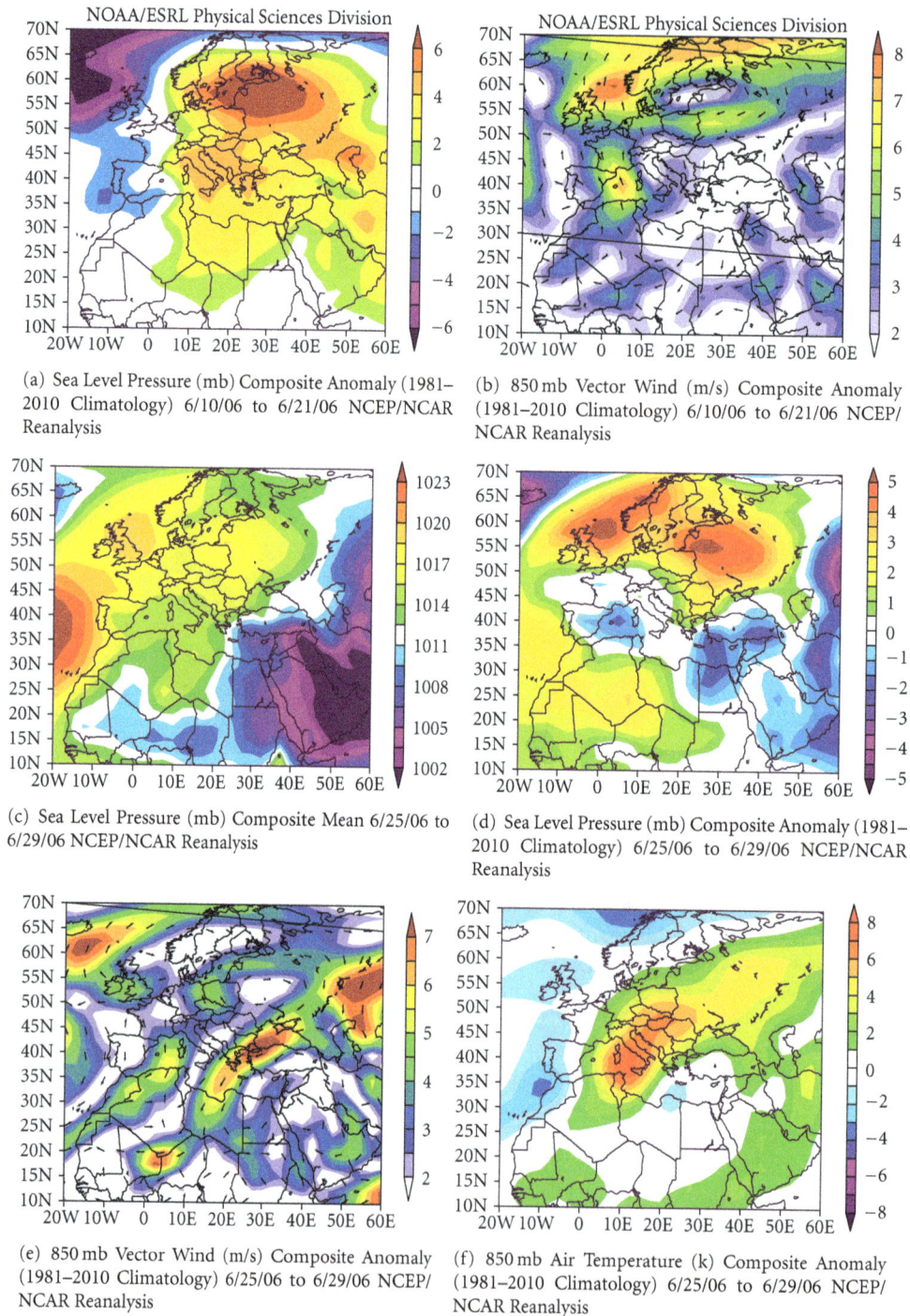

(a) Sea Level Pressure (mb) Composite Anomaly (1981–2010 Climatology) 6/10/06 to 6/21/06 NCEP/NCAR Reanalysis

(b) 850 mb Vector Wind (m/s) Composite Anomaly (1981–2010 Climatology) 6/10/06 to 6/21/06 NCEP/NCAR Reanalysis

(c) Sea Level Pressure (mb) Composite Mean 6/25/06 to 6/29/06 NCEP/NCAR Reanalysis

(d) Sea Level Pressure (mb) Composite Anomaly (1981–2010 Climatology) 6/25/06 to 6/29/06 NCEP/NCAR Reanalysis

(e) 850 mb Vector Wind (m/s) Composite Anomaly (1981–2010 Climatology) 6/25/06 to 6/29/06 NCEP/NCAR Reanalysis

(f) 850 mb Air Temperature (k) Composite Anomaly (1981–2010 Climatology) 6/25/06 to 6/29/06 NCEP/NCAR Reanalysis

FIGURE 2: Meteorological fields from NCEP/NCAR Reanalysis. The anomalies are evaluated on the climatology 1981–2010. (a) Sea level pressure anomaly for 10–21 June 2006, (b) 850 mb wind vector anomaly for 10–21 June 2006, (c) mean sea level pressure for 25–29 June 2006, (d) sea level pressure anomaly for 25–29 June 2006, (e) 850 mb wind vector anomaly for 25–29 June 2006, and (f) 850 mb air temperature anomaly for 25–29 June 2006.

the soil moisture content due to the positive surface, which favours the raising of dust.

Once the dust is raised above the boundary layer, a strong circulation between 850 and 700 hPa is necessary to transport the dust far from the emission area. Throughout the year, massive airborne plumes of desert dust from the Sahara and surrounding regions are exported to the tropical Atlantic Ocean and the Mediterranean Sea. The majority of the dust intrusions over the MB are usually associated with the passage of either a cold or a warm low-pressure system. Saharan depressions develop most readily when a Polar or Arctic air mass from the northwest (Maritime Polar or Maritime Arctic) or northeast (Continental Polar or Continental Arctic) flows behind the dry desert air [45].

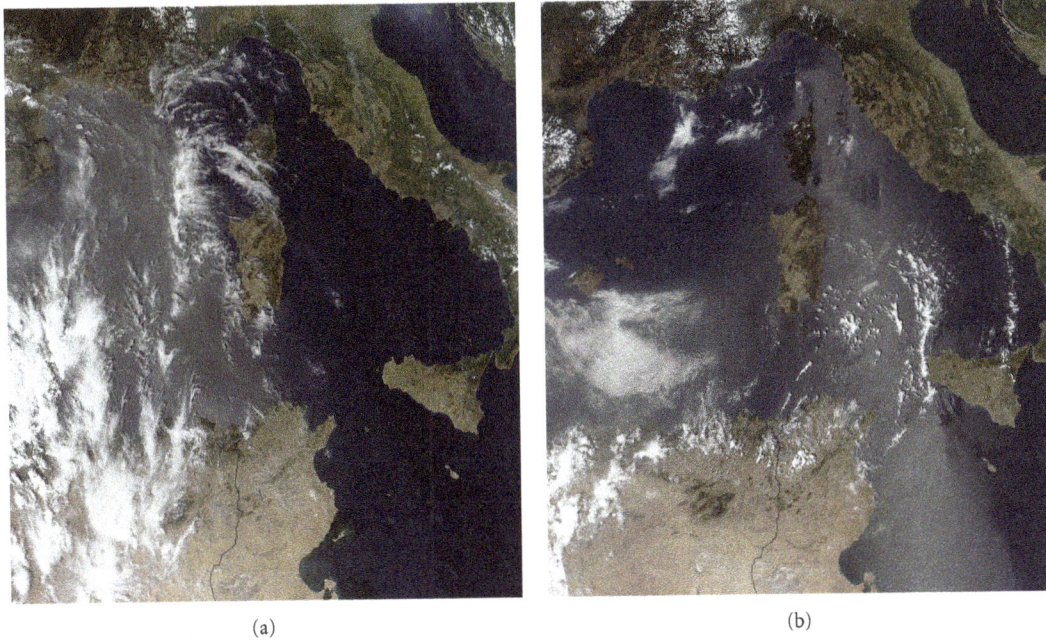

(a)   (b)

FIGURE 3: MODIS/AQUA images for 15 June (a) and 26 June (b) 2006.

To analyse the synoptic circulation over the Mediterranean area during the case study of June 2006, the sea level pressure, air temperature, and wind fields are investigated using a daily NCEP Reanalysis database provided by the NOAA/OAR/ESRL PSD (website at http://www.esrl.noaa.gov/psd/). The anomalies are calculated with respect to the long-term mean (LTM) period of 1981–2010. The screening of 6-hourly sea level pressures (SLPs), wind vectors, and air temperatures at different pressure levels reveals major dust outbreaks originating from the North African deserts towards Europe and provides insight into their temporal evolution. The means and anomalies of these 3 meteorological fields, compared with their LTMs, are calculated and discussed as follows.

From 10 to 21 June 2006, the average SLP distribution over the MB highlights a pronounced high-pressure system over the West Mediterranean Basin (4 mb in excess of the LTM for this period) and a low-pressure system over southern Algeria (Figure 2(a)). The West Mediterranean (WM), exposed to the influence of both systems, is subjected to the generation of SE gradient winds at 850 hPa, resulting in this episode being 6-7 m s^{-1} stronger than the normal wind flow over this region (Figure 2(b)). On 23 June, an Atlantic perturbation advances toward the Central Mediterranean area and weakens the SW flux and the dust transport over Italy. The following period, from 25 to 29 June 2006, is characterised by an extension of the Azorean ridge deep into northern Europe together with a weakening of the African Monsoon over southern Algeria (Figure 2(c)). This synoptic configuration yields a negative anomaly of 1-2 hPa over the whole MB (Figure 2(d)) with one of its two lowest pressure cores centred over the North Western African coast. This low-pressure system induces a cyclonic flow at shallow atmospheric layers over the WM, characterised by anomalous strong Westerlies over this region accompanied

by stronger than usual SE winds over Sardinia and Corsica (Figure 2(e)). This counterclockwise flow led to a positive anomaly in air temperature over the WM and the Central Mediterranean (Figure 2(f)).

The satellite observations, as shown in Figure 3(a) for 15 June, and the TOMS aerosol index reveal air masses transporting significant amounts of dust over large parts of the WM originating from southern Algeria and SW Africa. The TOMS data and the AQUA satellite observations (Figure 3(b)) for 27 June 2006 show the plume of dust originating from SW Algeria migrating in a cyclonic direction path through Sicily and Central Italy, crossing Sardinia and Corsica on its way to the Genoa Bay and eastern Spain.

The GOCART dust aerosol optical depth (AOD) daily maps are also used to follow the evolution of the dust outbreak on a daily basis [20]. As an example, Figure 4 represents the AOD for 15, 21, 24, and 26 June, showing the dust transport over the MB.

Finally, the Lagrangian trajectory HYSPLIT (Hybrid Simple Particle Lagrangian Integrated Trajectory) model is used to generate backward trajectories to trace back the sources of the air masses at different levels and for different hours of the day. Figure 5 shows the trajectories ending in Florence on 29 June 2006 at three different levels of A.G.L. (500, 1000, and 1500 m) starting from Northern Africa.

3.2. Numerical Simulations of Mineral Dust Events. The model chain is performed from 1 June to 5 July 2006. An analysis of the model results, based on the daily vertically integrated dust concentrations for the particle size 1–20 μm (as a sum of clay and small silt soil types), is presented. From 6 June, the Saharan desert dust is transported to the Iberian Peninsula, and in the following days, the dust reaches Northern Europe (France, UK, and Norway). From 16 to 22 June, a large area ranging from the West MB

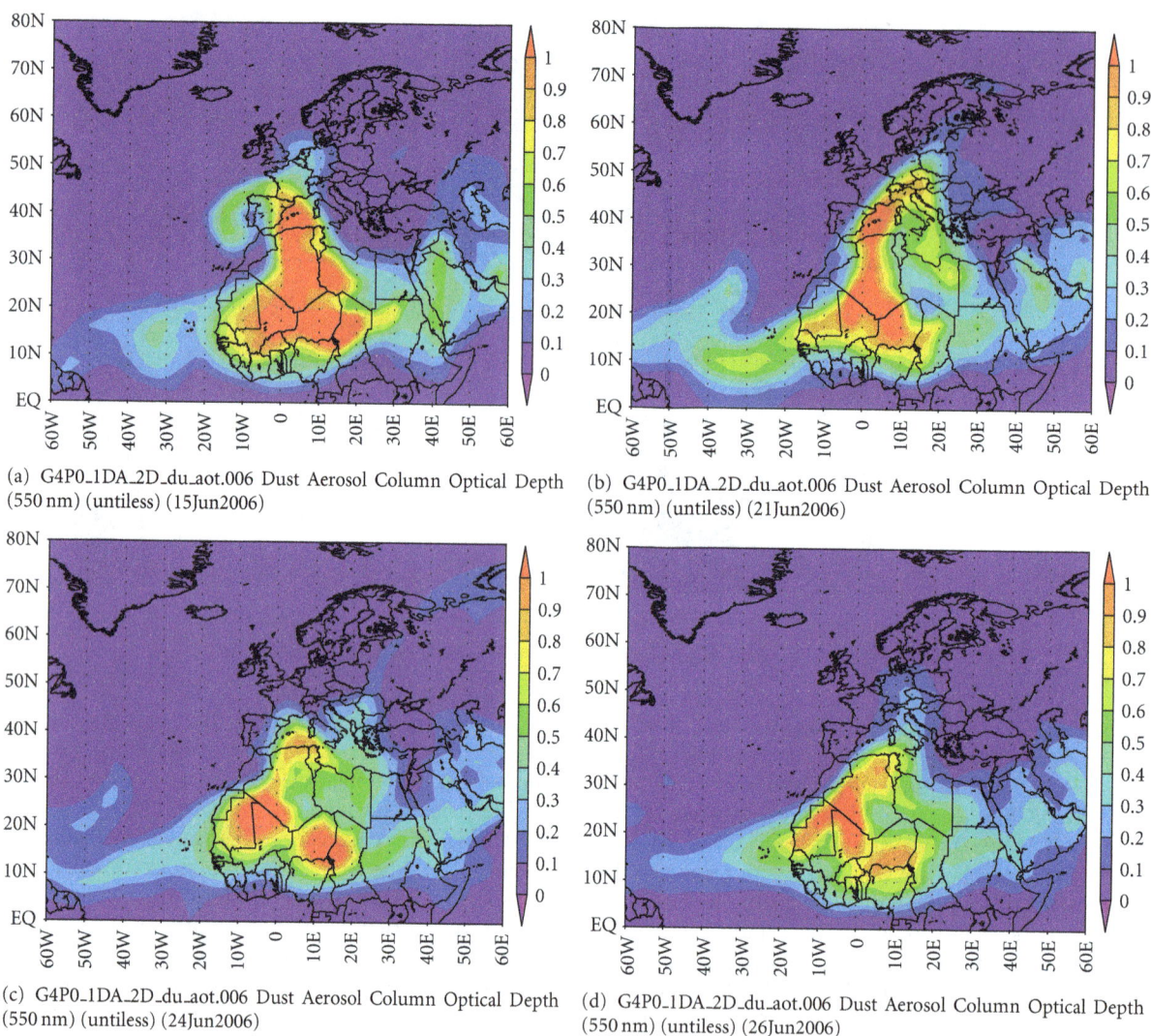

(a) G4P0_1DA_2D_du_aot.006 Dust Aerosol Column Optical Depth (550 nm) (untiless) (15 Jun 2006)

(b) G4P0_1DA_2D_du_aot.006 Dust Aerosol Column Optical Depth (550 nm) (untiless) (21 Jun 2006)

(c) G4P0_1DA_2D_du_aot.006 Dust Aerosol Column Optical Depth (550 nm) (untiless) (24 Jun 2006)

(d) G4P0_1DA_2D_du_aot.006 Dust Aerosol Column Optical Depth (550 nm) (untiless) (26 Jun 2006)

FIGURE 4: Dust Aerosol Column Optical Depth (550 nm) from GOCART aerosol model data V006, for the 15th of June (a), 21st of June (b), 24th of June (c), and 26th of June (d) 2006.

to the Scandinavian Peninsula and Russia is affected by a mineral dust intrusion that also reached Italy. In particular, during 17 and 18 June, Spain, France, Switzerland, Germany, Italy, and the Balkan Peninsula are affected by elevated dust concentrations. During 23 and 24 June, the Italian peninsula experienced a decrease of dust concentration due to a newly formed cyclone coming from the Atlantic Ocean, which interrupted the southwestern flux over Italy. Afterwards, the transport mechanism proceeds again until the first days of July, although the concentration values are lower than those in the previous period. Figure 6 shows the daily vertically integrated dust concentrations for 15, 21, 24, and 26 June 2006. The qualitative comparison between these simulation results, the satellite retrievals, and the TOMS index data show a good agreement with the location of the zones that encountered dust intrusions. Furthermore, the GOCART dust aerosol optical depth daily maps are compared with simulation's results for the whole period, from 1 June to 5 July, to evaluate whether the model chain correctly reproduces the dust event. There is temporal and

spatial agreement between the CAMx and GOCART results, as shown in Figures 4 and 6.

The RAMS/DUSTEM/CAMx model chain, configured with a finer resolution than GOCART, provides more detailed information on the vertical distribution of dust concentration and on the deposition at the ground level. In fact, the vertical sections of the mean daily dust concentration, for example at latitude $43.78°$ N (Florence), show dust advancing towards Spain and France, at both low and high levels up to 5000–7000 meters. The greatest dust concentration over Europe is reached during the period of 15–22 June between 1000 and 8000 metres, whereas under the boundary layer, the concentrations are lower (30–40 $\mu g/m^3$ versus 200–500 $\mu g/m^3$) (Figure 7). During 23-24 June, the Atlantic cyclonic flow temporarily interrupts the dust transport over Tuscany. On 25 June, the dust is transported only above the boundary layer without deposition. Observing the daily modeled dust concentration at the ground level, a strong decrease is evident, as also proved by ground measurements (next paragraphs). Such information cannot be obtained

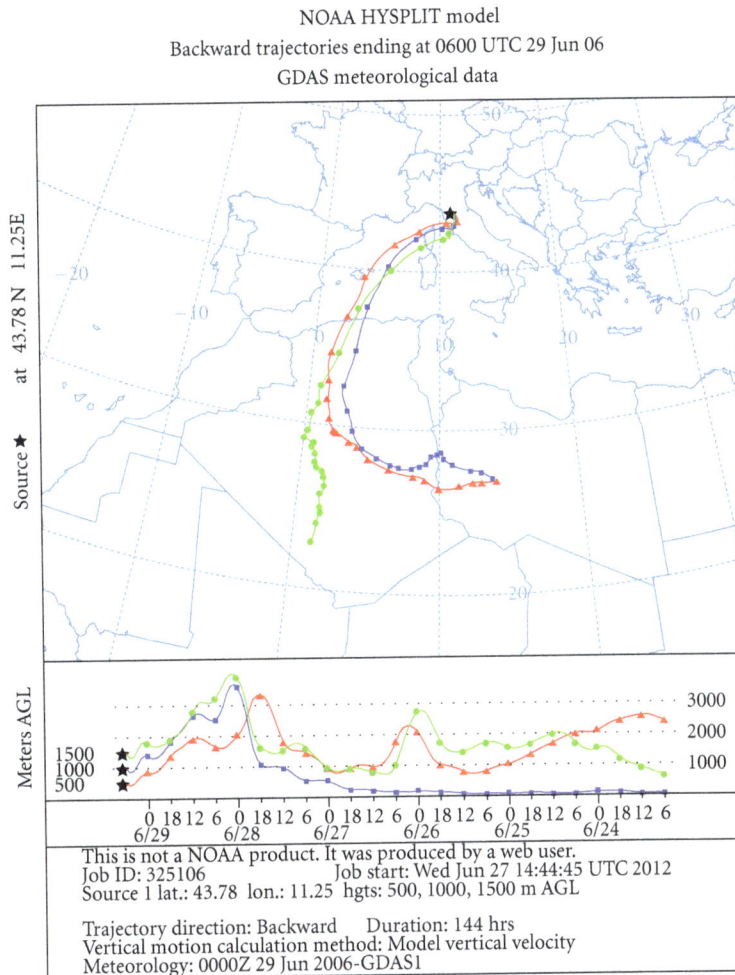

FIGURE 5: 144 hr back trajectories calculated by the HYSPLIT transport model (NOAA Air Resource Laboratory). The trajectories end in Florence at three different levels above the ground (10, 100, and 500 m) on the 29th of June 6 : 00 UTC.

from the results based on the daily vertically integrated dust concentrations or from an analysis of the optical depth maps. Finally, in the last part of the episode (26 June–2 July), the vertical extent of the dust concentration is lower, reaching 5000–6000 meters, and the dust event is less intense (Figure 7).

3.3. Chemical and Physical Analysis. During the month of June 2006, the concentrations of all the main crustal elements (Na, Mg, Al, Si, K, Ca, Ti, Mn, Fe, and Sr) show a clear increase together with the WM Saharan intrusion simultaneously in all sampling sites in Tuscany (Figure 8). The concentrations start to increase from 18 June and return to background values on 2 July. The time series show a two-peak shape, with a central minimum (24-25 June). As mentioned previously, the samplers are moved from three sampling sites to the other three every 15 days: in June, these displacements occurred on the 14th and 29th. Consequently, most of the episode (18–28 June) is measured in Florence, Capannori-Lucca, and Grosseto, one day (28 June) is missed, and the last part of the event (30 June–2 July) is observed in Arezzo, Livorno, and Prato. The average concentration values of the crustal elements during the episode are 2–4

times higher than their background concentrations. Al, Si, and Ti show a higher increase: their maximum concentration values during the episode ($\sim$2.5 μg/m^3, $\sim$6.0 μg/m^3, and $\sim$0.2 μg/m^3) are approximately 6 times higher than their background values.

Elemental ratios among the crustal elements also show significant changes, thus further reinforcing the Saharan intrusion hypothesis. For example, the Si/Al ratio mean and standard deviation during the episode is 2.3 $\pm$ 0.1 in Florence, 2.4 $\pm$ 0.1 in Capannori-Lucca, and 2.2 $\pm$ 0.1 in Grosseto, while the same ratios calculated for the other days (i.e., excluding the days of the episode) are $\sim$15% higher (2.7 $\pm$ 0.2, 2.8 $\pm$ 0.2, and 2.5 $\pm$ 0.2, resp.); the Ti/Fe ratio during the episode is 0.08 $\pm$ 0.01 in Florence, 0.08 $\pm$ 0.01 in Capannori-Lucca, and 0.09 $\pm$ 0.01 in Grosseto, while it is $\sim$50–70% lower during the other days (0.04 $\pm$ 0.01, 0.05 $\pm$ 0.01, and 0.06 $\pm$ 0.01, resp.); the Al/Ca ratio during the episode is 0.41$\pm$0.07 in Florence, 0.49 $\pm$ 0.11 in Capannori-Lucca, and 0.57 $\pm$ 0.12 in Grosseto, while it is $\sim$40–80% lower during the other days (0.29 $\pm$ 0.06, 0.28 $\pm$ 0.07, and 0.32 $\pm$ 0.08, resp.). Similar results are found at two remote sites of Northern and Central Italy by Bonelli et al. [36]. In both studies, a decrease of the Si/Al ratio and an

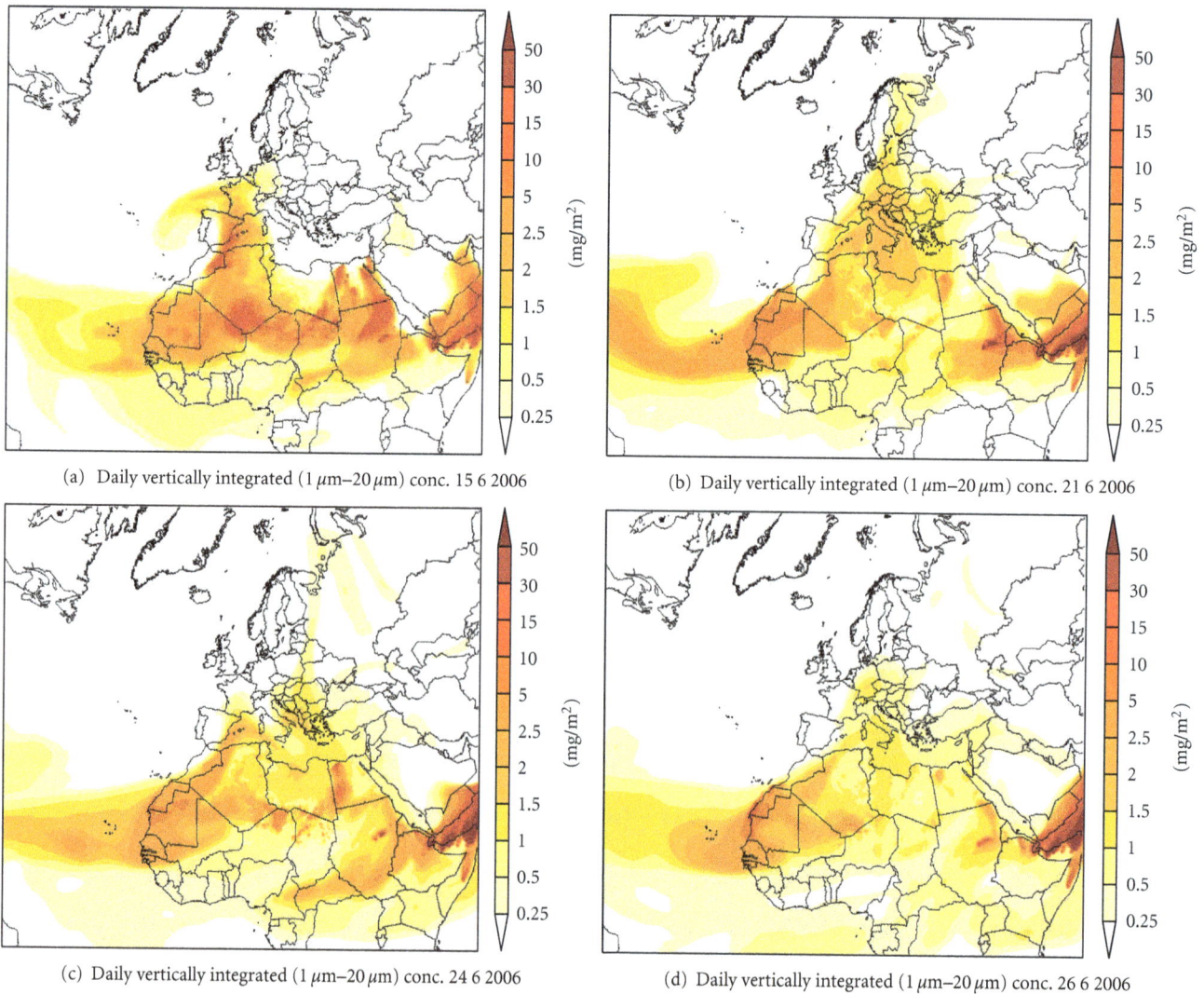

(a) Daily vertically integrated ($1\,\mu m$–$20\,\mu m$) conc. 15 6 2006

(b) Daily vertically integrated ($1\,\mu m$–$20\,\mu m$) conc. 21 6 2006

(c) Daily vertically integrated ($1\,\mu m$–$20\,\mu m$) conc. 24 6 2006

(d) Daily vertically integrated ($1\,\mu m$–$20\,\mu m$) conc. 26 6 2006

FIGURE 6: Daily vertically integrated dust concentration for 15th of June (a), 21st of June (b), 24th of June (c), and 26th of June 2006 (d).

increase of the Ti/Fe and Al/Ca ratios are indicated as highly representative fingerprints of African dust transport events. Their reported values for the Si/Al ratio during these events (and during days not affected by Saharan intrusions), that is, 2.3-2.4 (2.7-2.8), match well with those found in this study. The Si/Al ratio decrease may be explained by a higher contribution during the Saharan intrusions of agglomerated clay minerals with respect to bulk crustal material, while the increase of the Ti/Fe and Al/Ca ratios may be ascribed to an iron and Ca enrichment in local soil dust.

An estimation of the soil dust component may be calculated as the sum of the contributions of the main oxides of all the crustal elements (Na_2O_2, MgO, SiO_2, Al_2O_3, TiO_2, K_2O, CaO, and Fe_2O_3):

$$[\text{soil dust}] = 1{,}7\,[\text{nssNa}] + 1{,}67\,[\text{nssMg}] + 1{,}89\,[\text{Al}] + 2{,}14\,[\text{Si}]$$

$$+ 1{,}2\,[\text{K}] + 1{,}4\,[\text{Ca}] + 1{,}43\,[\text{Fe}] + 1{,}67\,[\text{Ti}], \tag{1}$$

where [nssNa] and [nssMg] are the concentrations of "non-sea-salt" Na and Mg (i.e., excluding the sea-salt contribution).

This expression may overestimate soil dust due to the contribution of other sources, such as biomass burning (for K), traffic, and other anthropogenic sources for Fe, and Ca. During the studied period the enrichment factors of K, Fe, and Ca (with respect to Al using the crust composition reported in Mason [18]) turn out to be low, suggesting the absence of strong anthropogenic contributions. Consequently, no correction is applied to the calculation of their contribution to soil dust. The sea-salt fractions of Na and Mg are calculated using the IC data, as the measured Na^+ and Mg^{2+} soluble ion concentrations may actually be considered good estimations for these contributions.

The daily time series of soil dust concentration, calculated as reported above, are shown in Figure 8. During the Saharan intrusion, values of 20–30 $\mu g/m^3$ are reached, to be compared with background values of the order of a few $\mu g/m^3$. The Saharan dust contribution, for all the days of the episode, is estimated by subtracting the background soil dust concentration, which is calculated, site by site, as an average from mid-May to the end of July, excluding the Saharan intrusion days. The results together with the PM_{10}

(a) Florence latitude cross-section dust (1 μm–20 μm) daily average on 15 6 2006

(b) Florence latitude cross-section dust (1 μm–20 μm) daily average on 21 6 2006

(c) Florence latitude cross-section dust (1 μm–20 μm) daily average on 24 6 2006

(d) Florence latitude cross-section dust (1 μm–20 μm) daily average on 26 6 2006

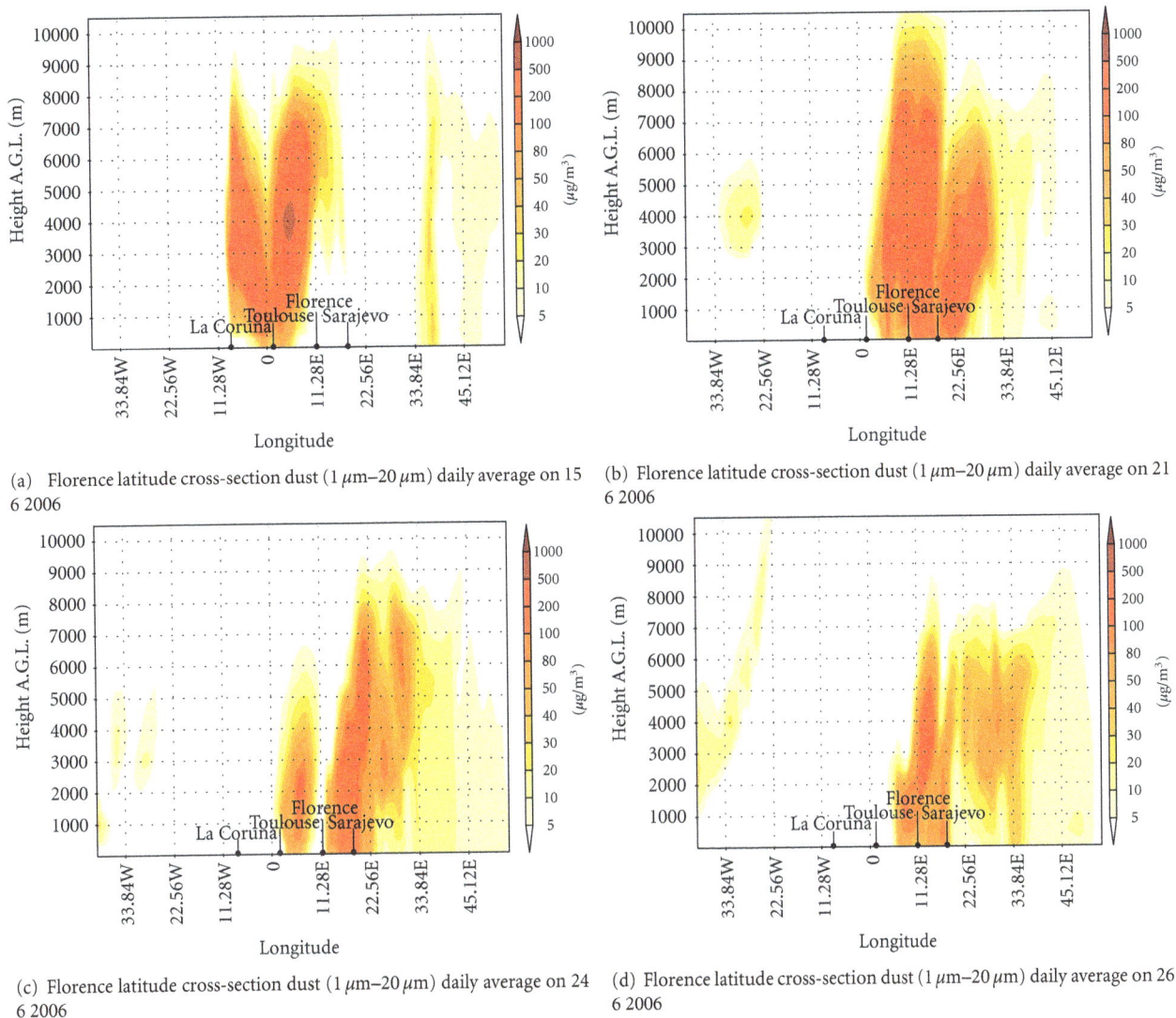

FIGURE 7: Latitude cross-section at 43.78°N of the daily average of dust concentration for 15th of June (a), 21st of June (b), 24th of June (c) and 26th of June 2006 (d).

concentration values are reported in Table 2. The Saharan dust contribution is quite high, with concentrations up to ~20 μg/m³, and it can be considered the main cause of the PM_{10} 50 μg/m³ limit value exceedances (values shown in bold in the table) occurring in this period.

3.4. Comparison between Model and Measurements. The comparison with in situ measurements plays a fundamental role for an evaluation of the model chain effectiveness in reproducing the deposition mechanism.

As explained in Section 3.3, the in situ measurements identify the dust outbreak from 18 June to 1 July, with a small residual contribution on 2 July. For the comparison, only three sites are used (Firenze, Capannori-Lucca, and Grosseto), as the relative sampling periods cover almost the entire extent of the episode.

In Figure 9, the time series of the daily simulated dust concentration at the ground level are compared with the Saharan dust contribution to the PM_{10} obtained by the measurements. Concerning the period extension, in the

inland cities (Florence and Capannori-Lucca), the model simulation identifies the dust episode from 16 June to 2 July. The onset of the dust outbreak is thus anticipated in 2 days with respect to the measured data, while in the coastal city (Grosseto), the simulation and the measurements identify the same temporal range.

At all the sites, both the simulated and observed time series show a good agreement in the representation of the concentration decrease on 24-25 June and the following increasing phase up to the end of the episode.

Because the DUSTEM emission model provides particle sizes of 1–20 μm, the comparison with PM_{10} measurements is limited to a qualitative analysis and may lead to simulated concentration values that are higher than the soil dust concentrations obtained from the measurements. This behaviour may actually be observed in the inland cities. However, in the coastal cities, the simulation gives lower values than the measurements, which could be attributed to the model representation of the boundary layer's vertical extent, which could be underestimated near the coast,

TABLE 2: Saharan dust and PM$_{10}$ concentrations, in μg/m^3, during the period of the Saharan intrusion of June 2006, for the six sampling sites ("—" indicates a contribution below the minimum detection limit).

	Firenze		Capannori		Grosseto	
	Saharan dust	PM$_{10}$	Saharan dust	PM$_{10}$	Saharan dust	PM$_{10}$
18/06/2006	2	22	2	24	—	18
19/06/2006	6	30	4	32	3	24
20/06/2006	14	42	9	38	5	27
21/06/2006	19	46	12	44	15	40
22/06/2006	**23**	**56**	14	48	**22**	**51**
23/06/2006	10	41	9	42	14	42
24/06/2006	8	42	6	45	3	31
25/06/2006	8	41	7	46	5	32
26/06/2006	12	44	8	40	9	37
27/06/2006	**13**	**50**	14	49	11	44
28/06/2006	**21**	**57**	**17**	**51**	**18**	**50**
	Arezzo		Prato		Livorno	
	Saharan dust	PM$_{10}$	Saharan dust	PM$_{10}$	Saharan dust	PM$_{10}$
30/06/2006	8	36	11	44	**19**	**50**
01/07/2006	4	15	2	25	6	40
02/07/2006	—	11	—	15	2	31

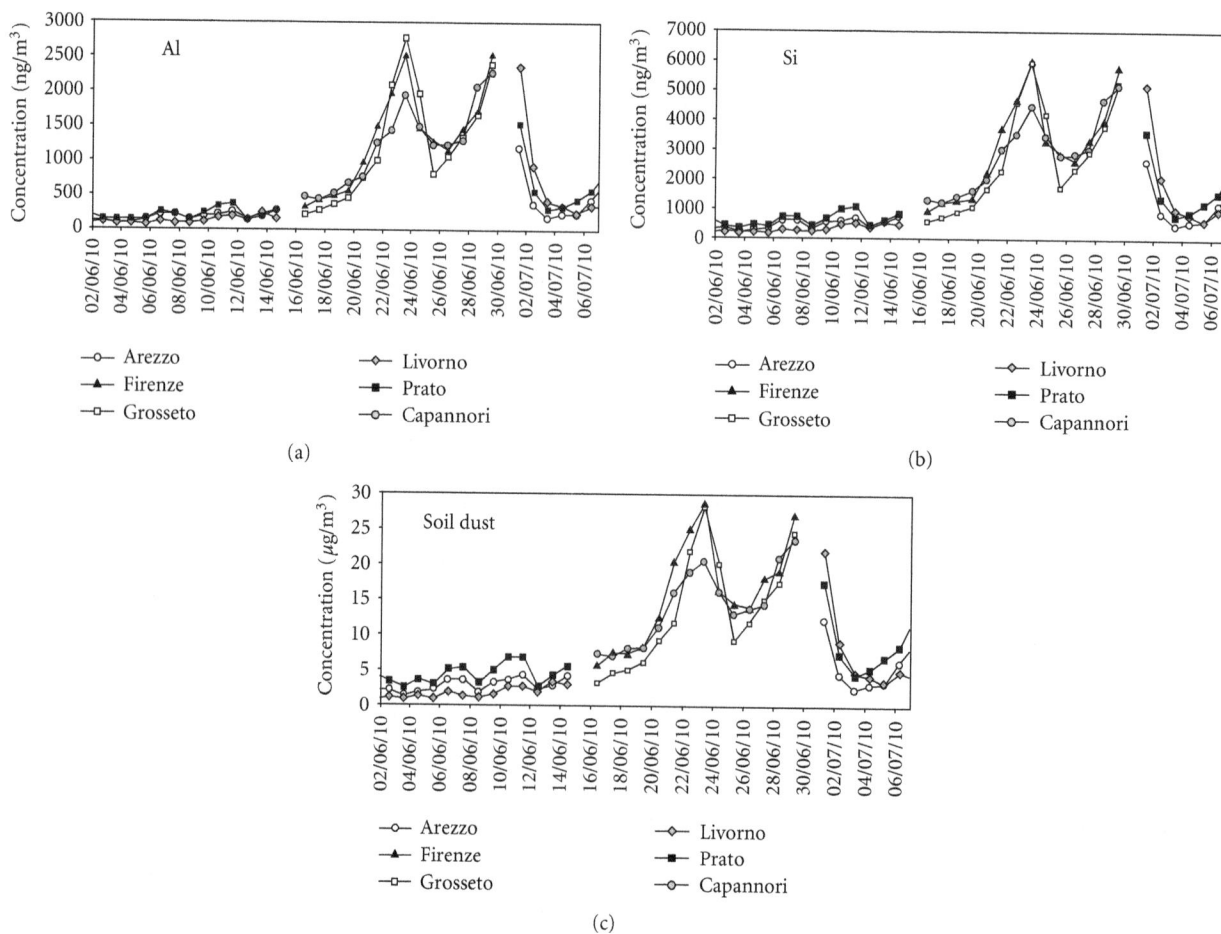

(a)

(b)

(c)

FIGURE 8: Daily concentrations of Al (ng/m^3), Si (ng/m^3), and soil dust (μg/m^3) in PM$_{10}$ samples collected in June 2006 in six sampling sites in Tuscany (Italy).

(a)

(b)

(c)

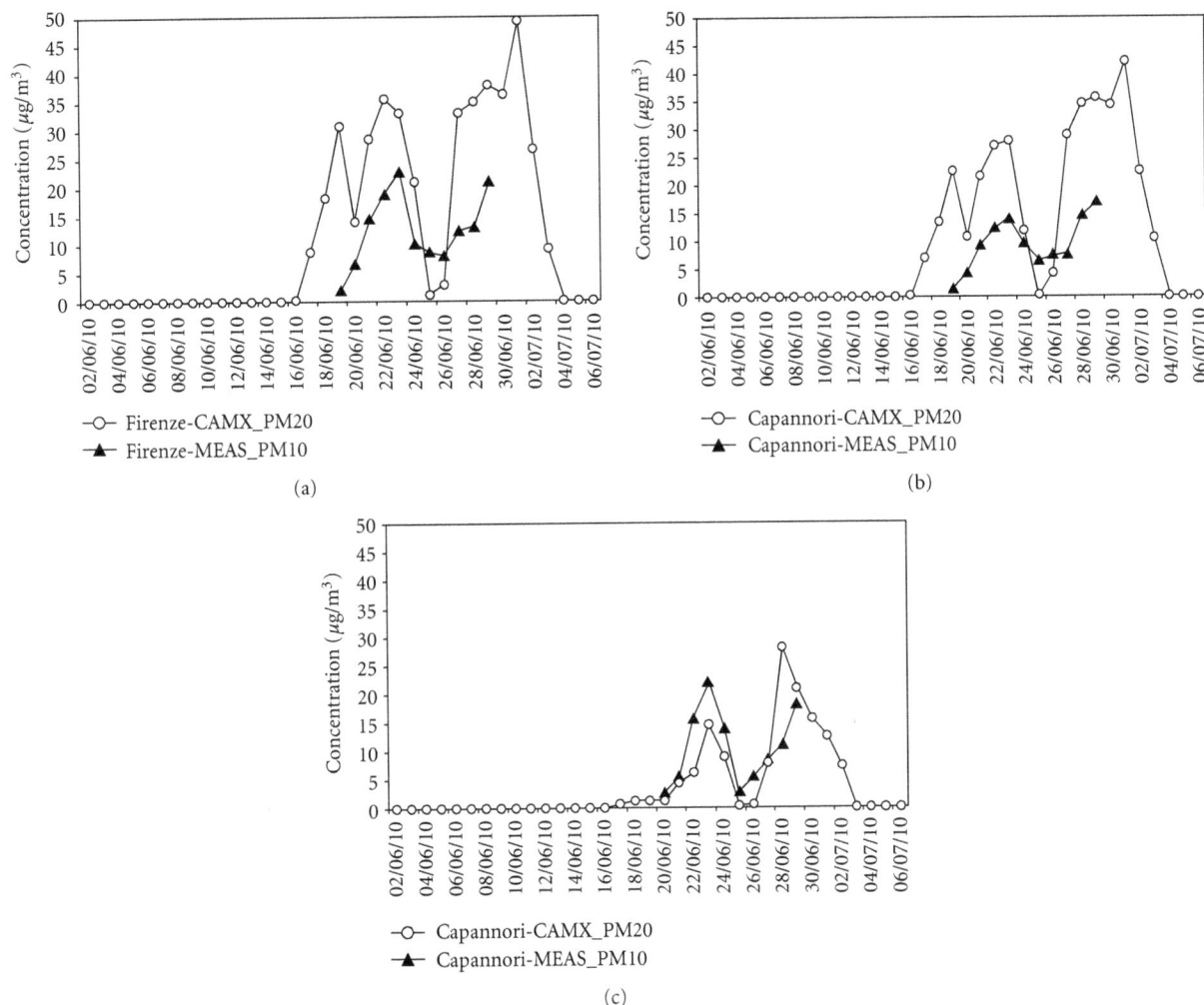

FIGURE 9: Comparison between mean daily modeled dust concentrations (CAMX_PM20) and Saharan dust contribution to PM_{10} obtained by ground measurements (MEAS_PM10), in Florence, Capannori-Lucca, and Grosseto (Tuscany, Italy).

obstructing the dust intrusions from higher to lower levels. The meteorological model RAMS has been configured with a horizontal and vertical resolution useful to reproduce the raising and transport phenomena, but it is not fine enough to describe the boundary layer near the coastline. Perhaps it will be necessary to introduce a finer nested grid into the meteorological model (and, consequently, the CAMx model) to describe the boundary layer with a better resolution over the target area.

4. Conclusions

Within the framework of the European Commission's guidelines to assess the natural contribution to PM_{10}, the aim of this work is to study the Saharan dust intrusions over the MB to quantify the dust impact at the ground level. The selected case study considers the large dust outbreak over the MB of June 2006. After a preliminary synoptic analysis, an ad hoc model chain is configured and physical-chemical analyses are conducted on PM_{10} samples collected in Central Italy. The model chain properly reproduces the dust emission

and transport dynamics, as proved by comparison with available model maps. Furthermore, it describes the dust distribution on vertical profiles and the deposition at the ground level, adding detail to the satellite observations and optical thickness data analysis. Regarding ground-based measurements, the specific assessment of all the crustal elements by PIXE gives a more accurate quantitative estimate of the desert dust contribution than that obtained solely by PM_{10} mass concentration. The comparison between simulation results and specific insitu measurements shows a satisfying overall agreement, especially for the temporal evolution of the studied episode. Nevertheless, in the sampling site near the coast, the model simulation gives lower values than the measured ones. Near the coast, the model representation of the boundary layer's vertical extent could be underestimated, obstructing the dust intrusions to the ground level. To improve the performance of the model chain, it may be useful to introduce a nested grid in the meteorological model (and, consequently, in the CAMx model) with a higher resolution over the target area, where the measurement sites are located. However, the comparison with PM_{10} measurements is limited to a qualitative analysis because the DUSTEM model provides emission rates for particle sizes of 1–20 μm. As a further development, the DUSTEM will be improved, taking into account a better soil type description to provide PM_{10} emission rates.

Acknowledgments

The authors would like to acknowledge the Tuscany Regional Government (Italy), which supported the PATOS project and provided the experimental data used here, and the support received within the CNR-Italy Short Term Mobility Program 2010. The authors would also like to acknowledge the NOAA Air Resources Laboratory for the provision of the HYSPLIT model results and NASA for the TOMS maps, satellite images, and the Giovanni online data system, developed and maintained by the NASA GES DISC for the dust maps used in this study.

References

[1] IPCC, "Changes in atmospheric constituents and in radiative forcing," in *Climate Change 2007: the Physical Basis*, P. Forster, V. Ramaswamy, P. Artaxo et al., Eds., pp. 129–234, Cambridge University Press, New York, NY, USA.

[2] J. Haywood and O. Boucher, "Estimates of the direct and indirect radiative forcing due to tropospheric aerosols: a review," *Reviews of Geophysics*, vol. 38, no. 4, pp. 513–543, 2000.

[3] J. E. Penner et al., "Aerosols, their direct and indirect effects," in *Climate Change 2001: the Scientific Basis*, J. T. Houghton, Y. Ding, D. J. Griggs et al., Eds., pp. 289–348, Cambridge University Press, Cambridge, UK, 2001.

[4] V. Ramaswamy et al., "Radiative forcing of climate change," in *Climate Change 2001: the Scientific Basis*, J. T. Houghton, Y. Ding, D. J. Griggs et al., Eds., pp. 349–416, Cambridge University Press, Cambridge, UK, 2001.

[5] U. Lohmann and J. Feichter, "Global indirect aerosol effects: a review," *Atmospheric Chemistry and Physics*, vol. 5, no. 3, pp. 715–737, 2005.

[6] IPCC, "Summary for policymakers," in *Climate Change 2007: the Physical Science Basis*, S. Solomon, D. Qin, M. Manning et al., Eds., Cambridge University Press, New York, NY, USA, 2007.

[7] N. M. Mahowald and L. M. Kiehl, "Mineral aerosol and cloud interactions," *Geophysical Research Letters*, vol. 30, no. 9, pp. 28-1, 2003.

[8] K. H. Rosenlof, S. J. Oltmans, D. Kley et al., "Stratospheric water vapor increases over the past half-century," *Geophysical Research Letters*, vol. 28, no. 7, pp. 1195–1198, 2001.

[9] T. D. Jickells, Z. S. An, K. K. Andersen et al., "Global iron connections between desert dust, ocean biogeochemistry, and climate," *Science*, vol. 308, no. 5718, pp. 67–71, 2005.

[10] G. S. Okin, N. Mahowald, O. A. Chadwick, and P. Artaxo, "Impact of desert dust on the biogeochemistry of phosphorus in terrestrial ecosystems," *Global Biogeochemical Cycles*, vol. 18, no. 2, Article ID GB2005, 9 pages, 2004.

[11] J. M. Prospero, P. Ginoux, O. Torres, S. E. Nicholson, and T. E. Gill, "Environmental characterization of global sources of atmospheric soil dust identified with the Nimbus 7 Total Ozone Mapping Spectrometer (TOMS) absorbing aerosol product," *Reviews of Geophysics*, vol. 40, no. 1, pp. 1–31, 2002.

[12] X. Tie, S. Madronich, S. Walters et al., "Assessment of the global impact of aerosols on tropospheric oxidants," *Journal of Geophysical Research D*, vol. 110, no. 3, pp. 1–32, 2005.

[13] J. M. Prospero, "Saharan dust transport over the North Atlantic Ocean and Mediterranean: an overview," in *The Impact of Desert Dust Across the Mediterranean*, S. Guerzoni and R. Chester, Eds., pp. 133–151, Kluwer Academic Publishers, 1996.

[14] U. Dayan, J. Heffter, J. Miller, and G. Gutman, "Dust intrusion events into the Mediterranean basin," *Journal of Applied Meteorology*, vol. 30, no. 8, pp. 1185–1199, 1991.

[15] X. Querol, J. Pey, M. Pandolfi et al., "African dust contributions to mean ambient PM10 mass-levels across the Mediterranean Basin," *Atmospheric Environment*, vol. 43, no. 28, pp. 4266–4277, 2009.

[16] S. Rodríguez, X. Querol, A. Alastuey, G. Kallos, and O. Kakaliagou, "Saharan dust contributions to PM10 and TSP levels in Southern and Eastern Spain," *Atmospheric Environment*, vol. 35, no. 14, pp. 2433–2447, 2001.

[17] M. Escudero, X. Querol, J. Pey et al., "A methodology for the quantification of the net African dust load in air quality monitoring networks," *Atmospheric Environment*, vol. 41, no. 26, pp. 5516–5524, 2007.

[18] B. Mason, *Principles of Geochemistry*, Wiley, New York, NY, USA, 1966.

[19] R. A. Pielke, W. R. Cotton, R. L. Walko et al., "A comprehensive meteorological modeling system-RAMS," *Meteorology and Atmospheric Physics*, vol. 49, no. 1-4, pp. 69–91, 1992.

[20] J. G. Acker and G. Leptoukh, "Online analysis enhances use of NASA Earth Science Data," *Eos*, vol. 88, no. 2, pp. 14–17, 2007.

[21] M. Chin, P. Ginoux, S. Kinne et al., "Tropospheric aerosol optical thickness from the GOCART model and comparisons with satellite and sun photometer measurements," *Journal of the Atmospheric Sciences*, vol. 59, no. 3, pp. 461–483, 2002.

[22] D. A. Gillette, "Major contributions of natural primary continental aerosols: source mechanisms," *Annals of the New York Academy of Sciences*, vol. 338, pp. 348–358, 1980.

[23] D. A. Gillette, "Production of dust that may be carried great distances," in *Desert Dust: Origin, Characteristics, and Effect on Man*, T. L. Pewe, Ed., vol. 186, pp. 11–26, Geological Society of America Special, 1981.

[24] B. Marticorena and G. Bergametti, "Modeling the atmospheric dust cycle: 1. Design of a soil-derived dust emission scheme," *Journal of Geophysical Research*, vol. 100, no. 8, pp. 16–430, 1995.

[25] F. Meneguzzo, M. Pasqui, G. Menduni et al., "Sensitivity of meteorological high-resolution numerical simulations of the biggest floods occurred over the Arno river basin, Italy, in the 20th century," *Journal of Hydrology*, vol. 288, no. 1-2, pp. 37–56, 2004.

[26] G. Messeri, A. Pellegrini, M. Pasqui et al., "Weather forecast using RAMS model. A case study," in *Proceedings of the Italian Physical Society Conference*, vol. 80, pp. 55–68, 2002.

[27] M. Pasqui, J. Lichtenegger, F. Meneguzzo, and G. Messeri, "Validation of Model rams with ERS-2 SAR observations in convectives storm events over the Mediterranean sea," in *Mediterranean Storms*, R. Deidda, A. Mugnai, and F. Siccardi, Eds., publicatioin no.2560, GNDCI, 2002.

[28] M. Pasqui, L. Bottai, C. Busillo et al., "Dust sandstorm dynamics analysis in northern China by means of atmospheric, emission, dispersion modeling," in *Proceeding of the 8th International Conference on Development of Drylands*, Beijing, China, 2006.

[29] M. Kanamitsu, W. Ebisuzaki, J. Woollen et al., "NCEP-DOE AMIP-II reanalysis (R-2)," *Bulletin of the American Meteorological Society*, vol. 83, no. 11, pp. 1631–1643, 2002.

[30] S. Nickovic, G. Kallos, A. Papadopoulos, and O. Kakaliagou, "A model for prediction of desert dust cycle in the atmosphere," *Journal of Geophysical Research D*, vol. 106, no. 16, pp. 18113–18129, 2001.

[31] I. Tegen and I. Fung, "Modeling of mineral dust in the atmosphere: sources, transport, and optical thickness," *Journal of Geophysical Research*, vol. 99, no. 11, pp. 22–914, 1994.

[32] E. Bartholomé and A. S. Belward, "GLC2000: a new approach to global land cover mapping from earth observation data," *International Journal of Remote Sensing*, vol. 26, no. 9, pp. 1959–1977, 2005.

[33] FAO, IIASA, ISRIC, ISSCAS, and JRC, "Harmonized world soil database," (version 1. 1), FAO, Rome, Italy and IIASA, Laxenburg, Austria, 2009.

[34] S. A. Slinn and W. G. N. Slinn, "Predictions for particle deposition on natural waters," *Atmospheric Environment. Part A*, vol. 14, no. 9, pp. 1013–1016, 1980.

[35] J. H. Seinfeld and S. N. Pandis, *Atmospheric Chemistry and Physics. from Air Pollution to Climate Change*, John Wiley and Sons, New York, NY, USA, 1998.

[36] P. Bonelli, G. M. Braga Marcazzan, and E. Cereda, "Elemental composition and air trajectories of African dust transported in Northern Italy," in *The Impact of Desert Dust Across the Mediterranean*, S. Guerzoni and R. Chester, Eds., pp. 275–283, Kluwer Academic Publishers, Dordrecht, The Netherlands, 1996.

[37] I. Borbely-Kiss, A. Z. Kiss, E. Koltay, G. Szabo, and L. Bozo, "Saharan dust episodes in Hungarian aerosol: elemental signatures and transport trajectories," *Journal of Aerosol Science*, vol. 35, no. 10, pp. 1205–1224, 2004.

[38] G. M. B. Marcazzan, P. Bonelli, E. Della Bella, A. Fumagalli, R. Ricci, and U. Pellegrini, "Study of regional and long-range transport in an Alpine station by PIXE analysis of aerosol particles," *Nuclear Inst. and Methods in Physics Research, B*, vol. 75, no. 1-4, pp. 312–316, 1993.

[39] M. Chiari, F. Lucarelli, F. Mazzei et al., "Characterization of airborne particulate matter in an industrial district near Florence by PIXE and PESA," *X-Ray Spectrometry*, vol. 34, no. 4, pp. 323–329, 2005.

[40] P. Formenti, M. O. Andreae, L. Lange et al., "Saharan dust in Brazil and Suriname during the Large-Scale Biosphere-Atmosphere Experiment in Amazonia (LBA)—cooperative LBA Regional Experiment (CLAIRE) in March 1998," *Journal of Geophysical Research D*, vol. 106, no. 14, pp. 14919–14934, 2001.

[41] G. Calzolai, M. Chiari, I. García Orellana et al., "The new external beam facility for environmental studies at the Tandetron accelerator of LABEC," *Nuclear Instruments and Methods in Physics Research, Section B*, vol. 249, no. 1-2, pp. 928–931, 2006.

[42] J. L. Campbell, N. I. Boyd, N. Grassi, P. Bonnick, and J. A. Maxwell, "The Guelph PIXE software package IV," *Nuclear Instruments and Methods in Physics Research, Section B*, vol. 268, no. 20, pp. 3356–3363, 2010.

[43] S. Becagli, C. Ghedini, S. Peeters et al., "MBAS (Methylene Blue Active Substances) and LAS (Linear Alkylbenzene Sulphonates) in Mediterranean coastal aerosols: sources and transport processes," *Atmospheric Environment*, vol. 45, no. 37, pp. 6788–6801, 2011.

[44] R. Washington, M. C. Todd, S. Engelstaedter, S. Mbainayel, and F. Mitchell, "Dust and the low-level circulation over the Bodélé Depression, Chad: observations from BoDEx 2005," *Journal of Geophysical Research D*, vol. 111, no. 3, Article ID D03201, 2006.

[45] C. Moulin, C. E. Lambert, F. Dulac, and U. Dayan, "Control of atmospheric export of dust from North Africa by the North Atlantic Oscillation," *Nature*, vol. 387, no. 6634, pp. 691–694, 1997.

15

An Integrative Approach to Understand the Climatic-Hydrological Process: A Case Study of Yarkand River, Northwest China

Jianhua Xu, Yiwen Xu, and Chunan Song

The Research Center for East-West Cooperation in China, The Key Lab of GIScience of the Education Ministry PRC, East China Normal University, 500 Dongchuan Road, Minhang, Shanghai 200241, China

Correspondence should be addressed to Jianhua Xu; jhxu@geo.ecnu.edu.cn

Academic Editor: Luis Gimeno

Taking the Yarkand River as an example, this paper conducted an integrative approach combining the Durbin-Watson statistic test (DWST), multiple linear regression (MLR), wavelet analysis (WA), coefficient of determination (CD), and Akaike information criterion (AIC) to analyze the climatic-hydrological process of inland river, Northwest China from a multitime scale perspective. The main findings are as follows. (1) The hydrologic and climatic variables, that is, annual runoff (AR), annual average temperature, (AAT) and annual precipitation (AP), are stochastic and, no significant autocorrelation. (2) The variation patterns of runoff, temperature, and precipitation were scale dependent in time. AR, AAT, and AP basically present linear trends at 16-year and 32-year scales, but they show nonlinear fluctuations at 2-year and 4-year scales. (3) The relationship between AR with AAT and AP was simulated by the multiple linear regression equation (MLRE) based on wavelet analysis at each time scale. But the simulated effect at a larger time scale is better than that at a smaller time scale.

1. Introduction

The hydrological response to climate change is an important science issue. To well understand this issue, the coupled system of climatic-hydrological process should be thoroughly studied at different spatial and temporal scales.

Theoretically, a process can be evaluated to determine if they comprise an ordered, deterministic system, an unordered, random system, or a chaotic, dynamic system, and whether change patterns of periodicity or quasi-periodicity exist. However, it is difficult to achieve a thorough understanding of the mechanism of climatic-hydrological processes [1]. To date, these questions have not received satisfactory answers [2].

Case studies in different countries and regions have suggested that the climatic-hydrological process is a complex system [3–6]. Therefore, more studies are required to explore the mechanism of climatic-hydrological process from different perspectives and using different methods. As a result, the climatic-hydrological process has been explored using various analytical methods, including the fractal theory [7–9], self-organized criticality [10], wavelets analysis [11–13], and artificial neural networks [14, 15]. Although there were several effective methods available to reveal the variations in climatic-hydrological process [16–19], it has proven difficult to achieve a thorough understanding of the mechanism of climatic-hydrological process in inland river [2].

In the last 20 years, studies have been conducted to evaluate climate change and hydrological and ecological processes in the arid and semiarid regions in northwestern China [18–25]. Some studies have indicated that there was a visible transition in the hydroclimatic processes in the past half-century [24, 26–28]. This transition was characterized by a continual increase in temperature and precipitation, added river runoff volumes, increased lake water surface elevation and area, and elevated groundwater level. This transition may present a series of questions if these changes represent a localized transition to a warm and wet climate type in response to global warming, or merely reflect a centennial periodicity in hydrological dynamics. To date, these questions have not received satisfactory answers; therefore, more studies are required to explore the nonlinear characteristics of hydroclimatic process from different perspectives and using different methods [2, 15, 29].

Though some studies have shown that the inland river in northwest China (NW China), such as the Yarkand River, is mainly recharged by snowmelt, the main climatic factors affecting the streamflow are temperature and precipitation [20, 30, 31]. But due to the complexity of hydroclimatic system, it is difficult to understand the mechanism of climatic-hydrological process thoroughly [2]. For the above reasons, this paper did not involve the complex physical mechanisms but conducted an integrative approach combining statistics and wavelet analysis to understand the variation of annual runoff and its response to climatic factors at different time scales.

2. Materials and Methods

2.1. Study Area and Data. The Yarkand River is a typical representative of inland rivers, which is located in the Tarim River Basin of Xinjiang Uygur Autonomous Region, northwestern China (Figure 1), with a length of 1097 km. The Yarkand River ($35°40'$ ~ $40°31'$N, $74°28'$ ~ $80°54'$E) has a total basin area of 9.89×10^4 km^2, including 6.08×10^4 km^2 as the mountain area, which accounts for 61.5%, and 3.81×10^4 km^2 as the plain area, which takes up 38.5% [31]. The main stream of Yarkand River originates from Karakoram Pass in the north slope of Karakoram Mountain, which is full of towering peaks and glaciers, as well as the extremely rare precipitation in plain. Due to the special geographical conditions, the accumulation of ice and snow in high mountain is the only supply source for runoff. Therefore, the Yarkand River is a typical ice-snow supply river, in which the multiyear average runoff in Kaqun hydrometric station consists of 64.0% from mean volume of glacial ablation, 13.4% from rain and snow supply, and 22.6% from groundwater supply, respectively [32, 33].

For the Yarkand River is an inland river, no water recharges in the plain area, and its stream flow mainly comes from mountainous area, that is, the Pamir Mountains. In other words, the streamflow of the Yarkand River is mainly fed by glacier and snowmelt in the Pamir Mountains. Therefore, the climatic factors, especially temperature and precipitation, directly affect the annual changes in the runoff. So we use the runoff as well as temperature and precipitation data to analyze the climatic-hydrological process in Yarkand River. The runoff data were from the Kaqun hydrologic station, and temperature and precipitation data were from Tash Kurghan meteorological station. The two stations are located in the source areas of the river; the amount of water used by humans is minimal compared to the total discharge. Therefore, the observed hydrological and meteorological records reflect the natural conditions.

Long-term climate changes can alter the runoff production pattern, the timing of hydrological events, and the frequency and severity of floods, particularly in arid or semiarid regions. Therefore, a small change in precipitation

and temperature may result in marked changes in runoff. To investigate the runoff and its related climatic effect, this study used the time series data of annual runoff (AR), annual average temperature (AAT), and annual precipitation (AP) from 1957 to 2008.

2.2. Methods. In order to study the variations of streamflow with regional climate change at different time scales, this paper conducted a comprehensive method including the Durbin-Watson statistic test (DWST), multiple linear regression (MLR), wavelet analysis (WA), coefficient of determination (CD), and Akaike information criterion (AIC). Firstly, the DWST was used to explore the stochastic characteristic of hydrologic and climatic variables. Secondly, the WA was used to reveal the variation patterns of annual runoff (AR) and its related climatic factors at different time scales. Thirdly, the relationship between AR with AAT and AP was simulated by MLR based on WA at different time scales. Finally, the estimated effect of multiple linear regression equation (MLRE) at each time scale was tested by CD and AIC.

2.2.1. Durbin-Watson Statistic Test. The Durbin-Watson statistic is a test statistic used to detect the presence of autocorrelation (a relationship between values separated from each other by a given time lag) in the residuals (prediction errors) from a regression analysis [34, 35].

For a variable y, the Durbin-Watson statistic is

$$\text{DW} = \frac{\sum_{i=2}^{n} (e_i - e_{i-1})^2}{\sum_{i=1}^{n} e_i^2}, \tag{1}$$

where $e_i = y_i - \hat{y}_i$, and y_i and $\hat{y}_i$ are, respectively, the observed and predicted values of the response variable for individual i; n is the number of observations.

To test for positive autocorrelation at significance α, the test statistic DW is compared to lower and upper critical values (d_L and d_U): if DW $< d_L$, there is statistical evidence that the error terms are positively autocorrelated; if DW $> d_U$, there is no statistical evidence that the error terms are positively autocorrelated; if $d_L <$ DW $< d_U$, the test is inconclusive.

To test for negative autocorrelation at significance α, the test statistic $4 - $ DW is compared to lower and upper critical values (d_L and d_U): if $(4 - $ DW$) < d_L$, there is statistical evidence that the error terms are negatively autocorrelated; if $(4 - $ DW$) > d_U$, there is no statistical evidence that the error terms are negatively autocorrelated; if $d_L < (4 - $ DW$) < d_U$, the test is inconclusive.

Using the Durbin-Watson statistic, we checked the autocorrelation of hydrological and climatic variables, such as temperature, precipitation, and runoff.

2.2.2. Wavelet Analysis. Wavelet transformation has been shown to be a powerful technique for characterization of the frequency, intensity, time position, and duration of variations in climate and hydrological time series [11, 12, 16, 36]. Wavelet analysis can also reveal the localized time and frequency information without requiring the time series to

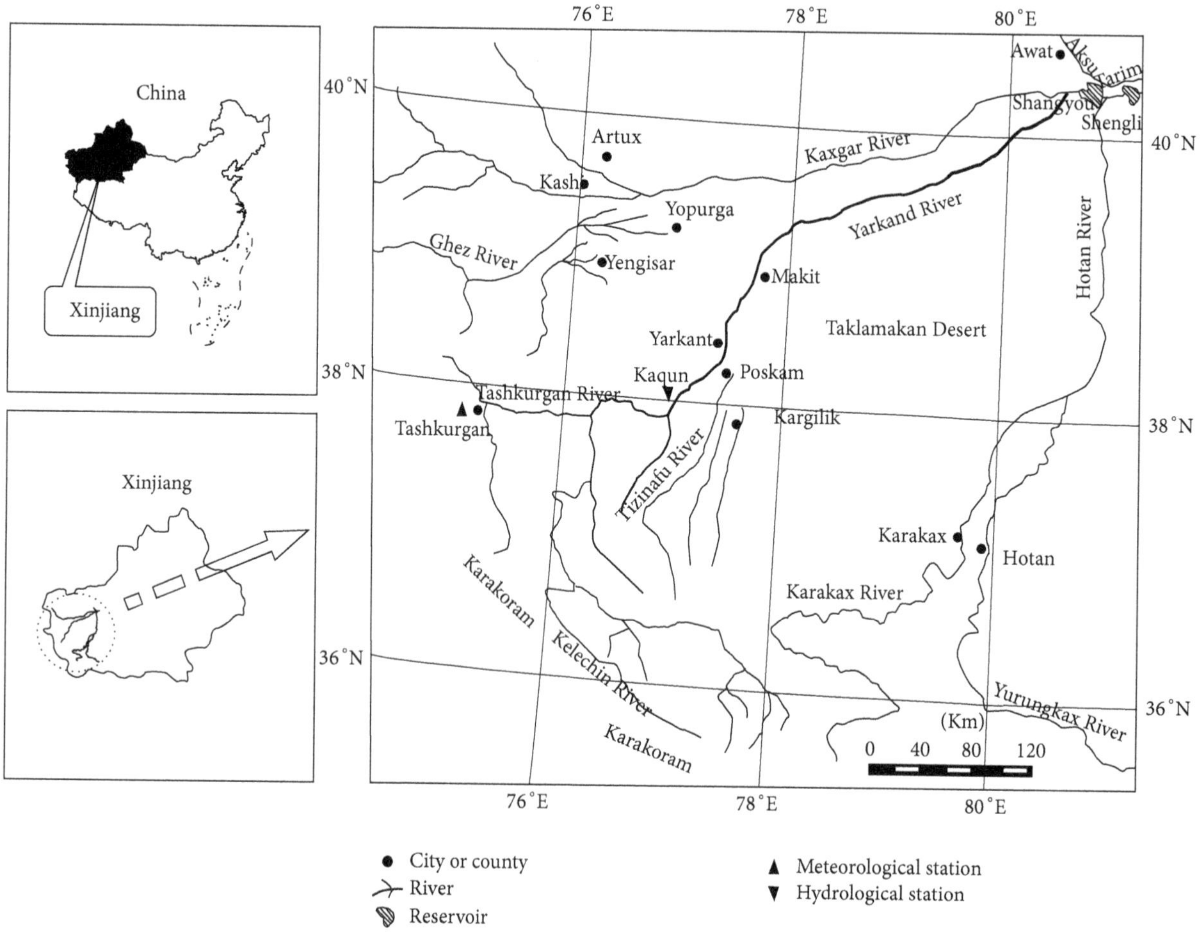

FIGURE 1: Location of the Yarkand River.

be stationary, as required by the Fourier transform and other spectral methods [37].

A continuous wavelet function $\Psi(\eta)$ that depends on a nondimensional time parameter η can be written as [36]

$$\Psi(\eta) = \Psi(a,b) = |a|^{-1/2} \Psi\left(\frac{t-b}{a}\right), \quad (2)$$

where t denotes time, a is the scale parameter, and b is the translation parameter. $\Psi(\eta)$ must have a zero mean and be localized in both time and Fourier space [38]. The continuous wavelet transform (CWT) of a discrete signal, $x(t)$, such as the time series of runoff, temperature, or precipitation, is expressed by the convolution of $x(t)$ with a scaled and translated $\Psi(\eta)$,

$$W_x(a,b) = |a|^{-1/2} \int_{-\infty}^{+\infty} x(t) \Psi^*\left(\frac{t-b}{a}\right) dt, \quad (3)$$

where $*$ indicates the complex conjugate, and $W_x(a,b)$ denotes the wavelet coefficient. Thus, the concept of frequency is replaced by that of scale, which can characterize the variation in the signal, $x(t)$, at a given time scale.

Selecting a proper wavelet function is a prerequisite for time series analysis [39, 40]. The actual criteria for wavelet selection include self-similarity, compactness, and smoothness [41]. Because the symlets are nearly symmetrical, orthogonal, and biorthogonal wavelets proposed by Daubechies as modifications to the db family [41], this study chose the symlets 8 to analyze the variation patterns of runoff and its related climatic factors in the computing environment of MATLAB.

For a time series, $x(t)$, it can be analyzed at multiple scales through wavelet decomposition on the basis of the discrete wavelet transform (DWT). The DWT is defined taking discrete values of a and b. The full DWT for signal, $x(t)$, can be represented as [42]

$$x(t) = \sum_k \mu_{j_0,k} \phi_{j_0,k}(t) + \sum_{j=1}^{j_0} \sum_k \omega_{j,k} \psi_{j,k}(t), \quad (4)$$

where $\phi_{j_0,k}(t)$ and $\psi_{j,k}(t)$ are the flexing and parallel shift of the basic scaling function, $\phi(t)$, and the mother wavelet function, $\psi(t)$, and $\mu_{j_0,k}$ $(j < j_0)$ and $\omega_{j,k}$ are the scaling coefficients and the wavelet coefficients, respectively. Generally, scales and positions are based on powers of 2, which is the dyadic DWT.

Once a mother wavelet is selected, the wavelet transform can be used to decompose a signal according to scale, allowing separation of the fine-scale behavior (detail) from the large-scale behavior (approximation) of the signal [43]. The relationship between scale and signal behavior is designated as follows: low scale corresponds to compressed wavelet as well as rapidly changing details, namely, high frequency; whereas high scale corresponds to stretched wavelet and slowly changing coarse features, namely, low frequency. Signal decomposition is typically conducted in an iterative fashion using a series of scales such as $a = 2, 4, 8, \ldots, 2^L$, with successive approximations being split in turn so that one signal is broken down into many lower resolution components.

The wavelet decomposition and reconstruction were used to approximate the variation patterns of AR and its related factors over the entire study period at the selected different time scales.

2.2.3. Multiple Linear Regression. For understanding the relationship between annual runoff with its related climatic factors at different time scales, we employed multiple linear regression (MLR) based on wavelet analysis. This method fits multiple linear regression equation (MLRE) between AR with AAT and AP by using multiple linear regression (MLR) based on the results of wavelet approximation [44].

The multiple linear regression model is

$$y = a_0 + a_1 x_1 + a_2 x_2 + \cdots + a_k x_k, \tag{5}$$

where y is dependent variable, x_i is the independent variables; a_i is the regression coefficient, which is generally calculated by method of least squares [45]. In this study, the dependent variable is the annual runoff (AR) and the independent variables are related climatic factors, such as the annual average temperature (AAT) and annual precipitation (AP).

2.2.4. Coefficient of Determination and Akaike Information Criterion. In order to identify the uncertainty of the estimated model for a given time scale, the coefficient of determination, also known as the goodness of fit, was calculated as follows:

$$R^2 = 1 - \frac{\text{RSS}}{\text{TSS}} = 1 - \frac{\sum_{i=1}^{n} (y_i - \widehat{y}_i)^2}{\sum_{i=1}^{n} (y_i - \overline{y})^2}, \tag{6}$$

where R^2 is the coefficient of determination; $\widehat{y}_i$ and y_i are the simulated value and actual data of runoff, respectively; $\overline{y}$ is the mean of y_i $(i = 1, 2, \ldots, n)$; RSS $= \sum_{i=1}^{n} (y_i - \widehat{y}_i)^2$ is the residual sum of squares; TSS $= \sum_{i=1}^{n} (y_i - \overline{y})^2$ is the total sum of squares.

The coefficient of determination is a measure of how well the simulated results represent the actual data. A bigger R^2 indicates a higher certainty and lower uncertainty of the estimates [45].

To compare the relative goodness between the ANN and multiple linear regression (MLR) fit for a given timescale, we

also used the measure of Akaike information criterion (AIC) [46]. The formula of AIC is as follows:

$$\text{AIC} = 2k + n \ln \left(\frac{\text{RSS}}{n} \right), \tag{7}$$

where k is the number of parameters estimated in the model; n is the number of samples; RSS is the same as in formula (6). A smaller AIC indicates a better model.

For small sample sizes (i.e., $n/K \leq 40$), the second-order Akaike information criterion (AIC_c) should be used instead:

$$\text{AIC}_c = \text{AIC} + \frac{2k(k+1)}{n - k - 1}, \tag{8}$$

where n is the sample size. As the sample size increases, the last term of the AIC_c approaches zero, and the AIC_c tends to yield the same conclusions as the AIC [47].

3. Results

3.1. Check for Variable's Autocorrelation. The premise of statistics indicates that models imply an assumption; that is, variables are stochastic and no significant autocorrelation is present. Is it really? This can be demonstrated by statistical check for variable's autocorrelation (Table 1).

For (1^*), (2^*), and (3^*) in Table 1, their degree of freedom is, respectively, k equals 2 and N (i.e., $n-k-1$) equals 51. Upper and lower critical values of the Durbin-Watson Statistic (DW) are d_L equals 1.509 and d_U equals 1.58 when significance level (α) equals 0.01. Because the values of DW are between d_U and $4 - d_U$, it is obvious that annual runoff (AR), annual average temperature (ATT), and annual precipitation (AP) indicate no autocorrelation.

For (4^*) in Table 1, its degree of freedom is, respectively, k equals 4 and N equals 51. Upper and lower critical values of DW are d_L equals 1.25350 and d_U equals 1.49384 when significance level (α) equals 0.01. For (5^*), its degree of freedom is, respectively, k equals 6 and N equals 51. Upper and lower critical values of DW are d_L equals 1.17372 and d_U equals 1.58811 when significance level (α) equals 0.01. Thereby they indicate no autocorrelation either.

In fact, it can be determined that variables and model reveal non-autocorrelation because the value of DW is close to 2 for each regression equation shown in Table 1. Therefore, the assumption of our model is logical.

3.2. Variation Patterns of Climatic-Hydrological Process at Different Time Scales. Our previous study indicated that [44] the annual average temperature and annual precipitation are the most important factors that related with the annual runoff. The result was also supported by the other studies for the headwaters of the Tarim River Basin [20, 30–33].

The raw data of AR, AAT, and AP showed fluctuation. It is difficult to identify any patterns simply based on the raw data. In order to show the scale-dependent with time for the climatic-hydrological process of the Yarkand River, the wavelet analysis was used. The nonlinear variation for the annual runoff process and the related climate factors were

TABLE 1: Statistical check for variable's autocorrelation.

Dependent variable	Independent variable	Function		R^2	F	DW
AR_t	AR_{t-1}	$AR_t = 76.852 - 0.156AR_{t-1}$	(1*)	0.024	1.125	1.981
AAT_t	AAT_{t-1}	$AAT_t = 2.526 - 0.290AAT_{t-1}$	(2*)	0.093	5.014	1.835
AP_t	AP_{t-1}	$AP_t = 81.338 - 0.118AP_{t-1}$	(3*)	0.013	0.659	1.851
AR_t	AAT_t, AP_t, AR_{t-1}	$AR_t = 66.677 + 3.191AAT_t - 0.025AP_t - 0.145AR_{t-1}$	(4*)	0.091	1.572	2.023
AR_t	$AAT_t, AP_t, AR_{t-1},$ AAT_{t-1}, AP_{t-1}	$AR_t = 62.516 + 2.051AAT_t - 0.048AP_t - 0.186AR_{t-1} + 2.702AAT_{t-1} + 0.042AP_{t-1}$	(5*)	0.123	1.266	1.931

Notes. AR: annual runoff, AAT: annual average temperature, and AP: annual precipitation; the subscripts, t and $t - 1$, represent time.

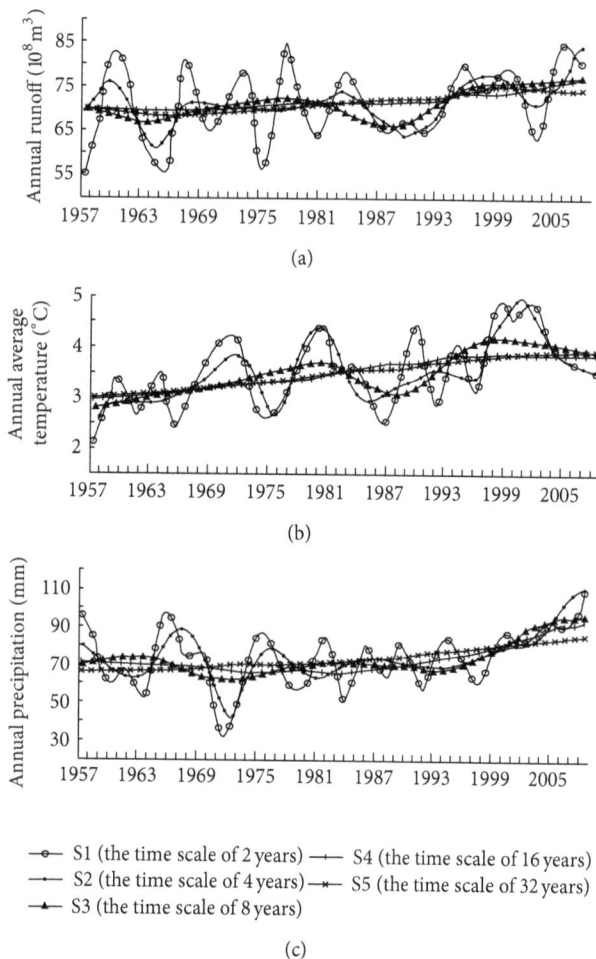

(a)

(b)

— S1 (the time scale of 2 years)   —▲— S4 (the time scale of 16 years)
— S2 (the time scale of 4 years)   —✳— S5 (the time scale of 32 years)
—▲— S3 (the time scale of 8 years)

(c)

FIGURE 2: Variation patterns at different time scales of (a) annual runoff, (b) annual average temperature, and (c) annual precipitation.

analyzed at multiple-year scales through wavelet decomposition on the basis of the discrete wavelet transform (DWT).

The wavelet decomposition for the time series of annual runoff at five time scales resulted in five variation patterns (Figure 2(a)). The S1 curve retains a large amount of residual from the raw data, and drastic fluctuations exist in the period from 1957 to 2008. These characteristics indicate that although the runoff varied greatly throughout the study period, there was a hidden increasing trend. The S2 curve still

retains a considerable amount of residual, as indicated by the presence of 4 peaks and 4 valleys. However, the S2 curve is much smoother than the S1 curve, which allows the hidden increasing trend to be more apparent. The S3 curve retains much less residual, as indicated by the presence of 2 peaks and 2 valleys. Compared to S2, the increase in runoff over time is more apparent in S3. Finally, the S5 curve presents an ascending tendency, whereas the increasing trend is obvious in the S4 curve.

Accordingly, Figures 2(b) and 2(c) provide us with a method for comparing the variation patterns of annual average temperature and annual precipitation at different time scales. The wavelet decomposition for the time series at five time scales resulted in five variation patterns, respectively. These five time scales are also designated as S1 to S5. The curves present an ascending tendency despite drastic fluctuations in S1 and S2. Then, the curves are getting much smoother and the increasing trend becomes even more obvious as the scale level increases.

The upper analysis showed that the nonlinear variations of runoff, temperature, and precipitation of the Yarkand River basin were dependent on time scales. The annual runoff, annual average temperature and annual precipitation at five time scales resulted in five patterns of nonlinear variations, respectively.

3.3. Relationship between Streamflow and Climate Factors. Based on the raw data of AAT, AP, and AR, multiple linear regression equation (MLRE) was developed as follows:

$$AR = 3.5AAT - 0.037AP + 56.75,$$
$$R^2 = 0.1983, \qquad F = 2.517, \qquad \alpha = 0.1. \tag{9}$$

Equation (9) reveals a positive correlation between the annual runoff and the annual average temperature. These results are readily supported by the fact that the majority of streamflow comes from glacial melt and snowmelt, which have been occurring at increased rates as the temperature increases. These results have been confirmed by other studies [48]. However, (9) also indicates the existence of a weak, negative correlation between the annual runoff and the annual precipitation, which does not seem reasonable. Indeed, this finding conflicts with the results of other studies, which have suggested that both the temperature and precipitation series in the Tarim Basin have been increasing in a pattern similar to that of annual runoff over the past 50 years [20, 30].

Additionally, the coefficient of determination of (9) is as low as 0.093. Furthermore, the average absolute error and average relative error for predicted results is as high as $9.014 \times 10^8 \, \text{m}^3$ and 14.11%, respectively. All this means that the regression (9) is not authentic. What is the reason for this? It is possible that this inconsistency is caused by randomicity in the raw time-series data, which should be filtered out via wavelet decomposition based on the discrete wavelet transform [17, 18].

To understand the response of the runoff to regional climatic change at different time scales, based on the results of wavelet decomposition (Figure 2), multiple linear regression equation (MLRE) at each time scale was fitted for describing the relationship among annual runoff, annual average temperature, and annual precipitation (Table 2).

3.4. Comparison of the Estimated Results at Different Time Scales. Though all MLREs at each time scale in Table 2 got across the statistical test at the significant level of 0.01 or 0.001, the predicted error of MLRE at the chosen time scales is different. Figure 3 shows the comparison for the simulated value by MLREs and original data of AR at different time scales. The predicted error of MLRE at the time scale of S1 and S2 (i.e., 2-year and 4-year scales) is large, that at the time scale of S3 (i.e., 8-year scale) is also fairish, and wee predicted error of MLRE at the time scale of S4 and S5 (i.e., 4-year and 5-year scales) appears. These results show that MLRE only can well fit the relationship between runoff and climate factors at large time scale such as at 16-year and 32-year scales.

By comparing the R^2 and AIC value in Table 3, we can know the estimated effect (good or bad) of models at different time scales.

Table 3 tells us that the R^2 for MLRE at the time scale of S1 and S2 (i.e., 2-year and 4-year scales) is lower (only 0.361 and 0.416, resp.) and that at the time scale of S3 (i.e., 8-year scales) is higher, reaching 0.894. Only the MLRE at the time scale of S4 and S5 (i.e., 16-year and 32-year scales) has the high coefficient of determination, which is as high as 0.975 and 0.996, respectively.

The lower AIC value indicates better model, which tells us that the MLRE at time scale of S5 is the best, that at time scale of S4 is better, that at time scale of S3 is moderate, that at time scale of S2 is the penult, and that at time scale of S1 is the worst.

Overall, the relationship between AR with AAT and AP was simulated by the multiple linear regression (MLR) based on wavelet analysis at different time scales, but the simulated effect at a larger time scale is better than that at a smaller time scale.

4. Discussion and Conclusions

Many studies indicated that the climatic-hydrological process is a complex system with nonlinearity, but it is difficult to understand the mechanism of climatic-hydrological process thoroughly [2]. Our results showed the following fact: the simulated effect at large time scale is better than that at small time scale, and the estimated precision at large time scale is

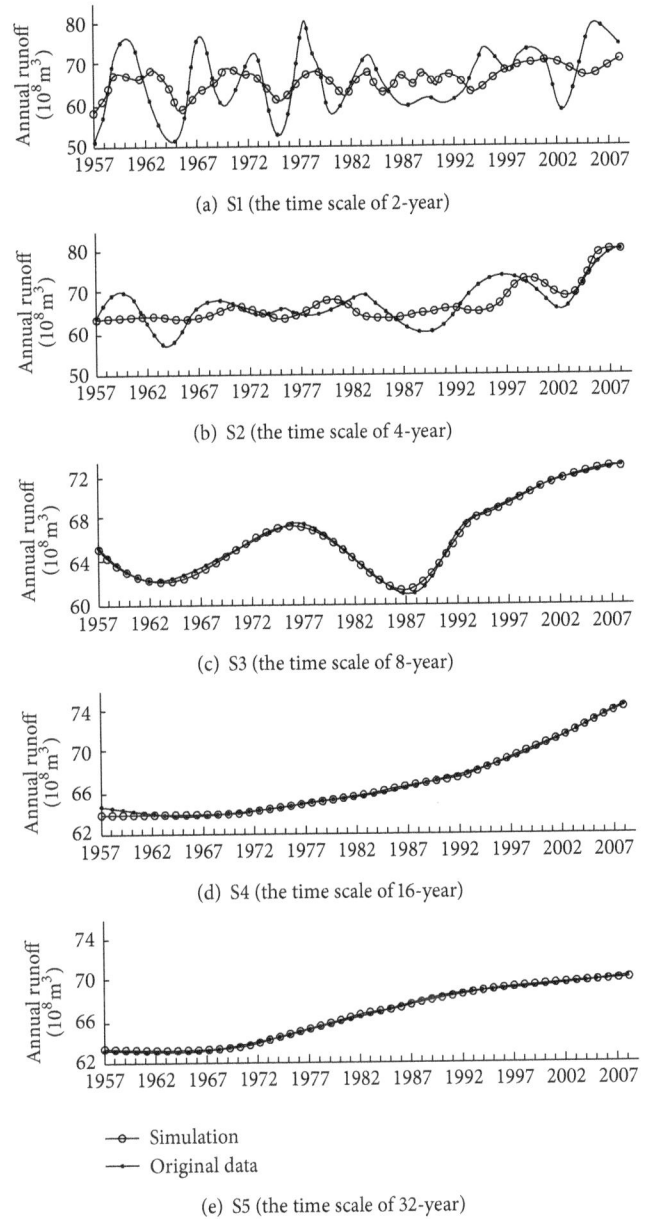

(a) S1 (the time scale of 2-year)

(b) S2 (the time scale of 4-year)

(c) S3 (the time scale of 8-year)

(d) S4 (the time scale of 16-year)

(e) S5 (the time scale of 32-year)

FIGURE 3: The simulated value for AR and its original data.

higher than that at small time scale. What are the causes for this? It is difficult to thoroughly answer the question because of the nonlinear complicated climatic-hydrological process, which is essentially difficult to precisely predict [15].

Our study revealed that the climatic-hydrological process at larger time scale (e.g., 16-year or 32-year scales) basically presented a linear process, but that at smaller time scale (e.g., 2-year or 4-year scales) is essentially nonlinear process with complicated causations. Because the time series of runoff are essentially monotonic trends related to long-term climatic changes at large time scale (e.g., 16-year and 32-year scales), the estimated precision is much higher. Otherwise,

TABLE 2: MLREs for climatic-hydrological process at different time scales.

Time scale	Regression equation	R^2	F	Significance level α	Average absolute error	Average relative error
S1	AR = 3.243AAT + 54.89	0.361	4.763	0.01	6.266	9.665%
S2	AR = 2.883AAT + 0.173AP + 43.62	0.4157	15.6492	0.001	3.250	4.959%
S3	AR = 5.792AAT + 0.116AP + 37.276	0.894	205.729	0.001	0.880	1.354%
S4	AR = 3.332AAT + 0.189AP + 40.618	0.975	943.228	0.001	0.329	0.502%
S5	AR = 2.555AAT + 0.206AP + 42.353	0.996	6701.914	0.001	0.109	0.165%

Notes. AR: annual runoff, AAT: annual average temperature, and AP: annual precipitation.

TABLE 3: R^2 and AIC for MLREs at different time scales.

Time scale	R^2	AIC
S1	0.361	209.924
S2	0.416	143.263
S3	0.894	12.960
S4	0.975	−96.714
S5	0.996	−209.172

due to the difficulty to accurately predict nonlinear climatic-hydrological process at small time scales (e.g., 2-year or 4-year scale), the estimated precision and simulated effect are not satisfactory.

Nevertheless, the comprehensive method conducted by this paper provides a method to understand the climatic-hydrological process in the Yarkand River from the perspective of multiscale, which may be used to explore the climatic-hydrological process in other inland rivers of northwest China.

The main conclusions of this work can be summarized as follows.

(1) The hydrologic and climatic variables, that is, annual runoff (AR), annual average temperature (AAT), and annual precipitation (AP) are stochastic and show no significant autocorrelation.

(2) The variation pattern of runoff, temperature, and precipitation was scale dependent with time. The annual runoff (AR), annual average temperature (AAT), and annual precipitation (AP) basically present linear trends at 16-year and 32-year scales, but they show nonlinear fluctuations at 2-year, 4-year, and 8-year scales.

(3) The relationship between AR with AAT and AP was simulated by the multiple linear regression equation (MLRE) based on wavelet analysis at each time scale. The results showed that the AR is basically monotonic trend related to long-term climatic changes at a larger time scale (e.g., 16-year or 32-year scales), and the estimated precision is much higher. But due to an essentially nonlinear climatic-hydrological process at a smaller time scale (e.g., 2-year or 4-year scales), the estimated precision is lower than that at a larger time scale.

Acknowledgments

This work was supported by the Director Fund of the Key Lab of GIScience of the Education Ministry PRC. The authors are grateful to the editor and referees whose comments helped us in improving the article's quality.

References

[1] A. J. Cannon and I. G. McKendry, "A graphical sensitivity analysis for statistical climate models: application to Indian monsoon rainfall prediction by artificial neural networks and multiple linear regression models," *International Journal of Climatology*, vol. 22, no. 13, pp. 1687–1708, 2002.

[2] J. H. Xu, Y. N. Chen, W. H. Li, Q. Nie, Y. L. Hong, and Y. Yang, "The nonlinear hydro-climatic process in the Yarkand River, northwestern China," *Stochastic Environmental Research and Risk Assessment*, vol. 27, no. 2, pp. 389–399, 2013.

[3] R. Ibbitt and R. Woods, "Re-scaling the topographic index to improve the representation of physical processes in catchment models," *Journal of Hydrology*, vol. 293, no. 1–4, pp. 205–218, 2004.

[4] W. G. Strupczewski, V. P. Singh, S. Weglarczyk, K. Kochanek, and H. T. Mitosek, "Complementary aspects of linear flood routing modelling and flood frequency analysis," *Hydrological Processes*, vol. 20, no. 16, pp. 3535–3554, 2006.

[5] B. Sivakumar, "Nonlinear determinism in river flow: prediction as a possible indicator," *Earth Surface Processes and Landforms*, vol. 32, no. 7, pp. 969–979, 2007.

[6] J. H. Xu, Y. N. Chen, W. H. Li, and S. Dong, "Long-term trend and fractal of annual runoff process in mainstream of Tarim River," *Chinese Geographical Science*, vol. 18, no. 1, pp. 77–84, 2008.

[7] B. P. Wilcox, M. S. Seyfried, and T. H. Matison, "Searching for chaotic dynamics in snowmelt runoff," *Water Resources Research*, vol. 27, no. 6, pp. 1005–1010, 1991.

[8] J. W. Kantelhardt, D. Rybski, S. A. Zschiegner et al., "Multifractality of river runoff and precipitation: comparison of fluctuation analysis and wavelet methods," *Physica A*, vol. 330, no. 1-2, pp. 240–245, 2003.

[9] J. H. Xu, Y. N. Chen, W. H. Li, M. H. Ji, and S. Dong, "The complex nonlinear systems with fractal as well as chaotic dynamics of annual runoff processes in the three headwaters of the Tarim River," *Journal of Geographical Sciences*, vol. 19, no. 1, pp. 25–35, 2009.

[10] Z. L. Wang and C. Y. Huang, "Self-organized criticality of rainfall in central China," *Advances in Meteorology*, vol. 2012, Article ID 203682, 8 pages, 2012.

[11] L. C. Smith, D. L. Turcotte, and B. L. Isacks, "Streamflow characterization and feature detection using a discrete wavelet

transform," *Hydrological Processes*, vol. 12, no. 2, pp. 233–249, 1998.

[12] C. M. Chou, "Efficient nonlinear modeling of rainfall-runoff process using wavelet compression," *Journal of Hydrology*, vol. 332, no. 3-4, pp. 442–455, 2007.

[13] J. H. Xu, Y. N. Chen, M. H. Ji, and F. Lu, "Climate change and its effects on runoff of Kaidu River, Xinjiang, China: a multiple time-scale analysis," *Chinese Geographical Science*, vol. 18, no. 4, pp. 331–339, 2008.

[14] C. H. Hu, Y. H. Hao, T. C. J. Yeh, B. Pang, and Z. N. Wu, "Simulation of spring flows from a karst aquifer with an artificial neural network," *Hydrological Processes*, vol. 22, no. 5, pp. 596–604, 2008.

[15] J. H. Xu, Y. N. Chen, W. H. Li et al., "Combining BPANN and wavelet analysis to simulate hydro-climatic process—a case study of the Kaidu River, NW China," *Frontiers of Earth Science*, 2013.

[16] J. H. Xu, Y. N. Chen, W. H. Li, M. H. Ji, S. Dong, and Y. L. Hong, "Wavelet analysis and nonparametric test for climate change in Tarim River Basin of Xinjiang during 1959–2006," *Chinese Geographical Science*, vol. 19, no. 4, pp. 306–313, 2009.

[17] J. H. Xu, W. H. Li, M. H. Ji, F. Lu, and S. Dong, "A comprehensive approach to characterization of the nonlinearity of runoff in the headwaters of the Tarim River, western China," *Hydrological Processes*, vol. 24, no. 2, pp. 136–146, 2010.

[18] J. H. Xu, Y. N. Chen, F. Lu, W. H. Li, L. J. Zhang, and Y. L. Hong, "The nonlinear trend of runoff and its response to climate change in the Aksu River, western China," *International Journal of Climatology*, vol. 31, no. 5, pp. 687–695, 2011.

[19] J. H. Xu, Y. N. Chen, W. H. Li, Y. Yang, and Y. L. Hong, "An integrated statistical approach to identify the nonlinear trend of runoff in the Hotan River and its relation with climatic factors," *Stochastic Environmental Research and Risk Assessment*, vol. 25, no. 2, pp. 223–233, 2011.

[20] X. M. Hao, Y. N. Chen, C. C. Xu, and W. H. Li, "Impacts of climate change and human activities on the surface runoff in the Tarim River Basin over the last fifty years," *Water Resources Management*, vol. 22, no. 9, pp. 1159–1171, 2008.

[21] Y. N. Chen and Z. X. Xu, "Plausible impact of global climate change on water resources in the Tarim River Basin," *Science in China D*, vol. 48, no. 1, pp. 65–73, 2005.

[22] Y. N. Chen, K. Takeuchi, C. C. Xu, Y. P. Chen, and Z. X. Xu, "Regional climate change and its effects on river runoff in the Tarim Basin, China," *Hydrological Processes*, vol. 20, no. 10, pp. 2207–2216, 2006.

[23] Y. N. Chen, C. C. Xu, X. M. Hao et al., "Fifty-year climate change and its effect on annual runoff in the Tarim River Basin, China," *Quaternary International*, vol. 208, no. 1-2, pp. 53–61, 2009.

[24] J. Wang, H. Li, and X. Hao, "Responses of snowmelt runoff to climatic change in an inland river basin, Northwestern China, over the past 50 years," *Hydrology and Earth System Sciences*, vol. 14, no. 10, pp. 1979–1987, 2010.

[25] Q. Zhang, C. Y. Xu, H. Tao, T. Jiang, and Y. D. Chen, "Climate changes and their impacts on water resources in the arid regions: a case study of the Tarim River basin, China," *Stochastic Environmental Research and Risk Assessment*, vol. 24, no. 3, pp. 349–358, 2010.

[26] Y. F. Shi, Y. P. Shen, E. S. Kang et al., "Recent and future climate change in northwest china," *Climatic Change*, vol. 80, no. 3-4, pp. 379–393, 2007.

[27] B. F. Li, Y. N. Chen, Z. S. Chen, and W. H. Li, "Trends in runoff versus climate change in typical rivers in the arid region of northwest China," *Quaternary International*, vol. 282, pp. 87–95, 2012.

[28] B. F. Li, Y. N. Chen, and X. Shi, "Why does the temperature rise faster in the arid region of northwest China?" *Journal of Geophysical Research*, vol. 117, no. 16, Article ID 16115, 2012.

[29] Y. Yang, J. H. Xu, Y. L. Hong, and G. H. Lv, "The dynamic of vegetation coverage and its response to climate factors in Inner Mongolia, China," *Stochastic Environmental Research and Risk Assessment*, vol. 26, no. 3, pp. 357–373, 2012.

[30] Y. N. Chen, K. Takeuchi, C. C. Xu, Y. P. Chen, and Z. X. Xu, "Regional climate change and its effects on river runoff in the Tarim Basin, China," *Hydrological Processes*, vol. 20, no. 10, pp. 2207–2216, 2006.

[31] B. G. Sun, W. Y. Mao, Y. R. Feng et al., "study on the change of air temperature, precipitation and runoff volume in the Yarkant River basin," *Arid Zone Research*, vol. 23, no. 2, pp. 203–209, 2006 (Chinese).

[32] M. Sabit and A. Tohti, "An analysis of water resources and it's hydrological characteristic of Yarkend River Valley," *Journal of Xinjiang Normal University*, vol. 24, no. 1, pp. 74–78, 2005 (Chinese).

[33] T. L. Liu, Q. Yang, R. Qin, Y. P. He, and R. Liu, "Climate change towards warming-wetting trend and its effects on runoff at the headwater region of the Yarkand River in Xinjiang," *Journal of Arid Land Resources and Environment*, vol. 22, no. 9, pp. 49–53, 2008 (Chinese).

[34] J. Durbin and G. S. Watson, "Testing for serial correlation in least squares regression. I," *Biometrika*, vol. 37, no. 3-4, pp. 409–428, 1950.

[35] J. Durbin and G. S. Watson, "Testing for serial correlation in least squares regression. II," *Biometrika*, vol. 38, no. 1-2, pp. 159–178, 1951.

[36] D. Labat, "Recent advances in wavelet analyses: part 1. A review of concepts," *Journal of Hydrology*, vol. 314, no. 1–4, pp. 275–288, 2005.

[37] C. Torrence and G. P. Compo, "A practical guide to wavelet analysis," *Bulletin of the American Meteorological Society*, vol. 79, no. 1, pp. 61–78, 1998.

[38] M. Farge, "Wavelet transforms and their applications to turbulence," *Annual Review of Fluid Mechanics*, vol. 24, no. 1, pp. 395–457, 1992.

[39] J. B. Ramsey, "Regression over timescale decompositions: a sampling analysis of distributional properties," *Economic Systems Research*, vol. 11, no. 2, pp. 163–183, 1999.

[40] J. H. Xu, Y. Lu, F. L. Su, and N. S. Ai, "R/S and wavelet analysis on the evolutionary process of regional economic disparity in China during the past 50 years," *Chinese Geographical Science*, vol. 14, no. 3, pp. 193–201, 2004.

[41] I. Daubechies, "Orthonormal bases of compactly supported wavelets," *Communications on Pure and Applied Mathematics*, vol. 41, pp. 909–996, 1988.

[42] S. G. Mallat, "Theory for multiresolution signal decomposition: the wavelet representation," *IEEE Transactions on Pattern Analysis and Machine Intelligence*, vol. 11, no. 7, pp. 674–693, 1989.

[43] L. M. Bruce, C. H. Koger, and J. Li, "Dimensionality reduction of hyperspectral data using discrete wavelet transform feature extraction," *IEEE Transactions on Geoscience and Remote Sensing*, vol. 40, no. 10, pp. 2331–2338, 2002.

[44] J. H. Xu, Y. N. Chen, M. H. Ji, and F. Lu, "Climate change and its effects on runoff of Kaidu River, Xinjiang, China: a multiple time-scale analysis," *Chinese Geographical Science*, vol. 18, no. 4, pp. 331–339, 2008.

[45] J. H. Xu, *Mathematical Methods in Contemporary Geography*, Higher Education Press, Beijing, China, 2002, (in Chinese).

[46] D. R. Anderson, K. P. Burnham, and W. L. Thompson, "Null hypothesis testing: problems, prevalence, and an alternative," *Journal of Wildlife Management*, vol. 64, no. 4, pp. 912–923, 2000.

[47] K. P. Burnham and D. R. Anderson, *Model Selection and Multimodel Inference: A Practical Information-Theoretic Method*, Springer, New York, NY, USA, 2nd edition, 2002.

[48] X. Wang, Z. C. Xie, S. Y. Liu, D. H. Shangguan, J. J. Tao, and Y. L. Yang, "Prediction on the variation trend of glacier system in the source region of Tarim River responding to climate change," *Journal of Mountain Research*, vol. 24, no. 6, pp. 641–646, 2006 (Chinese).

16

Impact of Asian Dust Aerosol and Surface Albedo on Photosynthetically Active Radiation and Surface Radiative Balance in Dryland Ecosystems

X. Xi and I. N. Sokolik

School of Earth and Atmospheric Sciences, Georgia Institute of Technology, 311 Ferst Drive, Atlanta, GA 30332-0340, USA

Correspondence should be addressed to I. N. Sokolik, isokolik@eas.gatech.edu

Academic Editor: Dimitris G. Kaskaoutis

We investigated the extent to which Asian dust can affect vegetation in dryland ecosystems through altering photosynthetically active radiation (PAR) and shortwave and longwave radiation components of the surface energy balance. Results show that dust decreases the surface radiative balance and total PAR. The diffuse component of PAR, however, increases with increasing dust load but then decreases after reaching a maximum at a certain optimum condition. The forcing efficiency ranges from -67.7 to $-82.2\,\mathrm{Wm^{-2}}\,\tau_{0.5}^{-1}$ in total PAR and from -68.8 to $-122.1\,\mathrm{Wm^{-2}}\,\tau_{0.5}^{-1}$ in surface radiative balance. The ratio of total PAR to downwelling shortwave flux remains nearly constant ($0.45 \pm 4\%$) similar to other aerosol types, while the ratio for the diffuse faction of PAR exhibits significant variations. The impact of dust on the gross photosynthetic rate varies among different types of crops. C4 plants such as corn tend to be less sensitive to the dust optical properties compared to C3 plants such as soybean and wheat.

1. Introduction

There has been a growing interest in the impact of atmospheric aerosols upon the terrestrial ecosystems and their role in land-atmosphere interactions in the context of earth system science. These interactions are thought to involve multiple, interrelated processes and various feedbacks that remain poorly constrained [1]. Here we address the impact of mineral aerosol (dust) that involves the radiative transfer processes, focusing on aerosol-induced changes in photosynthetically active radiation (PAR, 0.4–0.7 μm) and surface radiative balance (SRB, 0.3–20 μm). Light is a vital factor governing the plant photosynthetic activities, and hence changes in PAR caused by aerosols can influence the plant-air carbon/water exchanges and ecosystem functioning. Changes in the land surface energy balance are important because they affect the surface evapotranspiration, sensible and latent heat, soil temperature and moisture, and major land-atmosphere exchange processes that along with light availability are all important to the ecosystems.

Dust can affect both the shortwave (SW, 0.3–2.5 μm) and longwave (LW, 2.5–20 μm) components of the radiative energy balance, but in opposing ways [2]. Based on regional model simulations, Mallet et al. [3] found that dust aerosol decreased the SW radiation by up to $-137\,\mathrm{Wm^{-2}}$ (regional mean) in North Africa, resulting in a significant decrease in surface temperature and sensible heat. Based on a study with a coupled aerosol transport-radiation model, Takemura et al. [4] reported that dust caused a monthly mean SW surface forcing of $-2.0\,\mathrm{Wm^{-2}}$ over East Asia, where the surface forcing was defined as the difference in SW fluxes between clean and aerosol-laden conditions. Using satellite observations in conjunction with modeling, Huang et al. [5] found that dust aerosol caused a daily-mean surface SW forcing of up to $-41.9\,\mathrm{Wm^{-2}}$ over the Taklamakan desert.

In contrast to the reduction in SW radiation, dust significantly increases the LW radiation reaching the surface. For instance, based on ship-based measurements during the ACE-Asia campaign, Vogelmann et al. [6] demonstrated that Asian dust contributed to a surface LW forcing of up to

$10\,\mathrm{Wm}^{-2}$; Markowicz et al. [7] found that the LW forcing compensated about 20% of the SW cooling. Huang et al. [5] found that about one-third of the SW surface cooling caused by Asian dust was compensated by its LW warming effect. Thus, dust-induced changes in both the SW and LW radiation should be accounted for in assessing the surface energy balance.

Although there have been numerous studies of the dust SW and LW radiative impact, we are not aware of any study that comprehensively addressed the impact of dust on PAR. Past studies, however, explored the influence of several other aerosol types on PAR, including volcanic aerosol [8, 9], urban pollution aerosol [10–12], and biomass burning smoke [13]. Several studies reported reductions in the plant photosynthetic rate and primary production due to less incoming PAR as a result of aerosol attenuation [10, 11]. However, while reducing the total PAR, aerosols can enhance the diffuse fraction of PAR and lead to a higher gross photosynthetic rate. The underlying reasoning is believed to be due to the redistribution of light between the sunlit and shaded leaves within the plant canopy: the aerosol absorption and scattering causes reductions in the PAR received by the sunlit leaves with no change or some reductions in the photosynthesis, while more scattered diffuse PAR becomes available to the majority of light-limited shaded leaves, such that the gross photosynthetic rate increases. This so-called "diffuse radiation fertilization" effect due to aerosols is well demonstrated in modeling [12, 14] and observational [9, 13, 15] studies.

Given that changes in total PAR and its diffuse fraction are controlled by the aerosol type, especially by aerosol burden and optical properties and because of distinct differences between the optical properties of dust and other aerosols, it is important to understand how dust aerosol can affect PAR. Furthermore, dust-induced changes in both PAR and SRB need to be addressed to understand the net radiative impact of dust on terrestrial ecosystems. Indeed, concurrent with the modifications of PAR, aerosols can alter the surface net radiation, latent/sensible heat, soil/leaf temperature, and atmospheric humidity, which can also affect the photosynthesis and respiration processes and the net primary production (NPP) [16, 17].

The focus of this paper is on dust in East Asia. This region is vulnerable to the dust impact because of vast and prodigious dust sources. Each year large quantities of windblown dust are being emitted from arid and semiarid regions in China and Mongolia and transported downwind over thousands of kilometers, potentially affecting various ecosystems [18]. The most intense dust outbreaks occur in the spring season coinciding with the vegetation growing season that further amplifies the importance of dust in dryland ecosystems.

Assessment of PAR along with SW and LW components of the surface energy balance in dust-laden conditions necessitates a consistent representation of the dust optical characteristics across the wide spectral range (i.e., from the UV to the IR). Current measurement capabilities cannot provide this information so that optical modeling must be performed to compute the required spectral optical characteristics. However, computation of the optical characteristics, such as

extinction coefficient, single scattering albedo, and scattering phase function, is subject to large uncertainties due to the complex nature of mineral aerosols. Dust particles exhibit various nonspherical shapes, mineralogical compositions, and size spectra that depend on dust sources and physicochemical changes (aging) of dust particles during the transport in the atmosphere [19]. To perform optical modeling for radiative budget assessments, past studies often considered dust as a single generic species and used a spectral refractive index reported for limited dust bulk samples collected in a few geographical regions. For instance, the dust refractive index used by Yoshioka et al. [20] was a combination of data from Patterson [21] for the visible spectrum, from Sokolik et al. [22] for the near-infrared (near-IR), and from Volz [23] for the IR, despite the fact that these datasets represent three completely different dust samples—one from Central Asia and the other two from Northern Africa. The OPAC (Optical Properties of Aerosols and Clouds) library, which is widely used in the radiation/climate modeling, also consists of the bulk dust models that are based on the Patterson and Volz refractive indices [24]. To overcome these limitations, here we explicitly consider the size-resolved mineralogical composition of dust aerosol that allows the use of the spectral refractive indices of the major minerals [2, 25, 26]. One key advantage of this approach is that it is possible to incorporate recent data of dust mineralogical composition, providing the improved representation of region-specific dust optical characteristics from the UV to the IR.

The radiative impact of aerosols also depends on the properties and state of the underlying land surfaces that control surface albedo and emissivity. In dryland regions, surface albedo exhibits large spatiotemporal and spectral variability. In particular in the case of barren and sparsely vegetated surfaces, albedo depends on a combination of several factors such as soil type, composition, and moisture [27, 28]. For vegetated surfaces, the variability of the surface albedo is controlled by the plant phenology [29]. Shrublands and grasslands exhibit the largest albedo variations in space and time compared to other land categories [29]. Although surface albedo varies with wavelength, it is commonly represented in regional and global climate models by a wavelength-independent constant (often called SW broadband albedo), which is then prescribed to a certain land category as part of the land surface module. Studies demonstrated that, however, neglecting the spectral dependence of surface albedo (e.g., in the visible versus near-IR) can lead to significant errors in modeled climate variables [30]. Therefore, it is important to account for the spectral surface albedo and its spatiotemporal variability to better understand the regional and temporal (e.g., seasonal) dynamics of the dust radiative impact.

The goal of this study is to assess the extent to which Asian dust can impact the PAR and surface radiative energy balance considering the regional specifics of Asian dust properties and spectral surface albedo in the dryland ecosystems of East Asia. The specific objectives are to (1) compute and examine the size- and composition-dependent spectral (i.e., from the UV to the IR) behavior of Asian dust optical characteristics, (2) determine the spectral surface albedo of the major dryland ecosystems in East Asia that are frequently affected by dust transport, (3) examine the dust radiative impact on the total PAR, diffuse fraction of PAR, and surface energy balance in these ecosystems, and (4) explore implications of the dust radiative impact on the ecosystem functioning using several light use efficiency models. Our approach is to perform a comprehensive one-dimensional radiative transfer modeling constrained by ground-based and satellite observations of dust aerosol and land surface properties.

The paper is organized as follows. Section 2 introduces the data and methodology used in this study. Analysis of computed dust spectral optical properties is presented in Section 3. This section also presents our results of the dust impact on the PAR and surface radiative balance in different dryland ecosystems and assessments of associated changes in the ecosystem LUE and photosynthetic rates. Section 4 summarizes our major findings and addresses the implications.

2. Methodology and Data

We used a one-dimensional radiative transfer code SBDART (Santa Barbara DISORT Atmospheric Radiative Transfer, [31]) to compute radiative fluxes. SBDART solves the radiative transfer equation in a vertically inhomogeneous plane-parallel atmosphere taking into account scattering, absorption, and emission by major gases and aerosols. The SBDART code was modified to allow for the incorporation of a new module to treat the aerosol vertical profile and spectral dust optical characteristics that were computed in this study (see Section 3.1). In addition, spectral surface albedos were reconstructed for the dominant dryland ecosystems in East Asia (see Section 2.2) and incorporated into SBDART. Radiative transfer calculations were performed for cloud-free conditions with a spectral resolution of $0.05\,\mu m$ in the SW and $20\,cm^{-1}$ in the LW. Spectral radiative fluxes were integrated over the certain wavelength intervals to calculate SW, LW, and PAR fluxes at the surface. We also computed and analyzed the diffuse component of PAR (PAR_{dif}), the diffuse fraction (α) of PAR, and the surface radiative balance (SRB) as follows:

$$PAR_{dif} = PAR - PAR_{dir}, \qquad (1)$$

$$\alpha = \frac{PAR_{dif}}{PAR}, \qquad (2)$$

$$SRB = \left(SW_{dn} - SW_{up}\right) + \left(LW_{dn} - LW_{up}\right), \qquad (3)$$

where SW_{dn}, SW_{up}, LW_{dn}, and LW_{up} denote the surface downwelling SW flux, upwelling SW flux, downwelling LW flux, and upwelling LW flux, respectively. Here, PAR is the total photosynthetically active radiation incident at the surface and PAR_{dir} is its direct component.

We also computed and analyzed the efficiency of dust radiative forcing in the PAR (δ_{PAR}) and in the surface radiative balance (δ_{SRB}) defined as the differential change in these quantities per change in the dust aerosol optical depth at $0.5\,\mu m$, $\tau_{0.5}$:

$$\delta_{PAR} = \frac{dRF_{PAR}}{d\tau_{0.5}},$$

$$\delta_{SRB} = \frac{dRF_{SRB}}{d\tau_{0.5}}, \qquad (4)$$

where RF_{PAR} is the radiative forcing in PAR, $RF_{PAR} = PAR_{dust} - PAR_{clean}$, and RF_{SRB} in the radiative forcing in the surface radiative balance, $RF_{SRB} = SRB_{dust} - SRB_{clean}$. Here, the subscript "clean" denotes clean (aerosol-free) atmospheric condition.

2.1. Selection of Mineralogical Composition and Particle Size Distributions Representative of Asian Dust. We used a Mie code to compute the dust optical characteristics over the wide spectral range (from the UV to thermal IR) that is required for this study. As a necessity, dust particles are assumed to be spheres. A number of studies have demonstrated the validity of the spherical shape assumption in radiative flux calculations. For instance, Fu et al. [32] showed that this assumption caused less than 5% error in radiative fluxes compared to the spheroidal-shape approximation. Yi et al. [33] reported the 5–10% error in surface radiative fluxes over land by comparing the results for spheres versus ellipsoids.

The dust composition is represented by a mixture of quartz, calcite, and clay-iron oxide aggregates based on recent measurements of the size-resolved mineralogical composition of Asian dust [25, 26]. Quartz and calcite both have negligible absorption in the SW but exhibit significant absorption in the LW [2]. Clays are often aggregated with iron oxides in such a way that these aggregates have much higher light absorption than individual minerals. In particular, illite is found to be the most abundant type of clay in Asian dust, while goethite and hematite are two most important iron oxides [25]. The clay-iron oxide aggregates considered in our modeling are illite-hematite (IH) and illite-goethite (IG) following Lafon et al. [25]. The effective refractive indices of the aggregates were computed using the Bruggeman approximation [2].

Past studies show that mineralogical composition varies with particle size. We used measurements reported by Lafon et al. [25] to constrain the composition of fine and coarse particle size modes. The number fractions of quartz, calcite, and iron oxide-clay aggregates in the fine mode are 16%, 25%, and 59%, respectively, and they are 28%, 29%, and 43% in the coarse mode. In both size modes, IG is assumed to constitute 70% of the aggregates (IG plus IH). An important factor that can significantly affect the light absorption by dust is the volume fraction (ν) of iron oxides (in this case hematite or goethite) in the aggregates. Here, we use the values representative of Asian dust as recommended by Lafon et al. [25], with ν being equal to 3.0% and 6.7% for the fine and coarse modes, respectively.

Selection of representative dust particle size distributions was performed by examining measurements from several AERONET sites located in East Asia, including Dunhuang, Inner Mongolia, Yulin, and Beijing. These sites are located

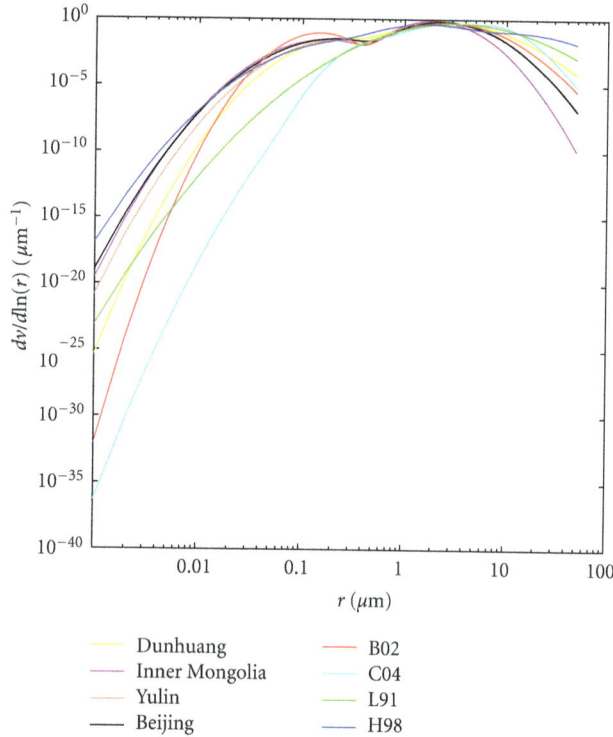

FIGURE 1: Normalized dust volume size distributions for the dust cases considered in this study (see text for details). The associated size distribution parameters are given in Table 1.

TABLE 1: Parameters of dust volume size distributions from AERONET sites in East Asia (Dunhuang, Inner Mongolia, Yulin, and Beijing) and past studies. The lognormal parameters shown are volume median radius (μm), geometric standard deviation, and mass fraction (%).

Size Distribution	Source	Fine mode(s)		Coarse mode
Dunhuang	AERONET	0.254 μm		2.561 μm
		1.697		2.024
		4.0%		96.0%
Inner Mongolia	AERONET	0.169 μm		1.932 μm
		1.757		1.623
		3.8%		96.2%
Yulin	AERONET	0.235 μm		2.23 μm
		1.786		1.758
		3.9%		96.1%
Beijing	AERONET	0.209 μm		2.331 μm
		1.815		1.736
		4.9%		95.1%
B02	[36]	0.149 μm		2.538 μm
		1.52		1.84
		9.1%		90.9%
C04	[37]	0.53 μm	2.751 μm	7.099 μm
		1.46	1.85	1.50
		1.8%	69.4%	28.8%
L91	[38]	3.228 μm		
		2.20		
		100%		
H98	[24]	0.267 μm	1.648 μm	11.02 μm
		1.95	2.00	2.15
		3.4%	76.1%	20.5%

along the transport routes of dust originating from the Taklamakan and Gobi deserts. Because these stations are located at different distances from the dust sources, selected aerosol size distributions help to examine how dust optical properties can change during transport and implications to the dust radiative impact.

AERONET size distributions are retrieved from measurements of the sun and sky radiances and reported in terms of the standard parameters of a bimodal lognormal function including the volume median radius and geometric standard deviation [36]. AERONET also measures the aerosol optical depth (τ) at several wavelengths. These measurements are used to derive the Ångström exponent (β). Given that dust events are commonly associated with relatively large aerosol optical depth and the presence of coarse particles, we selected several representative dust size distribution cases by examining daily averaged $\tau_{0.5}$ and the Ångström exponent during spring 2001, when the AERONET sites in East Asia provided intensive observations as part of the ACE-Asia field campaign [18]. The β is inversely related to particle size such that $\beta < 0.5$–0.7 is often used to identify dust events.

For the considered time period, we found that the AERONET sites frequently showed high $\tau_{0.5}$ and low β, and the retrieved size distributions contained large amounts of coarse particles. The relative proportion of the fine and coarse modes, however, showed significant temporal variations at all sites. To address this observed dynamics in aerosol size spectra, for our modeling we selected four representative cases shown in Table 1. In addition, we considered four dust

size distributions from past studies. B02 denotes the size distribution averaged from 8-year AERONET data at the Bahrain site, which was suggested as a representative size distribution of dust aerosol [36]. C04 was measured by Clarke et al. [37] during the ACE-Asia campaign on a strong dust event. L91 is the single-mode particle size distribution of d'Almeida [38], which is included in the OPAC library [24]. L91 is thought to represent the long-range transported dust. We also considered another dust size distribution from OPAC consisting of three size modes (hereafter H98). Table 1 presents all the considered size distributions in terms of their parameters of the lognormal size distribution expressed as

$$\frac{dV(r)}{d(\log r)} = \sum_j \frac{V_j}{\sqrt{2\pi} \log \sigma_j} \exp \frac{\left(\log r - \log r_{0v,j}\right)^2}{2\left(\log \sigma_j\right)^2}, \quad (5)$$

where j denotes the jth size mode with the volume concentration (V_j), volume median radius ($r_{0v,j}$), and geometric standard deviation (σ_j).

These size distributions are further compared in Figure 1. Various similarities and differences are apparent. Some

Impact of Asian Dust Aerosol and Surface Albedo on Photosynthetically Active Radiation and Surface Radiative Balance in Dryland Ecosystems

161

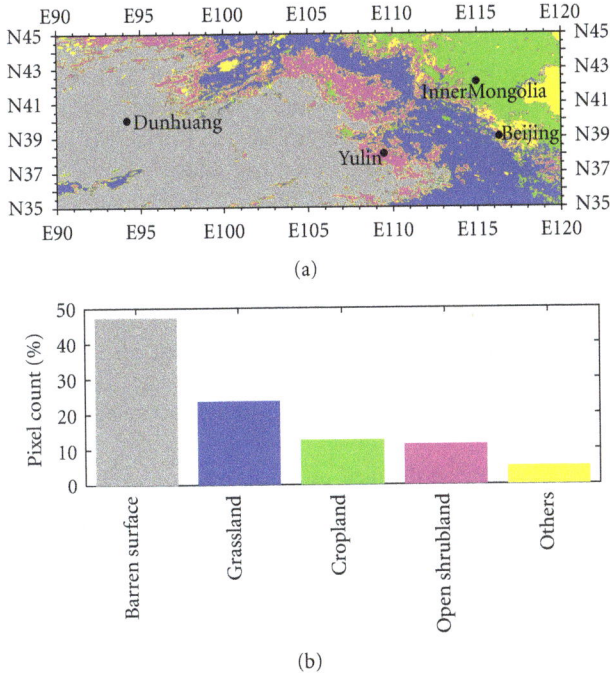

FIGURE 2: (a) MODIS land cover map of the study domain. The locations of four AERONET sites (Dunhuang, Inner Mongolia, Yulin, and Beijing) are also shown. (b) Dominant ecosystem types identified from MODIS land cover statistics: barren surface, cropland, grassland, and open shrubland.

differences in measured size distributions could be due to the variability in size spectra controlled by the dust emission and transport processes, sedimentation of large particles, and mixing of dust with other types of aerosols. For example, at the Beijing site, dust aerosol can be mixed with fine particles originating from urban and industrial sources, so that the size distribution has a larger fine mode than at other sites (see Table 1). B02 has a larger fine mode than the size distributions from the Asian AERONET sites. This might be due to the multiyear averaging or because of an actual difference in dust sizes between Persian Gulf and East Asia. In addition, considered size distributions were derived by different means, either from the limited sampling at local sources, such as C04 and OPAC, or from column-averaged optical inversion, such as AERONET.

We first computed the spectral optical characteristics of each mineral species in each size mode of a certain size distribution, including the normalized extinction $K^*_{\text{ext},i,j}(\lambda)$ and scattering $K^*_{\text{sc},i,j}(\lambda)$ coefficients and asymmetry parameter $g_{i,j}$, where i denotes the ith mineral species and j denotes the jth size mode. Then, for the jth mode, the normalized extinction coefficient is calculated by summing up the normalized extinction coefficient of each species weighted by its number fraction, $f_{i,j}$:

$$K^*_{\text{ext},j}(\lambda) = \sum f_{i,j} K^*_{\text{ext},i,j}(\lambda). \tag{6}$$

For instance for the fine mode, (6) is

$$K^*_{\text{ext},f} = f_{\text{cal},f} K^*_{\text{cal},f} + f_{\text{qtz},f} K^*_{\text{qtz},f} + f_{\text{IG},f} K^*_{\text{IG},f} + f_{\text{IH},f} K^*_{\text{IH},f}, \tag{7}$$

where the subscripts cal, qtz, IG, and IH denote calcite, quartz, IG aggregate, and IH aggregate, respectively. $K^*_{\text{ext},j}(\lambda)$ $(\text{km}^{-1}/\text{cm}^{-3})$ is then weighted by the number concentration N_j of each mode to give the extinction coefficient of the dust mixture:

$$K_{\text{ext}}(\lambda) = \sum N_j K^*_{\text{ext},j}(\lambda). \tag{8}$$

The $K_{\text{ext}}(\lambda)$ can also be expressed in terms of the particle mass concentration. The N_j is related to the total mass concentration (M) as $N_j = m_j M / M^*_j$. Here m_j is the mass fraction of the jth mode and $M^*_j = 4/3 \pi \rho r^3_{0v,j} \exp(-9(\log \sigma_j)^2/2)$ $(\mu\text{g m}^{-3}/\text{cm}^{-3})$, where $r_{0v,j}$ is the volume median radius (see (5)) and ρ is the particle density (in our case, 2.5 g cm^{-3}). The scattering coefficient $K_{\text{sc}}(\lambda)$ can be calculated in a similar fashion, so the single scattering albedo ω_0 of the dust mixture is

$$\omega_0(\lambda) = \frac{K_{\text{sc}}(\lambda)}{K_{\text{ext}}(\lambda)}. \tag{9}$$

By performing the Mie calculations for the selected dust size distributions (shown in Table 1) and comparing the results to the aerosol optical depth measured at AERONET sites, we selected $M = 250$, 500, and 750 μg m^{-3} to represent low, moderate, and high dust loadings, respectively.

The computed spectral optical characteristics of Asian dust were then incorporated into the SBDART code. The dust vertical profile was specified based on the CALIPSO (Cloud-Aerosol Lidar and Infrared Pathfinder Satellite Observations) lidar data. We examined the CALIPSO vertical feature mask during spring 2007 in East Asia to select the representative cases. We found that Asian dust aerosol often extends from the ground up to 4 km in the dust source regions and downwind, although dust layers aloft reaching up to 8 or 9 km were also observed. Here we consider two different profiles: one with a uniform dust layer between 0 and 4 km, (hereafter the mixed layer) and another one with an elevated layer between 5 and 9 km (hereafter the elevated layer).

2.2. Reconstruction of Spectral Surface Albedo. We used the MODIS land products to obtain the spectral surface albedo for the dryland ecosystems that are affected by the dust in East Asia. We first examined the MODIS (Moderate Resolution Imaging Spectroradiometer) CMG (Climate Modeling Grid) land cover product (MOD12C1), which contains fractions of each IGBP (International Geosphere-Biosphere Program) land type in 1 km resolution pixels. These fractions are used to identify pure (>95% coverage) land cover grids for each land type. Figure 2(a) shows a land cover map of the region (N35°–N45°, E90°–E120°) for which we identified four dominant land types: barren surface, cropland, grassland, and open shrubland (see Figure 2(b)).

To assign the albedo to different ecosystems, the land cover is collocated with the MODIS CMG albedo (MCD43C)

(a)

(b)

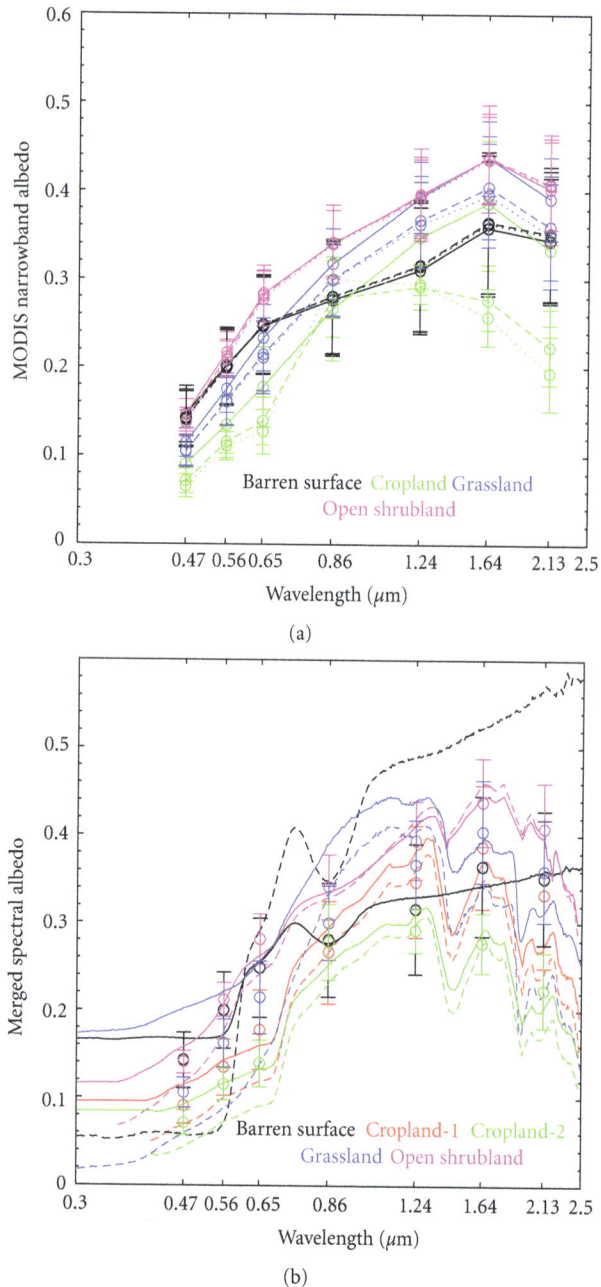

FIGURE 3: (a) Mean value (unfilled circles) and standard deviation (error bars) of MODIS narrowband albedo for different dryland ecosystems (barren surface, cropland, grassland, and open shrubland). Data are averaged for the production periods of April 23 (solid curve), May 1 (dash curve), and May 9 (dotted curve) in 2001. (b) Reconstructed spectral surface albedo (solid curves) by merging MODIS albedo with the USGS spectroscopy dataset (dotted curves).

data. The MODIS CMG albedo is generated every 8 days on a geographic latitude/longitude projection at 0.05 degree resolution [29]. The CMG albedo is reported at seven narrowbands centered at 0.47, 0.56, 0.65, 0.86, 1.24, 1.64, and 2.13 μm. It consists of white-sky albedo and black-sky albedo that represent contributions from the diffuse and direct solar

radiation components, respectively. The white-sky albedo is used in our calculations in accordance with the Lambertian surface assumption in SBDART. Pixels with best retrieval quality were selected to compute a spatial mean value and standard deviation of the surface albedo during the growth season (April 23–31, May 1–8, and May 9–17 in 2001). Figure 3(a) shows that albedo values of barren surface and open shrubland change little with time. The cropland albedo exhibits the largest temporal variability due to the vegetation phenology and land management. This is consistent with previous findings by Gao et al. [29]. For our analyses, we selected five cases of the surface albedo: cropland (hereafter cropland-1) for April 23–31, and barren surface, grassland, open shrubland, and cropland (hereafter cropland-2) for May 1–8.

To provide the detailed spectral dependence of surface albedo, we merged the MODIS narrowband albedo with the USGS spectroscopy dataset. The USGS spectroscopy library (http://speclab.cr.usgs.gov/spectral.lib06/ds231/datatable .html) provides detailed wavelength dependence of albedo for various land targets. This allows expansion of the MODIS narrowband albedo across the entire solar spectrum. To do so, we matched MODIS land types with corresponding land targets from the spectroscopy dataset. Then, we selected the spectroscopy data that best relate to MODIS albedo in the seven narrowbands and applied a least-square fitting to calculate values of the corresponding spectral surface albedo. The resulting spectral surface albedos serve as input into SBDART and are shown in Figure 3(b). These albedos capture the spectral behavior of MODIS narrowband albedo and contain the detailed spectral information from the USGS spectroscopy dataset.

3. Results

3.1. Examination of Spectral Optical Characteristics of Asian Dust. Using the size distributions presented in Table 1, we computed the dust spectral optical characteristics from the UV to the IR. Figures 4(a) and 4(b) show the extinction coefficient (K_{ext}) for the dust loading $M = 250\,\mu\mathrm{g\,m^{-3}}$ and the single scattering albedo (ω_0) in the SW. For comparison, we also show the OPAC bulk dust optical characteristics (denoted by H98). All shown cases have the same composition, except the Inner Mongolia_agg dust and H98, so the strong influence of size distribution on the magnitude and spectral behavior of K_{ext} and ω_0 in the solar and IR (not shown) is apparent. The Inner Mongolia_agg dust has the same size distribution as Inner Mongolia (see Table 1) but consists only of clay-iron oxide aggregates. This case, associated with the largest light absorption compared to other Asian dust cases shown in Figure 4 helps to demonstrate the influence of the mineralogical composition on dust optics. This composition difference has little effect on K_{ext}, but it causes a significant decrease in ω_0, for example, from 0.93 to 0.84 at 0.5 μm with even larger decrease across the PAR region towards the shorter wavelength. Comparing to our Asian dust cases, the OPAC bulk dust model (H98) has relatively high K_{ext} values but the lowest ω_0.

Impact of Asian Dust Aerosol and Surface Albedo on Photosynthetically Active Radiation and Surface Radiative
Balance in Dryland Ecosystems

163

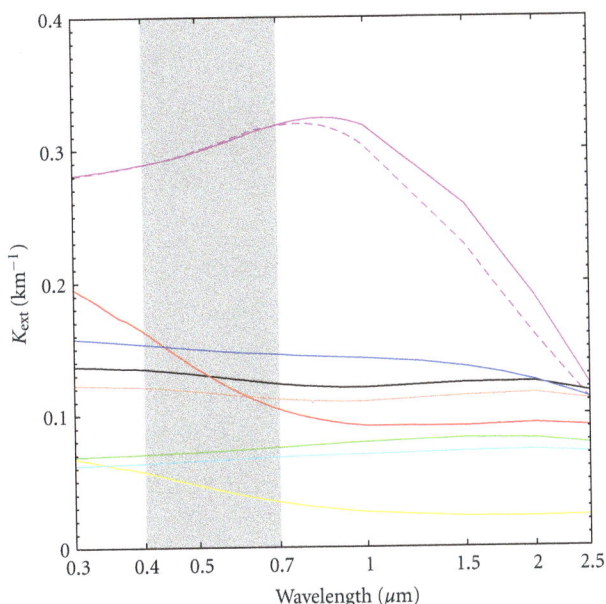

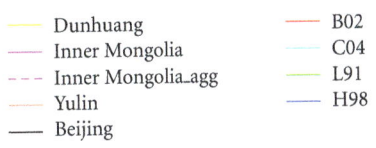

Dunhuang	B02
Inner Mongolia	C04
Inner Mongolia_agg	L91
Yulin	H98
Beijing	

(a)

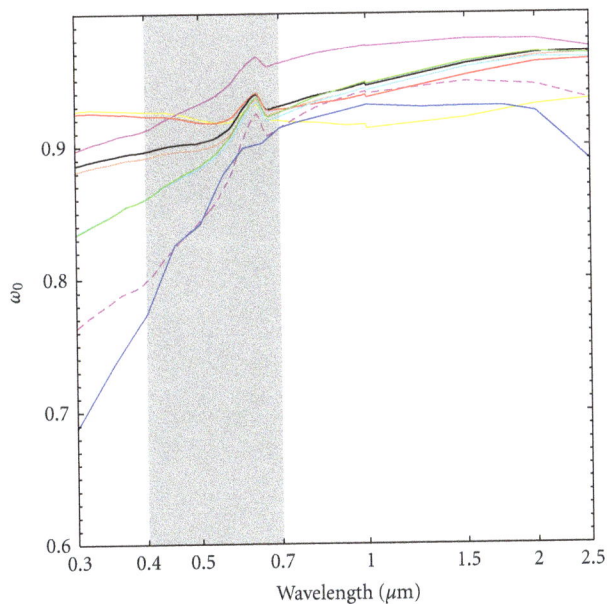

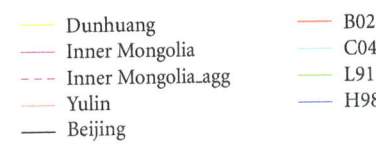

Dunhuang	B02
Inner Mongolia	C04
Inner Mongolia_agg	L91
Yulin	H98
Beijing	

(b)

FIGURE 4: (a) Extinction coefficient K_{ext} for dust loading $M = 250\,\mu g\,m^{-3}$, and (b) single scattering albedo (ω_0) computed for Asian dust and OPAC bulk dust (H98) cases. Shaded areas highlight the PAR spectral region.

Even for the same composition, the dust ω_0 strongly varies among the considered size distributions, for example, ω_0 ($0.5\,\mu m$) ranges from 0.88 to 0.93. These values of ω_0 ($0.5\,\mu m$) are comparable to those reported by past studies for Asian dust-laden environments: 0.88 ± 0.07 ($0.55\,\mu m$) at an urban site in Seoul, Republic of Korea [39], 0.919 ± 0.056 at Gosan [40], 0.957 ± 0.031 ($0.55\,\mu m$) from the in situ measurements during ACE-Asia [41], and 0.97 ± 0.01 ($0.55\,\mu m$) from the aircraft measurements of an Asian dust plume over the Pacific [42].

The large spread of K_{ext} values at $0.5\,\mu m$ is clearly seen in Figure 4(a). One important implication is that, for a given dust loading, this results in large differences in the aerosol optical depth, $\tau_{0.5}$. Figure 5 shows $\tau_{0.5}$ computed for three different dust loads for all dust cases shown in Figure 4(a). Significant differences seen in $\tau_{0.5}$ values suggest that caution must be exercised in interpreting linkages between $\tau_{0.5}$ and dust loading, which will be further addressed below.

In terms of the spectral behavior, K_{ext} varies slightly with wavelength in the SW, except for the Dunhuang, Inner Mongolia, and B02 dust cases, which are associated with very small Ångström exponents. For the Inner Mongolia case, K_{ext} increases with wavelength up to $0.8\,\mu m$ and rapidly decreases in the near-IR, while for Dunhuang and B02 cases, K_{ext} decreases with wavelength up to $1.0\,\mu m$ and then levels off. The value of ω_0 increases with wavelength in the PAR region, except for the Dunhuang and B02 cases. In the latter cases, ω_0 slightly decreases with wavelength in the PAR region and then increases in the near-IR. The spike at $0.66\,\mu m$ is due to the imaginary part of the refractive index of goethite. Note that ω_0 computed for considered Asian dust cases has spectral behavior similar to that of Saharan dust reported by Otto et al. [43] (see their Figure 6).

Overall, examination of Figure 4 clearly shows that both K_{ext} and ω_0 strongly depend on the presence of both fine and coarse modes. This finding is in agreement with past studies (e.g., [43]) and reinforces the need for taking into account the broad range of particle sizes and having realistic representation of the fine/coarse mode ratio.

3.2. Impact of Dust on Total and Diffuse PAR. To assess the extent to which dust can affect the total PAR and its partitioning into direct and diffuse components, we performed radiative transfer modeling considering different combinations of dust optical cases and dryland ecosystems. PAR is absorbed by green vegetation and converted to biomass through photosynthesis. In ecological and land surface models, PAR is a key factor controlling the biophysical processes that govern photosynthesis and stomatal regulation of water, energy, and biogeochemical cycles. Figure 6(a) shows how PAR varies with $\tau_{0.5}$ for the considered dust cases. In each case, results are shown for three $\tau_{0.5}$ values corresponding to dust loadings of 250, 500, and $750\,\mu g\,m^{-3}$. These results are for the mixed layer dust profile and sun elevation angle of 90°.

Under clean (dust-free) conditions, PAR varies from 490.0 to $495.6\,Wm^{-2}$ among the analyzed ecosystems (surface albedos). Note that Figure 6(a) only shows the results for cropland-2 and grassland, which represent the lowest and

164

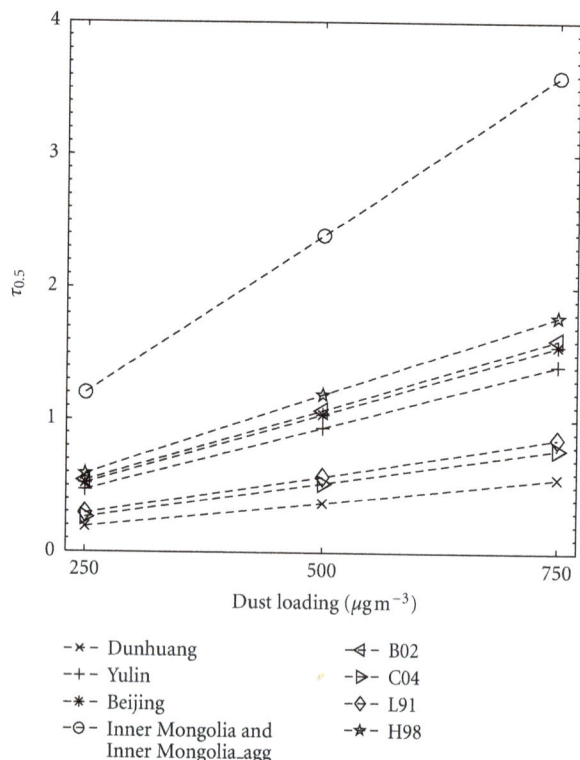

FIGURE 5: Dust optical depth at 0.5 μm ($\tau_{0.5}$) computed for all dust cases at loadings of $M = 250, 500,$ and $750\,\mu g\,m^{-3}$.

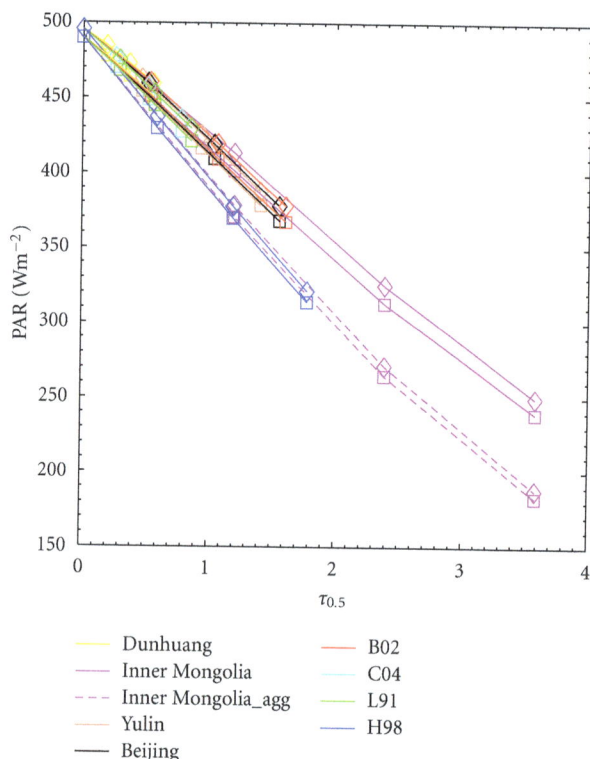

(a)

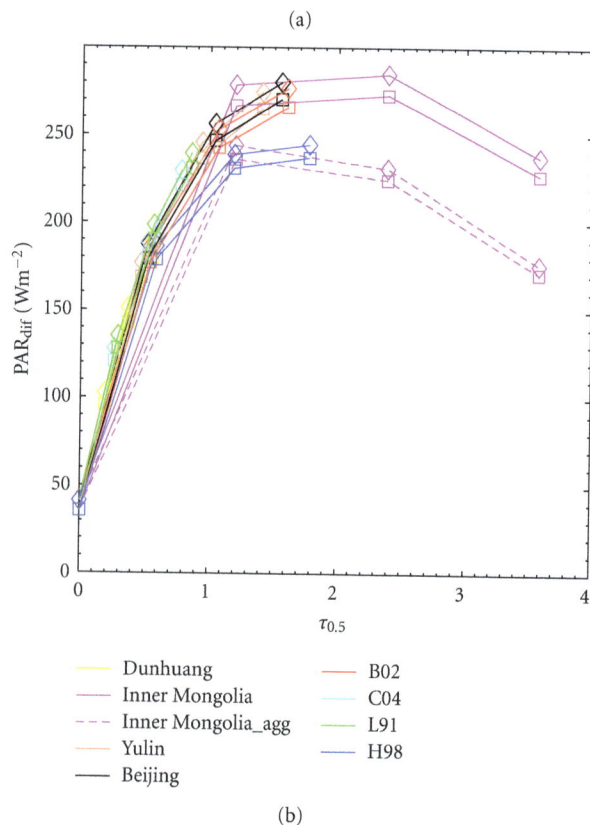

(b)

FIGURE 6: (a) Surface downwelling PAR as a function of dust optical depth ($\tau_{0.5}$); (b) same as (a) but for the diffuse PAR component (PAR$_{dif}$). Squares denote the use of cropland-2 surface albedo, and diamonds are for grassland. Sun elevation angle is 90°. Dust vertical profile is for mixed layer case.

highest values of surface albedo, respectively. For a given dust loading M, the case with the Inner Mongolia size distribution gives the highest K_{ext} in the PAR region and hence the largest τ, causing the largest reduction in PAR compared to other cases. The Dunhuang size distribution gives the lowest K_{ext} values and hence the highest available PAR. For instance, for $M = 250\,\mu g\,m^{-3}$ and cropland-2 albedo, the reduction in PAR is $12.6\,Wm^{-2}$ for Dunhuang ($\tau_{0.5} = 0.19$) and $88.9\,Wm^{-2}$ for Inner Mongolia ($\tau_{0.5} = 1.2$), while the PAR reduction is $119.9\,Wm^{-2}$ for Inner Mongolia_agg. The latter indicates that the composition change alone contributes to a reduction in PAR of $31.0\,Wm^{-2}$. Due to its low ω_0 values and relatively high K_{ext}, H98 dust causes a larger reduction in PAR than the other dust cases for the same loading, with the exception of the Inner Mongolia and Inner Mongolia_agg dust.

As seen in Figure 6(a), PAR is nearly linearly related to $\tau_{0.5}$ that supports the validity of using the dust forcing efficiency (δ_{PAR}) (see (4)). The δ_{PAR} mainly depends on the dust ω_0 and to a lesser extent on the surface albedo. Computed δ_{PAR} values range from $-67.7\,Wm^{-2}\,\tau_{0.5}^{-1}$ for the Inner Mongolia dust case to $-82.2\,Wm^{-2}\,\tau_{0.5}^{-1}$ for the L91 case over cropland-2. The dust vertical profile is found to have a negligible effect on δ_{PAR}.

Our modeling results are in good agreement with observations. In particular, we used radiation measurement data reported by Bush and Valero [44] in dust-laden conditions at Gosan, Republic of Korea, to compare with our modeling results. Based on their data, we estimated an efficiency δ_{PAR}

Impact of Asian Dust Aerosol and Surface Albedo on Photosynthetically Active Radiation and Surface Radiative Balance in Dryland Ecosystems

165

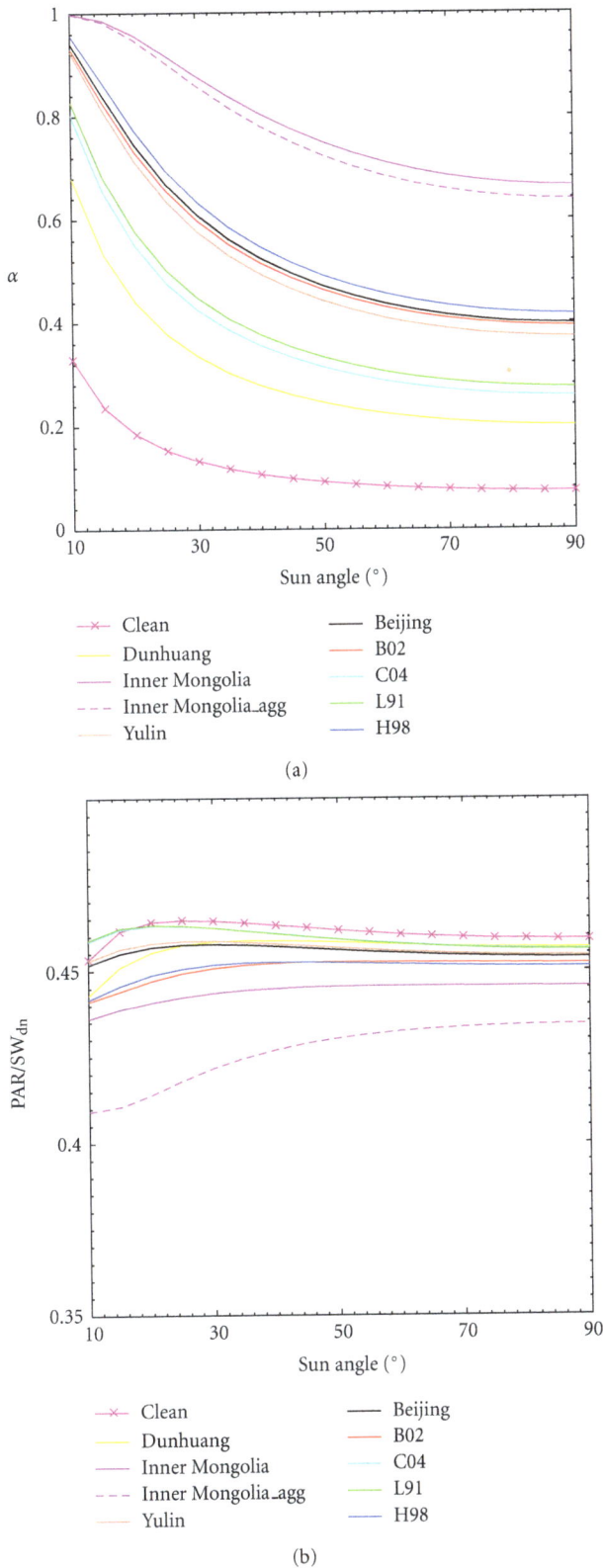

FIGURE 7: (a) Diffuse fraction (α) of PAR, and (b) the ratio of PAR to downwelling shortwave radiation, PAR/SW$_{dn}$, as a function of sun elevation angle. Dust loading is $M = 250\,\mu g\,m^{-3}$. Dust vertical profile is for the mixed layer case. The surface albedo is for the cropland-2 case.

of $-93.6 \pm 12.9\,Wm^{-2}\,\tau_{0.5}^{-1}$. This value is slightly higher but in general agreement with our results.

As demonstrated in past studies, diffuse PAR (PAR$_{dif}$) is an important factor in assessing the plant gross photosynthetic rate due to the fact that PAR$_{dif}$ is often associated with higher light use efficiency than PAR$_{dir}$. Figure 6(b) shows that PAR$_{dif}$ drastically increases in the presence of dust, although changes in PAR$_{dif}$ differ between dust cases. For $M = 250\,\mu g\,m^{-3}$, PAR$_{dif}$ reaches $95.3\,Wm^{-2}$ for the Dunhuang case and $266.5\,Wm^{-2}$ for the Inner Mongolia case over cropland-2. These differences in PAR$_{dif}$ changes are mainly due to the large difference in the dust optical depth between dust cases (see Figure 5). The compositional effect is seen by comparing the Inner Mongolia with Inner Mongolia_agg cases. The former causes $30.6\,Wm^{-2}$ more PAR$_{dif}$ than the latter due to the low ω_0 associated with Inner Mongolia_agg case. Overall, PAR$_{dif}$ tends to increase when M increases from 250 to $750\,\mu g\,m^{-3}$, except for Inner Mongolia and Inner Mongolia_agg cases, for which PAR$_{dif}$ begins to decrease when M increases from 500 to $750\,\mu g\,m^{-3}$. When the dust profile was changed to the elevated-layer case, PAR$_{dif}$ exhibited negligible changes.

The diurnal variation of aerosol loading and solar radiation can affect the ecosystem carbon uptake on a daily basis. To this end, we examined the solar angle dependence of the diffuse fraction of PAR (α, see (2)). Figure 7(a) shows that α remains low ($\sim$10%) during most of the day under dust-free conditions. Values of α become significantly higher in the presence of dust, especially at low sun elevation angles (measured as the angle above horizon). Here, we show the results for $M = 250\,\mu g\,m^{-3}$ and mixed layer dust profile. For all dust cases, α decreases with increasing sun angle before leveling off around 90° (i.e., local noon). For a given dust loading, considered dust cases have very different optical depth values and, as a result, significant differences in α are clearly seen in Figure 7(a). For instance at 90°, α ranges from 20.0% for the Dunhuang case to 66.4% for the Inner Mongolia case. Due to solely the difference in ω_0, α is reduced down to 63.7% for the Inner Mongolia_agg case compared to the Inner Mongolia case. We also found that α depends weakly (within 1.2%) on the ecosystem type (i.e., surface albedo), but the sensitivity increases with dust loading due to the effect of multiple scattering.

Because of the lack of direct measurements of PAR, the common way to estimate its value is from measurements of downwelling SW radiation, SW$_{dn}$ [45]. In the models, PAR is often computed as a fraction of SW$_{dn}$. Although the PAR/SW$_{dn}$ ratio remains fairly constant for a given location at daily or longer time scales, it is sensitive to the presence of clouds and aerosols [45]. Clouds generally increase the PAR/SW$_{dn}$ ratio due to the fact that clouds absorb much more in the near-IR than in the PAR spectrum. Aerosols can affect this ratio in a more complicated way, depending on the spectral dependence of the aerosol optical characteristics [45]. Figure 7(b) shows that the PAR/SW$_{dn}$ ratio remains nearly constant at 0.45 ($\pm$4%) for clean and dusty conditions, except for the Inner Mongolia_agg case. This is due to the fact that both the extinction coefficient and single scattering albedo of the Inner Mongolia_agg case have contrasting

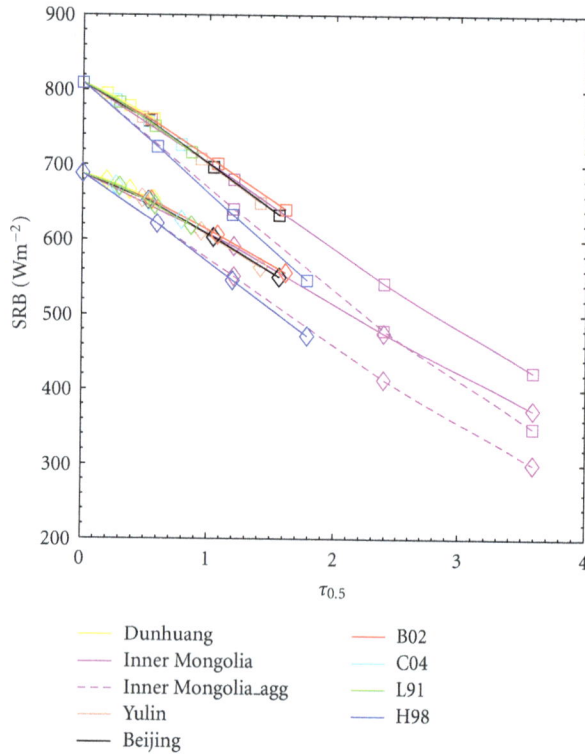

FIGURE 8: Surface radiative balance (SRB) over the cropland-2 (squares) and grassland (diamonds) for considered dust cases (at three dust loadings).

spectral behaviors in the PAR versus near-IR. Our PAR/SW$_{dn}$ value (0.45) is slightly lower than that (0.45–0.50) reported by Frouin and Pinker [45] and is close to the value (0.443–0.445) reported by Jacovides et al. [46] for soot aerosol.

3.3. Impact of Dust on Surface Radiative Balance.

It is well known that dust can reduce the surface net radiation, which controls the available energy for surface turbulent fluxes and affects the surface temperature and boundary layer dynamics [1, 3]. Figure 8 shows surface radiative balance (SRB, (3)) as a function of $\tau_{0.5}$ for the mixed layer profile case. Under clean conditions, SRB ranges from 688.4 Wm^{-2} over grassland to 809.1 Wm^{-2} over cropland-2. We used the same surface emissivity and temperature for all ecosystems, so this difference is chiefly due to differences in surface albedo. For dust loading of $M = 250\,\mu g\,m^{-3}$, the reduction in SRB over cropland-2 ranges from -14.6 Wm^{-2} for the Dunhuang dust case to -128.8 Wm^{-2} for the Inner Mongolia case and to -168.3 Wm^{-2} for the Inner Mongolia_agg case. The forcing efficiency in SRB (δ_{SRB}) ranges from -91.4 Wm$^{-2}\,\tau_{0.5}^{-1}$ for the Dunhuang case to -122.1 Wm$^{-2}\,\tau_{0.5}^{-1}$ for the Inner Mongolia_agg case. As expected, the large differences in the dust forcing in the SRB (RF$_{SRB}$) and δ_{SRB} are due to the differences in the dust optical characteristics.

When the SW and LW radiation components were examined separately, we found that the reduction in SW net radiation ranges from -18.8 Wm^{-2} for the Dunhuang dust case to -140.6 Wm^{-2} for the Inner Mongolia case and to

-182.7 Wm^{-2} for the Inner Mongolia_agg case. The forcing efficiency in the SW ranges from -110.1 Wm$^{-2}\,\tau_{0.5}^{-1}$ for the Dunhuang case to -139.4 Wm$^{-2}\,\tau_{0.5}^{-1}$ for the L91 case. These values are higher than those (-73.0 ± 9.6 Wm$^{-2}\,\tau_{0.5}^{-1}$) reported by Bush and Valero [44]. We also found that the LW positive forcing compensates the SW reduction by about 7.9–26.4%. In the case of elevated layer, the dust LW forcing decreased by 1.7–5.6 Wm^{-2} because of lower temperatures associated with the elevated dust layer.

3.4. Impact of Dust on Vegetation Light Use Efficiency.

In Section 3.2, we demonstrated that Asian dust can exert a substantial impact on the total PAR and its direct/diffuse partitioning. To explore the potential implications for croplands, here we consider several light use efficiency (LUE) models for different crop types (wheat, soybean, and corn) [8, 12, 34, 35, 47].

It is well established that the biomass accumulation is nearly linearly related to PAR intercepted (or absorbed) by the vegetation canopy (APAR) so that the carbon assimilation rate of the canopy (A_c) can be expressed as [48]

$$A_c = LUE * APAR = LUE * fPAR * PAR, \quad (10)$$

where the ratio of A_c to APAR is called the light use efficiency (LUE, mol CO$_2$ (mol APAR)$^{-1}$) and fPAR is the fraction of PAR absorbed by the canopy. fPAR can be readily estimated from the leaf characteristics and is currently operationally retrieved from satellite observations, for instance, MODIS (see http://modis-land.gsfc.nasa.gov/lai.html).

Observations showed that LUE increases with the increasing PAR$_{dif}$ within the canopy [16]. Past studies also demonstrated that LUE increases more or less linearly with the diffuse fraction, α [34, 35, 47]. For the case of wheat canopy with a leaf area index (LAI) of 2.9, Choudhury [34] derived the LUE-α relationship as follows:

$$LUE = 0.0356\alpha + 0.0108. \quad (11)$$

Based on model results, Anderson et al. [35] parameterized LUE as a function of α for both C3 and C4 plants. For soybean (C3 plant):

$$LUE = 0.025[1 + 0.8(\alpha - 0.5)] \quad (12)$$

and for corn (C4 plant):

$$LUE = 0.04[1 + 0.3(\alpha - 0.5)]. \quad (13)$$

Based on the modeling results of Choudhury [47] for a variety of crop and forest canopies in different climatic zones, Roderick et al. [8] derived the following LUE-α relationship:

$$LUE = 0.024\alpha + 0.012. \quad (14)$$

Given that the latter is based on an ensemble of modeling cases for various vegetation types, this LUE model is not for a specific vegetation type but more generic in nature. Although LUE is expressed solely as a function of diffuse fraction of PAR, these LUE models do account (implicitly) for likely

Impact of Asian Dust Aerosol and Surface Albedo on Photosynthetically Active Radiation and Surface Radiative Balance in Dryland Ecosystems

167

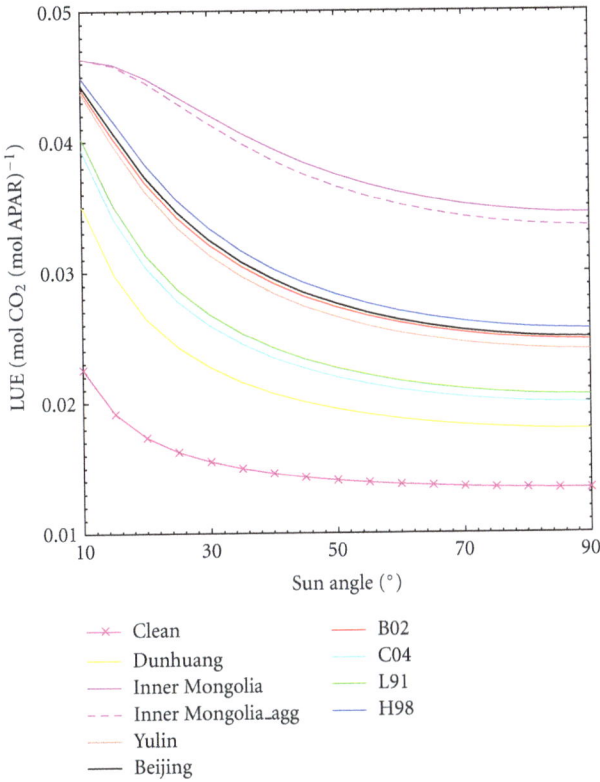

(a)

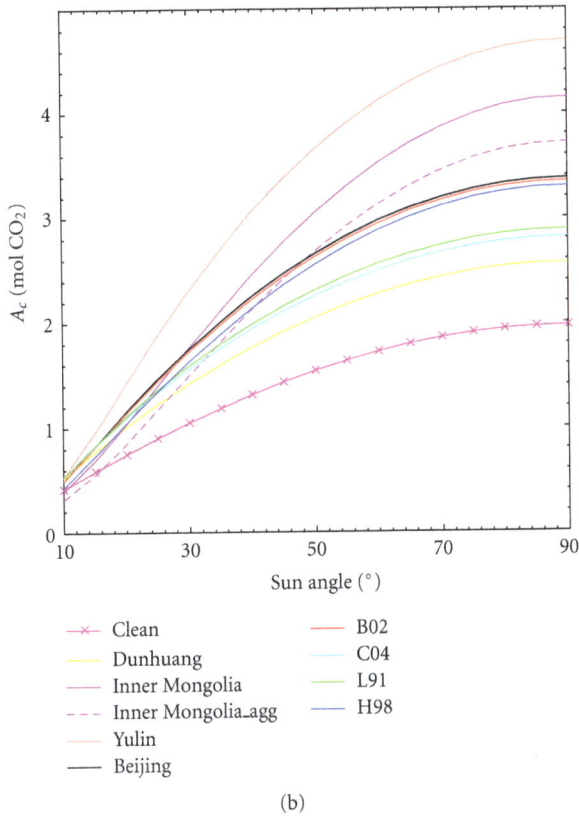

(b)

FIGURE 9: (a) Light use efficiency (LUE), and (b) carbon assimilation rate (A_c) computed with the Choudhury [34] model (for wheat) as a function of sun elevation angle. Dust loading $M = 250\,\mu g\,m^{-3}$. Surface albedo is for the cropland-2 case.

TABLE 2: Comparison of past studies on the aerosol impact on PAR and plant photosynthesis in cloud-free condition.

Study	Data and method	Aerosol (max. $\tau_{0.5}/\omega_0$)[a]	Ecosystem (s)	τ_{ct}
Cohan et al. [12]	Crop models	Urban pollution (0.5/0.9)	Crop	~0.6
Niyogi et al. [15]	Eddy fluxes; AERONET	— (0.8/−)	Forest, crop, grassland	>0.8
Yamasoe et al. [13]	Eddy fluxes; AERONET	Biomass burning (3.0/0.93)	Tropical rainforest	1.5–2.0
Jing et al. [49]	Eddy fluxes; AERONET	— (1.2/−)	Semiarid grassland	Not found, but >1.2

[a]Values in the parentheses are maximum τ and ω_0 at $0.5\,\mu m$.

changes in surface radiative balance, given that the models were developed for specific environmental conditions (temperature, soil moisture, etc.).

Using the LUE models (11)–(14) and our radiation modeling results, we assessed the potential dust impact on the vegetation carbon uptake. Figure 9 presents the LUE and A_c computed for wheat (11) under clean and dusty ($M = 250\,\mu g\,m^{-3}$) conditions. Apparently, LUE has the same dependence on sun elevation angle as α (see Figure 7). As the sun angle increases, LUE decreases and begins to level off when sun angle >70°. In contrast, Figure 9(b) shows that the gross photosynthetic rate A_c increases with the sun angle. Generally, A_c is larger in all dust cases compared to clean conditions, except that at low sun elevation angles when the diffuse fraction is already very high, the reduction in PAR by dust can lead to a decrease in A_c, in this case the Inner Mongolia_agg dust case for sun angle <15°. Although the Inner Mongolia dust case leads to the largest LUE, the largest A_c is associated with the Yulin case. This is due to the fact that the Inner Mongolia case causes a large reduction in PAR, whose effect dominates the increase in α and LUE.

Previous studies reported that the plant gross photosynthesis can reach a maximum at a critical aerosol loading corresponding to a certain optical depth, τ_{ct} (see Table 2). When the aerosol optical depth drops below τ_{ct}, the majority of shaded leaves receive very low sunlight while the sunlit leaves are light saturated. On the other hand, when the aerosol optical depth exceeds τ_{ct}, most sunlight is attenuated such that the plant is light-starved. Using the LUE model of Roderick et al. [8], we calculated A_c as a function of dust loading for all dust cases, as shown in Figure 10. There are noticeable differences in the behavior of A_c: it decreases with increasing dust loading for the Inner Mongolia and Inner Mongolia_agg dust cases, while in the OPAC case, A_c increases with dust loading (up to $M = 500\,\mu g\,m^{-3}$) and then decreases. For all other dust cases, A_c increases with dust loading. These can be explained by the fact that the aerosol impact on the plant carbon uptake depends on a balance between the reduction in PAR and the increase in PAR$_{dif}$. For example, the Inner Mongolia dust case has a much

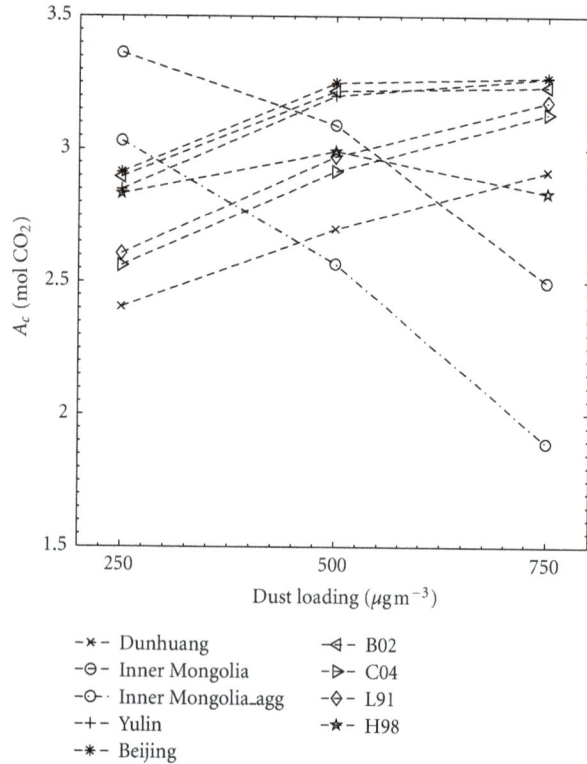

FIGURE 10: Carbon assimilation rate calculated with the LUE model of Roderick et al. [8] as a function of dust loading for all dust cases.

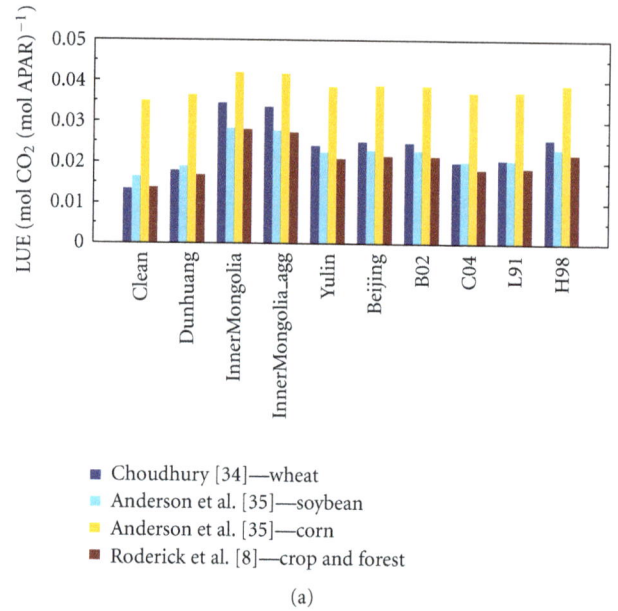

(a)

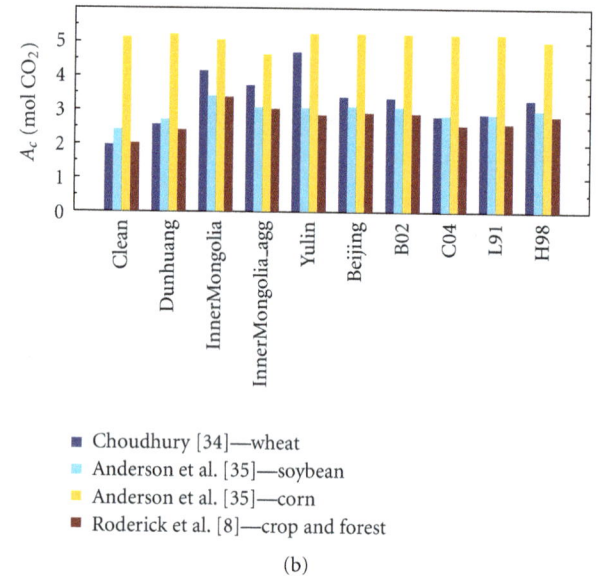

(b)

FIGURE 11: (a) Light use efficiency (LUE), and (b) carbon assimilation rate (A_c) computed for clean and dust conditions (sun elevation angle = 90°) using the LUE models of [8, 34, 35].

larger extinction coefficient than other cases, such that at dust loading $M = 250\,\mu g\,m^{-3}$, resultant dust optical depth exceeds the optimum value for gross photosynthesis.

Figure 11 presents LUE and A_c computed with the LUE models using (11)–(14) at sun elevation angle = 90° for dust loading $M = 250\,\mu g\,m^{-3}$. Due to the fact that corn (C4 plant) has a higher light saturation point and is less sensitive to changes in PAR_{dif} than C3 plants [47], the LUE and A_c values for corn are higher and less sensitive to the dust optical properties compared to other plant types. In addition, for certain crop type A_c peaks at different values depending on the dust case, for example, A_c of soybean and wheat reaches the maximum for the Inner Mongolia (3.4 mol CO_2 (mol APAR)$^{-1}$) and Yulin (5.2 mol CO_2 (mol APAR)$^{-1}$) dust cases, respectively. Thus, our results demonstrate that the dust impact on the PAR and plant photosynthesis depends on both the dust properties and ecosystem type (e.g., C3 versus C4 crops).

4. Conclusions and Discussions

In this study, we investigated the dust impact on the PAR and surface radiative balance (SRB) considering conditions representative of dryland ecosystems in East Asia. The spectral (from the UV to thermal IR) optical characteristics of Asian dust were computed based on representative size-resolved mineralogical composition and remote sensing retrievals of dust size distributions. MODIS narrowband albedo products along with the USGS spectroscopy library data were used to reconstruct the spectral surface albedo of different ecosystems in East Asia. These albedo data and dust optical models were incorporated in the radiative transfer model SBDART, which was used to investigate the dust-induced changes in PAR, and SRB in terms of forcing and its efficiency. The radiative transfer modeling results were analyzed with several LUE models to examine the implications of the dust radiative impact on the plant gross photosynthetic rate.

For a representative Asian dust composition, our results demonstrate significant variations in the optical characteristics in terms of both the magnitude and spectral dependence caused by variations in size distribution. For instance, at $0.5\,\mu m$, the normalized extinction coefficient ranges from 1.9×10^{-4} to $1.2 \times 10^{-3}\,km^{-1}$ $(\mu g\,m^{-3})^{-1}$ and

Impact of Asian Dust Aerosol and Surface Albedo on Photosynthetically Active Radiation and Surface Radiative
Balance in Dryland Ecosystems

169

the single scattering albedo ranges from 0.88 to 0.93 for Asian dust cases (Figure 4). The highest absorption case (Inner Mongolia_agg), which only contains iron-oxide clay aggregates, gives the lowest ω_0 (0.5 μm) of 0.88 and so does the OPAC dust model. However, the OPAC dust has too low ω_0 values across the SW spectrum compared to Asian dust cases. Comparison with the OPAC bulk dust optical model stresses the limitations of this model in representing regional dust optics, in particular, Asian dust. This also demonstrates the need and advantage of representing the dust mineralogy and size distribution covering fine and coarse modes in assessments of the dust radiative impact.

The dust-induced changes in the total PAR, diffuse PAR and SRB are found to exhibit large variations over the dryland ecosystems, depending on the dust optical properties and the surface albedo. The estimated range of the forcing efficiency of Asian dust in SRB is from -68.8 to $-122.1\,\mathrm{Wm^{-2}}\,\tau_{0.5}^{-1}$ (Figure 8), while in the total PAR it ranges from -67.7 to $-82.2\,\mathrm{Wm^{-2}}\,\tau_{0.5}^{-1}$. The OPAC and Inner Mongolia_agg give the largest absolute value (about $-110\,\mathrm{Wm^{-2}}\,\tau_{0.5}^{-1}$ in total PAR) caused by higher absorption in these two cases. They also give the smallest increase in the diffuse component of PAR compared to the Asian dust cases (Figure 6(b)). Similar to other aerosol types, the ratio of total PAR to downwelling shortwave flux remains nearly constant (0.45 $\pm$ 4%) (Figure 7(b)). However, the diffuse faction of PAR exhibits significant variations among considered Asian dust cases (Figure 7(a)).

Using the light use efficiency (LUE) models for several types of crops (wheat, soybean, and corn), we estimated the influence of dust-induced changes in PAR on the plant photosynthesis. We found that the dust impact on the vegetation gross photosynthetic rate is also a strong function of dust optical properties but differs among crop types. The plant photosynthetic rate was enhanced under a low dust loading, but was decreased when dust loading exceeded a certain optimal level. This behavior is consistent with previous studies of other types of aerosols that identified a critical aerosol optical depth. We demonstrate, however, that the critical optical depth depends on both the loading and size distribution of dust. In particular, the relative proportion of fine and coarse modes is a key factor controlling the normalized $K_{ext}(\lambda)$ so that the same dust loading will result in different optical depth depending on the size distribution considered (see Figure 5). Thus, in the case of dust aerosol, both loading and size distribution will need to be considered in determining the optimal regime of plant photosynthesis.

Given that the diffuse radiation fertilization is due to the fact that more scattered sunlight reaches shaded leaves, the extent of this effect on vegetation also depends on LAI. The lower the LAI the less the effect of the dust-enhanced diffuse radiation. Wohlfahrt et al. [49] provided observational evidence by showing that temperate mountain grassland is less sensitive to the diffuse radiation when the green area is low. They suggested that biomes with LAI <2 such as desert shrublands exhibit little sensitivity to diffuse PAR. Jing et al. [50] also showed that semiarid grassland exhibits no fertilization effect to the aerosol-enhanced diffuse PAR that is likely to be due to the low LAI and low light saturation point

of grassland. However, dust can be transported downwind for thousands of kilometers affecting large regions with vegetation having higher LAI values.

In addition to diffuse PAR and dust-induced changes in SRB, there are a number of important factors that can significantly affect the vegetation functioning including photosynthesis, respiration, and transpiration processes. Under aerosol-laden conditions, concurrent variations in leaf/soil temperature and humidity may occur that can amplify the diffuse PAR effect [16]. Specifically, due to less incoming solar radiation, lowering leaf/soil temperature could depress the leaf/soil respiration, while a lower vapor pressure deficit tends to enhance the stomatal conductance and leaf-air exchanges. These environmental changes can exert either significant [13, 14, 17, 49] or negligible [50] effects on the canopy photosynthesis. Under certain conditions, the changes in the environmental factors can overcome the effect of diffuse PAR. Steiner and Chameides [17] showed that under high-irradiance condition, the presence of aerosol reduces the incoming sunlight and leaf temperature down to an optimum level and thus enhances the photosynthesis, in which case the effect of increased diffuse PAR is negligible. Accounting for these different mechanisms will require an earth system framework that couples biosphere with the physical climate system. We suggest that improved representation of dust that takes into account size-resolved composition of fine and coarse modes will be needed to provide more accurate assessments of how dust-induced changes in the radiation regime affect the ecosystem functioning and the role of these processes in overall land-atmosphere interactions.

Acknowledgment

This work was partially funded by the NASA LCLUC Program.

References

[1] K. S. Carslaw, O. Boucher, D. V. Spracklen et al., "A review of natural aerosol interactions and feedbacks within the earth system," *Atmospheric Chemistry and Physics*, vol. 10, no. 4, pp. 1701–1737, 2010.

[2] I. N. Sokolik and O. B. Toon, "Incorporation of mineralogical composition into models of the radiative properties of mineral aerosol from uv to ir wavelengths," *Journal of Geophysical Research D*, vol. 104, no. 8, pp. 9423–9444, 1999.

[3] M. Mallet, P. Tulet, D. Serca et al., "Impact of dust aerosols on the radiative budget, surface heat fluxes, heating rate profiles and convective activity over west africa during march 2006," *Atmospheric Chemistry and Physics*, vol. 9, no. 18, pp. 7143–7160, 2009.

[4] T. Takemura, T. Nakajima, A. Higurashi, S. Ohta, and N. Sugimoto, "Aerosol distributions and radiative forcing over the Asian Pacific region simulated by spectral radiation-transport model for aerosol species (SPRINTARS)," *Journal of Geophysical Research D*, vol. 108, no. 8659, 10 pages, 2003.

[5] J. Huang, Q. Fu, J. Su et al., "Taklimakan dust aerosol radiative heating derived from calipso observations using

the fu-liou radiation model with Ceres constraints," *Atmospheric Chemistry and Physics*, vol. 9, no. 12, pp. 4011–4021, 2009.

[6] A. M. Vogelmann, P. J. Flatau, M. Szczodrak, K. M. Markowicz, and P. J. Minnett, "Observations of large aerosol infrared forcing at the surface," *Geophysical Research Letters*, vol. 30, no. 1655, 4 pages, 2003.

[7] K. M. Markowicz, P. J. Flatau, A. M. Vogelmann, P. K. Quinn, and E. J. Welton, "Clear-sky infrared aerosol radiative forcing at the surface and the top of the atmosphere," *Quarterly Journal of the Royal Meteorological Society*, vol. 129, no. 594, pp. 2927–2947, 2003.

[8] M. L. Roderick, G. D. Farquhar, S. L. Berry, and I. R. Noble, "On the direct effect of clouds and atmospheric particles on the productivity and structure of vegetation," *Oecologia*, vol. 129, no. 1, pp. 21–30, 2001.

[9] L. Gu, D. D. Baldocchi, S. C. Wofsy et al., "Response of a deciduous forest to the mount pinatubo eruption: enhanced photosynthesis," *Science*, vol. 299, no. 5615, pp. 2035–2038, 2003.

[10] W. L. Chameides, H. Yu, S. C. Liu et al., "Case study of the effects of atmospheric aerosols and regional haze on agriculture: an opportunity to enhance crop yields in china through emission controls?" *Proceedings of the National Academy of Sciences of the United States of America*, vol. 96, no. 24, pp. 13626–13633, 1999.

[11] M. H. Bergin, R. Greenwald, J. Xu, Y. Berta, and W. L. Chameides, "Influence of aerosol dry deposition on photosynthetically active radiation available to plants: a case study in the yangtze delta region of china," *Geophysical Research Letters*, vol. 28, no. 18, pp. 3605–3608, 2001.

[12] D. Cohan, J. Xu, R. Greenwald, M. Bergin, and W. Chameides, "Impact of atmospheric aerosol light scattering and absorption on terrestrial net primary productivity," *Global Biogeochemical Cycles*, vol. 16, no. 1090, 12 pages, 2002.

[13] M. Yamasoe, C. Von Randow, A. Manzi, J. Schafer, T. Eck, and B. Holben, "Effect of smoke and clouds on the transmissivity of photosynthetically active radiation inside the canopy," *Atmospheric Chemistry and Physics*, vol. 6, no. 6, pp. 1645–1656, 2006.

[14] T. Matsui, A. Beltrán-Przekurat, D. Niyogi, R. A. Pielke Sr., and M. Coughenour, "Aerosol light scattering effect on terrestrial plant productivity and energy fluxes over the eastern united states," *Journal of Geophysical Research D*, vol. 113, Article ID D14S14, 17 pages, 2008.

[15] D. Niyogi, H. Chang, V. Saxena et al., "Direct observations of the effects of aerosol loading on net ecosystem CO_2 exchanges over different landscapes," *Geophysical Research Letters*, vol. 31, Article ID L20506, 5 pages, 2004.

[16] L. Gu, D. Baldocchi, S. B. Verma et al., "Advantages of diffuse radiation for terrestrial ecosystem productivity," *Journal of Geophysical Research D*, vol. 107, no. 4050, 23 pages, 2002.

[17] A. L. Steiner and W. L. Chameides, "Aerosol-induced thermal effects increase modelled terrestrial photosynthesis and transpiration," *Tellus B*, vol. 57, no. 5, pp. 404–411, 2005.

[18] R. Arimoto, Y. J. Kim, Y. P. Kim et al., "Characterization of asian dust during ACE-Asia," *Global and Planetary Change*, vol. 52, no. 1–4, pp. 23–56, 2006.

[19] I. N. Sokolik, D. Winker, G. Bergametti et al., "Introduction to special section: outstanding problems in quantifying the radiative impacts of mineral dust," *Journal of Geophysical Research D*, vol. 106, no. 16, pp. 18015–18027, 2001.

[20] M. Yoshioka, N. Mahowald, A. Conley et al., "Impact of desert dust radiative forcing on sahel precipitation: relative importance of dust compared to sea surface temperature variations, vegetation changes, and greenhouse gas warming," *Journal of Climate*, vol. 20, no. 8, pp. 1445–1467, 2007.

[21] E. Patterson, "Optical properties of the crustal aerosol-relation to chemical and physical characteristics," *Journal of Geophysical Research*, vol. 86, pp. 3236–3246, 1981.

[22] I. N. Sokolik, A. Andronova, and T. C. Johnson, "Complex refractive index of atmospheric dust aerosols," *Atmospheric Environment A*, vol. 27, no. 16, pp. 2495–2502, 1993.

[23] F. Volz, "Infrared optical constants of ammonium sulfate, sahara dust, volcanic pumice, and flyash," *Applied Optics*, vol. 12, no. 3, pp. 564–568, 1973.

[24] M. Hess, P. Koepke, and I. Schult, "Optical properties of aerosols and clouds: the software package OPAC," *Bulletin of the American Meteorological Society*, vol. 79, no. 5, pp. 831–844, 1998.

[25] S. Lafon, I. N. Sokolik, J. Rajot, S. Caquincau, and A. Gaudichet, "Characterization of iron oxides in mineral dust aerosols: implications for light absorption," *Journal of Geophysical Research D*, vol. 111, Article ID D21207, 19 pages, 2006.

[26] G. R. Jeong and I. N. Sokolik, "Effect of mineral dust aerosols on the photolysis rates in the clean and polluted marine environments," *Journal of Geophysical Research D*, vol. 112, Article ID D21308, 18 pages, 2007.

[27] E. A. Tsvetsinskaya, C. B. Schaaf, F. Gao, A. H. Strahler, and R. E. Dickinson, "Spatial and temporal variability in moderate resolution imaging spectroradiometer-derived surface albedo over global arid regions," *Journal of Geophysical Research D*, vol. 111, Article ID D20106, 10 pages, 2006.

[28] D. Waggoner and I. N. Sokolik, "Seasonal dynamics and regional features of MODIS-derived land surface characteristics in dust source regions of East Asia," *Remote Sensing of Environment*, vol. 114, no. 10, pp. 2126–2136, 2010.

[29] F. Gao, C. B. Schaaf, A. H. Strahler, A. Roesch, W. Lucht, and R. Dickinson, "MODIS bidirectional reflectance distribution function and albedo Climate Modeling Grid products and the variability of albedo major global vegetation types," *Journal of Geophysical Research D*, vol. 110, Article ID D01104, 13 pages, 2005.

[30] A. Roesch, M. Wild, R. Pinker, and A. Ohmura, "Comparison of spectral surface albedos and their impact on the general circulation model simulated surface climate," *Journal of Geophysical Research D*, vol. 107, no. 4221, 18 pages, 2002.

[31] P. Ricchiazzi, S. Yang, C. Gautier, and D. Sowle, "SBDART: a research and teaching software tool for plane-parallel radiative transfer in the earth's atmosphere," *Bulletin of the American Meteorological Society*, vol. 79, no. 10, pp. 2101–2114, 1998.

[32] Q. Fu, T. J. Thorsen, J. Su, J. M. Ge, and J. P. Huang, "Test of mie-based single-scattering properties of non-spherical dust aerosols in radiative flux calculations," *Journal of Quantitative Spectroscopy and Radiative Transfer*, vol. 110, no. 14–16, pp. 1640–1653, 2009.

[33] B. Yi, C. N. Hsu, P. Yang, and S. C. Tsay, "Radiative transfer simulation of dust-like aerosols: uncertainties from particle shape and refractive index," *Journal of Aerosol Science*, vol. 42, no. 10, pp. 631–644, 2011.

[34] B. J. Choudhury, "A sensitivity analysis of the radiation use efficiency for gross photosynthesis and net carbon accumulation by wheat," *Agricultural and Forest Meteorology*, vol. 101, no. 2-3, pp. 217–234, 2000.

[35] M. C. Anderson, J. M. Norman, T. P. Meyers, and G. R. Diak, "An analytical model for estimating canopy transpiration

Impact of Asian Dust Aerosol and Surface Albedo on Photosynthetically Active Radiation and Surface Radiative
Balance in Dryland Ecosystems

171

and carbon assimilation fluxes based on canopy light-use efficiency," *Agricultural and Forest Meteorology*, vol. 101, no. 4,
pp. 265–289, 2000.

[36] O. Dubovik, B. Holben, T. Eck et al., "Variability of absorption
and optical properties of key aerosol types observed in
worldwide locations," *Journal of the Atmospheric Sciences*, vol.
59, no. 3, pp. 590–608, 2002.

[37] A. D. Clarke, Y. Shinozuka, V. N. Kapustin et al., "Size distributions and mixtures of dust and black carbon aerosol in
asian outflow: physiochemistry and optical properties," *Journal of Geophysical Research D*, vol. 109, Article ID D15S09, 20
pages, 2004.

[38] G. d'Almeida, P. Koepke, and E. Shettle, *Atmospheric Aerosols:
Global Climatology and Radiative Characteristics*, A. Deepak,
Hampton, VA, USA, 1991.

[39] J. Jung, Y. J. Kim, K. Y. Lee et al., "Spectral optical properties
of long-range transport Asian dust and pollution aerosols over
Northeast Asia in 2007 and 2008," *Atmospheric Chemistry and
Physics*, vol. 10, no. 12, pp. 5391–5408, 2010.

[40] T. Nakajima, M. Sekiguchi, T. Takemura et al., "Significance
of direct and indirect radiative forcings of aerosols in the East
China Sea region," *Journal of Geophysical Research D*, vol. 108,
no. 8658, 16 pages, 2003.

[41] S. J. Doherty, P. K. Quinn, A. Jefferson, C. M. Carrico, T.
L. Anderson, and D. Hegg, "A comparison and summary of
aerosol optical properties as observed in situ from aircraft,
ship, and land during ACE-Asia," *Journal of Geophysical Research D*, vol. 110, no. 4, Article ID D04201, 35 pages, 2005.

[42] A. D. Clarke, W. G. Collins, P. J. Rasch et al., "Dust and pollution transport on global scales: aerosol measurements and
model predictions," *Journal of Geophysical Research D*, vol.
106, no. 23, pp. 32555–32569, 2001.

[43] S. Otto, M. De Reus, T. Trautmann, A. Thomas, M. Wendisch,
and S. Borrmann, "Atmospheric radiative effects of an in-situ
measured saharan dust plume and the role of large particles,"
Atmospheric Chemistry and Physics, vol. 7, no. 18, pp. 4887–
4903, 2007.

[44] B. C. Bush and F. P. J. Valero, "Surface aerosol radiative
forcing at gosan during the ACE-Asia campaign," *Journal of
Geophysical Research D*, vol. 108, no. 8660, 8 pages, 2003.

[45] R. Frouin and R. T. Pinker, "Estimating photosynthetically
active radiation (PAR) at the earth's surface from satellite
observations," *Remote Sensing of Environment*, vol. 51, no. 1,
pp. 98–107, 1995.

[46] C. Jacovides, F. Tymvios, D. Asimakopoulos, K. Theofilou,
and S. Pashiardes, "Global photosynthetically active radiation
and its relationship with global solar radiation in the eastern
mediterranean basin," *Theoretical and Applied Climatology*,
vol. 74, no. 3-4, pp. 227–233, 2003.

[47] B. J. Choudhury, "Modeling radiation-and carbon-use efficiencies of maize, sorghum, and rice," *Agricultural and Forest
Meteorology*, vol. 106, no. 4, pp. 317–330, 2001.

[48] J. L. Monteith, "Solar radiation and production in tropical
ecosystems," *Journal of Applied Ecology*, vol. 9, pp. 747–766,
1972.

[49] G. Wohlfahrt, A. Hammerle, A. Haslwanter, M. Bahn, U. Tap-
peiner, and A. Cernusca, "Disentangling leaf area and environmental effects on the response of the net ecosystem CO_2
exchange to diffuse radiation," *Geophysical Research Letters*,
vol. 35, Article ID L16805, 5 pages, 2008.

[50] X. Jing, J. Huang, G. Wang et al., "The effects of clouds and
aerosols on net ecosystem CO_2 exchange over semi-arid loess
plateau of northwest china," *Atmospheric Chemistry and Physics*, vol. 10, no. 17, pp. 8205–8218, 2010.

17

The Dynamics of the Skin Temperature of the Dead Sea

Roni Nehorai,[1,2] **Nadav Lensky,**[2] **Steve Brenner,**[1] **and Itamar Lensky**[1]

[1] *Department of Geography and Environment, Bar-Ilan University, 52900 Ramat-Gan, Israel*
[2] *Geological Survey of Israel, 30 Malkhe Israel Street, 95501 Jerusalem, Israel*

Correspondence should be addressed to Itamar Lensky; itamar.lensky@biu.ac.il

Academic Editor: Lian Xie

We explored the dynamics of the temperature of the skin layer of the Dead Sea surface by means of in situ meteorological and hydrographic measurements from a buoy located near the center of the lake. The skin temperature is most highly correlated to air temperature (0.93–0.98) in all seasons. The skin temperature is much less correlated to the bulk surface water temperature in the summer (0.80), when the lake is thermally stratified, and uncorrelated in the winter, when the Dead Sea is vertically mixed. Low correlations were found between the skin temperature and the solar radiation and wind speed in all seasons. The skin, with its low thermal inertia, responds immediately to the atmospheric forcing. Heat fluxes across the sea surface are also presented. The high correlation of skin temperature to air temperature with minimal time lag is a result of the nearly immediate response of the thin skin layer to the surface heat fluxes, primarily the sensible heat flux.

1. Introduction

Sea surface temperature (SST) is a critically important parameter in the study of ocean-atmosphere interactions. SST has a major role in atmospheric models, weather forecasting, climate change models, and energy balance calculations. SST can be measured from satellites and represents a very thin boundary layer (~10 μm skin layer) between the turbulent ocean and atmospheric layers. At this boundary layer, exchanges of sensible and latent heat occur, and long-wave radiation is emitted and absorbed [1]. Different processes act on the skin layer and on the water body beneath it (bulk layer), resulting in a difference between the skin and bulk temperatures.

Saunders [2] presented a simple theory in which the difference between bulk temperature and skin temperature, commonly termed "the skin effect" (ΔT), is proportional to the heat flux (including sensible, latent, and long wave radiative heat fluxes from ocean to atmosphere) and inversely proportional to the kinematic stress (wind friction); the theory is limited to conditions of negligibly low solar radiation and excludes very low wind intensity. One of the predictions of this model is that the ocean is usually covered with a "cool skin". The skin effect is estimated using measured in situ bulk

temperature and long wave radiation from which the skin temperature is calculated [3].

The effects of wind, waves, and the upper layer mixing on the boundary layer have been investigated [3–5]. These studies have shown that wind mixes the upper layer, cooling the skin layer, and that breaking waves momentarily destroy the skin layer, which reestablishes itself within less than one second [6].

Physical processes that control the skin effect vary throughout the seasonal and diurnal cycles. Emery et al. [1] described three mixing regimes in the water body affecting the skin effect: free convection, forced convection driven by wind stress, and forced convection driven by microscale wave breaking. They used different models to represent the physics of the skin layer and applied the models to measured data sets. Although these models reproduced the overall variability of the measured skin effect, nevertheless most of the variance was not explained ($R^2 = 0.28$ was the highest of all models.)

Over the last decade there has been an improvement in the calibration of satellite measured SST to in situ measurements. Much of this work is coordinated through the Group for High Resolution Sea Surface Temperature (GHRSST), using radiative transfer models [7, 8] or regression-based retrieval [9–11].

Little is known about the dynamics of the skin layer and the skin effect of the Dead Sea (Figure 1(a)). The Dead Sea is a hypersaline lake with a reduced evaporation rate due to the low water activity (the vapor pressure of the brine is about 0.7 of that of pure water at the same temperature) [12] and is the warmest large water body on Earth. Nehorai et al. [13] characterized the Dead Sea surface temperature using sequences of 15-minute interval satellite images and in situ measurements of wind speed, solar radiation, and air temperature. They found that at night the SST over the Dead Sea is relatively uniform, whereas during daytime the spatial variability is much larger. They concluded that the horizontal uniformity of the Dead Sea surface temperature during nighttime is due to the strong night winds that cause a vertical mixing of the upper few meters. During the day, the skin temperature rises due to intense solar radiation and the low wind speeds. The skin layer is very sensitive to wind intensity [6]. In the Dead Sea, even weak winds (<5 m/s) during daytime locally destroy the skin layer, causing the observed nonuniformity in SST [13]. In contrast, during nighttime strong winds (>5 m/s) associated with the Mediterranean Sea breeze intensify vertical mixing over the entire Dead Sea surface, which results in uniform SST over the Dead Sea. SST in closed and stratified lakes can be influenced by wind driven upwelling events, as indicated in thermistor chains and thermal infrared images when the cooler epilimnion or metalimnion reaches the surface (e.g., in Lake Tahoe, [14]). Thermistor chain data in the Dead Sea [15–18] and satellite thermal images [13] show that the upper layer is well mixed, the thermocline depth throughout the stratified season is ~25 m, and cooler hypolimnion water does not reach the sea surface during typical wind events. The circulation in the Dead Sea was previously studied with a few short term, sporadic observations. Neev and Emery [19] and Emery and Csanady [20] measured currents in 1959, when the Dead Sea was meromictic and had two connected basins, a deep northern basin and a shallow southern basin. Since then the northern and southern basins have disconnected due to the dropping sea level, and the Dead Sea has switched to holomictic conditions due to the increasing salinity of the upper layer.

In this paper we use in situ measurements of skin, bulk and air temperatures, solar radiation, and wind speed measured from a buoy located near the center of the Dead Sea to explore the diurnal and seasonal cycles and the major forcing of the skin temperature.

2. Data and Methods

Meteorological data, including air temperature (T_a), wind speed (Ws), and solar radiation (Ra), were collected every 20 minutes at 3.7 m above the sea surface from a hydro-meteorological buoy operated by Israel Oceanological and Limnological Research (IOLR), located ~5 km offshore of Ein Gedi, near the center of the lake (Figures 1(a) and 1(b)). All measuring instruments were manufactured by Aanderaa, as presented in [16, 21].

Bulk temperature (T_b) was measured using a thermistor placed at a depth of 5 cm. The thermistor was tied to a small

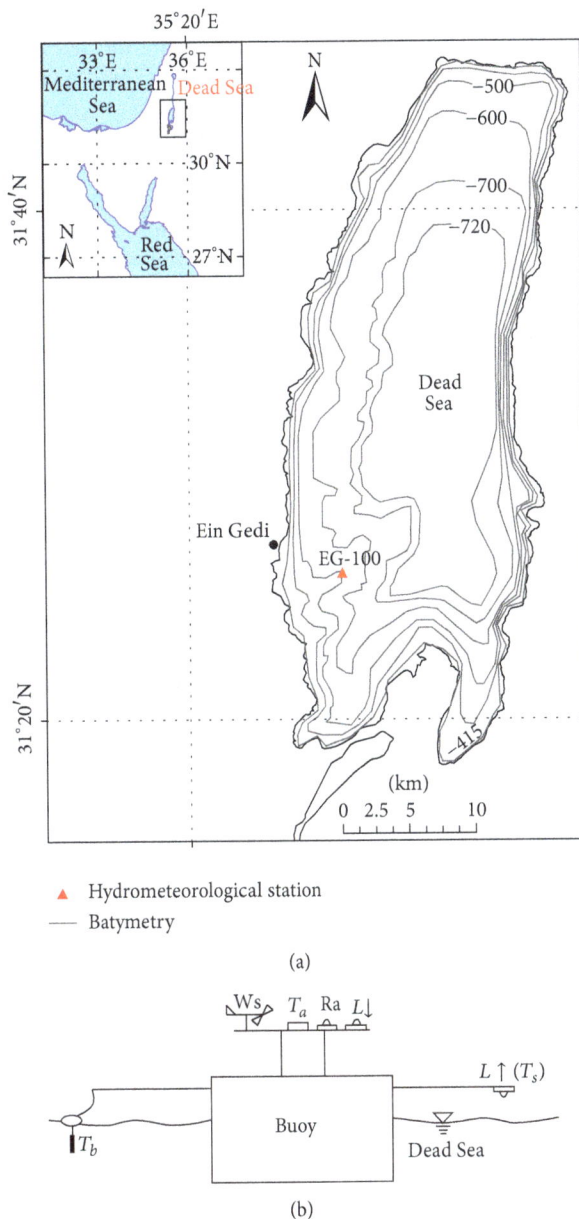

Figure 1: (a) Map of the Dead Sea, including the hydrometeorological station (red triangle) mentioned in the text. (b) Schematic diagram of the hydrometeorological buoy showing the relevant instruments. This buoy measured: T_s: skin layer temperature, Ws: wind speed; Ra: short wave radiation; T_a: air temperature; T_b: bulk temperature; L: long wave radiation.

buoy 2 m away from the hydrometeorological buoy to avoid the influence of the hydrometeorological buoy on the thermal structure of the top 5 cm (Figure 1(b)). We used a thermistor (Solinst model 3001, levelogger junior) with an accuracy of ±0.1°C and a temperature resolution of 0.1°C. Another thermistor was located at a depth of 1 m.

The skin temperature (T_s) was measured using two long-wave radiometers (Kipp & Zonen, CGR4) installed on the buoy. The CGR4 radiometer is sensitive to infrared radiation in a wavelength range from 4.5 to 42 μm (it has an extremely low window heating offset and diamond-like coating for optimal protection against environmental influences and low temperature dependence of sensitivity). The downward directed radiometer was placed at the edge of an extension arm 2 m away from the buoy and one meter above the water surface, minimizing the atmospheric effects on the measurements and the effect of the buoy on the fine thermal structure. The second radiometer was placed next to the meteorological instruments (3.7 m above the water surface), directed upward. It received the downward long-wave radiation emitted from the atmosphere ($L\downarrow$). The downward directed radiometer received the total long-wave radiation flux ($L\uparrow$) consisting of the radiation emitted from the sea surface and the radiation reflected upwards from the sea surface. To calculate T_s we used the following equation:

$$L\uparrow - L\downarrow (1-\varepsilon) = \varepsilon \times \sigma \times T_s^4, \qquad (1)$$

where $\varepsilon = 0.975$ is the water emissivity (and absorption); $(1-\varepsilon)$ is the water reflectance, and σ is the Stefan-Boltzmann constant.

We used cross correlation to analyze the correlation and time lag between T_s and the other measured quantities (T_a, T_b, Ws, and Ra). Since the skin layer is very thin, with a very short thermal response time (seconds), we expected that the measured quantity that showed the minimum time lag and the highest correlation to be indicative of the major forcing of the skin layer.

Heat fluxes were calculated following Lensky et al. [12], which include the required adaptations to the unique conditions of the Dead Sea. In this work we measured the net radiation, Q_{RN}, using four radiometers; short wave and long wave directed upward and downward. The contradiction will be explained in the next point. Latent heat (Q_L) and sensible (Q_S) heat were calculated using bulk formulas ((2) and (3), resp.), with the numeric values suitable for the Dead Sea, accounting for the heat and water balances:

$$Q_L = f(w) \cdot (e_{br} - e_a), \qquad (2)$$

$$Q_S = c_b \cdot \frac{P}{1000} \cdot f(w) \cdot (T_s - T_a), \qquad (3)$$

where e_{br}, is vapor pressure of the brine, calculated using the water activity of the brine ($\beta \sim 0.7$), e_a is the atmospheric vapor pressure, c_b is the Bowen constant (0.61 hPa/°C), P is the air pressure (hPa), and $f(w)$ is the wind function (4) with the parameterization accounting for energy and mass balance of the Dead Sea [12, 15]:

$$f(w) = 0.483 \cdot \left(9.2 + \left(0.46 \cdot Ws^2 \right) \right). \qquad (4)$$

The Bowen ratio, B, (5) is also presented in the results:

$$Q_S = Q_L \cdot B. \qquad (5)$$

The net heat (Q_n) is calculated using two independent approaches: as the residual heat flux, and from the change in the heat storage of the lake ((6) and (7), resp.):

$$Q_n = Q_L + Q_S + Q_{RN}, \qquad (6)$$

$$Q_n = -C_p \cdot m_t \cdot \frac{\Delta T}{\Delta t}, \qquad (7)$$

where C_p is the specific heat capacity, m_t is the total mass, and $\Delta T/\Delta t$ is the rate of temperature change.

3. Results

3.1. The Diurnal and Seasonal Cycles of the Skin Effect. The major finding of this study is that the diurnal cycle of T_s is most highly correlated to T_a (0.93–0.98) with a minimal time lag in all seasons. It is much less correlated to T_b with a larger time lag. T_s is even less correlated to Ws, and Ra with even higher time lags. Figure 2 presents the time series of all measured quantities of four representative days in winter, summer, and autumn. Figure 3 presents scatter diagrams of T_s versus T_a, T_b, Ws and Ra in the three seasons. The correlations are computed for five, 45, and 17 days in the summer, winter, and autumn, respectively. Figures 4 and 5 present the cross correlations of these pairs. Figure 6 is a schematic diagram of the diurnal and seasonal cycles of temperature profiles, from the air, through the skin to the different layers of the main water body. Figure 7 presents the heat fluxes time series.

3.2. Correlation of T_s to T_b and T_a. During the winter, convection fully mixes the Dead Sea; therefore, the water temperature is almost uniform throughout the entire water column (300 m). The diurnal amplitude of T_b is ~0.2°C, whereas T_s and T_a show much larger amplitudes of 2–4°C. Throughout the day T_b is higher than T_a by 1–5°C, leading to a continuous cooling of the entire water column through the skin at a rate of ~0.02°C/day. The skin layer is cooler and saltier due to evaporation and with higher density than the layers underneath and is therefore unstable. This difference drives the winter convection and cooling. Accordingly, T_s and T_b exhibit practically no correlation in winter (−0.17), whereas the correlation between T_a and T_s is very high (0.97), with practically no time lag (see Figures 4(a) and 5).

In the summer, the Dead Sea is stratified with an upper mixed layer above a sharp thermocline at a depth of 20–30 m. The upper diurnal layer with a thickness of ~5 m develops due to the solar heating and mixes at night with the layer beneath it due to density changes (Figure 6) [16]. As in the winter, T_s is most highly correlated to T_a (0.93) with no significant time lag (Figures 4(b) and 5). The correlation between T_b and T_s is highest in the summer (0.79) with a time lag of one hour. This is higher than the correlation of T_s – Ra and T_s – Ws (Figures 4 and 5). The diurnal amplitude of T_b is smaller than T_a and T_s (±1°C and ±2-3°C, resp., Figure 2(b)), but still significantly higher than in the winter, representing stronger coupling between the sea and the atmosphere during the stratified period.

In the autumn, the Dead Sea is still stratified, but the stability of the upper layer decreases together with the decrease of water temperature and its reduced diurnal amplitude. T_s highly correlates to T_a (0.98) with no time lag and correlates less to T_b (0.68, Figures 4(c) and 5). The correlation of T_b to T_s is higher in the summer than in autumn (Figures 4(b) and 4(c)), probably due to the weaker incoming solar radiation (Figures 2(e) and 2(f)), which results in the reduced diurnal amplitude of T_b in autumn. The cooling rate of T_b in the autumn is ~0.18°C/day, which is ten times larger than in

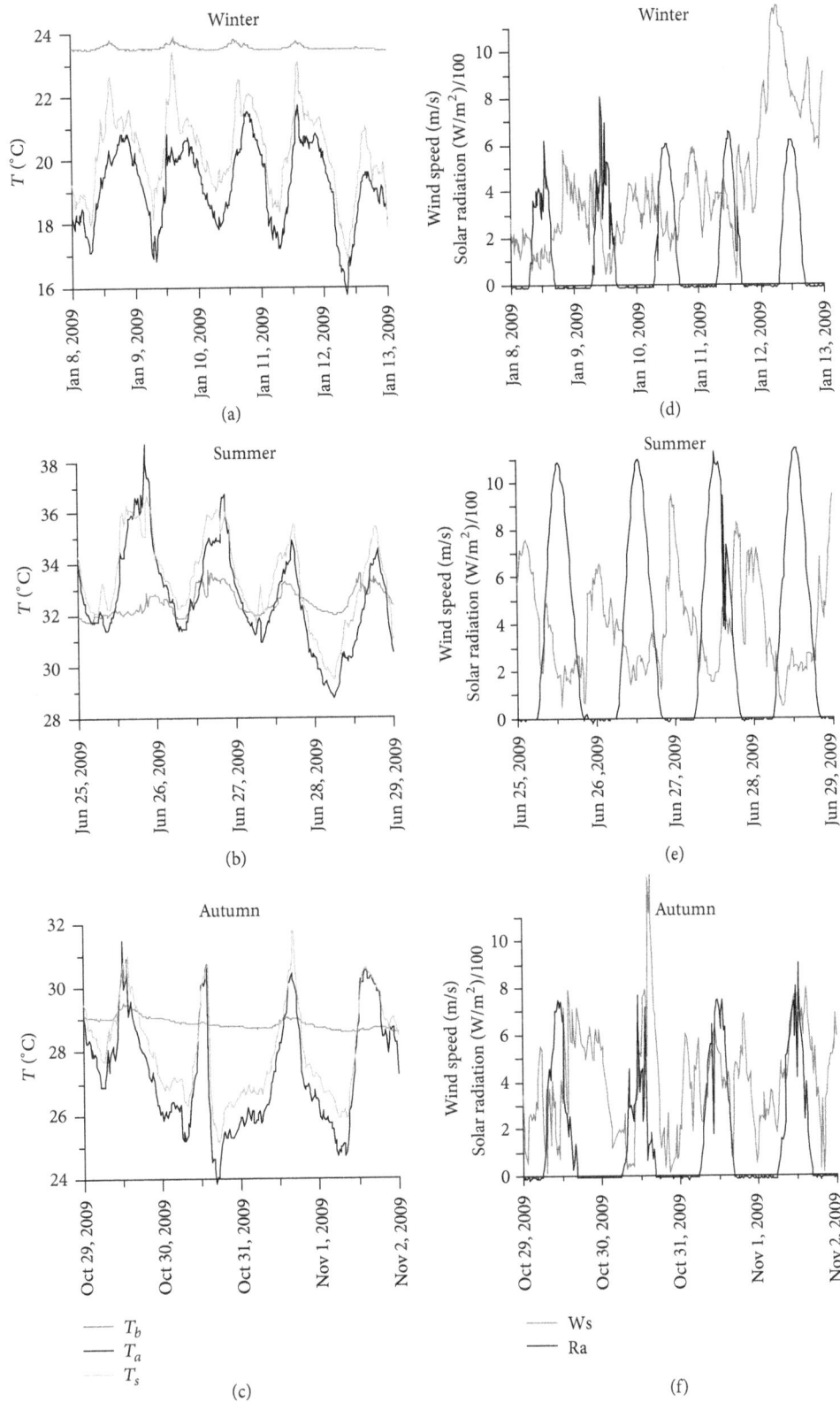

FIGURE 2: Time series of bulk (T_b), skin (T_s), air temperatures (T_a), wind speed (Ws), and solar radiation (Ra) during four representative days in winter (a), (d), when the water column is homogeneous; and summer and autumn (b), (e) and (c), (f), respectively, when the water column is stratified.

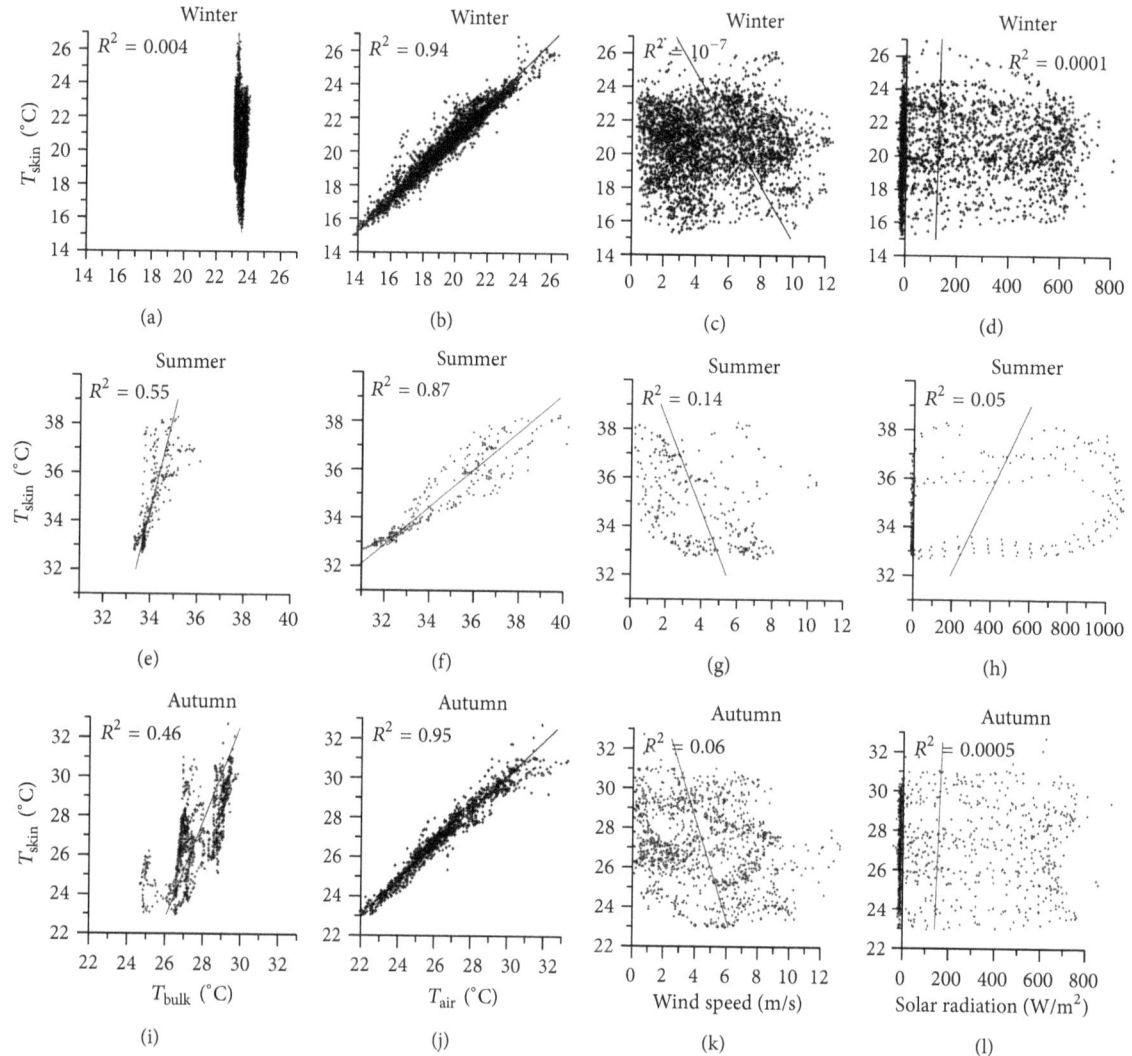

FIGURE 3: Correlations between surface, bulk water temperature, air temperature, wind speed, and solar radiation in winter ((a), (b), (c), (d))—(22 Dec 2008–14 Jan 2009); summer ((e), (f), (g), (h))—(25–29 Jun 2009); and autumn ((i), (j), (k), (l))—(27 Oct –10 Nov 2009).

winter. The more efficient cooling of T_b in autumn relative to winter is due to the difference in the thickness of layer interacting with the atmosphere, which in the winter is ten times thicker than in autumn (~300 m and ~30 m, resp.).

3.3. Correlation of T_s to Ra and Ws.

Low correlations and significant time lags were found between T_s and the atmospheric forcing (Ws and Ra) in all seasons (Figures 3, 4, and 5). Ra and T_s both show a diurnal cycle with high values during daytime and low values at nighttime (Figures 2(d), 2(e), and 2(f)). However, there is a lag of 4-5 hours in which Ra precedes T_s. In the scatter diagram this is seen as a counterclockwise cycle, as is best demonstrated in Figures 3(h) and 2(e). From sunrise (no solar radiation and $T_s \sim 33°C$), the solar radiation increases to ~1100 Wm^{-2} with minor changes in T_s. At noontime, T_s increases rapidly to about 37°C, gradually decreases to 36°C at sunset, and falls back to 33°C at nighttime.

The wind speed is even less correlated to T_s (0 to −0.71). The correlation of T_s – Ws is negative, with a time lag of 2-3.3

hours. Negative shifts in T_s – Ws occur while the T_s – T_a shift is positive. This is due to strong dry winds at nighttime, which cool the skin layer by increased evaporation.

3.4. Heat Fluxes.

Heat fluxes were calculated for 45 days during the winter. Figure 7 presents a time series of the heat flux components and the main meteorological parameters determining the fluxes. The net radiation (Figure 7(a)) has a typical diurnal cycle with high net radiation during the day and a repeating cycle with a very weak trend throughout the period. The latent and sensible heat fluxes are presented in Figure 7(b). The latent heat flux is characterized by periods of high fluxes (high evaporation rate), which is mostly related to the wind intensity (Figure 7(f)). The sensible heat flux has a clear diurnal cycle with higher flux during the night, when the air temperature is lower than the bulk water temperature by up to 8°C (Figure 7(e)). Note that the skin and air temperatures are highly correlated, as mentioned above, whereas the bulk water temperature shows a negligible

Correlation

(a)

Correlation

(b)

Correlation

$$T_s - T_a \qquad\qquad T_s - Ws$$
$$T_s - T_b \qquad\qquad T_s - Ra$$

(c)

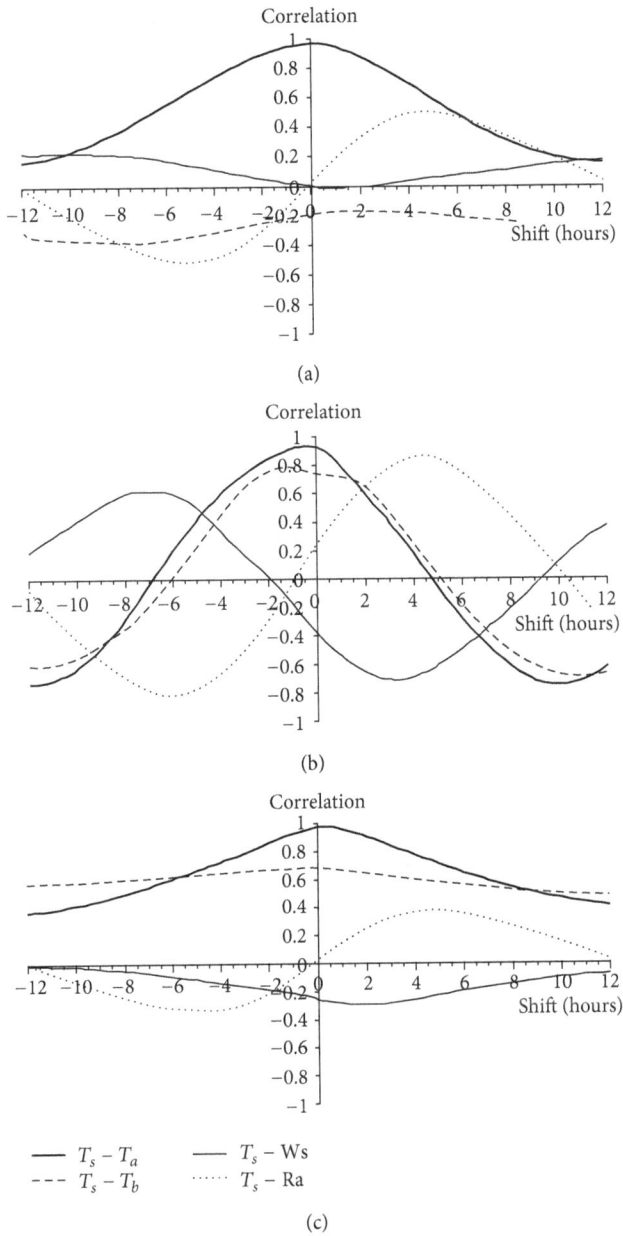

FIGURE 4: Cross correlation between the skin temperature (T_s) and the following measured quantities in winter, summer, and autumn: air temperature (T_a), bulk water temperature (T_b), wind speed (Ws), and solar radiation (Ra).

$$\square\ T_a \qquad\qquad \boxtimes\ Ws$$
$$\boxplus\ T_b \qquad\qquad \boxminus\ Ra$$

FIGURE 5: The maximum correlations from Figure 4 with the corresponding time lags. The three seasons are denoted by S: summer, W: winter, A: autumn.

diurnal cycle. Figure 7(c) presents the net heat flux calculated through two independent approaches, by calculating the residual or net heat flux and by calculating the changes in the heat storage (see the end of Section 2). Since the Dead Sea is fully mixed during this period of the year (holomictic conditions), the heat storage calculation is rather straightforward. The general trend of the bulk water temperature decreasing with time (Figure 7(e), ~0.015°C/day) suggests that the net heat flux is directed upwards from the lake to the atmosphere, which is also confirmed by the calculated

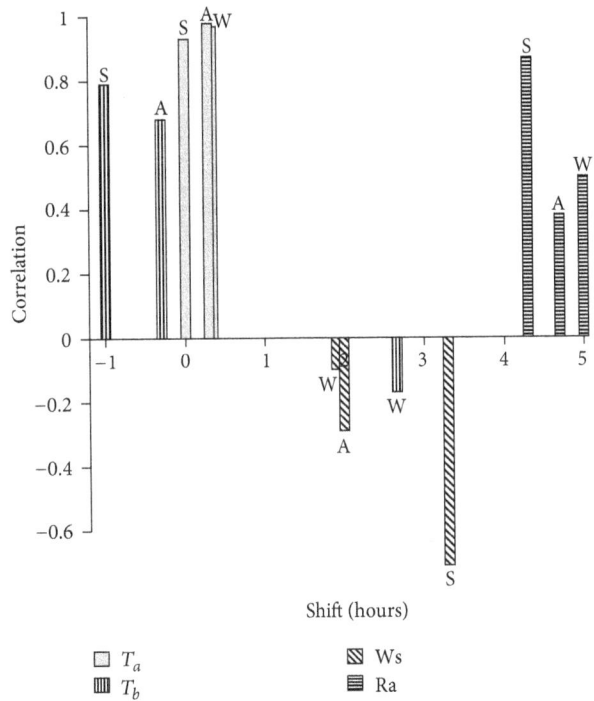

net flux using the residual approach. The cooling of the Dead Sea during this period is the "engine" that drives the vertical convection, which leads to a uniform water column. Note the general agreement of the calculated net heat flux in the two independent approaches. Figures 7(d), 7(e), and 7(f) present the Bowen ratio, the air, bulk and skin temperatures, and the wind speed. The Bowen ratio is around 0.2 (Figure 7(d)), with diurnal changes ranging from 0 to 0.5, with one event with a negative value (when air temperature was higher than bulk water temperature). Figure 7(f) presents the wind intensity through the period, influencing the latent and sensible heat fluxes (Figure 7(b) and (2)–(4)). Throughout the year T_s is nearly uncorrelated with the latent heat flux (0.06 and 0.08 in summer and winter, resp.). This is not surprising since the water temperature affects the evaporation only indirectly through the computation of the saturation vapor pressure. However, T_s is highly correlated with the sensible heat flux in summer (0.93), but somewhat less correlated in winter and autumn (0.62 and 0.66, resp.). The sensible heat flux, according to (3), is the product of the nonlinear wind speed factor and the air-sea temperature difference, which is a linear function of the water surface temperature. From Figure 3 it is clear that T_s and T_a are highly correlated in all seasons and thus suggesting that the sensible heat flux is the main component of the atmospheric forcing of T_s. The lower correlations between T_s and the sensible heat flux in winter and autumn are therefore related to the weaker dependence of Q_S on the wind speed in these seasons (Figure 3).

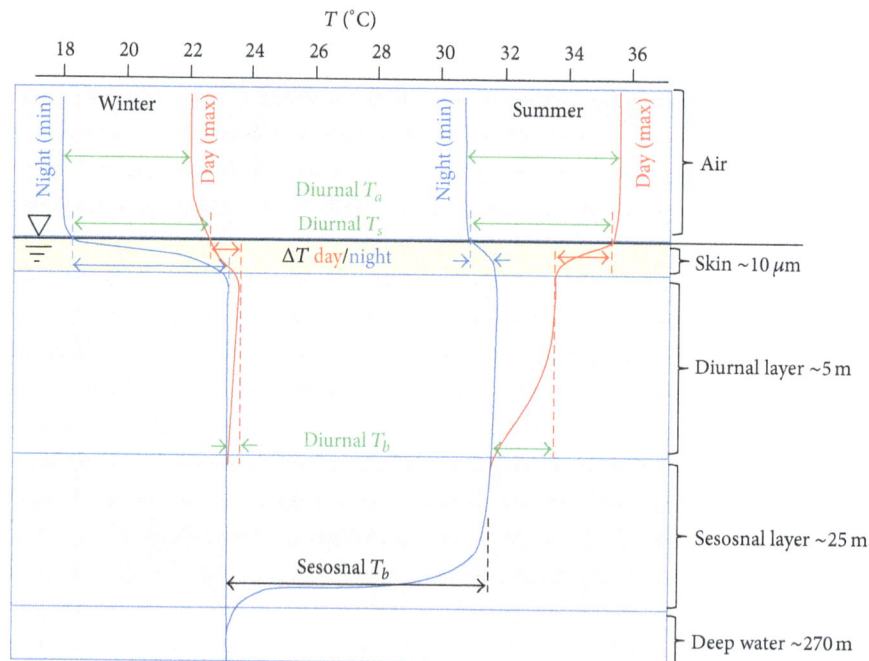

FIGURE 6: A schematic diagram showing the vertical temperature profiles including the major physical layers (air, skin, and the water column). The diurnal and seasonal variations are shown. Dotted arrows stand for diurnal variations between day (solid gray curves) and night (black curves) temperature profiles. Solid gray and solid black arrows show the day and night skin effect, respectively. The dashed arrow represents the amplitude of the seasonal thermocline.

4. Discussion

The magnitude of the skin effect is influenced by the diurnal cycle of solar radiation and winds [3, 4, 22]. Neohrai et al. [13] have shown that the solar radiation and wind intensity control the spatial variance of the Dead Sea SST in the diurnal and seasonal cycles. Here we show that in the Dead Sea the skin temperature is mostly correlated to the air temperature and much less to bulk water temperature, wind speed and solar radiation. Measurements were conducted from a buoy located towards the center of the lake at water depth of 60 m and 5 km offshore (Figure 1), far enough from significant influence of boundary effects (such as upwelling, downwelling, or shallow water effects) on the sea surface. Surface cooling due to wind induced upwelling events was not observed during the earlier [17] or during the more recent [15, 18] measurement periods; in the winter, the lake is fully mixed, and bulk water temperature is practically uniform, while, in the summer, the upper warm mixed layer reaches depth of ~25 m, and the cooler hypolimnion or metalimnion does not approache the surface (see also [15–17]). While SST in closed and stratified lakes can be influenced by wind driven upwelling events (e.g., in Lake Tahoe, [14]), this does not appear to be a significant factor in the Dead Sea as indicated by thermistor chains [15–18] and by satellite thermal images [13]. Furthermore, Sirkes [17] concluded that oscillations due to internal seiches, and their weak surface manifestation in summer, are not a result of direct wind forcing.

Figure 6 presents day/night and summer/winter schematic temperature profiles, summarizing our findings. Whereas the amplitude of the air and skin temperature diurnal cycles is similar in the summer and winter (four upper dotted arrows), the amplitude of the bulk temperature diurnal cycle is very different between summer and winter (two lower dotted arrows). The skin effect is low during summer nights and winter days while during summer days and winter nights it increases (solid black and solid gray arrows). In the summer, a daily thermocline builds up, resulting in significant amplitude of the diurnal cycle, whereas in the winter the amplitude of the diurnal cycle is very small, due to the vertical mixing. Therefore, the correlation between air and skin temperature is high in all seasons, whereas the correlation between bulk and skin temperature is higher in the summer, and no correlation was found in the winter. The highest skin effect $(T_b - T_s)$ occurs during winter nights (left black arrow), when the bulk water temperature is ~24°C, whereas the air and skin temperatures drop below 19°C.

Figure 5 summarizes the correlations and time lags of the skin temperature and the other measured parameters. The wind intensity and the solar radiation are correlated to the skin temperature, but with a time lag of a few hours. Since the response time of the skin layer is very short (<second), it implies that solar radiation and wind intensity have an indirect effect on the skin layer. High correlations and negligible time lags between air and skin temperatures suggest that the air temperature plays a major role in the forcing of the skin temperature. This forcing is accomplished through the sensible heat flux, which is a direct linear function of the air-sea temperature difference. This is also manifested in summer evenings when high air and skin temperatures decrease from

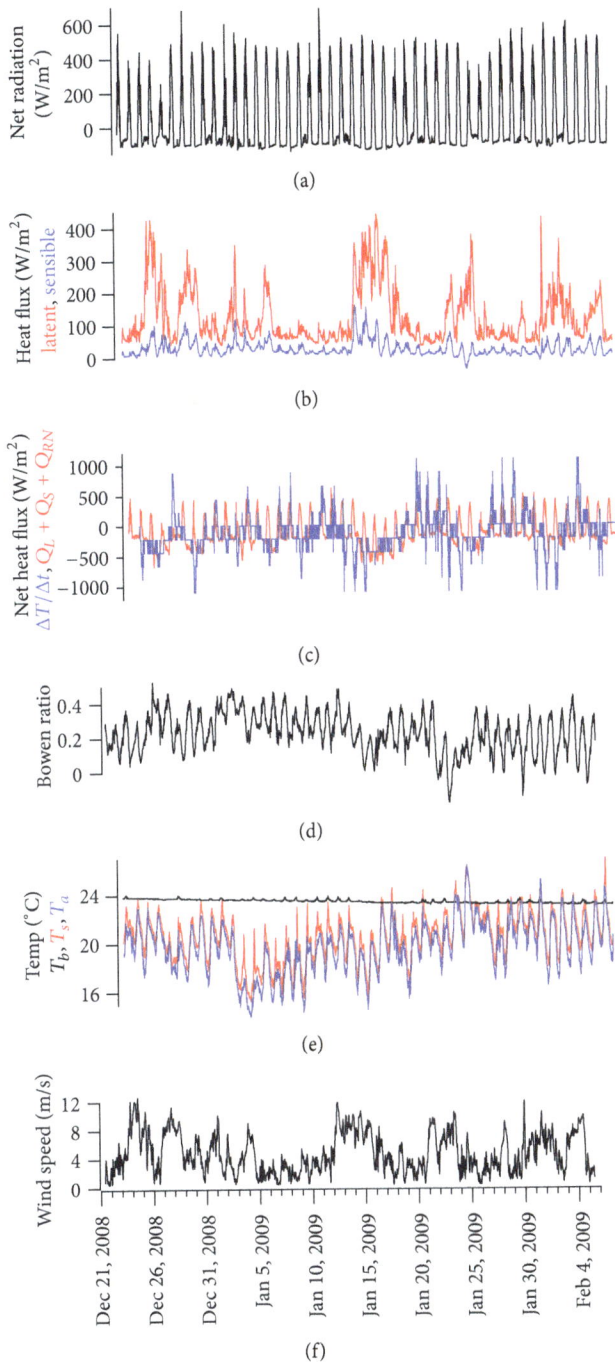

FIGURE 7: Time series of the heat fluxes at the Dead Sea surface and the governing meteorological factors, measured during wintertime.

The skin layer of the Dead Sea can be classified into two mixing regimes:

(i) unstable conditions in winter when the skin temperature is controlled by free convection (where $T_s < T_b$). The spatial variations of SST are low in such conditions [13]. The skin layer temperature is less affected by wind since it is unstable and it continuously sinks and rebuilds. To some extent this is also the case in summer nights, when night cooling takes place,

(ii) stable conditions during the daytime in summer when the skin is affected by the stable thermal layering due to heating by solar radiation. The stable structure of the upper water layer is very sensitive to wind gusts that cause significant spatial variations [13].

5. Conclusion

We found that the skin temperature in the Dead Sea is most highly correlated to the air temperature in all seasons (0.93–0.98). In the summer, when the Dead Sea is stratified, the skin temperature is also correlated to the bulk water temperature of the surface (0.80). In the winter, however, when the Dead Sea is vertically mixed, the amplitude of the skin temperature diurnal cycle is ~4°C, whereas the bulk water shows an amplitude of ~0.2°C, and therefore there is no correlation between the skin and bulk temperatures in the winter. Low correlations were found between the skin temperature and the solar radiation and wind speed in all seasons. The skin, with its low thermal inertia, responds immediately to the governing forcing. Thus the air temperature with its highest correlation and minimal time lag is considered to be the most important factor in the forcing of the skin layer's temperature, which is accomplished primarily through the sensible heat flux.

Acknowledgments

The authors thank Isaac Gertman for supplying the in situ measurements from the buoy, which enabled this research, and for his critical reading. The authors thank Raanan Bodzin for the assistance in the field work, critical reading, and helpful discussions. Ittai Gavrieli, Vladimir Lyakhovsky, and Gerald Stanhill are acknowledged for fruitful discussions. The authors also thank Uri Malik and Shabtai Cohen for helping with the installation and calibration of the instruments, and the late Moti Gonen, Silvy Gonen, and the crew of the "Taglit" for cruise services. The authors thank Tal Ozer and Boris Katsanelson for assistance in the field. Ali Arnon is acknowledged for statistical analysis of the time series. The research was supported by the Earth Science Research Administration, the Ministry of National Infrastructures (Israel).

References

[1] W. J. Emery, S. Castro, G. A. Wick, P. Schluessel, and C. Donlon, "Estimating sea surface temperature from infrared

the peak temperature only a few hours after sunset and after the bulk temperature has decreased (Figures 2(b) and 2(e)). This happens when air from the Mediterranean Sea breeze (Figure 2(e)) is heated adiabatically while descending from the Judean Mountains to the Dead Sea, delaying the evening cooling to a few hours after sunset. There is no other explanation for the high temperature of the skin layer at night other than the influence of the air temperature.

satellite and in situ temperature data," *Bulletin of the American Meteorological Society*, vol. 82, no. 12, pp. 2773–2785, 2001.

[2] P. M. Saunders, "The temperature at the ocean-air interface," *Journal of the Atmospheric Sciences*, vol. 24, pp. 269–273, 2011.

[3] C. J. Donlon, P. J. Minnett, C. Gentemann et al., "Toward improved validation of satellite sea surface skin temperature measurements for climate research," *Journal of Climate*, vol. 15, no. 4, pp. 353–369, 2002.

[4] D. C. Oesch, J.-M. Jaquet, A. Hauser, and S. Wunderle, "Lake surface water temperature retrieval using advanced very high resolution radiometer and Moderate Resolution Imaging Spectroradiometer data: validation and feasibility study," *Journal of Geophysical Research C*, vol. 110, no. 12, Article ID C12014, 17 pages, 2005.

[5] C. J. Merchant, M. J. Filipiak, P. Le Borgne et al., "Diurnal warm-layer events in the western Mediterranean and European shelf seas," *Geophysical Research Letters*, vol. 35, no. 4, Article ID L04601, 2008.

[6] A. T. Jessup, C. J. Zappa, M. R. Loewen, and V. Hesany, "Infrared remote sensing of breaking waves," *Nature*, vol. 385, no. 6611, pp. 52–55, 1997.

[7] C. J. Merchant, L. A. Horrocks, J. R. Eyre, and A. G. O'Carroll, "Retrievals of sea surface temperature from infrared imagery: origin and form of systematic errors," *Quarterly Journal of the Royal Meteorological Society*, vol. 132, no. 617, pp. 1205–1223, 2006.

[8] C. J. Merchant, P. Le Borgne, A. Marsouin, and H. Roquet, "Optimal estimation of sea surface temperature from split-window observations," *Remote Sensing of Environment*, vol. 112, no. 5, pp. 2469–2484, 2008.

[9] S. L. Castro, G. A. Wick, P. J. Minnett, A. T. Jessup, and W. J. Emery, "The impact of measurement uncertainty and spatial variability on the accuracy of skin and subsurface regression-based sea surface temperature algorithms," *Remote Sensing of Environment*, vol. 114, no. 11, pp. 2666–2678, 2010.

[10] A. T. Jessup and R. Brance, "Integrated ocean skin and bulk temperature measurements using the Calibrated Infrared in Situ Measurement System (CIRIMS) and through-hull ports," *Journal of Atmospheric and Oceanic Technology*, vol. 25, no. 4, pp. 579–597, 2011.

[11] E. J. Noyes, P. J. Minnett, J. J. Remedios, G. K. Corlett, S. A. Good, and D. T. Llewellyn-Jones, "The accuracy of the AATSR sea surface temperatures in the Caribbean," *Remote Sensing of Environment*, vol. 101, no. 1, pp. 38–51, 2006.

[12] N. G. Lensky, Y. Dvorkin, V. Lyakhovsky, I. Gertman, and I. Gavrieli, "Water, salt, and energy balances of the Dead Sea," *Water Resources Research*, vol. 41, no. 12, Article ID W12418, 13 pages, 2005.

[13] R. Nehorai, I. M. Lensky, N. G. Lensky, and S. Shiff, "Remote sensing of the Dead Sea surface temperature," *Journal of Geophysical Research C*, vol. 114, no. 5, Article ID C05021, 2009.

[14] T. E. Steissberg, S. J. Hook, and S. G. Schladow, "Characterizing partial upwellings and surface circulation at Lake Tahoe, California-Nevada, USA with thermal infrared images," *Remote Sensing of Environment*, vol. 99, no. 1-2, pp. 2–15, 2005.

[15] I. Gavrieli, N. G. Lensky, M. Abelson et al., "Red Sea to Dead Sea Water Conveyance (RSDSC) study: Dead Sea research team," Tech. Rep. GSI/10/2011, Tahal Group and Geological Survey of Israel, Jerusalem, Israel, 2011.

[16] I. Gertman and A. Hecht, "The Dead Sea hydrography from 1992 to 2000," *Journal of Marine Systems*, vol. 35, no. 3-4, pp. 169–181, 2002.

[17] Z. Sirkes, "Surface manifestations of internal oscillations in a highly stratified saline lake (the Dead Sea)," *Limnology and Oceanography*, vol. 32, pp. 76–82, 1987.

[18] N. G. Lensky, I. Gertman, Z. Rosentraub et al., "Alternative dumping sites in the Dead Sea for harvested salt from pond 5," Tech. Rep. GSI/05/2010, Geological Survey of Israel, Jerusalem, Israel, 2010.

[19] D. Neev and K. O. Emery, "The Dead Sea: depositional processes and environments of evaporates," Geological Survey of Israel Bulletin 41, 147 pp., 1967.

[20] K. O. Emery and G. T. Csanady, "Surface circulation of lakes and nearly land-locked seas," *Proceedings of the National Academy of Sciences of the United States of America*, vol. 70, no. 1, pp. 93–97, 1973.

[21] A. Hecht and I. Gertman, "Fungal life in the Dead Sea," *Mycological Research*, vol. 108, pp. 1106–1106, 2004.

[22] I. J. Barton, "Interpretation of satellite-derived sea surface temperatures," *Advances in Space Research*, vol. 28, no. 1, pp. 165–170, 2001.

18

Weather Support for the 2008 Olympic and Paralympic Sailing Events

Yan Ma,[1] Rongzhen Gao,[1] Yunchuan Xue,[1] Yuqiang Yang,[2] Xiaoyun Wang,[3] Bin Liu,[4] Xiaoliang Xu,[1] Xuezhong Liu,[1] Jianwei Hou,[1] and Hang Lin[1]

[1] *Qingdao Meteorological Bureau, 4 Fulong Shan, Shinan District, Qingdao, Shandong 266003, China*
[2] *Hangzhou Meteorological Bureau, Hangzhou, Zhejiang 310008, China*
[3] *Department of Synthetic Observing, China Meteorological Administration, Beijing 100081, China*
[4] *Department of Marine, Earth, and Atmospheric Sciences, NC State University, Raleigh, North Carolina 27695, USA*

Correspondence should be addressed to Yan Ma; qdyanma@163.com

Academic Editor: Huiwang Gao

The Beijing 2008 Olympic and Paralympic Sailing Competitions (referred to as OPSC hereafter) were held at Qingdao during August 9–23 and September 7–13 2008, respectively. The Qingdao Meteorological Bureau was the official provider of weather support for the OPSC. Three-dimensional real-time information with high spatial-temporal resolution was obtained by the comprehensive observation system during the OPSC, which included weather radars, wind profile radars, buoys, automated weather stations, and other conventional observations. The refined forecasting system based on MM5, WRF, and statistical modules provided point-specific hourly wind forecasts for the five venues, and the severe weather monitoring and forecasting system was used in short-term forecasts and nowcasts for rainstorms, gales, and hailstones. Moreover, latest forecasting products, warnings, and weather information were communicated conveniently and timely through a synthetic, speedy, and digitalized network system to different customers. Daily weather information briefings, notice boards, websites, and community short messages were the main approaches for regatta organizers, athletes, and coaches to receive weather service products at 8:00 PM of each day and whenever new updates were available. During the period of OPSC, almost one hundred people were involved in the weather service with innovative service concept, and the weather support was found to be successful and helpful to the OPSC.

1. Introduction

The Beijing 2008 Olympic and Paralympic Sailing Competitions (termed as OPSC hereafter) were held at Qingdao, Shandong province, during August 9–23 and September 7–13 2008, respectively. Approximately 400 athletes coming from 62 countries and regions competed in 11 sports of Olympic sailing games, and 80 athletes from 25 countries and regions participated in 3 sailing games in the Paralympic sailing competition. The weather support group of Qingdao Meteorological Bureau (QMB) provided weather information for the games. Specialized operational weather service supplied from the national weather support system was the key element for the games, since they are sensitive to local weather situations [1–6]. The Olympic forecasting experience provided an exciting insight into capabilities for future National Weather Services forecast operations. But the atypical weather around the venue as frequent thunderstorm and persistent light winds proved challenging to forecasters, athletes, and Olympic management officials alike [7]. It is a great challenge for weather forecasting to provide detailed wind prediction for sailing competitions. Operational wind forecasting system is more used to support power system at present discussed by Zhu and Genton [8]. A distinguished aspect of OPSC support system is the high-resolution wind forecasting system, which gives a good demonstration for the weather support to large outdoor activities.

The planning for the weather support system began in 2001, shortly after the International Olympic Committee elected Beijing as the site of the 2008 Olympics and

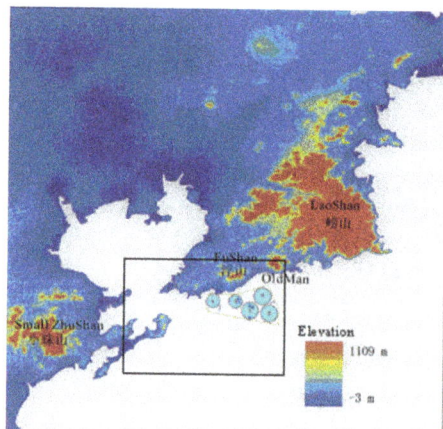

FIGURE 1: Topography around Qingdao OPSC venues. The region of rectangle is the position of observational system.

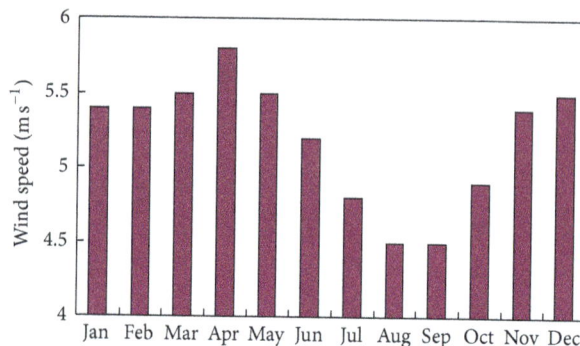

FIGURE 2: Monthly wind speed of Qingdao based on data of 1971–2007.

Paralympics and Qingdao as the site of the 2008 Olympic and Paralympic Sailing Competitions. The weather support system including monitoring, forecasting, and networking systems was then constructed, evaluated, and improved through weather support for the 2006, 2007 international regattas and the 2008 international regatta for the handicapped.

It is well known that the changes of wind speed and wind direction are of crucial importance for sailing events. First of all, the setting for sailing routes is dependent upon the current wind direction. If the wind direction wasto shift to larger than 50 degrees in a round, the game would have to be canceled. In addition, the required wind speed range is 3–20 m s^{-1}. In the situations of wind speed less than 3 m s^{-1}, the competition cannot go on because the wind is too weak and, therefore, not suitable for driving the sails. During the 2004 summer Olympic games, there were three days of "no wind" appeared at Athens Metropolitan area. On the other hand, when the wind speed exceeds 20 m s^{-1}, it is too dangerous for the athletes to sail. Additionally, a visibility of greater than 1500 m and no possible thunderstorm weather threats are also essential safety conditions for the sailing events. The weather support required for QMB was hourly wind speed and wind direction forecasts aimed at five Olympic venues distributed at the scope of 50 km^2 (Figure 1) as well as any severe weather warnings issued one hour prior to the competition. At the same time, QMB must meet the diverse requirements such as providing flooding and thunderstorm warnings and providing weather support for cleaning the sailing venues due to the breaking out of *Enteromorpha prolifera*. QMB faced an unprecedented challenge in the weather support for 2008 OPSC.

2. Qingdao Climate Analyses

Qingdao, as a coastal city facing the Yellow Sea, lies on the southern tip of Shandong Peninsula. The climate of Qingdao has both monsoon and marine climate characteristics. It is worth noting that the period of OPSC (August 9–September

13) falls within the flood season of Qingdao. Typhoon, heavy rainfall, thunderstorms with gusts, heavy fog, and other high impact weathers occur from time to time. For example, on August 10, 2007, during the period of the 2007 International Sailing Regatta, more than 282.7 mm of precipitation fell on the city of Qingdao from 21:00 Local Standard Time (LST) 10 to 06:00 LST 11 August, with wind gusts of 24.1 m s^{-1}. The loss was estimated to be greater than 2.8 billion Chinese Yuan in the process.

The Olympic sailing field consisted of five venues: A, B, C, D, and E and was set up at an area of 50 km^2 from Fushan Bay to the south of Old Man Bay (Figure 1). Several mountains including Taipingshan (with elevation of 348 m), Fushan (368 m), Wushan (298 m), and Laoshan (1133 m) are all located to the north of the sailing fields. The complicated offshore terrain disrupts the accuracy of wind forecasts, especially for the weak wind conditions. Analyses based on observations at the Qingdao station from 1971 to 2007 indicate that the mean wind speeds in the months of August and September are the smallest, only 4.5 m s^{-1} (Figure 2). The prevailing wind direction in August is southeasterly while northerly and southerly wind in September. Moreover, the sea-land breeze is the main local atmospheric feature along the coast of Qingdao. Usually, the sea breeze appears during the period of 10:00 and 13:00 LST and lasts 6 to 9 hours. For the land breeze, it appears around 3:00 to 5:00 LST and lasts 2 to 4 hours or so. During the daytime in August and September, wind speed rises to over 3.5 m s^{-1} after 9:00 LST, and the maximum wind speed occurs during 13:00–15:00 LST with an hourly average wind speed of 4.3 m s^{-1}. In September, the maximum wind speed occurs between 14:00 LST and 15:00 LST with a maximum hourly speed of 4.4 m s^{-1}. After that time, wind speed decreases gradually and reaches 3.5 m s^{-1} at 21:00 LST (Figure 3). The 2008 Olympic sailing events were begun after 13:00 LST as planned, though there were several competitions delayed or forwarded because of the situations of weak wind speed.

The complicated topography around Olympic sailing fields plays an important role in the strength and direction of wind changes. The sea-land breeze circulation is the most important local wind system in Qingdao. The observations during 1971–2007 indicate that the sea breeze prevails the

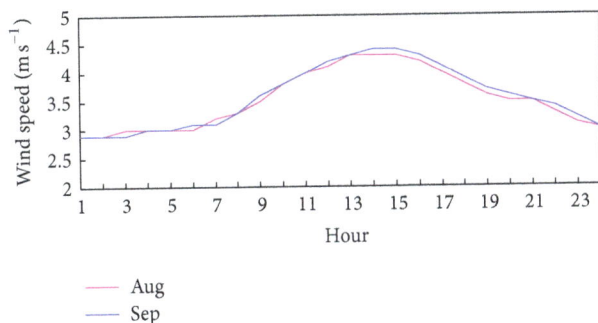

FIGURE 3: Time series of wind speed in August and September for Qingdao based on data of 1971–2007.

offshore of Qingdao during the daytime with the frequency of occurrence at 35% in August and 25% in September. Figure 3 gives the wind rose of buoy B at 08:00 LST and 16:00 LST in August of 2005–2008. It is shown that the prevailing wind direction is easterly to northwesterly in the morning at the venues, while in the afternoon southerly to southeasterly wind with apparent sea breeze characteristics prevails. The predominant wind speed was 3–6 m s^{-1}. Analyses also show that the strength of sea breeze is related to environmental circulation. While the large-scale geostrophic wind is in the direction of the easterlies, the strength of the sea breeze is stronger (larger than 6 m s^{-1}) just offshore Qingdao; while the large-scale geostrophic wind is northerly, the wind speed of sea breeze is larger than 3 m s^{-1}.

3. Weather Support System for OPSC

Research and assessment have been underway at QMB to improve the understanding and accuracy of prediction of the atmospheric circulation around the sailing venues since 2001. QMB has contributed to the effort in aspects of improvements to the temporal-spatial comprehensive monitoring system, the introduction of high-resolution real-time wind forecast system, and the development of modernized network system for the OPSC weather support. Setting up special projects such as developing high-resolution multimodel real-time mesoscale modeling system and Model Output Statistics (MOS) methods based on these model outputs provided strong technological support for the wind forecasts, especially for the weak wind forecasts under the condition of complex terrain.

3.1. Monitoring System. Under the base of the original monitoring system of QMB, a number of monitoring devices have been installed and have been successively operational in Qingdao since 2001. For example, 107 four-parameter Automatic Weather Stations (AWSs) were installed from 2003 to 2008 (Figure 4 is the schematic of the observational system) which monitored weather processes effectively, for the sailing competition. It consists of Doppler radars, wind profile radars, laser wind-detection radars, and portable weather stations. In addition, five buoys A, B, C, D, and E, are located around the sailing venues. Buoys A, B, and D

were set up by the State Oceanic Administration, China, and QMB shared the observed information during the period of competition. Buoys C and E were installed by China Meteorological Administration. Buoy C, with the diameter of 3 m, has been supporting the sailing games since July 30, 2007, and buoy E, with the diameter of 10 m, was put into operation on July 12, 2008. Variables of wind speed, wind direction, temperature, and humidity with an interval of 10 minutes, variables of sea surface waves with an interval of 30 minutes, and current with an interval of 10 minutes were collected from the buoy stations. At the same time, observations from eight AWSs, which are either located offshore or on an island, strengthen the weather support around the sailing venues. All of the high-resolution and multisource observed information was provided to both the committee of OPSC and the athletes through the weather service center in forms of figures and tables. The observed information was also ingested into QMB real-time mesoscale model by data assimilation techniques.

3.2. High-Resolution Wind Forecast System. The multimodel high-resolution sea surface wind forecast system has been constructing to meet high requirements of the sea surface wind prediction for the sailing games since 2001. It has been continuously modified and improved through the application in the 2006, 2007 International Sailing Regattas and the 2008 International Sailing Regatta for the handicapped. Five dynamical models and two statistical modules were included in the forecast system. Table 1 shows the information of five numerical models, including the QMB operational numerical model based on the Pennsylvania State University, National Center for Atmospheric Research fifth generation Mesoscale Model (MM5) [9–11], MM5RUC, and three Weather Research and Forecasting (WRF) systems [12, 13]: WRF (500 m), WRF (3 km), and WRF (5 km) with different horizontal and vertical resolutions and different initial fields. MM5 has been running since 2004 at QMB, and the initial time is at 20:00 LST everyday with a forecast duration of 84 h. Based on this operational modeling system, the model referred to as MM5RUC has been developed by assimilating observations from AWSs, buoys, wind profile radar, Doppler radar, and gradient wind tower around Qingdao, in order to get more accurate initial condition for the model. By the way, the observations were also ingested into the WRF (3 km) modeling system which was provided by Beijing Meteorological Bureau. The WRF (3 km) was updated every 3 hours for a forecasting duration of 24 hours. The WRF model with a horizontal resolution of 500 m uses a high-resolution Qingdao land use data, which was derived from Landsat remote sensing data of 2005, reflecting the influence of detailed land-use characteristics on the sea surface wind around the offshore. The WRF (500 m) finescale simulations are used to investigate impacts of urban processes and urbanization on a localized, summer, heavy rainfall in Beijing presented by Miao et al. [14]. Evaluation using radar and gauge data shows that this configuration of WRF with three-dimensional variational data assimilation of local weather and GPS precipitable water data can simulate storm generally well. The modeling system WRF (5 km) was provided to

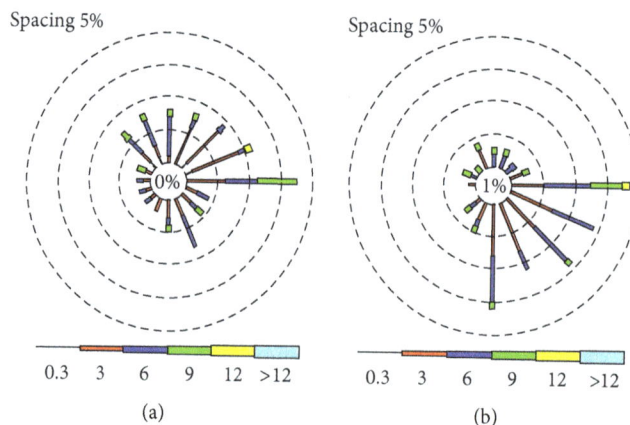

FIGURE 4: Wind rose of buoy B at 08:00 LST (a) and 16:00 LST (b) based on hourly wind in August of 2005–2008 (Color scale: mean wind speed; ray direction: wind direction; ray length: frequency of wind direction; value in the circular: frequency of calm wind).

QMB by the National Meteorological Center of China, while postprocessing was carried out at QMB. MM5-based model output statistical modules, MOS (MM5) and DAMOS, provided hourly forecasts of wind speed and wind direction at the five venues. During the 2002 Olympic and Paralympic Winter Games, a similar method of MOS was used to provide hourly forecasts for the sites [15]. During the 2008 Olympic and Paralympic sailing games, MOS (MM5) was developed at QMB by considering daily wind observations and 10-year wind diurnal fluctuation. DAMOS was supported by a team from the Department of Marine, Earth, and Atmospheric Sciences and Department of Statistics of North Carolina State University, which considers pre-30-day observations and QMB real-time MM5 output. The two modules were run operationally at QMB once a day during the Olympic game. All products were shown in the form of graphs and tables and then sent to the weather support office before 8:00 AM everyday during the period of the competition. On the basis of these forecast products, QMB weather forecasters then issued the wind prediction around the sailing venues after considering field observations and the local weather situation. Figures 5 and 6 is an example of hourly forecasts for 10 m wind speed and wind direction on August 18, 2008, for Buoy B, in which the ensemble forecast was carried out by considering the prediction from five dynamic models and two statistic modules through a kind of statistical method. The dynamic models (MM5, MM5RUC, and WRF 500 m) showed that there was the west-northwest wind with the force of 7–10 m s^{-1} at the field of OPSC in the morning of August 18 and turned to the southwest wind with the force of 3-4 m s^{-1} in the afternoon (Figures 7 and 8). The competition group of OPSC adopted suggestions from QMB and shifted the games to be held at 11:00 AM. The competition ratio at that day was 81% based on strong support from high-resolution wind forecasting system. This was also a transformation process between land breeze and sea breeze and the wind forecasting system reproduced the characteristics very well. Moreover, the statistical analyses for the four buoy stations A, B, C, and D during August 9–21, 2008, are shown in Table 2. It is shown that the forecasters' forecasts have the highest accuracy, with

TABLE 1: Overview of dynamical models.

Model	Grid	Initial conditions
MM5	4-nested grids (45/15/5/1.67 km)	AVN analyses
MM5RUC	2-nested grids (5/1.67 km)	5 km MM5 forecast
WRF (500 m)	500 m	1.67 km MM5RUC forecast
WRF (3 km)	3-nested grids (27/9/3 km)	AVN analyses
WRF (5 km)	2-nested grids (15/5 km)	T213 forecast

TABLE 2: Statistical analyses for four buoy stations during the period of August 9–21, 2008.

	Wind speed MAE (ms^{-1})	Wind direction MAE (°)	Wind speed RMSE (ms^{-1})	Wind direction RMSE (°)
Correction from forecasters	1.1	34	1.4	42
Ensembled forecasts	1.2	36	1.4	45
MOS (MM5)	1.3	37	1.6	45
WRF (500 m)	1.5	41	1.8	52
MM5RUC	1.6	38	1.9	49
DAMOS (MM5)	1.7	46	1.9	54
WRF (5 km)	1.9	49	2.2	57
WRF (3 km)	2.0	50	2.4	63

MAE: mean absolute error; RMSE: root mean square error.

the mean RMSE for the four buoy stations being 1.4 m s^{-1} for wind speed and 42 degree for wind direction, respectively.

3.3. Severe Weather Monitoring and Forecasting System. The QMB severe weather monitoring and forecasting system used throughout the period of OPSC consists of two components. The first component refers to the Severe Weather Integrated

TABLE 3: Itinerary of Olympic and Paralympic Games.

Date	9	10	11	12	13	14	15	16	17	18	19	20	21	22	23	8	9	10	11	12	13
Expected schedule	4	7	15	12	15	15	23	29	21	20	8	8	2			6	9		9	9	7
Actual completion	4	7	15	12	11	0	11	14	11	18	4	8	2	flexible date	Closing ceremony	6	9	flexible date	3	7	6
Competition ratio	100% Rate of fulfillment			rate of fulfillment reached 93%												100%			93.9%		

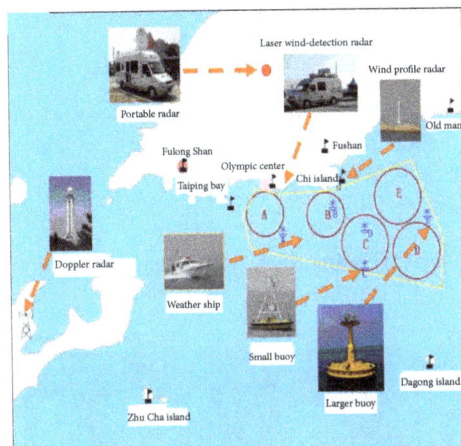

FIGURE 5: Schematic of the observational system for the sailing competition.

Forecasting Tools (SWIFT), which was originally developed by the Guangdong provincial Meteorological Bureau and was modified and improved by considering operational MM5 forecasts, AWS observations, and GPS/MET data in QMB. Thunderstorms, heavy rainfall, and gale can be detected and forecasted from this system. The second component focuses on lightning monitoring and forecasting by considering the following data: lightning location, surface electric field, and radar around the Qingdao region. The region and probability of lightning can be predicted by this system one hour in advance. Moreover, the sailing field was set as a special monitoring region, and the probability, location, and density of lighting for this region were updated every 15 minutes.

3.4. Weather Information Display and Distribution System. In the display and distribution platform, the latest data monitored by different sensors, historical data, numerical model results, thunderstorm prewarning products, and time series of statistical analyses are all displayed and updated every 10 minutes. The integrated display system is comprised of a multiple screen graphics workstation by using techniques of Oracle database, network communication, and net station designing. China Meteorological Administration, Beijing Meteorological Bureau, Shandong Province Meteorological Bureau, the weather service office at the sailing field, and other departments can access the information through high speed internet, mobile, and satellite. This effective network system guaranteed the successful transmission of multisource data quickly, especially for short-time nowcasts and updated forecasts.

4. Weather Service for OPSC

The weather service group, including almost 100 people, provided OPSC weather support from June 25 to September 13, 2008. The weather service was involved with supporting the cleaning of the *Enteromorpha Prolifera* around the sailing field, the sailing competition, the daily city running, and the opening and closing ceremonies of OPSC. Nine meteorologists who have extensive forecasting experience were responsible for providing detailed wind forecasts and severe weather information for the five venues directly. The refined weather service products were provided to regatta organizers, athletes, and coaches through daily weather information briefings, notice boards, internet, and community short messages. The TV platforms installed in buses, mobile phone text messages, and electronic information displays distributed at arterial streets all played a significant role in improving the coverage of weather information.

Based on fully understanding the demands of different branches, the weather service group used a different service strategy for the games. The government officials were concerned as to whether the sailing games would run smoothly or not. Therefore, the weather service conclusion provided to them would be that, for example, "the current weather situation is unsuitable for the competition under the condition of weak and unsteady winds." In contrast, the competition officials paid attention to the accomplishment ratio. Therefore, the service also provided forecasting and probability of what may happen. For the athletes, the weather service group simply provided them with forecast products since they are concerned about the accuracy of forecasting in order to form their strategy of competition. The International Olympic Committee adjusted the schedules of 11 days according to the weather forecasting in the 18 competition days, and the accomplishment ratio of 2008 Olympic Sailing events is 93% under the condition of two flexible days unused (Table 3). The number in the table is the number of competitions.

5. Conclusions

The Beijing 2008 Olympic and Paralympic Sailing Competitions (OPSC) drew to a close and accurate weather forecast and active weather service assured that the sailing games went on smoothly and successfully. The mean absolute forecast error at four buoy stations for the wind speed was $1.1 \, \text{m s}^{-1}$ and 34 degrees for the wind direction. The extensive experience for offshore wind forecasting gained from the OPSC will help to improve the understanding of sea-land

(a)

(b)

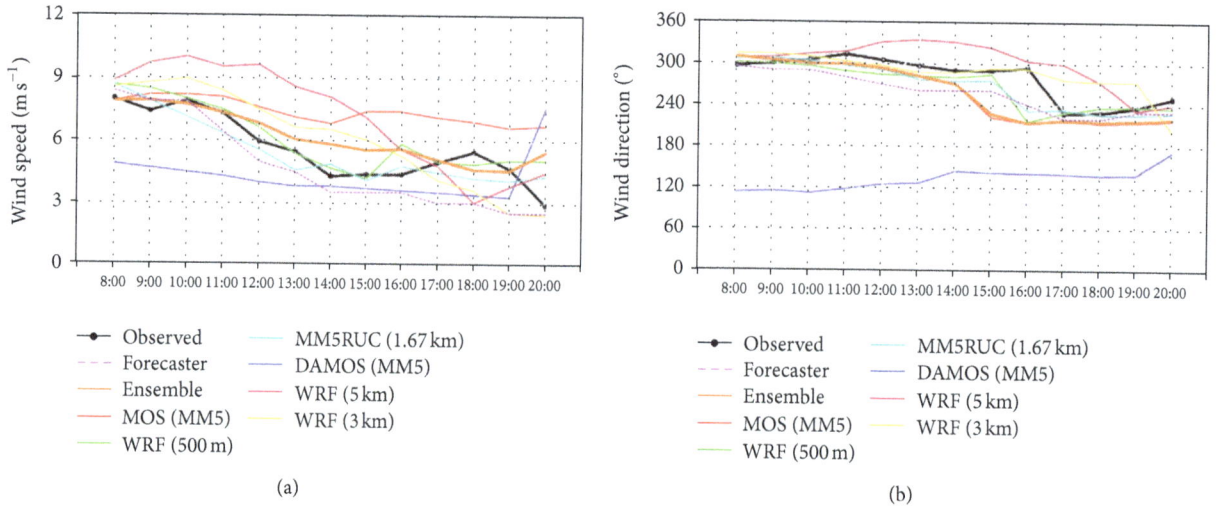

FIGURE 6: Hourly forecasts of 10 m wind speed (a) and wind direction (b) on August 18, 2008 for buoy B.

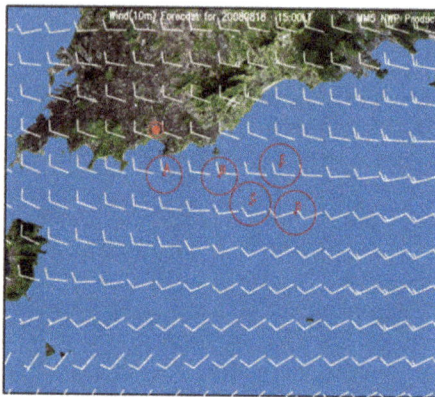

FIGURE 7: Distribution of wind vector at the height of 10 m at 15:00 LST of 18 August, 2008.

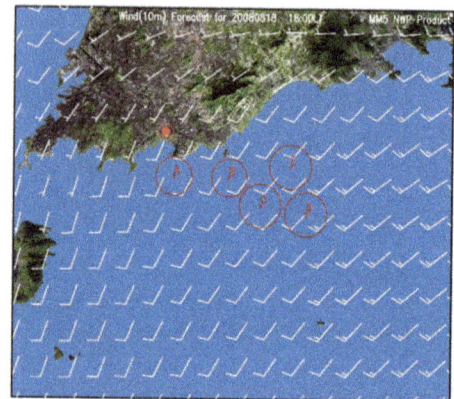

FIGURE 8: Distribution of wind vector at the height of 10 m at 18:00 LST of 18 August, 2008.

breeze circulation in complex terrain, on the aspects of both operational and research models. And the idea of synthetic observation network and service will play a significant role in the weather service of public activities.

Aiming at specific user groups and the public, QMB carried out a customer satisfaction survey on the 2007 international sailing regatta and 2008 Olympic and Paralympic sailing games according to Customer Satisfaction Index of Weather Service (CSIWS). The survey showed that the averaged satisfaction index rose 15% in two years and the satisfaction index reached up to 96.3% in 2008.

Acknowledgments

The authors are grateful for the English proofreading by Ms. Katie Costa. The authors also wish to acknowledge the helpful discussion with Dr. Liangbo Qi from Shanghai Meteorological Bureau and Mr. Hongwei Zhang from Dongying Meteorological Bureau.

References

[1] J. S. Snook, P. A. Stamus, J. Edwards, Z. Christidis, and J. A. McGinley, "Local-domain mesoscale analysis and forecast model support for the 1996 Centennial Olympic Games," *Weather and Forecasting*, vol. 13, no. 1, pp. 138–150, 1998.

[2] J. Horel, T. Potter, L. Dunn et al., "Weather support for the 2002 winter olympic and paralympic games," *Bulletin of the American Meteorological Society*, vol. 83, no. 2, pp. 227–240, 2002.

[3] P. T. May, T. D. Keenan, R. Potts et al., "The Sydney 2000 Olympic Games Forecast Demonstration Project: forecasting observing network infrastructure, and data processing issues," *Weather and Forecasting*, vol. 19, no. 1, pp. 115–130, 2004.

[4] E. Spark and G. J. Connor, "Wind forecasting for the sailing events at the Sydney 2000 Olympic and Paralympic games," *Weather and Forecasting*, vol. 19, no. 2, pp. 181–199, 2004.

[5] A. N. Hahmann, Y. Liu, and T. T. Warner, "Mesoscale circulations over the Athens metropolitan area during the 2004 summer Olympic games," in *Proceedings of the 86th AMS Annual Meeting: AMS Forum: Managing our Physical and Natural Resources: Successes and Challenges, and 6th Symposium*

on the Urban Environment, Atlanta, Ga, USA, January-February 2006, https://ams.confex.com/ams/pdfpapers/105087.pdf.

[6] N. H. Andrea, L. Yubao, and T. W. Thomas, "Mesoscale circulations over the Athens metropolitan area during the 2004 summer Olympic games," in *Proceedings of the 6th Symposium on the Urban Environment and AMS Forum: Managing our Physical and Natural Resources: Successes and Challenges*, 2006.

[7] M. D. Powell and S. K. Rinard, "Marine forecasting at the 1996 centennial Olympic games," *Weather and Forecasting*, vol. 13, no. 3, pp. 764–782, 1998.

[8] X. X. Zhu and M. G. Genton, "Short-term wind speed forecasting system for power system opersations," *International Statistical Review*, vol. 80, no. 1, pp. 2–23, 2012.

[9] G. A. Grell, J. Dudhia, and D. R. Stauffer, "A description of the 5th generation Penn State/NCAR Mesoscale Model (MM5)," NCAR Tech Note NCAR/TN 398+STR, 1994.

[10] J. Dudhia, "Numerical study of convection observed during the winter monsoon experiment using a mesoscale two-dimensional model," *Journal of the Atmospheric Sciences*, vol. 46, pp. 3077–3107, 1989.

[11] J. Dudhia, "A nonhydrostatic version of the Penn State-NCAR mesoscale model: validation tests and simulation of an Atlantic cyclone and cold front," *Monthly Weather Review*, vol. 121, no. 5, pp. 1493–1513, 1993.

[12] W. C. Skamarock, J. B. Klemp, J. Dudhia et al., "A description of the Advanced Research WRF version 2," NCAR Tech Note TN-468+STR, 2005.

[13] W. Wang, D. Barker, J. Bray et al., "WRF version 2 modeling system user's guide (2007)," http://www.mmm.ucar.edu/wrf/users/docs/user_guide/.

[14] S. G. Miao, F. Chen, Q. Li, and S. Fan, "Impacts of urban processes and urbanization on summer precipitation: a case study of heavy rainfall in Beijing on 1 August 2006," *Journal of Applied Meteorology and Climatology*, vol. 50, no. 4, pp. 806–825, 2011.

[15] K. A. Hart, W. J. Steenburgh, D. J. Onton, and A. J. Siffert, "An evaluation of mesoscale-model-based model output statistics (MOS) during the 2002 Olympic and Paralympic winter games," *Weather and Forecasting*, vol. 19, no. 2, pp. 200–218, 2004.

19

Numerical Simulation of a Lee Wave Case over Three-Dimensional Mountainous Terrain under Strong Wind Condition

Lei Li,[1] P. W. Chan,[2] Lijie Zhang,[1] and Fei Hu[3]

[1] *Shenzhen National Climate Observatory, Meteorological Bureau of Shenzhen Municipality, Shenzhen 518040, China*
[2] *Hong Kong Observatory, 134A Nathan Road, Kowloon, Hong Kong*
[3] *Institute of Atmospheric Physics, Chinese Academy of Sciences, Beijing 100029, China*

Correspondence should be addressed to P. W. Chan; pwchan@hko.gov.hk

Academic Editor: Sven-Erik Gryning

This study of a lee wave event over three-dimensional (3D) mountainous terrain in Lantau Island, Hong Kong, using a simulation combining mesoscale model and computational fluid dynamics (CFD) model has shown that (1) 3D steep mountainous terrain can trigger small scale lee waves under strong wind condition, and the horizontal extent of the wave structure is in a dimension of few kilometers and corresponds to the dimension of the horizontal cross-section of the mountain; (2) the life cycle of the lee wave is short, and the wave structures will continuously form roughly in the same location, then gradually move downstream, and dissipate over time; (3) the lee wave triggered by the mountainous terrain in this case can be categorized into "nonsymmetric vortex shedding" or "turbulent wake," as defined before based on water tank experiments; (4) the magnitude of the wave is related to strength of wind shear. This study also shows that a simulation combining mesoscale model and CFD can capture complex wave structure in the boundary layer over realistic 3D steep terrain, and have a potential value for operational jobs on air traffic warning, wind energy utilization, and atmospheric environmental assessment.

1. Introduction

Lee wave is one of the key indicators of how mountains affect air flows [1]. It also has significant impact on aviation safety, wind energy utilization, and air pollution. Over the years, scholars have studied lee waves through different approaches, including theoretical analysis [2–7], field observations [8–11], and numerical simulation [12–14]. In the light of the previous studies, the formation of lee waves is closely related to atmospheric stratification and buoyancy force [15]. When mountain-crossing winds occur in stable stratification of the atmosphere, leeward flows will form waves and even lee wave rotors if stable stratification is strong.

Though there were some studies discussing or summarizing the lee wave phenomena related to three-dimensional (3D) terrain [12, 13, 15, 16], most of the previous studies mainly focused on two-dimensional (2D) or semi-2D ridge.

Compared with the 2D ridge, the 3D terrain is far more complex and the research conclusions for 2D issues may not be valid for 3D issues. Actually, the study on the lee wave phenomena related to 3D terrain is insufficient, and the characteristics and the mechanism of the lee wave triggered by 3D terrain are not fully understood till now. Thus, it is necessary to do more in-depth researches on the lee waves related to 3D terrain to improve the understanding in this field.

From analysis of Doppler radar data, Chan [17] noticed significant lee waves in Lantau Island, Hong Kong, under a typical strong wind condition, and he took it as "vortex/wave shedding." Lantau Island is a typical 3D steep mountainous area, as elevation change reaches more than 700 m within a 2 km range. Compared with simple and idealized mountains [12] or semi-2D ridges [14] discussed in previous numerical studies, the terrain of Lantau Island is far more complex and realistic, which makes the lee wave case observed in this area highly representative and worth further in-depth studies, especially these related to characteristics of lee waves such as spatial structure, impact range, and life cycle related to 3D terrain.

Based on the authors' previous studies, this paper leverages the advantageous application of computational fluid dynamics (CFD) in microscale numerical simulation to perform a detailed simulation on the "vortex/wave shedding" case in Lantau Island raised by Chan [17] and to analyze the process of lee wave formation triggered by 3D mountainous terrain under strong wind condition.

2. Brief Descriptions of Simulation Method

The simulation period for this study is chosen to be 18:30–19:30 local time on September 29, 2011. During this period, Typhoon Nesat was transiting Hong Kong, steady southeast wind covered the entirety of Lantau Island, and surface wind speed was keeping steady at above 10 m/s.

The simulation uses a mesoscale model, namely, Regional Atmospheric Modeling System (RAMS) [18], which is offline coupled with FLUENT, a CFD software [19]. Its flowchart is shown in Figure 1. The simulation first uses 20 km resolution data exported from the Hong Kong Observatory's Operational Regional Spectral Model (ORSM) to initiate a four nested grids RAMS simulation. It then exports 800 m resolution RAMS results as boundary conditions to initiate FLUENT calculations. This procedure is similar to those used in the authors' previous studies [20–22] and has proved to be reliable to illustrate flow field characteristics under complex underlying surfaces.

RAMS simulation has the same model setting as that of Chan [17, 23]. There are totally four grids for RAMS simulation, and grid 1 has a horizontal resolution of 4 km with 77×67 grids. The first vertical layer has a height of 40 m with the stretching ratio of 1.15. Grid 2 has a horizontal resolution of 800 m. The number of points and vertical levels is the same as grid 1. Mellor-Yamada scheme is used in grid 1 and Deardorff [24] scheme is used in grid 2. More details on the RAMS simulation can be found in [17, 23]. For comparison purpose, RAMS simulations in grid 3, namely, a horizontal resolution of 200 m, would be used. Deardorff [24] turbulence parameterization scheme is adopted in this grid for "large eddy simulation" mode. Cumulus parameterization is switched off. The first grid has a height of 20 m and a stretching ratio of 1.15 is maintained.

In FLUENT simulation, the atmosphere is assumed as incompressible, and the main governing equations are as follows.

Momentum equation:

$$\frac{\partial \overline{u_i}}{\partial t} + \overline{u_j}\frac{\partial \overline{u_i}}{\partial x_j} = -\frac{1}{\rho}\frac{\partial \overline{p}}{\partial x_i} + \frac{\mu}{\rho}\frac{\partial^2 \overline{u_i}}{\partial x_j \partial x_j} - \frac{\partial}{\partial x_j}\left(\overline{u_i' u_j'}\right) + \overline{f_i}. \tag{1}$$

Continuity equation:

$$\frac{\partial \overline{u_i}}{\partial x_i} = 0, \tag{2}$$

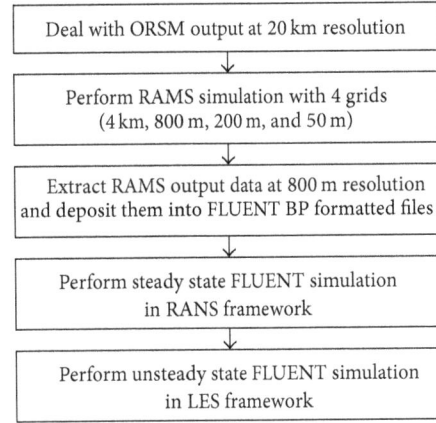

FIGURE 1: Flowchart of the simulation.

where $\overline{u_i}$ is the average velocity of air flow, p is the pressure, $\overline{u_i' u_j'}$ is the Reynolds stress, ρ is the density of atmosphere, μ is the molecular viscosity, and $\overline{f_i}$ is the body force.

The FLUENT simulation domain has a size of 22 km × 19 km, including Lantau Island, Hong Kong, and its surrounding areas. A computer aided design (CAD) model is built using terrain elevation data, as shown in Figures 2(a) and 2(b). Simulation domain is discretized by hexahedral meshing. The cell numbers for x and y directions are 140 and 120, respectively, which means that horizontal resolution is about 158 m. In vertical direction, simulation domain reached about 3000 m elevation and is divided into 22 layers. The size of each vertical grid cell sketches in a ratio 1 : 1.1 from the one below, which means that the vertical spaces of the bottom layer cells at sea level are about 45 m.

The FLUENT simulation is performed in two steps. The first step is a steady-state simulation in which time derivative terms and variables in dynamic equations are removed. In the first step of simulation, turbulent closure model uses the realizable k-e scheme in the Reynolds Averaged Navier-Stokes (RANS) framework. Once numerical solution reached convergence in steady state, we initiate the second simulation stage, a non-steady-state simulation. In the second stage simulation, variables depend on time and LES turbulent solution with Smagrinsky-Lilly scheme for sub-grid option are used in the simulation. This two-step simulation procedure guarantees that LES simulation yields higher quality and more physically meaningful initial value.

Considering there are steady southeast winds throughout the entire simulation period, the east, the north, and the south boundaries are set to velocity inlet type. Wind speed data at the three boundaries are taken from the second grid of the RAMS simulation output at September 29, 2011, 18:00, and are input into FLUENT computational model via the boundary profile (BP) module. For the west boundary, boundary conditions are set to outflow, which means that spatial derivatives at the boundary are zero. The top boundary is also set as the velocity inlet boundary, and the wind speed and direction are directly set as the RAMS output values. For all velocity inlet boundaries, the wind speed and direction are set as unchanged during the simulation. The boundary conditions

(a) Terrain in the simulation domain

(b) CAD model read in FLUENT

(c) Vertical profiles of the background atmosphere

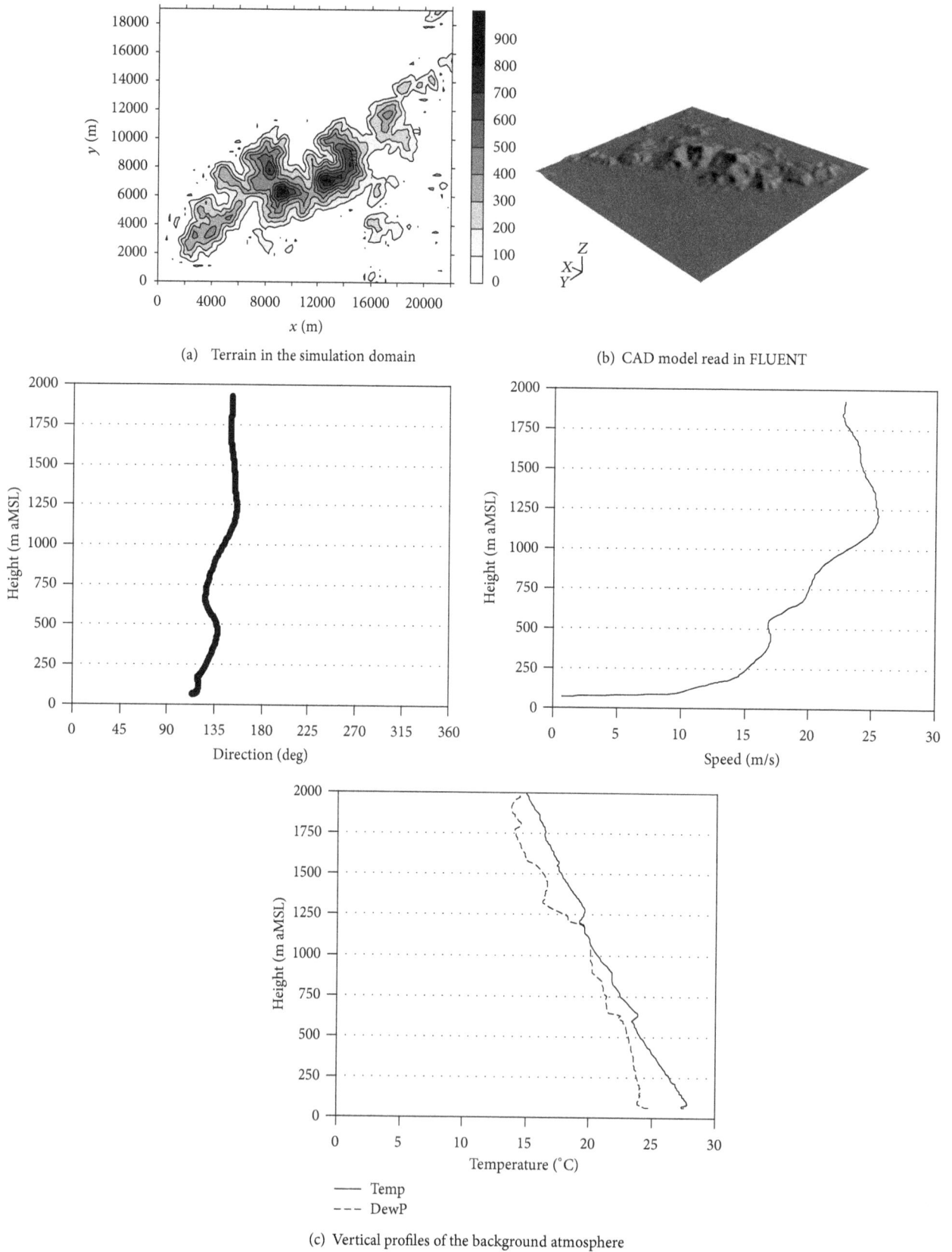

Figure 2: Terrain within FLUENT simulation is shown in (a) and (b). (c) gives the background atmospheric condition for the simulation as obtained from the radiosonde sounding in Hong Kong at 12 UTC, September 29, 2011. The figures include the wind direction profile, wind speed profile, and temperature and humidity profiles.

setting used in this study can provide strong forcing for the whole simulation domain and will help to ensure the dominant flow direction in the domain in accordance with the observed one. The bottom of the simulation region is set as no-slip boundary condition, with the roughness heights of the sea surface, and the airport in Lantau Island and the Lantau mountains are set as 0.0005 m, 0.005 m, and 0.5 m, respectively. The domain setting and the boundary conditions setting used in this study are similar to those used in the authors' previous studies [22] and have proven to be reliable to capture the characteristics of the flow field over mountainous area.

3. Analysis on Results

3.1. Radial Velocity. In order to help learn the background atmospheric condition of the vortex/wave event discussed in this study, the radiosonde sounding data in Hong Kong at 12 UTC, September 29, 2011, are shown in Figure 2(c). From Figure 2(c), it can be seen that during the transiting of Typhoon Nesat, the wind direction generally keeps as southeast and only slightly varies with height. The wind speed generally increases with height within the atmospheric boundary layer, and the average wind speed of the layer within the height of 2000 m is around 20.0 m/s. The temperature decreases with altitude, whilst there are ground inversion and two elevated inversion layers at approximately 650 m and 1200 m above sea level.

Detailed report on observation of lee wave during Typhoon Nesat can be referenced to Chan [17]. The major observations in Chan [17] are the shedding of wave/vortex from the mountains of Lantau Island from the radial velocity data from the weather radar. The radar observations are reproduced in Figure 3, with the shedded waves/vortices enclosed in ellipses. The warm colour (red, orange, yellow, and brown) means that is the radial velocity that away from the radar, and the cool colour (green, blue, and purple) means that the radial velocity is heading towards the radar. As a result, the general wind direction is east-southeasterly. The green features highlighted are embedded in an area with generally away-from-the-radar flow and are considered to be waves/vortices. Figure 3 showing 6 instances of Dropper radar near Lantau Island, with Figure 3(a) shows the observation at 18:56 local time and around 1-minute interval between each subsequent figure. As shown in Figure 3, there is a wave/vortex structure distinctly different with surroundings (indicated by green patch in red circle in figure) formed in mountain, labeled as "A," and then it shedded off and gradually moved downstream until dissipation. The entire procedure of this event lasts about 6 minutes.

Starting from an initial field obtained from a steady RANS simulation in the first step, the FLUENT LES simulation needs some time to adjust to get an equilibrium field. For this case, the adjusting time is about 25 minutes, and from the 27th minute of the simulation, the vortex/wave shedding event begins to appear. Thus, the simulation results starting from the 27th minute are used to analyze the characteristics of the simulated lee wave process. The radial velocity maps drawn by FLUENT simulation results are illustrated in

Figure 4. It can be seen that FLUENT successfully captures the lee waves observed by radar. There is a wave structure different with surrounding wind direction generated in northwest Lantau Island and continued to move downstream, and the entire process also lasts about 6 minutes. Figure 4 also shows that the wave structure extends about 3 km downwind from mountains, which is a typical microscale lee wave phenomenon.

Comparing Figures 3 and 4, FLUENT can accurately reproduce the lee wave observations, and the simulated evolution and impact areas of the wave structure are consistent with the observed ones.

Apart from the wave/vortex "A," which shedded off from Nei Lak Shan, there are also waves/vortices "C" and "B" that shedded off from Cheung Shan and Tai Tung Shan, respectively, as shown in Figure 3. Their sheddings are also reproduced successfully by FLUENT (see Figure 4).

3.2. Horizontal Wave Structure. Figure 5 depicts the simulated flow field at 220 m above sea level in northwest Lantau Island, and this elevation is about where radar observed lee waves. Figure 5 clearly shows that the green patch in Figures 3 and 4 is a wave structure, and FLUENT shows the process in which the wave structure detached from the mountain and moved downstream. Horizontal scale of each wave structure is about 3 km (in the east-west orientation) and about 1.5 km (in the north-south orientation) and corresponds to the horizontal cross-section of the mountain where it formed at 220 m elevation. The entire process of lee wave moving downstream lasts about 6 minutes for a distance about 3-4 km.

3.3. Vertical Wind Speed and Wave Structure. To analyze wave structure in vertical direction, we chose a cross section in simulation domain to analyze the changes in vertical wind speed and the wave structure. This cross section runs in southeast-northwest direction roughly consistent with dominant background wind direction and is through three control points within simulation domain, p1 ($x = 0$ m, $y = 15800$ m, $z = 0$ m), p2 ($x = 0$ m, $y = 15800$ m, $z = 3000$ m), and p3 ($x = 16900$, $y = 0$ m, $z = 3000$ m).

Figure 6 shows the vertical wind speed component (w) in the above-mentioned cross section and can find the following characteristics: (1) w alternates between positive and negative values along the background flow direction on the vertical cross section, indicating that vertical wind speed is clearly fluctuating in a typical wave-like fashion; (2) from the distance between negative w value regions (blue area), it can be determined that the wavelength of the lee wave is about 3 km; (3) the movement of negative w value regions over time shows that wave structures continue to move downstream and dissipate. Deduced from the changes from Figure 6(a) to 6(e), wave moves about 2.5 km downstream within 4 minutes; (4) Figure 6 also shows that wave not only moves downstream but also regenerates in original location and the whole process continuously repeats.

For better visualization of the shedded wave, Figure 6 is replotted by showing the vectors of the flow field as projected on the cross sectional plane, as shown in Figure 7.

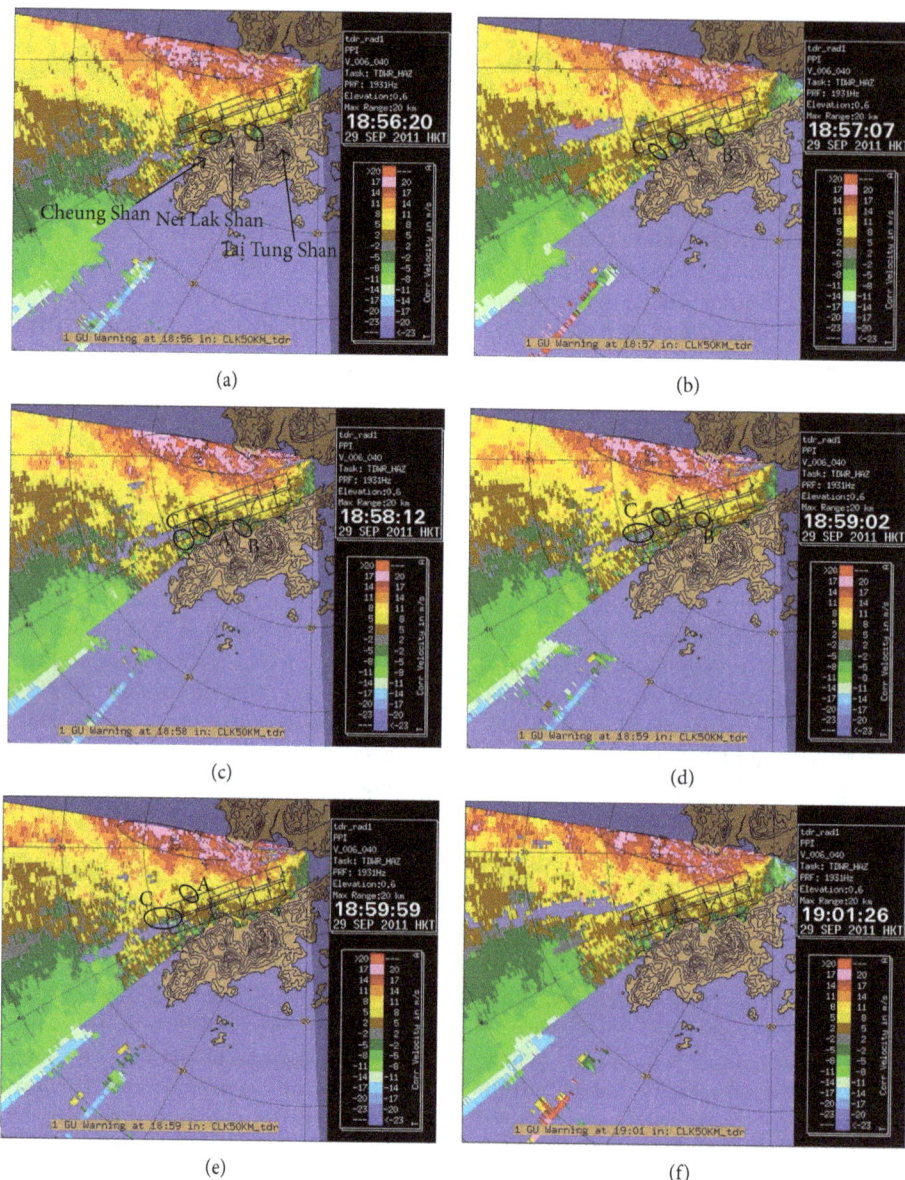

FIGURE 3: Radial wind speed observations from Doppler radar near Lantau Island ((a) represents 18:56 with 1-minute intervals between each figure).

The shedded wave is highlighted as a black arrow. It could be readily seen the shedding of the wave from the mountain from this wind field plot.

3.4. Pathline. Sections 3.2 and 3.3 describe the structure of shedded wave/vortex by using horizontal and vertical cross sections across the structure. In order to investigate the 3D structure of the wave/vortex, the "pathlines" of the flow (namely, following the trajectories of the air particles) are also analyzed, though the figure of the pathline is not provided in this paper.

It could be seen that, for structures "A" and "C" in Figure 3, the flows appear as oblique waves in the 3D space; that is, the wave axis does not lie purely horizontally or vertically, but appears as an oblique line in the 3D space. As

a result, when horizontal or vertical cross sections are cut through this structure, a wave shows up in the cross sectional planes.

On the other hand, the structure downstream of Tai Tung Shan, namely, feature "B" in Figure 3, appears to be more complicated, at least as given in the model simulation. There appears to be a rotor downstream of the mountain. The rotor has an axis generally parallel to the orientation of the mountain ranges of Lantau Island, that is, roughly northeast to southwest orientation.

3.5. Vorticity. As an alternative view of the flow field in the model simulation, vorticity plot is given in Figure 8 at a height of about 220 m above sea level. Downstream of the three mountains under consideration in this paper, there are

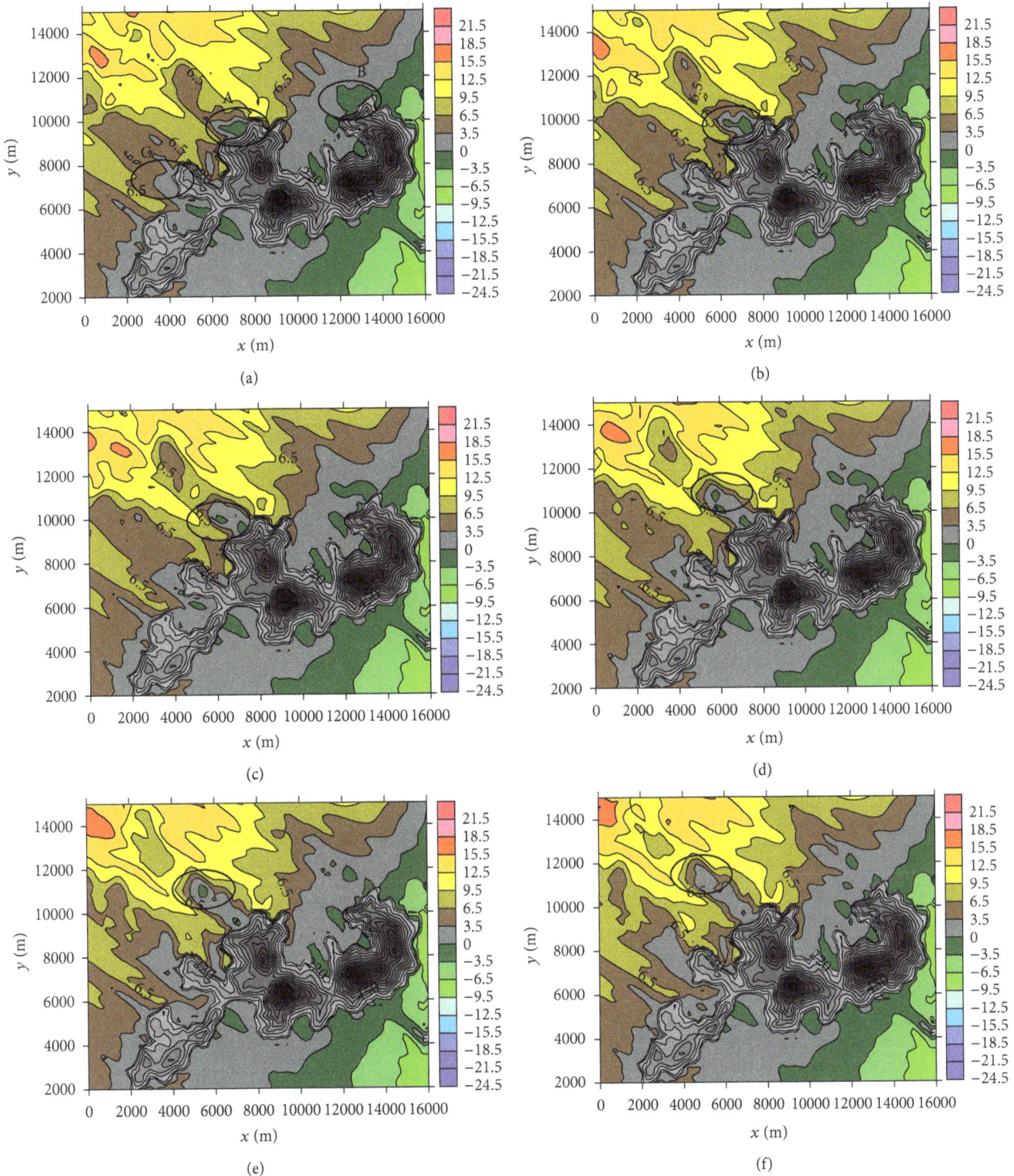

FIGURE 4: FLUENT simulated radial wind speed at 220 m elevation ((a) represents the 27th minute of the simulation period with 1-minute intervals between each figure).

areas of higher vorticity values, as coloured in green/yellow in Figure 8. The patterns are rather similar to the results of idealized model simulations of vortex shedding in, for instance, Schär and Smith [12]. The vorticity plot provides another view of the vortex/wave shedding from the three major mountains of Lantau Island.

From the simulation results including pathlines and vorticity, it could be seen that vortex/wave shedding occurs downstream of isolated mountains of Lautau Island. Each wave/vortex is associated with each mountain. Compared to the whole simulation domain, each mountain occupies a relatively small area only.

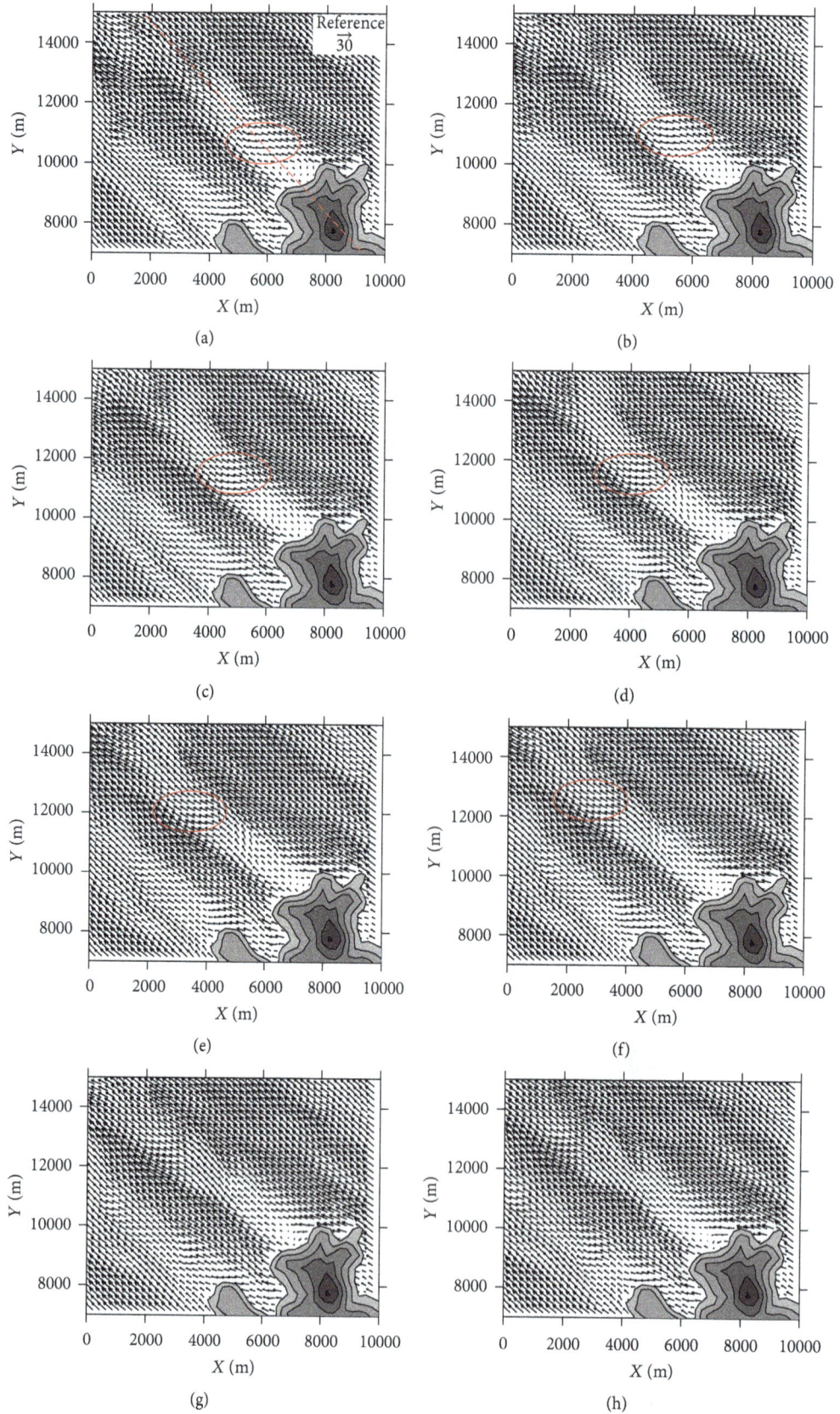

FIGURE 5: FLUENT simulated wind field in m/s at 220 m elevation ((a) represents the 27th minute of the simulation period with 1-minute intervals between each figure. The location of the vertical cross section in Figures 6 and 7 is shown as a red dotted line in (a)).

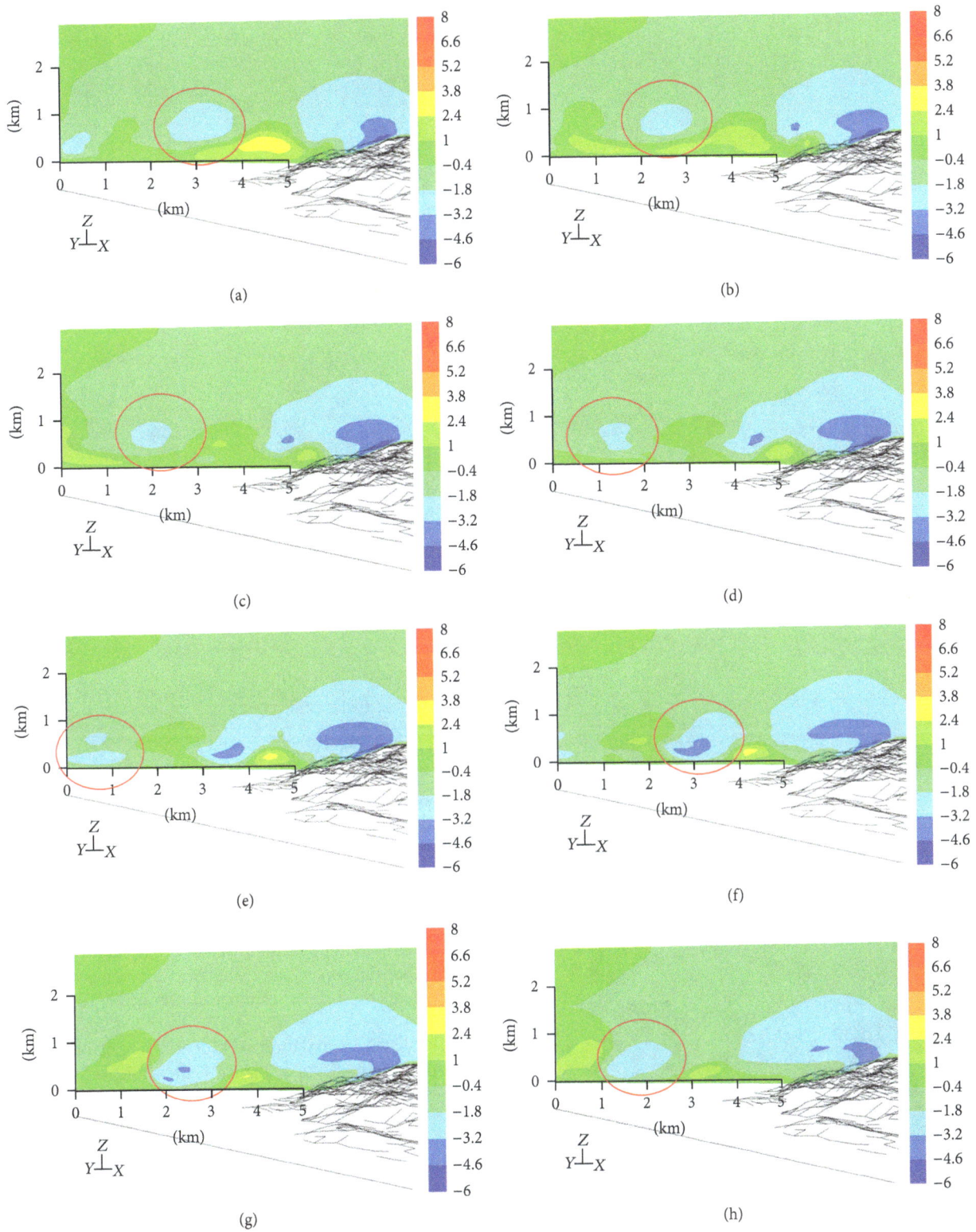

FIGURE 6: Vertical wind speed component at cross section (unit: m/s, (a) represents the 27th minute of the simulation period with 1-minute intervals between each figure, and the space scale is valid for the cross section plane).

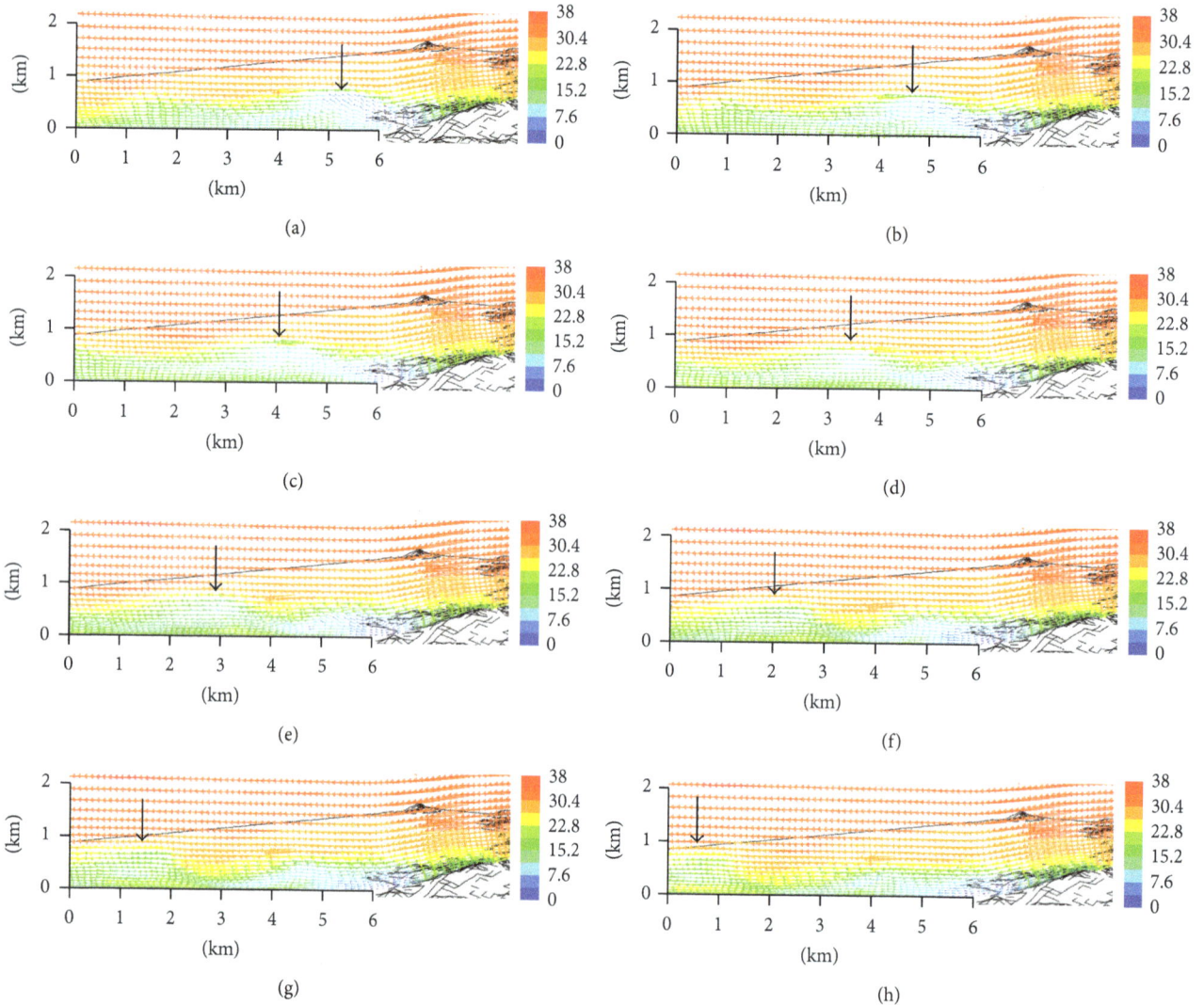

FIGURE 7: The wind vector projected on the vertical cross sectional planes. The wave shedding is highlighted in a black arrow (unit: m/s, time for each figure is same as that in Figure 5).

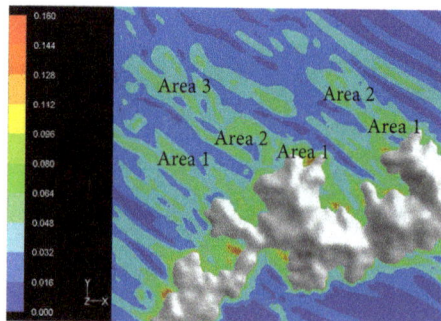

FIGURE 8: Vorticity plot for the model simulation result at 31 minutes (magnitude of vorticity is in s^{-1}).

4. Comparison with RAMS Simulation

Since there is RAMS simulation at horizontal resolution of 200 m, which is comparable with that of FLUENT simulation,

there is a question on the comparison of the results from these two model simulations. In particular, there is a question on whether the vortex/wave shedding features as observed in actual data could show up more clearly in FLUENT than in RAMS.

The RAMS simulation results at a height of 220 m are shown in Figure 9. As in Figure 4, the simulated wind fields are resolved along the measurement radials of the weather radar in order to reproduce the radar-observed features for ease of comparison. As shown in Figure 9 (highlighted in a black ellipse), vortex/wave shedding is also observed downstream of Nei Lak Shan in RAMS simulations at 200 m resolution. However, compared to the FLUENT simulation, the magnitude of the wave in RAMS simulation is smaller. In general, the flow field downstream of the mountains of Lantau Island appears to be smoother. No significant wave/vortex shedding could be reproduced downstream of Cheung Shan and Tai Tung Shan in RAMS. Though the 50 m resolution RAMS simulation can reproduce the wave shedding process,

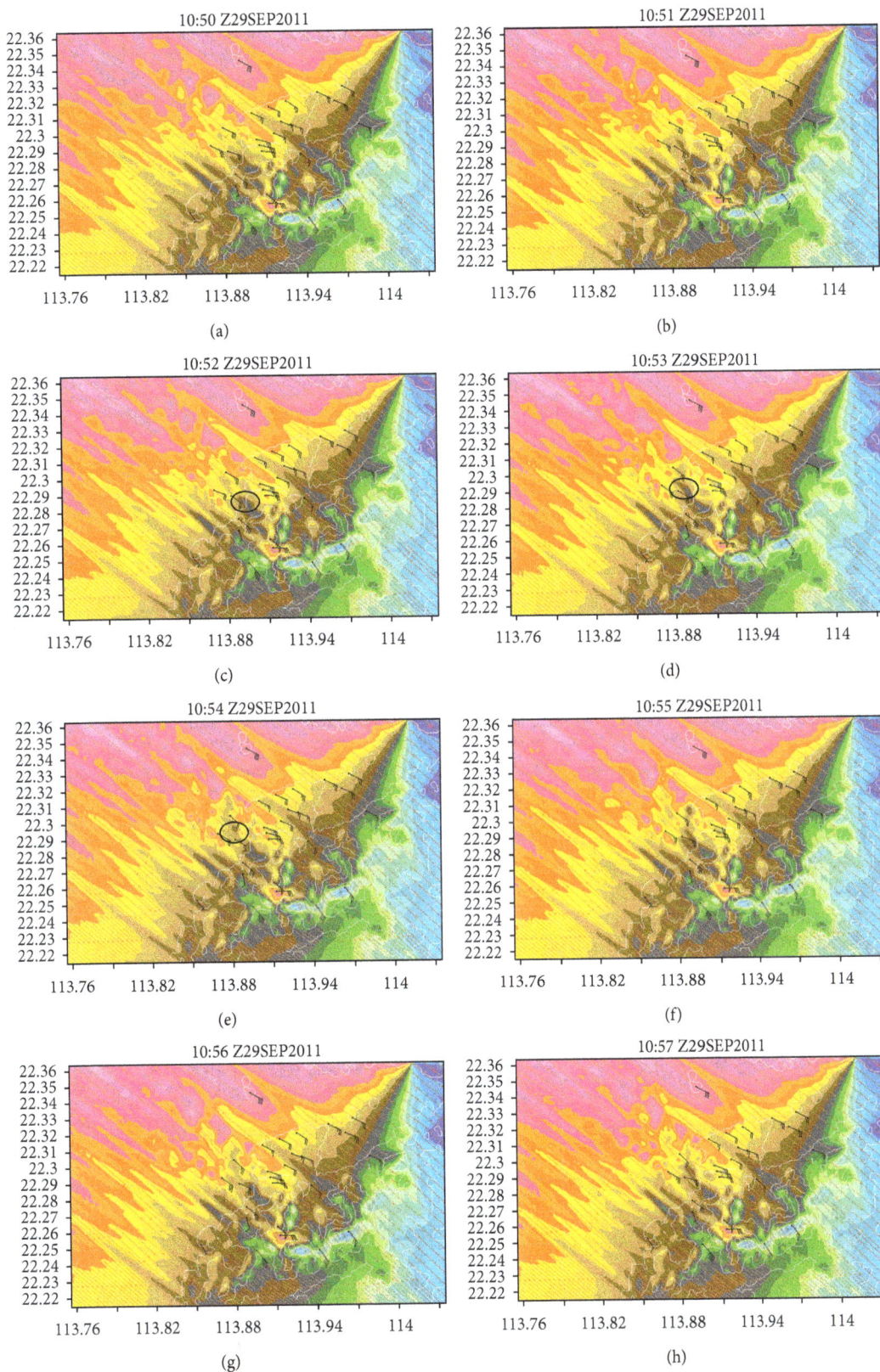

FIGURE 9: RAMS model simulation at a height of 220 m, with the flow field resolved in the direction of the measurement radials of the weather radar. A vortex/wave shedding feature is highlighted in a black ellipse.

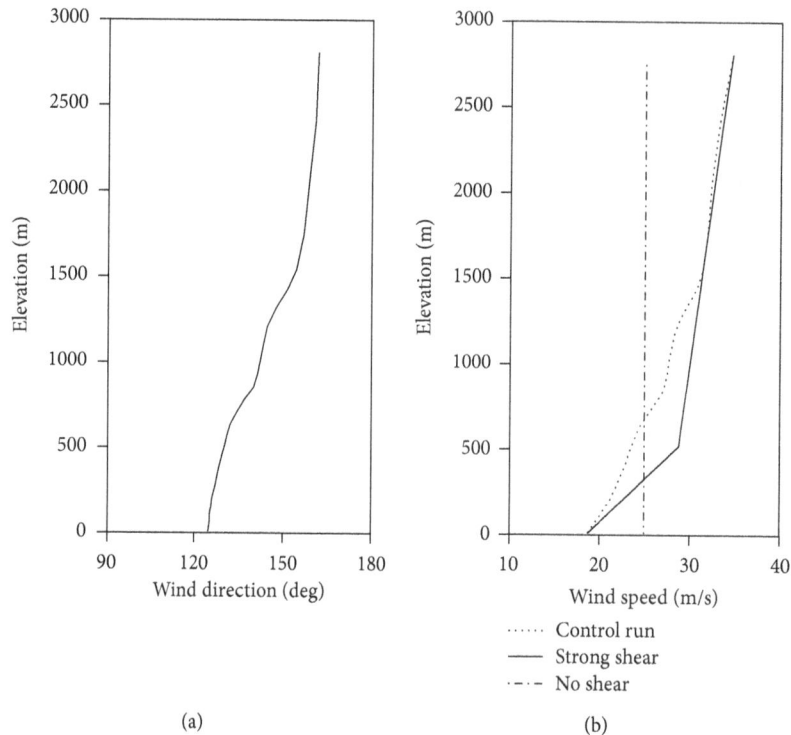

(a)   (b)

FIGURE 10: The vertical profile of wind direction (a) and the vertical profile of the three numerical simulations (b), namely, original (control, CTL), no shear (NS), and strong shear (SS).

it is too time-consuming and the wave structure is not as clear as that in FLUENT with a much coarser resolution. In terms of the computational time, the use of grid 4 (50 m resolution) would require around 10 times more about the time required for running the RAMS simulation up to grid 3 (200 m resolution) only. The use of FLUENT is computationally more efficient and it would be sufficient to simulate the vortex/wave shedding with a resolution of about 200 m by properly resolving the complex terrain near the airport.

From the comparison result, it appears that FLUENT is better in capturing the vortex/wave shedding features. On the other hand, the waves do not show up equally clearly in RAMS simulation. It shows that FLUENT has the potential of forecasting the occurrence of vortex/wave shedding in real time applications, for example, in providing an earlier alert about the occurrence of low-level windshear/turbulence arising from terrain-disrupted airflow over an airport.

5. Sensitivity of Wave Shedding to Wind Shear

In order to investigate the effect of vertical wind shear on the occurrence of wave/vortex shedding downstream of the mountains, two more numerical simulations are carried out with the modification of the vertical wind speed profile. The profiles in use are shown in Figure 10. They include the original profile (control, CTL), no wind shear (NS), and a stronger shear below 500 m above sea level (SS). For NS case, the average wind speed of CTL over the range of height under consideration is used.

The pathlines of the model simulation results for NS and SS shows that there are still waves/vortices and even rotors

downstream of the mountains (figures not shown in this paper). As such, the generation and shedding of the waves/vortices and the occurrence of rotors do not appear to be dependent on the vertical wind shear. It is believed that they are more related to the steep slopes associated with the mountains of Lantau Island under consideration.

However, the various simulations do show some differences. Figure 11 illustrates the vorticity plot on a horizontal plane. As could be expected, the SS case shows larger vorticity at a couple of kilometres downstream of the mountain. This is followed by CTL (maximum vorticity of about $0.8 \, s^{-1}$), and NS shows the smallest vorticity (vorticity generally in the order of 0.6 to $0.8 \, s^{-1}$ after a couple of kilometres downstream of the mountains, and the area of $0.8 \, s^{-1}$ vorticity is less extensive). Moreover, from the vertical cross section of the vertical velocity (figures not shown), SS gives the largest value of vertical velocity (in the order of 4 m/s or above), followed by CTL and NS (in the order of 2 to 3 m/s for the maximum value of the vertical velocity in the vertical plane). A larger vertical shear in the background flow is found to increase the vorticity and the upward/downward flow, which may be more hazardous to the aircraft.

6. Analysis of the Dynamics of the Vortex/Wave Shedding

In the present study, both observational data and numerical simulation suggest that lee waves can be formed in the lee of three-dimensional mountains under strong wind condition.

(a)

(b)

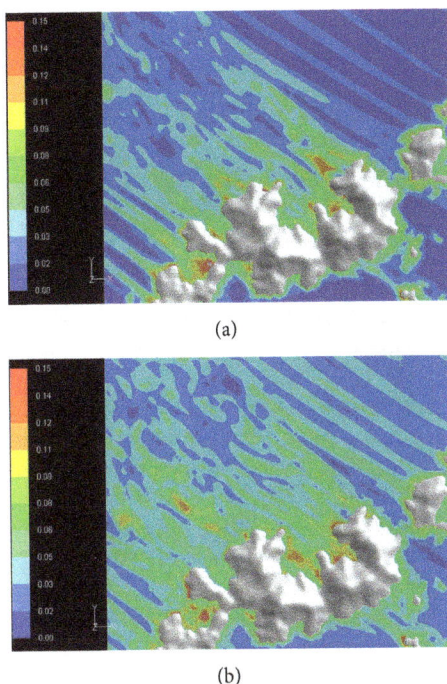

FIGURE 11: Vorticity field at a height of 220 m above sea level at 31 minutes for NS (a) and SS (b).

So, a new task for the present study is to explain the dynamics of the phenomenon.

A series of water tank experiments were performed by Lin et al. [25] to investigate the characteristics of the pattern for flows past a three-dimensional sphere. According to Lin et al. [25], the flow past sphere can be categorized into eight regimes, namely, steady two-dimensional attached vortices, unsteady two-dimensional attached vortices, two-dimensional vortex shedding, lee-wave instability, nonaxisymmetric attached vortex, symmetric vortex shedding, nonsymmetric vortex shedding, and turbulent wake. The eight regimes were located on an Fi-Re regime diagram, where Fi is the Froude number, and Re is the Reynolds number.

The study of Lin et al. [25] provides very valuable hints for explaining the dynamics of the lee wave/vortex shedding found in the present study. For the present case, based on the radiosonde data upstream of Lantau Island, the Brunt-Vaisala frequency, N, is in the order of 0.01 rad/s and U_0 is in the order of 20 m/s. As a result, the Froude number (Nh/U_0) is about 0.376, where h is the height of Nei Lak Shan (751 m) and U_0 is the average background wind speed. According to Lin et al., it is in the regime of "nonsymmetric vortex shedding" or "turbulent wake" triggered by 3D terrain.

Vortex shedding in tropical cyclone situation at Hong Kong International Airport has been discussed in Shun et al. [26]. Shun et al. [26] indicated that the vortex shedding may be related to "nonsymmetric vortex shedding" or "turbulent wake" with the Froude number in the order of 0.36, which is in accordance with the results in the present study. Nevertheless, the new result in this paper is that the vortex/wave shedding event is found to be successfully reproducible by

numerical simulation, which provides many more data (such as vertical cross section and vorticity field) to help understand the vortex/wave shedding event.

7. Conclusions

This study of a typical lee wave case in Lantau Island, Hong Kong, using RAMS/FLUENT combined simulation concludes the following.

(1) As a typical 3D steep mountainous terrain, Lantau Island can trigger small scale lee waves under strong wind condition. Both radar observation and CFD numerical simulation captured clear evidences of lee waves, and some waves are in a formation of rotor.

(2) The lee wave triggered by the mountainous terrain can be categorized into "nonsymmetric vortex shedding" or "turbulent wake," as defined before based on water tank experiments. The spatial dimension of the lee wave corresponds to the horizontal cross-section dimension of Lantau mountain, with a wavelength of about 3 km.

(3) In the present study, the life cycle of each lee wave is about 6 minutes. Wave structures will continuously form in roughly the same location, then gradually move downstream, and dissipate over time.

(4) The steep terrain is the major reason leading to the lee waves, though the magnitude of the wave is related to strength of wind shear.

Besides the above conclusions, this study once again shows that simulation combining mesoscale model and CFD can capture complex flow movement in boundary layer. CFD software, especially FLUENT, uses body-fitted mesh structure for domain discretization and uses finite volume method (FVM) in numerical calculation, which ensures CFD's capability to perform stable numerical calculations with nonconformal mesh. Therefore, CFD can obtain stable numerical solutions without the terrain being smoothed and ensures the accuracy of terrain. In addition, some special modules in CFD, such as user defined function (UDF, which has a grammar similar to C language in FLUENT) and boundary profile (BP), could help users code specific programs to describe boundary conditions and define some physical properties of fluid in simulations and thus provide an interface between the CFD and mesoscale models or observational data. The above special features make CFD have advantages over mesoscale models when describing wind field characteristics over steep terrain. The method of combining mesoscale model and CFD can ensure the accuracy of simulation by obtaining large-scale circulation data in mesoscale simulation and detailed microscale topographical description in CFD simulation simultaneously.

Finally, it can be expected that the simulation method of combining mesoscale model and CFD would be applied in operational air traffic warning, wind energy utilization, and atmospheric environmental assessment in a near future.

Acknowledgments

This study was supported by the National Natural Science Foundation of of China (nos. 51278308, 51008002, and

91215302) and the Open Science Foundation of the State Key of Laboratory of Atmospheric Boundary Physics and Atmospheric Chemistry.

References

[1] G. A. Corby, "The airflow over mountains. A review of the state of current knowledge," *Quarterly Journal of the Royal Meteorological Society*, vol. 80, pp. 491–521, 1954.

[2] R. S. Scorer and H. Klieforth, "Theory of mountain waves of large amplitude," *Quarterly Journal of the Royal Meteorological Society*, vol. 85, pp. 131–143, 1959.

[3] J. W. Miles, "Lee waves in a stratified flow. Part 1. Thin barrier," *Journal of Fluid Mechanics*, vol. 32, pp. 549–567, 1968.

[4] J. W. Miles, "Lee waves in a stratified flow. Part 2. Semi-circular obstacle," *Journal of Fluid Mechanics*, vol. 33, pp. 803–814, 1968.

[5] H. E. Huppert and J. W. Miles, "Leewaves in a stratified flow. Part 3. Semi-elliptical obstacle," *Journal of Fluid Mechanics*, vol. 35, pp. 481–496, 1969.

[6] R. R. Long, "Finite amplitude disturbances in the flow of inviscid rotating and stratified fluids over obstacles," *Annual Review of Fluid Mechanics*, vol. 4, pp. 69–92, 1972.

[7] D. R. Durran, "Two-layer solutions to Long's equation for vertically propagating mountain waves: how good is linear theory?" *Quarterly Journal of the Royal Meteorological Society*, vol. 118, no. 505, pp. 415–433, 1992.

[8] R. B. Smith, "The generation of lee waves by the Blue Ridge," *Journal of the Atmospheric Sciences*, vol. 33, no. 3, pp. 507–519, 1976.

[9] B. T. Gjevik, "Marthinsen, 1978: three-dimensional lee-wave pattern," *Quarterly Journal of the Royal Meteorological Society*, vol. 104, pp. 947–957.

[10] S. B. Vosper and S. D. Mobbs, "Lee waves over the English Lake District," *Quarterly Journal of the Royal Meteorological Society*, vol. 122, no. 534, pp. 1283–1305, 1996.

[11] L. S. Darby and G. S. Poulos, "The evolution of lee-wave-rotor activity in the lee of Pike's Peak under the influence of a cold frontal passage: implications for aircraft safety," *Monthly Weather Review*, vol. 134, no. 10, pp. 2857–2876, 2006.

[12] C. Schär and R. B. Smith, "Shallow-water flow past isolated topography. Part II: transition to vortex shedding," *Journal of the Atmospheric Sciences*, vol. 50, pp. 1401–1412, 1993.

[13] J. D. Doyle and D. R. Durran, "Rotor and subrotor dynamics in the Lee of three-dimensional terrain," *Journal of the Atmospheric Sciences*, vol. 64, no. 12, pp. 4202–4221, 2007.

[14] V. Grubišić and B. J. Billings, "The intense lee-wave rotor event of Sierra Rotors IOP 8," *Journal of the Atmospheric Sciences*, vol. 64, no. 12, pp. 4178–4201, 2007.

[15] R. B. Smith, "The influence of mountains on the atmosphere," *Advances in Geophysics*, vol. 21, pp. 87–230, 1979.

[16] R. S. Scorer and M. Wilkinson, "Waves in the lee of an isolated hill," *Quarterly Journal of the Royal Meteorological Society*, vol. 82, pp. 419–427, 1956.

[17] P. W. Chan, "Observation and numerical simulation of vortex/wave shedding for terrain-disrupted airflow at the Hong Kong International Airport during Typhoon Nesat in 2011," *Meteorological Applications*, 2012.

[18] R. A. Pielke, W. R. Cotton, R. L. Walko et al., "A comprehensive meteorological modeling system-RAMS," *Meteorology and Atmospheric Physics*, vol. 49, pp. 69–91, 1992.

[19] ANSYS, Inc., *Ansys Fluent 12.0 Theory Guide*, 2009.

[20] L. Li, F. Hu, J.-H. Jiang, and X.-L. Cheng, "An application of the RAMS/FLUENT system on the multi-scale numerical simulation of the urban surface layer—a preliminary study," *Advances in Atmospheric Sciences*, vol. 24, no. 2, pp. 271–280, 2007.

[21] L. Li, L. J. Zhang, N. Zhang et al., "Study on the micro-scale simulation of wind field over complex terrain by RAMS/FLUENT modeling system," *Wind and Structures*, vol. 13, no. 6, pp. 519–528, 2010.

[22] L. Lei and P. W. Chan, "Numerical simulation study of the effect of buildings and complex terrain on the low-level winds at an airport in typhoon situation," *Meteorologische Zeitschrift*, vol. 21, no. 2, pp. 183–192, 2012.

[23] P. W. Chan, "Atmospheric turbulence in complex terrain: verifying numerical model results with observations by remote-sensing instruments," *Meteorology and Atmospheric Physics*, vol. 103, no. 1–4, pp. 145–157, 2009.

[24] J. W. Deardorff, "Stratocumulus-capped mixed layers derived from a three-dimensional model," *Boundary-Layer Meteorology*, vol. 18, no. 4, pp. 495–527, 1980.

[25] Q. Lin, W. R. Lindberg, D. L. Boyer, and H. J. S. Fernando, "Stratified flow past a sphere," *Journal of Fluid Mechanics*, vol. 240, pp. 315–354, 1992.

[26] C. M. Shun, S. Y. Lau, and O. S. M. Lee, "Terminal doppler weather radar observation of atmospheric flow over complex terrain during tropical cyclone passages," *Journal of Applied Meteorology*, vol. 42, pp. 1697–1710, 2003.

Permissions

The contributors of this book come from diverse backgrounds, making this book a truly international effort. This book will bring forth new frontiers with its revolutionizing research information and detailed analysis of the nascent developments around the world.

We would like to thank all the contributing authors for lending their expertise to make the book truly unique. They have played a crucial role in the development of this book. Without their invaluable contributions this book wouldn't have been possible. They have made vital efforts to compile up to date information on the varied aspects of this subject to make this book a valuable addition to the collection of many professionals and students.

This book was conceptualized with the vision of imparting up-to-date information and advanced data in this field. To ensure the same, a matchless editorial board was set up. Every individual on the board went through rigorous rounds of assessment to prove their worth. After which they invested a large part of their time researching and compiling the most relevant data for our readers. Conferences and sessions were held from time to time between the editorial board and the contributing authors to present the data in the most comprehensible form. The editorial team has worked tirelessly to provide valuable and valid information to help people across the globe.

Every chapter published in this book has been scrutinized by our experts. Their significance has been extensively debated. The topics covered herein carry significant findings which will fuel the growth of the discipline. They may even be implemented as practical applications or may be referred to as a beginning point for another development. Chapters in this book were first published by Hindawi Publishing Corporation; hereby published with permission under the Creative Commons Attribution License or equivalent.

The editorial board has been involved in producing this book since its inception. They have spent rigorous hours researching and exploring the diverse topics which have resulted in the successful publishing of this book. They have passed on their knowledge of decades through this book. To expedite this challenging task, the publisher supported the team at every step. A small team of assistant editors was also appointed to further simplify the editing procedure and attain best results for the readers.

Our editorial team has been hand-picked from every corner of the world. Their multi-ethnicity adds dynamic inputs to the discussions which result in innovative outcomes. These outcomes are then further discussed with the researchers and contributors who give their valuable feedback and opinion regarding the same. The feedback is then collaborated with the researches and they are edited in a comprehensive manner to aid the understanding of the subject.

Apart from the editorial board, the designing team has also invested a significant amount of their time in understanding the subject and creating the most relevant covers. They scrutinized every image to scout for the most suitable representation of the subject and create an appropriate cover for the book.

The publishing team has been involved in this book since its early stages. They were actively engaged in every process, be it collecting the data, connecting with the contributors or procuring relevant information. The team has been an ardent support to the editorial, designing and production team. Their endless efforts to recruit the best for this project, has resulted in the accomplishment of this book. They are a veteran in the field of academics and their pool of knowledge is as vast as their experience in printing. Their expertise and guidance has proved useful at every step. Their uncompromising quality standards have made this book an exceptional effort. Their encouragement from time to time has been an inspiration for everyone.

The publisher and the editorial board hope that this book will prove to be a valuable piece of knowledge for researchers, students, practitioners and scholars across the globe.

List of Contributors

Bingui Wu, Yiyang Xie, Yi Lin, Jing Chen, Xiaobing Qiu and Yanan Wang
Tianjin Municipal Meteorological Bureau, Tianjin 300074, China

Xinxin Ye
Laboratory for Climate and Ocean-Atmosphere Studies, Department of Atmospheric and Oceanic Sciences, School of Physics, Peking University, Beijing 100871, China

Juan Bazo
Faculty of Sciences, Campus de Ourense, University of Vigo, 32004 Ourense, Spain
Peruvian National Meteorological and Hydrological Service (SENAMHI), Casilla 11 1308, Lima 11, Peru

María de las Nieves Lorenzo
Faculty of Sciences, Campus de Ourense, University of Vigo, 32004 Ourense, Spain

Rosmeri Porfirio da Rocha
Department of Atmospheric Sciences, Institute of Astronomy, Geophysics and Atmospheric Sciences, University of Sao Paulo, Sao Paulo, SP, Brazil

Jindrich Spicka and Jiri Hnilica
Department of Business Economics, University of Economics, Prague, W. Churchill Square 4, 130 67 Prague 3, Czech Republic

Fei Chen and Hans von Storch
Institute of Coastal Research, Helmholtz-Zentrum Geesthacht, Germany

N'Datchoh Evelyne Toure, Abdourahamane Konare and Siele Silue
Laboratoire de Physique de l'Atmosphere, Universite de Cocody, 22 BP 582 Abidjan 22, Cote d'Ivoire

Yiping Dou
Finance, eBay Inc., San Jose, CA 95125, USA

Nhu D. Le
BC Cancer Agency Research Center, Vancouver, BC, Canada V5Z 4E6

James V. Zidek
Department of Statistics, University of British Columbia, Vancouver, BC, Canada V6T 1Z2

Zekai Sen, Abdusselam Altunkaynak and Tarkan Erdik
Hydraulics Division, Civil Engineering Faculty, Istanbul Technical University, Maslak, 34469 Istanbul, Turkey

Zhiliang Wang and Chunyan Huang
College of Mathematics and Informatics, North China University of Water Conservancy and Hydroelectric Power, 36 Beihuan Road, Henan, Zhengzhou 450011, China

Kelin Zhuang
University of Arizona, Tucson, Arizona, AZ 85721, USA

John R. Giardino
Texas A&M University, College Station, Texas, TX 77843, USA

Vladimir V. Ivanov
Arctic and Antarctic Research Institute, St. Petersburg 199397, Russia
International Arctic Research Centre, University of Alaska, Fairbanks, AK 99775, USA
Scottish Marine Institute, Oban PA37 1 QA, UK

Vladimir A. Alexeev
International Arctic Research Centre, University of Alaska, Fairbanks, AK 99775, USA

Irina Repina
A.M. Obukhov Institute of Atmospheric Physics of RAS, Moscow 119017, Russia

Nikolay V. Koldunov
Institute of Oceanography, University of Hamburg, 20146, Hamburg, Germany

Alexander Smirnov
Arctic and Antarctic Research Institute, St. Petersburg 199397, Russia

J. Z. Wang, X. Y. Zhang, Y. Q. Yang, Q. Hou, C. H. Zhou and Y. Q. Wang
Center for Atmospheric Composition Observing & Service, Chinese Academy of Meteorological Sciences, Beijing 100081, China

S. L. Gong
Center for Atmospheric Composition Observing & Service, Chinese Academy of Meteorological Sciences, Beijing 100081, China
Air Quality Research Division, Science & Technology Branch, Environment Canada, 4905 Dufferin Street, Toronto, ON, Canada M3H 5T4

Lars Gidhagen and Magnuz Engardt
Swedish Meteorological and Hydrological Institute, 601 76 Norrkoping, Sweden

Boel Lovenheim
Environment and Health Administration, Box 8136, 104 20 Stockholm, Sweden

Christer Johansson
Environment and Health Administration, Box 8136, 104 20 Stockholm, Sweden
Department of Applied Environmental Science, Stockholm University, 106 91 Stockholm, Sweden

Duncan Ackerley
Department of Meteorology, University of Reading, Reading RG6 6BB, UK
Monash Weather and Climate, Monash University, VIC, Clayton 3800, Australia

Manoj M. Joshi
National Centres for Atmospheric Science (Climate), University of Reading, Reading RG6 6BB, UK

Claire L. Ryder, Eleanor J. Highwood and Jane Strachan
Department of Meteorology, University of Reading, Reading RG6 6BB, UK

Mark A. J. Harrison, David N. Walters and Sean F. Milton
Met Office, Exeter EX1 3PB, UK

F. Calastrini and F. Guarnieri
LaMMA Consortium, Laboratory of Monitoring and Environmental Modeling for the Sustainable Development, Via Madonna del Piano 10, 50019 Sesto Fiorentino, 50019 Sesto Fiorentino, Italy
IBIMET, National Research Council, Via G. Caproni 8, 50145 Florence, Italy

S. Becagli, R. Traversi and R. Udisti
Department of Chemistry, University of Florence, Via della Lastruccia 3, 50019 Sesto Fiorentino, Italy

C. Busillo
LaMMA Consortium, Laboratory of Monitoring and Environmental Modeling for the Sustainable Development, Via Madonna del Piano 10, 50019 Sesto Fiorentino, 50019 Sesto Fiorentino, Italy

M. Chiari and S. Nava
INFN, National Institute of Nuclear Physics, Via G. Sansone 1, 50019 Sesto Fiorentino, Italy

U. Dayan
Department of Geography, The Hebrew University of Jerusalem, Jerusalem 91905, Israel

F. Lucarelli
INFN, National Institute of Nuclear Physics, Via G. Sansone 1, 50019 Sesto Fiorentino, Italy
Department of Physics and Astronomy, University of Florence, Via G. Sansone 1, 50019 Sesto Fiorentino, Italy

M. Pasqui and G. Zipoli
IBIMET, National Research Council, Via G. Caproni 8, 50145 Florence, Italy

Jianhua Xu, Yiwen Xu and Chunan Song
The Research Center for East-West Cooperation in China, The Key Lab of GIScience of the Education Ministry PRC, East China Normal University, 500 Dongchuan Road, Minhang, Shanghai 200241, China

X. Xi and I. N. Sokolik
School of Earth and Atmospheric Sciences, Georgia Institute of Technology, 311 Ferst Drive, Atlanta, GA 30332-0340, USA

Roni Nehorai
Department of Geography and Environment, Bar-Ilan University, 52900 Ramat-Gan, Israel
Geological Survey of Israel, 30 Malkhe Israel Street, 95501 Jerusalem, Israel

Nadav Lensky
Geological Survey of Israel, 30 Malkhe Israel Street, 95501 Jerusalem, Israel

Steve Brenner and Itamar Lensky
Department of Geography and Environment, Bar-Ilan University, 52900 Ramat-Gan, Israel

Yan Ma, Rongzhen Gao, Yunchuan Xue, Xiaoliang Xu, Xuezhong Liu, Jianwei Hou and Hang Lin
Qingdao Meteorological Bureau, 4 Fulong Shan, Shinan District, Qingdao, Shandong 266003, China

Yuqiang Yang
Hangzhou Meteorological Bureau, Hangzhou, Zhejiang 310008, China

Xiaoyun Wang
Department of Synthetic Observing, China Meteorological Administration, Beijing 100081, China

Bin Liu
Department of Marine, Earth, and Atmospheric Sciences, NC State University, Raleigh, North Carolina 27695, USA

P. W. Chan
Hong Kong Observatory, 134A Nathan Road, Kowloon, Hong Kong

Lijie Zhang and Lei Li
Shenzhen National Climate Observatory, Meteorological Bureau of Shenzhen Municipality, Shenzhen 518040, China

Fei Hu
Institute of Atmospheric Physics, Chinese Academy of Sciences, Beijing 100029, China